NORTH ATLANTIC RIGHT WHALES

NORTH ATLANTIC

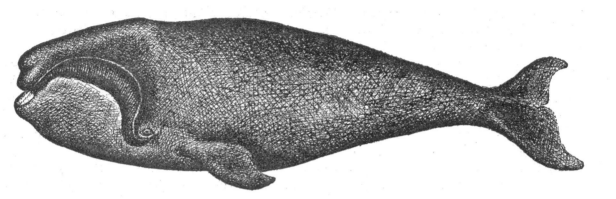

RIGHT
WHALES

FROM HUNTED LEVIATHAN
TO CONSERVATION ICON

David W. Laist

JOHNS HOPKINS UNIVERSITY PRESS ✳ BALTIMORE

JOHNS HOPKINS UNIVERSITY PRESS
2715 North Charles Street
Baltimore, Maryland 21218-4363
www.press.jhu.edu

Library of Congress Cataloging-in-Publication Data

Names: Laist, David W. (David Wallace), 1949– , author.
Title: North Atlantic right whales : from hunted leviathan
to conservation icon / David W. Laist.
Description: Baltimore : Johns Hopkins University Press, 2017. |
Includes bibliographical references and index.
Identifiers: LCCN 2016005168| ISBN 9781421420981 (hardcover : alk. paper) |
ISBN 9781421420998 (electronic) | ISBN 1421420988 (hardcover : alk. paper)
| ISBN 1421420996 (electronic)
Subjects: LCSH: Northern right whale. | Northern right whale—Conservation.
| Northern right whale—Pictorial works.
Classification: LCC QL737.C423 L35 2017 | DDC 599.5/273—dc23
LC record available at https://lccn.loc.gov/2016005168

A catalog record for this book is available from the British Library.

Special discounts are available for bulk purchases of this book.
For more information, please contact
Special Sales at 410-516-6936 or specialsales@press.jhu.edu.

Johns Hopkins University Press uses environmentally friendly
book materials, including recycled text paper that is composed of at least
30 percent post-consumer waste, whenever possible.

To my wife Judy,
and to the memory of John R. Twiss Jr.

CONTENTS

Preface ix
Acknowledgments xvii

1. Rescuing Nantucket	1
2. What's in a Name?	16
3. Foraging with a Smile	33
4. Evolution	55
5. The Origin of Whaling	72
6. Medieval Whaling in Northern Europe	83
7. Ghost Whalers	101
8. Basque Whaling in Terranova	120
9. The Dawn of International Whaling	149
10. A Fitful Start for Colonial Whalers	165
11. Long Island Whaling	179
12. Cape Cod Whaling	196
13. Nantucket, Martha's Vineyard, and Cape May	214
14. Whaling from the Carolinas to Florida	228
15. Estimating Pre-exploitation Population Size	251
16. A Second Chance	266
17. A Dedicated Recovery Program	276
18. Nobody *Wants* to Hit a Whale	300
19. Slow Speed Ahead	317
20. Entanglement	347
21. Oh, What a Tangled Web	367
22. Ten Thousand Right Whales	406

Appendix 419
Index 423

A GOOD CASE might be made for recognizing the oral history and rock art passed down through the generations of early cultures as the forerunners of all science. For untold millennia, primitive societies entrusted elders with responsibility for making major tribal decisions and serving as custodians of community legends, lore, traditions, and myths. Their stories, representing a collective wisdom molded by generations of observation, were every bit the functional equivalent of mathematical models used by today's scientists. These were the first attempts at making sense of a perceived world by drawing on past experience to predict and prepare for the future.

Today, historically based perspectives are often overlooked—or at least underutilized—by biologists and resource managers trying to ensure the survival of rare and endangered wildlife. Perhaps understandably, the urgency of looming threats compresses considered time frames to the immediate past and what might occur in the immediate future. Yet, often lost in this crisis mindset is the long-term view of a species' existence. As elders in today's indigenous societies are quick to point out, it is shortsighted not to consider traditional ecological knowledge accumulated over generations of study and reflection. Understanding shaped by events over a full spectrum of time scales—from evolutionary development through recorded history to living memory—can place today's pressing problems in a new light that offers invaluable insight and inspiration.

Countless books line the shelves of bookstores and libraries recounting the lives and achievements of famous and not-so-famous figures in human history. They routinely trace family lineages and examine every aspect of their subjects' lives from birth and pedigree, through interactions with contemporary figures and events, to death and legacy. Yet one would be hard pressed to find a single volume relating the full history of even the most iconic wildlife species making up our natural heritage. Because of humans' voracious appetite for natural resources and our pervasive effect on the natural world, the history of almost every species intersects with our own to at least some degree. Indeed, many do so in forgotten or discounted ways that have influenced the course of human events every bit as much as the world's most famous people have. For example, consider the prodigious run of spawning American shad that swam up Pennsylvania's Schuylkill

River in the spring of 1778, right under the noses of British troops ensconced in Philadelphia, toward a starving colonial army encamped at Valley Forge. Without that fortuitous run of shad so desperately needed by George Washington's beleaguered force, the American Revolution might have sputtered to an ignominious end then and there.[1]

Each species has a distinctive way of life and relationship to human endeavors and events. Often, this is just plain fascinating in its own right. Yet on another level, awareness of a species' history has the power to change the way we think about the animal or plant. It engenders a deeper respect for its place in today's world and how it has influenced our own lives. For some people, that history might even spark ideas for new lines of study or a new interest in ensuring a mutual coexistence with our own pursuits.

As the world's human population soars past 7 billion and its ever-widening sphere of influence reaches even the most remote corners of the planet, we are thoughtlessly and unwittingly exterminating species at warp speed. In this process we are losing living components of our natural heritage at a rate as astounding as it is sobering. The ongoing disappearance of species is on a track to rival the greatest mass extinctions since multicellular life first exploded across our planet some 600 million years ago. Not counting the ubiquitous array of bacteria and other microbes, current estimates place the number of species alive today at between 5 million and 11.5 million. To date, scientists have identified and described perhaps 1.5 million to 2 million species.[2] At their current rate of disappearance—estimated by some authorities to be between 11,000 and 58,000 per year, or 0.1 to 1 percent per *decade*—half of all species on Earth today will be lost before scientists can even identify them.[3] Among the relatively small group of vertebrates, we have lost at least 322 species since 1500, and nearly 30 percent of those that remain have declined over the past four decades.[4] The loss of each species saps the resilience of the natural ecosystems we all depend on in ways that are largely unrecognized or unappreciated.

With the ongoing planetary plunder pushing so many species toward extinction, one hardly knows where to begin to staunch the bleeding. It is clearly impossible to save all species, but that is no excuse for not trying to save all we can. Triage to stem the losses, however, raises several questions: Are some species more important to protect than others? Is an endangered whale more important than some other species? Are some individual species more important to protect than preserving habitats that might save many species? There are no simple or right answers to such questions, yet appeals on behalf of one species or another are constantly being made for a larger share of the meager resources available for such tasks. Almost all of them are justified; almost all are underfunded or unfunded.

There are certainly strong arguments for special attention to North Atlantic right whales, *Eubalaena glacialis*. After centuries of persecution at the hands of commercial whalers, the paltry number alive today—perhaps 450—makes them a candidate for the world's rarest large whale and one of the most endangered mammals anywhere, land or sea. Although whaling is no longer a threat, unintended entanglements in fishing gear and collisions with ships now account for roughly half of all North Atlantic right whale deaths, effectively doubling their mortality. Because of their low numbers, the few deaths inflicted today are enough to slow or even prevent recovery; they are functionally comparable to the thousands killed by whalers in the past that led to the species' present predicament. Considering the enormous size of right whales, the amount of prey they consume, and their vast oceanic range, restoring a healthy right whale population that is now effectively "ecologically extinct" would reestablish a major link in the nutrient cycle supporting marine food webs and help return regional ecosystems in many parts of the North Atlantic to a healthier, more productive state.[5]

Reasons for special attention to right whale recovery, however, are not purely scientific. A great number of people may simply care more about some species than others. In a democratic society, views and expectations shared by high numbers of people count. Large mammals in general, and whales in particular, regularly rank at the top of the list of "charismatic megafauna" people care about most. This interest and concern is amply apparent in the booming whale-watching business. In 2008 alone, more than 13 million people went whale watching and spent an estimated $2.1 billion worldwide.[6] The broad appeal of whales and their kin is also manifest in legislation, such as the US Marine Mammal Protection Act, which established national protection policies. At least part of the reason for this fascination undoubtedly lies in our shared mammalian body plan and life history, which has been miraculously modified in whales and other marine mammals for life in an ocean environment so alien to our familiar terrestrial world. Another reason is simply their enormous size. Approaching 60 feet in length and weighing over 70 tons, right whales are the third largest of all the great whales and one of the largest animals ever to live on Earth. Some would argue with good cause that if we can't protect one of the largest species ever to live on Earth, what species should we protect?

Still others view right whales as a living link to the romantic, if remorseless, era of Yankee whaling. For them, the remaining right whales represent a second chance to exercise the sense of responsible stewardship so clearly lacking in the sail-based whaling era of the 1700s and 1800s, and so poorly practiced in the age of modern whaling in the 1900s. These and other factors contribute to the broad success of campaigns to solicit funds and demand action for the protection of right whales. Although great progress has been made in restoring many of the world's depleted marine mammals, including whales, the North Atlantic right whale is a prime example of what remains to be done.

Nevertheless, efforts to protect North Atlantic right whales have been an exercise in patience and frustration. Today, they are victims of collateral damage, including unintended deaths due to incidental entanglement in commercial fishing gear and collisions with ships. From the perspective of the tens of thousands of fishermen and mariners plying their trade in right whale habitats, such interactions are extremely rare. Indeed, the vast majority have never been nor will likely ever be involved in an incident causing the death or serious injury of a right whale. And even if they were, they are unlikely to be aware of it. Entangled whales typically swim off with fishing gear, leaving fishermen with no clue as to what happened. Similarly, collisions between whales and ships almost always go unnoticed by mariners. The enormous size of today's oceangoing vessels dwarfs even the largest whales, making a collision no more perceptible to their crew than to passengers on a jumbo jet striking a small bird.

Extensive data on right whales collected by scientists over the past 40 years offer a powerful tool to guide conservation strategies. However, it would be shortsighted not to draw all we can from a more complete historical understanding of the species' existence. The capacity for history to stimulate interest in and sharpen perspectives on current issues has long been recognized. Aristotle counseled, "if you would understand anything, observe its beginning and development." Similar sentiments have echoed in lofty terms down through time in the writings of eminent thinkers in all walks of life. The French historian Alexis de Tocqueville warned, "when the past no longer illuminates the future, the spirit walks in darkness," and Mark Twain famously quipped, "the past may not repeat itself, but it rhymes." Of course, others with equal justification have adopted a more cynical view. In the early eighteenth century, Voltaire observed, "there is no history, only fictions of varying degrees of plausibility." George Bernard Shaw distilled this thought to an even pithier barb: "what we learn from history is that we learn nothing from history." Even historians routinely resort to self-deprecating humility. American historian W. Stull Holt likened history to "a damn dim candle over a damn dark abyss," while the highly regarded British historian Arnold Toynbee declared history is "just one damned thing after another." In the end, however, history is what we choose to make of it—an interesting diversion, a source of appreciation for our heritage, or inspiration for new or perhaps forgotten ideas for charting a better course ahead.

When the respected author David McCullough made the rounds to promote his book *The Greater Journey: Americans in Paris* (2011), he always made a point of citing a phrase penned by one of those Americans. That nineteenth-century pilgrim was Charles Sumner, a US senator from Massachusetts. When Sumner described his first impressions of a European culture that was punctuated at almost every turn by both grand and unassuming historical markers dating far back into antiquity, he remarked on the "prestige of age." For McCullough those words encapsulated the message of his book, if not the source of his own love affair with history and history's broad appeal to the public at large. The phrase was part of an entry in Sumner's journal in which he compared Europe's seasoned culture to the youthful American character forged by European immigrants who had been severed from a long historical perspective. Writing with a clear twinge of remorse, Sumner observed that in America, "there is none of the prestige of age about anything, . . . we are ready at any moment to lay our hands on any custom or mode of business and modify it; and though [we] may sometimes suffer by the proclivity, yet it is the means of keeping us constantly on the *qui vive* for improvement."[7] Concern over the implications of these opposing possibilities—change versus status quo—still lies at the heart of debate over many public policy issues, including wildlife conservation.

Precisely how history's lessons might manifest their influence is impossible to predict, but their power to do so is undeniable. The biological sciences have their own pantheon of great thinkers and big ideas about which many books have been written. But when it comes to the history of individual species, formal attention is largely restricted to the disciplines of evolution and a range of studies loosely termed "life history." Evolution, of course, maps the changes in a species across geological time; "life history" focuses on population statistics and demographic factors describing how species make their living over the course of a few generations (e.g., changes in population abundance, migratory patterns, how food is obtained, reproductive strategies, age/sex survival rates, and various other life cycle traits). Often overlooked—or reduced to broad generalizations—are efforts to add the dimension of a species' interactions with us across the full sweep of human history.

Many have investigated segments of species' histories in great detail, but few have tried to stitch those pieces together into a broad tapestry describing their existence across the full arc of time. Yet, to fully appreciate a species and its current status, one needs to consider all time scales, including long-term geological, archeological, and historic time frames; the shorter-term span of a person's memory or a species' generation; and even more myopic time frames measured in the months or years of regulatory and economic planning.

Few species emanate the "prestige of age" as much as North Atlantic right whales. From their ancient evolutionary lineage, to a potential life span far in excess of 100 years, to their surly, grizzled appearance that projects a longevity few species experience, right whales exude an antiquity deserving of respect. As the favored target of whalers for more than a thousand years, they were the first whale to be hunted commercially and the foundation on which all commercial whaling was built. North Atlantic right whales influenced the lifestyle of coastal residents from the end of the Iron Age through the European medieval era and Renaissance, and from the colonial period up to the present in eastern North America. In the process, they paid a heavy price. Eliminated throughout much of their former range, North Atlantic right whales now occur only along the east coast of North America in numbers that are likely smaller than the number killed in a single year at the height of their pursuit and smaller than all but a few other mammals.

This book presents a panorama of the North Atlantic right whale's existence, including how they evolved, how they arrived at their current imperiled status, and what is being done to safeguard their survival. Although the whales are rare today, their rich history and importance to coastal communities throughout time make them a true icon of the North Atlantic. As the pyramids symbolize both ancient and present-day Egypt, right whales serve as living monuments to the North Atlantic and the intrepid mariners who chased them on both sides of its expanse. This book tells the whale's epic story from evolutionary roots to the current conflicts that threaten tentative steps back from the brink of extinction. Their extraordinary path from small, obscure, terrestrial creatures to the giants of today, completely divorced of all ties to land, is one of the most remarkable examples of biological adaptability in the annals of life on Earth. Hunted continuously from at least the tenth century—and probably earlier—to the early twentieth century by a succession of Viking, Basque, English, Dutch, German, and American whalers, the North Atlantic right whale was one of the first marine species to be pushed to the edge of extinction by human hands.

For 35 years I have had the privilege of working for an independent federal agency called the Marine Mammal Commission. Created by the Marine Mammal Protection Act of 1972, the commission was established as something of an insurance policy to ensure that federal agencies consider the nation's special interest in marine mammals and adhere to the act's provisions for restoring and maintaining numbers at optimum sustainable levels. To do so, the commission provides science-based advice and recommendations on any federal actions or programs that might affect marine mammal populations. In the course of its efforts, it has played an instrumental role in forming and guiding a dedicated North Atlantic right whale conservation program.

Having joined the commission's staff just five years after it opened its doors and serving most of the years since then as its staff lead, responsible for developing its positions, recommendations, and reports on right whale protection issues, I have had a front-row seat for watching, and helping to guide, development of the right whale conservation program. During that time the program has grown from virtually nothing to a cooperative undertaking that involves thousands of people in various federal and state agencies, academic institutions, nongovernmental conservation groups, and industry organizations. In the course of participating on various committees, panels, and teams convened to weigh aspects of right whale conservation, I have often been amazed at the intensity of resistance to even modest suggestions for voluntary action. At least in part, that intransigence appeared rooted in underappreciated recognition of the species and of what was at stake with its possible loss. It was that thought that led to this book. By providing a historical perspective on North Atlantic right whales, it is hoped that readers—be they members of the public, stakeholders subject to regulation, managers charged with balancing conservation and development interests, or students contemplating a career in marine conservation—might place ongoing conservation needs and efforts on right whales' behalf in a new context. This context would be one that provides a more complete understanding and appreciation of the species, its significance, and the challenges facing its recovery. By the same token, it is hoped that this book might spark new ideas for research or conservation among scientists, resource managers, fishermen, and vessel operators.

NOTES

1. McPhee, J. 2002. *The Founding Fish*. Farrar, Straus and Giroux. New York, NY.

2. Pimm, SL, CN Jenkins, R Abell, TM Brooks, JL Gittleman, LN Joppa, PH Raven, et al. 2014. The biodiversity of species and their rate of extinction, distribution, and protection. Science 344(6187):987–997, 987.

3. Costello, MJ, RM May, NE Stork. 2013. Can we name Earth's species before they go extinct? Science 339:413–416.

4. Dirzo, R, HS Young, M Galetti, G Ceballos, NJB Isaac, and B Collen. 2014. Defaunation in the Anthropocene. Science 345:401–406.

5. McCauley, DJ, ML Pinsky, SR Palumbe, JA Estes, FH Joyce, and RR Warner. 2015. Marine defaunation: animal loss in the global ocean. Science 347(6219):247–254. Lavery, TJ, B Roudnew, J Seymour, JG Mitchell, V Smetacek, and S Nicol. 2014. Whales sustain fisheries: blue whales stimulate primary production in the Southern Ocean. Marine Mammal Science 30(3):888–904.

6. O'Connor, S, R Campbell, H Cortez, and T Knowles. 2009. *Whale Watching Worldwide: Tourism Numbers, Expenditures and Expanding Economic Benefits, a Special Report from the International Fund for Animal Welfare*. Prepared by Economists at Large. Yarmouth Port, MA, US, p. 8.

7. Pierce, EL (ed.). 1878. *Memoir and Letters of Charles Sumner*. Vol. 1, 1813–1838. Sampson Low, Marston, Searls, & Rivington. London, pp. 217–218 (quotation).

ACKNOWLEDGMENTS

WHILE WITH the Marine Mammal Commission, I have had the great fortune to meet and work with many of the world's foremost experts on North Atlantic right whales. When I began working on these whales in the early 1980s, there were perhaps a few dozen scientists knowledgeable about the species and even fewer who had ever seen one. Foremost among them were David and Melba Caldwell, Bill Schevill, Bill Watkins, Ed Mitchell, Howard Winn, John Prescott, Randy Reeves, Bob Brownell, and Peter Best; today there are hundreds of scientists, resource managers, and conservationists actively contributing in significant ways to understanding the species' biology and promoting their recovery. Without their dedication and efforts, this book would have been impossible.

I raise a glass in special thanks to those who gave generously of their time to review drafts of various sections, including Chris Clark, Bob Hofman, Bob Kenney, John King, Ed Lyman, Dave Mattila, Stormy Mayo, Bill McLellen, Jim Mead, Bill Perrin, Charlie Potter, Lori Quakenbush, Andy Read, Bruce Russell, John Strong, Hans Thewissen, Steve Thornton, Jim Valade, Dave Wiley, Doug Wartzok, and Sharon Young. Many other colleagues kindly helped directly or indirectly in the writing of this book by providing information, photos, suggestions, and other vital forms of assistance for which I am exceedingly grateful. In particular I would like to thank Dag Altin, Lutz Bachmann, Michael Barkham, Mark Baumgartner, Cathy Beck, Peter Best, Bob Bonde, Christina Brito, Moe Brown, Annalisa Burta, Phil Clapham, David Cottingham, Susan Crockford, Greg Donovan, Michael Dyer, Fernando Felix Grijalva, William Fitzhugh, Bill Fox, Cindy Gibbons, Marianna Hagbloom, Allison Henry, Mina Innes, Kate Jackson, Anne Jensen, Elizabeth Josephs, Christin Khan, Amy Knowlton, Scott Landry, Greta Lindquist, Judy Lund, Dan McKiernan, Richard Merrick, Michael Moore, Daniel Odess, Storrs Olson, Patrick Ramage, Randy Reeves, Jooke Robbins, Robert Rocha Jr., Vicky Rowntree, Sam Simmons, Pat Slattery, Tim Smith, Robert Suydam, Vicki Szabo, Peter Thomas, John Twiss, Leslie Ward-Geiger, Mike Webber, Alex Werth, and Øystine Wiig.

The resource specialists, curators, and other staff at various museums and libraries were a godsend in preparing this book. Their knowledge of collection holdings and their patience with endless requests were simply invaluable. My sincere gratitude goes to Eleanor Goodwin at Marine Mammal Commission; Richard Green, Leslie Overstreet, Erin Rushing, and Martha Rosen at the Smithsonian Museum of Natural History Library; Melanie Corriea, Mark Procknick, and Robert Rocca at the New Bedford Whaling Commission; Christopher Barfield at the Fort Lauderdale Historical Society; Zack Studenroth and Sundy Schermyer at the Southampton Town Archives; Tom Lisanti at the New York Public Library; Kim Anderson at the State Archives of North Carolina; Pam Morris at the Core Sound Wildlife Museum and Heritage Center; Marie "Ralph" Henke at the Nantucket Historical Society; Kathleen Monahan at the University of North Carolina Wilson Library; Dale Sauter at the J. Y. Joyner Library at East Carolina University; Margaret Corfuto at the North Carolina Museum of Natural Sciences; Jaclyn "Bam" Penny at the American Antiquarian Society; Jordan Goffin at the Providence Public Library; Soco Romano of the Musea Naval; and Steven Borkowski at the Provincetown Museum.

For their gracious help with putting this book together, I also thank Alain Harmon and Steve Thornton, who assisted with various figures and tables, and Linda White and Mina Innes for their yeomen's work preparing the various maps.

This book also would have been unachievable without the dedicated work of hundreds of people who have devoted much of their careers to various right whale research and management issues. The true heroes that made this book possible and have given this species its chance to survive, however, are the hundreds of people who spent countless hours photographing and archiving whale identification data; tagging and tracking whales to identify where they move and why; responding to reports of entangled whales and disentangling them; carrying out necropsies and compiling stranding records; deploying acoustic monitoring devices and analyzing recordings of whale calls; collecting and analyzing tissue samples and even fecal samples for genetic, contaminant, and chemical analyses; combing available data sets and generating models to elucidate biological and ecological trends; sifting archival records for historical accounts; participating in innumerable meetings to debate management needs; and developing and implementing regulatory strategies.

Finally, I would like to thank the Island Foundation, the International Fund for Animal Welfare, and the World Wildlife Fund for providing the funding that made this book possible.

NORTH ATLANTIC RIGHT WHALES

1

RESCUING NANTUCKET

THE CALL ARRIVED at the old cedar-shake house around noon on a warm June day in 1997. Other than a small sign tucked in the corner of a front window, there was little to suggest that its weather-beaten frame and a similar structure directly behind it had been appropriated for use as a small marine biological laboratory. Purchased just a few years earlier by a small cadre of newly minted biologists looking to ply their passion for marine life, the property had become the base of operations for an upstart research organization called the Provincetown Center for Coastal Studies. Just a few feet from its front door lay Commercial Street. Laid out nearly 300 years earlier as the town's main artery, the street retains its original no-nonsense name rooted in a colonial era when New England settlers cut straight to the chase with honest simplicity. Its location and street signs still stand in mute testimony to its maritime origins.

Only a sparse crowd of preseason tourists strolled by, soaking up the salty charm of the old fishing village. Now one of those quirky end-of-the-road destinations, Provincetown was incorporated in 1727 to take advantage of the rich fishing grounds lying at its doorstep. Before that, settlers on Cape Cod held a dim view of the peninsula's remote tip. To them it was known as "Hell Town" for the raucous collection of lawless, makeshift camps occupied by a hoard of questionable characters who descended seasonally on its sandy shores to drink, fish, and salvage what they could of dead whales cast upon its beaches. On a map, Provincetown is tucked in the crook of a barb on the tip of a scorpion's tail that juts 25 miles out into the cold, gray-green Atlantic from the Massachusetts mainland.

The friendly yet businesslike voice on the other end of the phone was a young coast guard officer calling to relay a radio message just received from a commercial gill-net fishermen on his way back to port. He reported the sighting of a right whale tangled in a mess of rope a mile off the beach near the entrance to Chatham

Harbor. This was 40 miles down the cape, where the peninsula bends sharply north. A coast guard patrol boat had already been dispatched and was standing by, awaiting arrival of the Provincetown Center's biologists for an attempt to free the whale from its burden of fishing gear. Under arrangements set up just a year earlier, in 1996, the Center had been sanctioned by the National Marine Fisheries Service to carry out the dangerous work of disentangling free-swimming whales caught in lines, buoys, traps, or other fishing paraphernalia.

A New Tool for the Tool Box

The mantle of responsibility for this task fell to the Center in the late 1980s as much by accident as design. For 15 years, two of its biologists—Charles Mayo III, known to everyone as "Stormy," and Dave Mattila—had been studying the humpback whales and right whales that returned each spring to forage in the productive waters surrounding the cape. Stormy, his wife, and a close college friend had invested their collective finances to form the Center in the early 1970s, and Stormy and Dave were part of a vanguard of dedicated young scientists with a passion for whales; these devotees were working out of a handful of research centers scattered around New England. They were all experimenting with a new research technique called photo identification—a method of identifying individual whales from scars and other unique marks etched on the animal's exterior. By laboriously tracking individuals one sighting at a time, they hoped to piece together their movements, routines, and life history. It wasn't long before researchers found that at least a few whales were seen every year with lines and floats from commercial fishing gear wrapped around tail stocks and flippers, or lodged in the narrow gaps between the filamentous baleen plates whales use to filter minute prey from their watery world. From the occasional dead whale found ashore with deep cuts and abrasions inflicted by chafing ropes, they knew entanglements could be a death sentence for whales unable to rid themselves of their encumbrances.[1]

Just a few years earlier and a thousand miles to the north, a young scientist named Jon Lien was seeking a degree at Memorial University in St. John's, Newfoundland, in the obscure field of biopsychology, and he had taken an interest in whale vocalizations. His odd choice of study had become a topic of gossip among the small seafaring community. It was therefore not entirely surprising when, in 1978, he received a call from a distraught fisherman asking for help in removing a humpback whale from his cod trap. A cod trap is a large rectangular box of net walls anchored to the sea floor and suspended from the surface by a line of floats. At one end of the box is an opening fitted with a pair of angled net walls to funnel fish into the trap. To lead fish in, a long, straight "leader" line of netting is hung along a line of poles at the surface and stretches to the bottom; the netting also extends a quarter mile from the trap's opening. When cod chase their prey—typically small schooling fish such as capelin—their quarry scoots through the mesh of the leader netting, forcing the larger cod to break off their pursuit and turn. Those turning the fisherman's way (toward the trap) follow the leader into the net enclosure, where they stay penned until scooped out every few days. Humpback whales also feed on schools of capelin, and the whale that Lien had been asked to help was likely trapped in much the same way.

When Jon accompanied the fisherman in his boat, they arrived at the scene and found the whale in the heart of the trap but otherwise free of entangling ropes. After surveying the situation, Jon suggested pushing the float line holding up one side of the trap's walls down toward the sea floor. When they did, the whale swam out, inflicting no damage to the trap's netting. At first, Jon considered this a unique opportunity to help both a fisherman and a whale, while also getting a close look at the subject of his selected area of interest. As word of his deed spread, Jon began receiving a handful of calls each summer from fishermen with the same problem. He soon learned that the frequency of such events had increased substantially over the preceding decade—a trend fishermen attributed to the suspension of commercial whaling for humpbacks in the early 1970s. With each entrapped whale threatening a fisherman's considerable investment in nets and their ability to make a living, the problem was becoming far more than a curious inconvenience. To help, Jon convinced the local and provincial governments to provide him with a small grant to start a whale-release program. Before long, he found himself besieged with calls; he was spending more time extracting whales from fish traps than on his own research. Within a few years, he had released several hundred whales.

When Stormy and Dave learned of Jon's exploits, they began wondering whether it might be possible to remove fishing lines from the free-swimming whales they were seeing. At the time, corralling a 50-foot whale in the open ocean for disentanglement was considered an impossible idea that others would have scoffed at and quickly discarded—assuming they had given it more than a passing thought in the first place. But Stormy and Dave were freethinkers willing to give harebrained notions a fair hearing. So before dismissing their fanciful musings entirely, they sought a second opinion from someone well versed in local lore about whales, fishing, and Cape Cod's fickle marine environment. For this, they had to look no further than Stormy's father, Charlie Mayo II, a retired fisherman and mariner in his late seventies, who knew the waters and marine life around the cape about as well as anyone.

Mayo roots on Cape Cod reached back to its first English settlers. The Reverend John Mayo, born in Dorsetshire, England, in 1598, brought his family to the Plymouth Colony aboard the *Truelove* in 1635. Arriving just 15 years after the *Mayflower*, the reverend and his family helped secure the first successful English colony north of Virginia. Ever since, the cape's sandy soils have felt the tread of Mayo footsteps, and most of the clan has pursued livelihoods in fishing or related marine pursuits.

Mulling over Stormy and Dave's odd proposition, the elder Mayo drew on his knowledge of traditional local practices and his own experience. Rather than dismissing the idea outright, he suggested taking a page from the old whalers' handbook. As recently as the early 1900s, fishermen had hunted whales off Cape Cod using small dories, handheld harpoons, and drogues (wooden floats) tied to the end of harpoon line. In the 1920s, Charlie himself had hunted pilot whales. The trailing drogue served as a drag to tire the whale until it became so exhausted it had to stop and rest at the surface. When it did, whalers seized their chance to pull in close and deliver a fatal series of stabbing blows with a long, bladed lance, a task far easier said than done. Stormy and Dave reasoned from this that they might use a similar ploy with the drogue, but in this case with the intent of saving a whale's life rather than taking it. If they could "keg" an entangled whale by attaching a large fishing float to a trailing line already caught on the animal, it might be possible to tire the animal until it came to a stop. Once "pinned" in this way, the rescuers could then move in to cut the entangling lines and set the whale free. It was a simple idea. Risky? You bet; but they were still young and foolish.

The first chance to put their plan into action came in October 1984. Stormy received a call from a colleague in Gloucester, 50 miles north across Massachusetts Bay on Cape Anne, advising that a whale had been found anchored in a gill net a mile off the town's breakwater. With Dave away at a meeting, Stormy and his assistant Carol Carlson loaded up their "new" research vessel, a vintage 36-foot lobster boat donated to the Center and rechristened with the lofty name RV (Research Vessel) *Halos* (figure 1.1). When their task was completed, they headed out to sea under a low midday sun, past the stubby profile of Woods End Light guarding Provincetown Harbor, and set course across the bay. They arrived in Gloucester late in the day. After checking into a motel, they met up with a few other researchers to learn the latest news of the whale's situation and discuss their approach to this novel task. The following morning, Stormy and Carol headed out to the anchored whale on an oily, calm sea accompanied by a small flotilla of researchers and curious onlookers who had heard plans of the bizarre undertaking.

When Stormy and Carol arrived, they found the whale still entangled and anchored in place. It was a humpback that they immediately recognized as an animal they had first identified as a calf five years earlier. Stormy and his colleagues had been giving whales names that conjure up images of marks on the tail or body to help them remember each animal. This whale was a female named "Ibis" for a dark squiggle on her fluke. The mark resembled the ancient Egyptian hieroglyph of the wading bird that represented the god of wisdom and learning, Thoth. The whale had netting and rope running through its mouth and trailing back to the tail. After becoming entangled, the trailing net had apparently snagged on a small mountain of trash that accumulated off the harbor entrance where vessels dumped old fish traps. It was all the whale could do to lift its head to the surface and take a quick chuffing breath before sinking back into the depths, pulled down by the weight of its attached gear.

Fearing the whale would drown, Stormy and Carol approached in the inflatable Zodiac they had brought with them, and attempted to attach a large float to the

flimsy mesh netting on the whale in hopes of keeping it at the surface where it could breathe. After an hour of futile effort, they abandoned the tactic and returned to the *Halos* for a set of tools to cut the attached lines. As they did, the repeated surfacing of the laboring whale suddenly stopped. Minutes ticked by as anxious eyes stared at the spot where the whale had been surfacing, but with no sign of the whale. After waiting about twenty minutes and fearing the whale must have drowned, Stormy flipped on their dual channel fish-finder and detected a large lump on the sea floor about the same size as Ibis. Disheartened and with nothing more they could do, they bid thanks and farewell to colleagues who had been standing by, and pointed the bow of the *Halos* back to Provincetown for a long, sullen trip home. Their first attempt to disentangle a whale had been a complete failure.

A month later on a crisp Thanksgiving Day, a second opportunity presented itself on their very doorstep. A group of Center colleagues and friends were preparing a turkey dinner at the lab in the best New England tradition, when a lingering group of humpback whales yet to turn south for their annual winter migration to the Caribbean was spotted just outside Provincetown Harbor. Enticed by good weather and unusually calm conditions, Stormy, Dave, Stormy's father, and a few lab workers decided to squeeze in a quick trip aboard the *Halos* to try their hand at recording a singing humpback whale. Although humpback whales were then thought to sing only in mating groups on tropical calving grounds, Stormy had recorded a singing humpback whale the previous spring off Provincetown. It was the first time anyone had heard a singing whale on a high-latitude feeding ground,

Figure 1.1. Researchers studying a humpback whale in 1983 aboard the Provincetown Center for Coastal Studies's first research vessel, a donated lobster boat refurbished and renamed the RV *Halos*. (Image courtesy of the Center for Coastal Studies, Provincetown, MA.)

and Stormy suspected the best time to hear them might be just before or just after their annual trips to and from their calving ground. With remaining whales about to head south for the winter, it was an ideal time to test the hypothesis.

And so on Thanksgiving Day 1984, Stormy, Dave, and a small group of Center colleagues found themselves aboard the *Halos* heading out of Provincetown Harbor with Stormy's father at the helm, a Zodiac in tow, and a set of hydrophones at the ready. Their plans, however, quickly changed. The first pair of whales was spotted barely a mile off the dock. As they approached for a closer look, a cry went up that one of the whales was entangled. Moments later, another cry rang out: "It's Ibis!" The whale they had given up for dead a month earlier off Gloucester had apparently pulled free of its snag and managed to swim underwater, towing its burden of net far enough to escape notice of all those watching the first disentanglement attempt.

As Stormy gazed intently at the whale to assess its condition, he noticed that it was distinctly thinner—undoubtedly the result of laboring to tow a heavy net and the consequent difficulty of feeding. The netting was still caught in the whale's mouth and trailed down both sides back to its tail. At the base of the tail a wrap of line had cut a deep gash into the skin. From there a twisted snarl of rope and netting extended a hundred feet behind the tail flukes. Not about to lose a second chance at disentangling Ibis, Stormy and Dave held a quick conference. With none of the tools they needed onboard, they decided Dave and a colleague would take the Zodiac back to the lab to gather up the necessary equipment, while Stormy and the others kept track of the whale in the *Halos*. Arriving back in port, it was a short dash to the Center where the two grabbed a second Zodiac, more lines, and a large polyball fishing float to keg the whale. Less than an hour later they rejoined the *Halos*, which was still close on the whale's heel.

After another quick conference to map out their approach, Dave realized he had forgotten the grapple used to catch hold of the trailing line and attach their kegging float. In a flash of inspiration, he suggested using a small boat anchor with retractable prongs that they had onboard. When locked in its unfolded position, the prongs could catch hold of the trailing line. With that, Stormy and Dave jumped back into the Zodiac with Dave at the helm and Stormy in the bow. As Ibis labored ahead, Stormy and Dave approached cautiously from behind. The whale accompanying Ibis hovered close by, clearly agitated as the Zodiac advanced. Stormy made a quick toss with the improvised grapple a short distance behind the flukes. It caught hold on the first try. When they pulled the line in, Stormy deftly tied on a tethered polyball float as the whale surged ahead in an attempt to escape. Retreating to see what would happen next, Stormy and Dave followed the bobbing float in their inflatable. Close behind was the *Halos* with the elder Mayo at the helm and an anxious crew ready to fish the would-be rescuers out of the icy water should something go awry.

Lo and behold, just as accounts of early whalers described—yet still a complete surprise—the exhausted whale came to a stop in just a few minutes. Drifting listlessly at the surface, its stamina was clearly sapped by weeks of towing the net. Stunned at their success and not quite sure what to do next, they decided to split up; Dave in one Zodiac would work on the lines caught around the head, while Stormy in the other would attack the lines wrapped around the tail. Grabbing a fillet knife as his only tool, Dave and a coworker eased their inflatable along one of the trailing lines until they were even with the whale's head. Leaning precariously over the side just a few feet from the whale's clenched jaw (figure 1.2), Dave cut the line as close to the mouth as he could. When that was done, they then dropped behind and began pulling up along the line trailing from the other side of the whale's mouth.

As Dave and his coworker proceeded with their task, Stormy pulled up to the whale's back end and began working on the ropes around the base of the tail (the tail stock) and flukes. In the whale's feeble state, its attempt to raise its seven-foot span of tail flukes succeeded in lifting them only a few feet before they flopped weakly onto the bow of the Zodiac directly into Stormy's lap. In a moment of eerie, frightening calm, Stormy's sense of curiosity took hold. Wondering if he could determine how much fat remained in the whale's blubber layer to fuel its journey south, Stormy reached forward and put his hand directly into the deep gash on the whale's tail stock. When he did, the wound gushed blood, yet the exhausted whale was too weak to react. As Stormy would recount the experience many times afterward, it was a moment of instant epiphany arousing a new appreciation and

Figure 1.2. Dave Mattila, knife in hand at the bow of an inflatable Zodiac, approaches the head of Ibis, an entangled humpback whale, in the first successful rescue to remove lines from a free-swimming whale. This occurred off Provincetown Harbor, Massachusetts, on Thanksgiving Day 1984. (Image courtesy of the Center for Coastal Studies, Provincetown, MA.)

awareness of whales as sentient, living beings. With his task unfinished, however, he resumed his work cutting the trailing lines.

Just as Stormy successfully cut the last line on the flukes, Dave had pulled up to the other side of the whale's head where a long length of line still trailed from her mouth. With only a short, bitter end now extending from the other side of the mouth, Dave tried to pull the remaining line through. The line held fast and all he accomplished was drawing the Zodiac flush up against the side of Ibis's mouth. With Stormy shouting words of encouragement from the other boat, a surge of adrenalin pumped through Dave's system. He again leaned over the side and began sawing the rope with his fish knife. As he did, the whale lowered her head and opened her mouth, and the remaining line pulled through. The two Zodiacs immediately dropped back and, with a stroke of the flukes, Ibis accelerated ahead, with the entangling net left behind. She was free. When they retrieved the netting, they found it strewn with rotting dogfish and other fish that it had apparently been set to catch. The condition of the degraded catch suggested that Ibis had become entangled only a short time before the first attempt to free her a month earlier.

By the time they returned to port and made their way to the Center, Thanksgiving dinner had long since been cleared from the table and relegated to the refrigerator for leftovers. Exhilarated by their accomplishment, the loss of a festive meal with friends and family seemed a small sacrifice. For the first time, a free-swimming whale had been successfully disentangled. Without fully realizing it, their ingenuity and commitment to try something novel had added a new technique to the sparse chest of tools available to recover endangered whales. Now convinced their harebrained scheme was not so harebrained after all, they set to work in the months and years that followed crafting a set of tools, including knives and snaphook line grabbers that could be fitted to poles, to be used to disentangle whales at a safer distance. Over the next 10 years the Center fielded dozens of calls a year reporting entangled whales in local waters and successfully removed all or at least some rope from about two dozen whales, mostly humpbacks. Ibis is still seen regularly today.

With experience came proficiency. And with a growing recognition that whale entanglements were far more common than previously thought, the government agency with lead responsibility for protecting whales—the National Marine Fisheries Service—formalized an agreement in the late 1980s authorizing the Center to carry out rescue work. All large whales in US waters are protected under the Marine Mammal Protection Act of 1972, and most are also listed as endangered under the Endangered Species Act of 1973. With those laws come requirements that anyone handling marine mammals, even those with the best intentions, must have a permit authorizing the activity. Because of risks to both people and whales, the Fisheries Service initially limited authorization for disentangling whales to the Center. It also provided a modicum of funding to set up a local hotline for fielding reports of entangled whales, to fabricate specialized equipment, and to train disentanglement teams to work in other areas.

Upping the Ante

When the coast guard's call arrived that late June day in 1997, it was received with far more than the normal sense of urgency. For this call involved an entangled North Atlantic right whale—one of the world's most endangered animals on land or in the sea. Up to that time, Stormy and Dave had worked on only a few right whales and the chance to save even a single member of such an imperiled species was not to be taken lightly. After nearly 20 years of photo-identification work on the species' calving grounds off Florida and Georgia, and their feeding grounds off New England and Canada, only some 325 individuals had been identified. Because almost all the right whales seen after the early 1990s had already been identified, it seemed likely that the total population size was not much larger and had not been increasing. Adding to their apprehension was the fact that encounters with entangled right whales had shown them to be far more cantankerous and unpredictable than the comparatively docile humpbacks.

When most entanglement reports are received, a response is impossible; sea conditions are too rough, the weather report is uncooperative, the whale is too far away, or the report arrives too late in the day to reach the animal. In other cases, coordinates are relayed, but reporting vessels are unwilling or unable to stay with the animal until the response team can arrive. These factors drastically reduce the odds of later locating a whale that can move tens of miles in any direction in a few hours. But on this June day, the winds were calm, it was still early, the whale was close to shore, and a coast guard vessel was standing by to lead them to the whale and provide backup. Even then, however, there was no time to waste.

Stormy, Dave, and a new member of the Center's team, Ed Lyman (who had joined the small but growing staff in 1994), figured their quickest route to Chatham Harbor was overland. So rather than taking the Center's vessel all the way around the tip of the cape and down the ocean side, they began loading their pickup truck for a trip down to Chatham. Into the back went a folded, 17-foot, one-piece inflatable (the same type used by US Navy SEALs on combat missions), a 25-horsepower outboard engine, grappling hooks, four 30-inch polyball floats, ropes, two different sizes of sea anchors, an assortment of specially crafted knives and aluminum poles, heavy gloves, life jackets, and protective sports helmets complete with face masks.

After an anxious hour-long drive down Route 6—the two-lane artery connecting the string of villages and hamlets along Cape Cod's outer arm—they arrived at Chatham Harbor. There they found John Our, the gill-net fisherman who had reported the whale to the coast guard. John had passed off the whale-sitting duty to the coast guard when the patrol boat arrived and was now waiting patiently in port in his 40-foot floating office, the fishing boat *Miss Fitz*. John filled in the Center's response team on the whale's condition and location and made an offer they couldn't refuse—his help and his boat. Grateful for the assistance, they wrestled the inflatable and assorted gear out of the truck. With the whale close to shore and the seas calm, they decided to launch the inflatable from the harbor rather than deploying at sea from the *Miss Fitz*. So they carried the Zodiac down a boat ramp and loaded their gear; with Stormy and Dave in the inflatable and Ed joining John on the *Miss Fitz*, they headed out of harbor to rendezvous with the coast guard patrol boat tracking the entangled whale.

When the response team arrived on scene at about 3 p.m., the coast guard vessel and the whale were a good two miles offshore. The whale had begun moving south at a steady 6-knot clip (10 mph) parallel to Monomoy Island, an uninhabited 12-mile sand spit dangling south of the cape's elbow, like the waddle on a turkey's chin. The shallow depth prevented the whale from diving deep, easing the task of following its course. Their first order of business was to assess the situation and devise their disentanglement plan (figure 1.3). With no two entanglements exactly alike, they had to weigh a host of factors. How were the lines caught on the animal? How thick were they? Where were the strategic points for cutting entangling lines? How much line might they be able to remove? How much time did they have before sundown? Which tools best fit the situation? Were any traps, floats, or anchors still attached to the trailing lines? How was the whale behaving? What was the sea condition and weather forecast?

Figure 1.3. A coast guard patrol boat providing safety support for the response team keeps pace with the entangled right whale off Monomoy Island, south of Cape Cod, in May 1997. (Image courtesy of the Center for Coastal Studies, Provincetown, MA.)

Moving the inflatable slowly around the whale at a safe distance while it swam south, the response team estimated its size at two-and-a-half times the length of their 17-foot boat. From what they could see, the line ran through the mouth trailing out of either side and back over the whale's long upper jaw like a bridle on a horse. About 10 feet back on the line exiting the left side was a surface buoy system with a 30-inch green polyball float and a high flyer (a pole with a flag or radar reflector fishermen use to help relocate their gear at sea). The polyball was essentially the same type of spherical float they used to keg whales and it didn't seem to be slowing down the pace of this one. Trailing from the opposite side of the mouth was a line leading back to the flukes but angling downward, suggesting it was dragging something heavy—possibly a trap or a string of traps. The gear was typical of that used by offshore lobstermen whose traps can be as big as a small coffee table.

After taking stock of the situation, the first step was to stop the whale's forward movement by latching hold of a trailing line and kegging the whale with additional floats to increase the drag and tire the animal. With Dave at the helm and Stormy in the inflatable's bow, they approached the whale's left side near its head as it plowed steadily south, occasionally lifting and shaking its head like a defiant stallion. Using a 15-foot aluminum pole with a snap hook fastened to the tip, Stormy deftly snatched hold of the line with the surface buoy, pulled it in, and lashed on a 30-foot length of line with two more 30-inch diameter floats. Dave and Stormy then dropped back to let the buoys do their work. But this whale had other ideas. Unfazed by the added drag, it continued moving south at the same, steady 6 knots. Keeping to the surface, the whale occasionally lifted its head out of the water as it dragged the traps, surface buoy system, and two additional polyballs bobbing and weaving halfway underwater some 10 feet behind its flukes.

After a half hour with no sign of the whale slowing, much less stopping, they reached the edge of Nantucket Shoals. With the whale still charging ahead and time to work on the animal diminishing, the response team decided on a risky move. They would try to cut the line with attached traps, which was caught on the right side of the animal's head, while the whale was still underway. With Stormy remaining on the bow and Dave at the helm of their inflatable, they made their move and approached cautiously up the right side of the whale with the *Miss Fitz* keeping pace on the opposite side in an attempt to keep the whale moving straight ahead. While holding the 15-foot pole now fitted with a V-shaped knife, Stormy leaned hard into the inflatable's low side, braced himself against the ocean chop, and extended the pole out toward the whale. Dropping the blade over the trailing line, he pulled hard to cut the half-inch line. As he worked on the line and the inflatable was drawn steadily closer to the whale's head by Stormy's repeated tugs, the whale suddenly turned toward the inflatable. In a flash their puny craft was directly alongside the whale's head. With Dave barely an arm's length from its mouth, he then tried to point the inflatable's bow in the opposite direction, away from the whale.

This was far closer than they wanted to be to a right whale that had shown no signs of tiring. Had the whale lifted its head sharply at that moment, the would-be rescuers and all their gear would have been tossed into the sea beneath a 40-foot whale and whatever it was dragging. Stormy nevertheless kept working feverishly on the line. As he did, the whale lurched sharply in the opposite direction in an attempt to escape. At that same moment, the line with its heavy load broke and sank instantly to the bottom. This left only the shorter length of line from the left side of the whale's mouth with the high-flyer system and their added kegging buoys still attached. Realizing this animal was in no mood to tolerate even the best-intentioned interlopers, Dave and Stormy dropped behind to regroup. With a determination not to be denied, the whale resumed its steady 6-knot pace, rounding the southern tip of Monomoy Island and setting a new heading of south-southwest toward Nantucket and a destination only it knew.

By then, the 22-foot coast guard patrol boat out of Chatham that was serving as their emergency support had to return to port, and a 41-foot coast guard vessel out of Nantucket arrived to assume the vital safety vigil. With line still on the animal and the rescue team not about to leave a job only half done, they plotted their next move. After considering their close call with the disgruntled whale on the first cut, they were not anxious to work any further until the whale's energy was spent. They therefore decided to bring out their big gun—a large sea anchor. This was essentially a 6-foot-wide underwater parachute designed to maximize drag and keep the bow of a 40-foot vessel headed straight into a heavy sea. Easing their inflatable up to within a few feet of the whale's churning flukes, Stormy snatched hold of the residual trailing line with the high flyer, polyball, and additional kegging floats, and tied on the sea anchor. Sure that his would either bring the whale to a quick stop or pull the remaining line free, Stormy kept hold of the line rather than dropping back. This whale, however, again had other plans; the rescue team's confident expectations soon turned to awe. Towing the sea anchor, three kegs, the surface system, and the inflatable, the whale's speed slowed only slightly to 4 knots. It continued to move steadily south-southwest toward Nantucket with its mouth clamped shut and the line jammed tight.

With light fading and the coast guard's new relief vessel nearing the point where it too would need to turn for home, they decided to up the ante. They would try hooking the 40-foot *Miss Fitz* to the trailing line with the sea anchor and floats to further increase the drag. With the whale maintaining a steady four knots, John Our skillfully nosed the *Miss Fitz* to within 30 feet of the whale's flukes. When the line was within reach, Ed leaned over the bow, caught hold of the line, and tied on. As he did, John cut the engine to idle (figure 1.4). Despite the added drag of the 40-foot gill-net boat, sea anchor, and three 30-inch polyballs, the whale continued plowing steadily ahead at the same 4-knot clip. After several minutes had elapsed, John eased the throttle into reverse, determined to stop the whale. With the *Miss Fitz* still being dragged ahead, the rope connecting it to the whale began to quiver

under the strain and then snapped. This left the whale still dragging the sea anchor, kegging floats, and surface system.

The sun was now dipping close to the horizon. They had been working on the whale for nearly four hours. Realizing time was about to run out, the group decided to make one final attempt to remove the remaining line. It was a risky option, but they resolved to again maneuver their inflatable up to the whale's head and try cutting the line that carried the surface buoy system and all their added paraphernalia as close to the mouth as they could. With Ed now joining Stormy and Dave in the Zodiac, Dave eased up to the whale. As he pulled even with the head and slowed to match the whale's speed, Stormy reached out with a 15-foot pole and a V-notched line cutter. Once he caught hold of the line a few feet beyond its exit point from the whale's mouth, Ed joined Stormy in grabbing hold of the pole, and both began pulling hard in hopes of cutting the half-inch thick line. As they pulled, the boat crept closer to the whale's head.

Just as the line was finally cut clean through, the whale lifted its enormous tail high out of the water. With a dexterity as amazing and unexpected as it was frightening, its 10-foot span of flukes weighing nearly as much as a small car swung swiftly and deftly left, up, and over its back toward their boat. As if swatting a fly, it smashed its flukes down on the ocean surface with a loud resounding crack, just feet from the inflatable. A huge wall of white water instantly shot 30 feet into the air and down over the boat, drenching the occupants in a frigid shower of seawater. With hearts pounding, Dave angled the inflatable smartly away from the whale to safety and dropped behind to pick up the severed surface system, sea anchor, and

Figure 1.4. The whale rescue team attempts to stop the entangled right whale by tying John Our's gill-net boat, the *Miss Fitz*, to a trailing line. (Image courtesy of the Center for Coastal Studies, Provincetown, MA.)

kegging floats before returning to their support vessels. The whale, now left with only about 8 to 10 feet of line trailing from its mouth, resumed its course toward Nantucket and faded from view into the setting sun.

By then it was 7:30 p.m. The whale had towed various configurations of floats, traps, sea anchors, the inflatable, and the *Miss Fitz* more than 15 nautical miles from the spot they first began working. They were nearly halfway to Nantucket when they last saw the whale, but they had removed most of the gear, giving it a chance for survival. With a new appreciation of the dangerous nature of their disentanglement work and a sharpened respect for the awesome power of a right whale, the vessels parted ways and the exhausted rescue team headed home aboard the *Miss Fitz*.

A comparison of pictures taken as they worked on the disentanglement to photos in a right whale identification catalogue kept by the New England Aquarium revealed the animal was an eight-year-old male first photographed as a calf off Florida. It had been assigned catalogue #1971, but given its steadfast determination to head for Nantucket during the rescue, it has since been known to researchers as "Nantucket." The following fall Nantucket was sighted again by a New England Aquarium research team 400 miles north of Cape Cod, in the Bay of Fundy, Canada. By then no line remained in its mouth and the whale appeared none the worse for its experience. Right whale researchers have continued to report sightings of Nantucket up to the time this was written, nearly 20 years later.

The National Marine Fisheries Service subsequently tested the line removed from Nantucket and found that its breaking strength was nearly 8,000 pounds. When recalling the day's events many years later, Stormy could still only shake his head and mutter, "the power of that animal was something else." Similar sentiments—probably in less polite words, but with no less awe—have undoubtedly been uttered by whalers in many languages for more than a thousand years. That such an encounter with a North Atlantic right whale is still possible is no minor miracle, as these whales are one of the world's most endangered. Hunted for a thousand years, North Atlantic right whales probably came as close to disappearing as any whale could and still survive.

Ironically, if you were to ask Stormy, Dave, or Ed today about their contributions to developing disentanglement capabilities, they would answer with mixed feelings. Although elated by the ability to save entangled whales otherwise facing dim prospects of survival, they are the first to emphasize that proper management should prevent entanglements before they occur and that we should not rely on the dangerous business of disentangling animals (figure 1.5). Disentanglement teams now exist in many parts of the world, and at least one practitioner has lost his life attempting to free whales. Many more have been seriously injured.

Nevertheless, the Fisheries Service now considers successful disentanglements that leave whales in good condition to be successful management; this absolves the agency of responsibility for actions that would otherwise be required to prevent whale entanglements. To this end, it recently adopted a policy to exclude

Figure 1.5. In March 2004, a whale disentanglement team off northeast Florida maneuvers danger-ously close to an entangled right whale named Kingfisher, in an attempt to remove a lobster trap buoy and associated lines. (Image courtesy of the Florida Fish and Wildlife Research Institute, taken under NOAA permit number 594-1759.)

successful disentanglements from the counts of seriously injured whales; the agency uses the counts to determine whether additional management actions are needed to prevent the animals from getting tangled in fishing gear.[2] Although there is some logic to this policy, for those believing that fisheries and fishery manag-ers should be responsible for preventing entanglements so that disentanglement workers don't have to risk their lives freeing animals, the new policy smacks of misplaced priorities.

NOTES

1. Johnson, T. 2005. *Entanglements: The Intertwined Fates of Whales and Fishermen.* University Press of Florida. Gainesville, FL, US, pp. 25–41.

2. The National Marine Fisheries Service uses an estimate of a population's "potential biolog-ical removal" level, or PBR, to assess the success of management measures designed to prevent whale entanglements. If entanglement deaths or serious injuries exceed the PBR, additional actions must be taken to prevent their occurrence. Under its new policy, freed whales judged to be in good condition are not counted against that standard, even if entanglements are initially judged threatening for whales. National Marine Fisheries Service. 2011. National policy for distinguishing between serious and non-serious injuries of marine mammals. 76 Fed. Reg. 42116–42118. (July 18.)

2

WHAT'S IN A NAME?

TODAY, THE BIOLOGY AND ECOLOGY of North Atlantic right whales is among the best known of any large whale. But to early whalers and naturalists, the species was as much a mystery as an economic bounty. As with any species, unraveling the mysteries of its existence begins with the deceptively difficult task of assigning a universally accepted name and description to distinguish it from all other creatures, regardless of how similar they appear or where they occur. Without an agreed-upon name and set of unique traits, one can never be sure if the description made by one observer in fact refers to the same kind of animal described by someone else. Deriving a universal name for a species as large as a whale might seem to be one of the easier challenges for early naturalists attempting to catalogue the world's fauna, yet for North Atlantic right whales it was anything but easy. Before receiving their present name, they were known by many colloquial pseudonyms in a wide array of languages. These names alluded to where they were found or characteristics used locally for identification, but provided an inadequate basis for confirming exactly what animal was being referred to.

To the enigmatic Basques who first hunted right whales in the Middle Ages from their homeland on the western neck of the Iberian Peninsula, these whales were known as sarde or sarda, a Basque word meaning school or schooling. Much like the origin of the Basques themselves, the source of those terms is uncertain. It might have referred to the whale's seemingly synchronous seasonal appearance off their shores, or perhaps a mistaken belief that it fed on small schooling fish that appeared at about the same time. Because right whales do not travel in large groups, it clearly was not based on any social bonds holding right whales together in large pods or schools. Centuries later, right whales caught in the same region were called "Biscay whales" after the coastal bight that sculpts the Basque's shoreline.

To the first Danish whalers working icy waters between Norway and Greenland, this whale was known as the svarthval, meaning black whale, a reference to its ebony hue. When Dutch, Danish, and German whalers discovered them along the northwestern coast of Norway, they were dubbed Noortkapers or Nordkapers, after North Cape, the name of Europe's northernmost point of land. To Iceland's early whalers, they were the sletbag, or smooth-backed whale, a reference to the whale's lack of a dorsal fin. For the same reason, German whalers called them the glattwal.[1]

When North Americans joined the hunt in the 1600s and early 1700s, this whale was variously called the black whale, the true whale, the whalebone whale, the seven-foot bone whale (a reference to the typical length of their baleen), the rocknose whale (a reference to the knobby callosities atop their head), and eventually the right whale (because its economic value was superior to that of other whales, and hence it was the right whale to chase and kill). Early American whalers even went so far as to use different names depending on a whale's age and oil yield. One-year-olds were called "short-heads"; cows still accompanied by calves were "dry-skins" because their fat reserves had been depleted by nursing; two-year-olds were "stunts" due to their relatively small size; and older animals were "skull fish," possibly because of either their enormous head or huge pectoral fins.[2]

It would take centuries to realize that all of those names and many others applied not only to the same kind of whale but often to the very same individuals who had migrated thousands of miles from one whaling ground to another. Indeed, debate over their taxonomic status persisted late into the twentieth century, when new genetic analyses in the 1990s revealed that right whales in the North Atlantic represented a separate species and established the current common and scientific names. The tortuous path to arrive at this conclusion and the names we use today is a fascinating tale offering insight into the struggles that many of the world's most famous naturalists had to go through to sort out confusion between right whales and bowhead whales, and thereby lift the veil of misunderstanding that had long shrouded the species' distribution, movements, and way of life.[3]

Finding the Right Whale Name

Whale descriptions in manuscripts written before the early 1500s are rarely adequate for marine mammal experts to link them to current species names with certainty. Most early citations offer little more than raw statistics on amounts of oil or baleen recovered, or else refer only to "whales." Despite that limitation, from what we now know of historical whaling methods, locations where whales were hunted, and the current distribution and movements of whale species, there is little doubt that most of those targeting large species in the North Atlantic before the early 1500s were hunting right whales. Other species, such as fin, blue, sei, and sperm whales, were too fast and too elusive for whalers still perfecting their craft. Even when whalers were successful, other kinds of whales were more likely to sink once killed, making the balance between risk and reward less appealing. However, whalers might have hunted the enigmatic North Atlantic gray whale with some success; this species is now extinct, but it survived until at least the early 1700s.[4] With no surviving accounts that can be ascribed unambiguously to gray whales, it seems that any catch of that species must have been limited and short lived.

Safety in the assumption that right whales were the primary target of early whalers (other than Arctic Natives who hunted bowhead whales) begins to unravel in the 1500s and early 1600s. By then right whales off Europe were already becoming scarce and whalers—then primarily the Basques—had expanded their range to whaling grounds off southern Ireland and Great Britain. In the early 1500s, perhaps first off Iceland or Greenland but certainly in the Strait of Belle Isle between Newfoundland and Labrador, the Basques encountered a new kind of whale. This one was very similar in form, yet even easier to catch and yielded larger quantities of oil and whalebone. They had reached the southern limit of the Arctic bowhead whale.

Because bowheads and right whales were so similar in appearance, with oil of identical quality, at first the Basques may have had little reason to call this new whale anything other than the familiar sarde they had hunted for centuries off Europe. However, by the late 1500s separate names began to appear for this new northern whale. The Basques used the name sardako baleak, the whale that goes in flocks,[5] possibly reflecting a perception that when one was found, others were usually close by. Later English whalers called it the "Grand Bay whale," an early English name for the Gulf of St. Lawrence or perhaps the Strait of Belle Isle where the Basques first exploited them in great numbers. Other early names included the "bearded whale," alluding to its long baleen plates reaching lengths of 13 to 14 feet; the "Greenland whale" for its common occurrence off Greenland; and the "Greenland right whale," which recognized both where it was found and its premium economic value.

Today this whale is called the "bowhead whale" or simply the "bowhead" (*Balaena mysticetus*)—a reference to the pronounced curvature of its mouth forming an even deeper, more distinctive arch than that of the right whale. When the various names for bowheads first began to appear, they were frequently used interchangeably with names for right whales. More often, however, catch records simply cited both species as "whales." Thus, with no economic reason to distinguish them, most catch records from the 1600s could have applied to either bowheads, right whales, or some combination of the two. For current whaling historians and biologists trying to deduce where, when, and how many of each species was killed, this has left a confused trail, frustrating efforts to parse early encounters by species.[6]

In the early 1700s when the name "right whale" came into common use, whalers initially applied it to bowheads, reserving colloquial names, such as Nordkapers or black whales, for today's right whales. However, recognizing their similar form and superior quality and quantity of oil and baleen compared to all other whales, the term "right whale" was soon applied to both species. For whalers the name identified all they needed to know—these were the whales producing the most profitable oil and baleen. All other kinds, including the finbacks, humpbacks, and blues, were the wrong whales to chase not only because they were harder to catch but because they yielded shorter baleen of inferior quality and less oil per whale.

For early naturalists, indiscriminate use of the term "right whale" blurred distinctions between the Arctic and temperate species. Indeed, its broad use made it unclear whether whalers were in fact hunting one kind of whale or perhaps even two or more. As "right whales" (then including both species) became increasingly scarce in the North Atlantic and adjacent Arctic seas, European whalers followed reports of their presence in other oceans. At first they were drawn to the South Atlantic, then the South Pacific, and eventually the North Pacific and the western Arctic Oceans. Again, so similar in appearance in all of those seas, the name "right whales" was applied regardless of which ocean they were found in. Indeed, early whalers may have thought they were merely chasing the same whales from ocean to ocean.

Enter the Naturalists

In 1735, a momentous step was taken to resolve confusion over names for all wildlife, including whales. That year a young Swedish medical doctor named Carl Linnaeus published an 11-page document called *Systema naturae*. It outlined an ambitious new system for naming and classifying any living organism. Up to then, naturalists had been using colloquial names, often stringing them together into long cumbersome chains of local descriptive terms to distinguish different kinds of plants and animals. To simplify this increasingly unwieldy practice and the bewildering array of names for individual plants and animals, Linnaeus proposed giving each kind of organism a descriptive Latin or Greek name referencing one of its key characteristics or attributes. Under his new approach, each species was to be identified worldwide by a one-word "genus" name and a one-word "species" name—a sort of family name and given name.

Linnaeus went on to propose grouping plants and animals under a descending hierarchy of ever narrowing categories—somewhat like a country, province, county, city, and street address—until eventually arriving at the genus and species levels, where individuals shared the greatest number of features. By this means, naturalists anywhere in the world would know precisely what kind of plant or animal another naturalist was discussing no matter what local vernacular names might be used. Moreover, they would know which other species were most similar to that species. In a dozen subsequent editions of *Systema naturae*, Linnaeus and his growing cadre of adherents expanded the list of named life forms to cover some 4,400 species of animals and 7,700 species of plants.

Because of its simplicity and elegance, the system caught on quickly and persists today as the cornerstone of modern taxonomy—the field of biological science responsible for assigning concise universal names to all the world's life forms and organizing them within the hierarchy of broader categories. Early naturalists struggled with decisions about how to classify marine mammals and some listed them with different groups of animals. For example, Konrad Gesner lumped all whales and porpoises together as a separate group of marine animals in the fourth volume of his monumental work *Historiae animalium*, published in 1558,[7] while the influential British naturalist John Ray listed them with quadrupeds in his 1713 compendium of animals.[8] Linnaeus initially followed the colloquial understanding that assumed whales were a type of fish and included them as such in his early editions of *Systema naturae*. However, in the tenth edition of his series, published in two volumes in 1758 and 1759,[9] Linnaeus moved whales to a separate "order" under the "class" of animals with all mammals, where they have remained ever since. As explained by Charles Knight in the nineteenth century, Linnaeus's explanation for doing this was due to "their warm bilocular [two chambered] heart, their lungs, their movable eyelids, their hollow ears, '*penem intrantem feminam mammis lactantem*' [females that produce milk from internal mammary glands], and '*ex lege nature jure meritoque*' [by virtue of the laws of nature it would be unjust to do otherwise]."[10]

Ever since Linnaeus's landmark work, researchers have wrestled with genus and species names for right whales. Today, most experts recognize North Atlantic right whales as members of the phylum Chordata (animals with backbones), class Mammalia (animals that nurse their young with milk), order Cetacea (with 87 living whales, porpoises, and dolphins), suborder Mysticeti (with all 14 species of living whales with baleen plates), family Balaenidae (including three species of right whales and the bowhead whale), genus *Eubalaena* (the "true whales" comprising the three right whale species), and species *glacialis* (ice-associated).

Arriving at this taxonomic address meant a circuitous route involving many of the world's foremost naturalists and zoologists of the eighteenth, nineteenth, and twentieth centuries. Indeed, although there is now broad agreement on the species name for North Atlantic right whales, matters are still unsettled higher on the taxonomic ladder, above the genus level. Recent fossil discoveries of extinct whales have provided fodder for merging the orders Cetacea and Artiodactyla (even-toed ungulates such as the hippopotamuses) into a single new order called Cetartiodactyla, and moving the old order Cetacea and suborder Mysticeti a notch down in the taxonomic pecking order to subgroups in the new order.[11]

Before the late 1900s, taxonomists relied almost entirely on physical traits and anatomical structures to decide whether differences between groups justified splitting them into separate species, or whether similarities dictated lumping them together under a single taxonomic name. Early naturalists, however, had few opportunities to examine whales directly. Bringing carcasses back to laboratories to lay out side-by-side for comparison was impractical. In fact, before the early 1800s, few naturalists attempting to sort out the different kinds of whales had ever even seen a whale, much less all of the varieties they sought to classify. Instead, to distinguish one kind from another, they relied largely on second- or thirdhand descriptions, the findings of other naturalists that were based on similarly poor information, the many common names used for whales, and where they were reportedly found.

Like the naturalists, artists hired to prepare etchings of whales for early texts rarely had any firsthand exposure to either living or stranded whales. They either created renderings from oral descriptions or copied earlier works of unknown accuracy, adding new details based on their own third-party sources or artistic inclinations. Thus, for all the painstaking care put into early illustrations of whales, they now seem like fanciful cartoon depictions of imagined animals—which is largely what they were. For example, in the fourth volume of Konrad Gesner's landmark natural history entitled *Historiae animalium*—a work revered by many scientists as the start of the discipline of zoology—woodcuts by the renowned German artist Albrecht Dürer depicted whales as frightful monsters bearing little resemblance to the actual animals. Many of the illustrations in Gesner's work, including those of whales, were embellishments of earlier artwork from the works of Olaus Magnus, a prolific Swedish author and Catholic official, between the 1530s and 1550s. In 1560, Gesner also published *Icones nomenclator*, an abridged version of his earlier volumes describing the world's animals and using the same wood cuts (figures 2.1, 2.2, and 2.3). Gesner himself cautioned that those of the whales were hardly believable. Not surprisingly, they offered little help to others trying to decide if the whales they were describing matched those discussed by Gesner.[12]

According to Joel Asaph Allen, the highly respected curator of birds and mammals at the American Museum of Natural History from 1885 until his death in 1921, credit for the first scientific description of a right whale under Linnaeus's system belonged to the Prussian naturalist Jacob Theodor Klein. In 1741 Klein classified the Nordkaper of northern Norway as a variety of a composite species he called *Balaena glacialis*. Klein's species name pooled features of right whales and bowhead whales and identified three "varieties": *australis*, *occidentalis*, and *borealis*.[13] The Nordkaper was described as a sub-Arctic resident under the varietal name *borealis*. Whether this variety included the bowhead whale or whether bowheads were included under the variety name *occidentalis* (the western whale) is unclear.

Klein almost had species identification for the entire family of balaenid whales organized as it is today. He was the first to use the species name *glacialis* and the first to recognize Nordkapers as boreal rather than Arctic residents. Credit for his work, however, has largely been ignored in taxonomic reviews because it predates the tenth edition of Linnaeus's taxonomic series published in 1758.[14] With that date now formally recognized as the start of modern taxonomy, rules for assigning scientific names stipulate that whatever name is given by the first person to correctly identify a species *after* 1758 should be the accepted one, even if another more fitting name was used before or after the first post-1758 name. Thus, taxonomists today have largely disregarded the works of naturalists before 1757, including Klein.

Figure 2.1. Three woodcuts by Albrecht Dürer from Konrad Gesner's 1560 *Icones nomenclator* of whales bearing down on ships. The illustrations suggest at least some contemporary reports of hazardous interactions between whales and ships. (Images courtesy of the Smithsonian Institution Libraries, Washington, DC.)

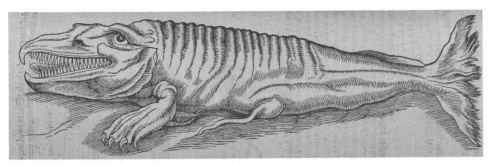

Figure 2.2. Many illustrations of whales in Konrad Gesner's 1558 *Historia animalium* and reproduced in his abridged 1560 edition *Icones nomenclator* bore little resemblance to any known animals. The *top* panel shows an "English" whale based on an earlier print of a stranded whale found along the Thames River in August 1532; the *bottom* panel is a "burrowing whale" based on earlier illustrations of a "sea hog" or "sea pig" that first appeared on the 1539 map *Carte marina*, by Olaus Magnus. (Images courtesy of the Smithsonian Institution Libraries, Washington, DC.)

As it turned out, views of the taxonomic status of right whales and their allies would wobble back and forth for over 150 years. With no direct way to compare one whale to another, early naturalists often assumed that any whale not previously described from a particular area was a new kind, regardless of similarity in appearance. However, bucking that trend, Klein's 1741 description was followed in 1756 by another from the respected French naturalist M. J. Brisson. He provided a brief, but more insightful, description that recognized Nordkapers as the same whales caught in Iceland and distinct from bowhead whales. Brisson therefore gave it full species status under the name *la baleine d'Islanda*.[15] This name was later translated into Latin as *Balaena islandica* by the German naturalist Johann Friedrich Gmelin, who published a list of whale varieties in Linnaeus's thirteenth edition of *Systema naturae* in 1788.[16] Interestingly, in the same volume Gmelin provides what may be the first reference to the common name "right whale"—and this wasn't used for the Nordkaper. Instead, it was listed as a common name for *Balaena mysticetus* (bowhead) and written in the German as *Der richte groenlaendifche Whallfifch*, or the right/true Greenland whale-fish.[17]

Between Brisson's initial publication and Gmelin's later Latin translation of the scientific name, a little known German-Danish zoologist named Otto Friderico Müller published a taxonomic review in 1776. Without reference to Klein's earlier work, Müller applied Klein's name *glacialis*, which Kline had used for both bowheads and right whales, to "Nord-Kapers" alone. He added that in Norway this

Figure 2.3. Most whale illustrations in Konrad Gesner's encyclopedia of wildlife were based on prints from earlier works by Olaus Magnus; the "horned whale" (*top*) is from Magnus's 1555 book, *History of Northern Peoples*; the "bearded whale" (*bottom*) is from Gesner's 1560 *Icones nomenclator*. (Images courtesy of the New Bedford Whaling Museum, New Bedford, MA; and the Smithsonian Institution Libraries, Washington, DC, respectively.)

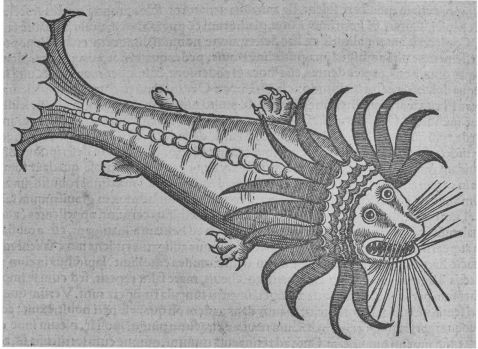

whale was also called the *Lille-Hval*, possibly a reference to a location in northern Norway, and the *Sild-Qval*, or herring whale. In doing so, Müller compounded confusion between different balaenid whales by giving the subarctic Nordkapers, which Kline called *borealis*, the very inappropriate scientific name *Balaena glacialis*, with *Balaena* meaning "whale" in Latin, and *glacialis* meaning "ice-associated."[18]

Most authorities of the day, however, including Linnaeus himself, scoffed at the idea that Nordkapers were a separate species and continued listing them as either identical to, or a variety of, bowhead whales. That view held until another French zoologist Pierre Joseph Bonnaterre once more conferred full species rank to "Nordkapers" in his 1789 review of whales included in *Tableau encyclopédique*.[19] Although Bonnaterre described Nordkapers as whales found in the seas of Norway and Iceland, he provided an illustration that looked like a bowhead, thus apparently still confusing the two whales, and assigned Nordkapers the scientific name *Balaena franche* (figure 2.4). Yet another Frenchman, one with the mountainous name Bernard Germain Étienne de la Ville-sur-Illon, comte de Lacépède, followed in 1804 with an illustrated review of the world's cetaceans based on a compilation of earlier accounts. Lacépède's influential work retained the Nordkaper's species status but gave it yet another scientific name—*Balaena nordkaper*.[20]

Other experts in the late 1700s and early 1800s refused to recognize Nordkapers as a separate species. Foremost among them was one of the most influential naturalists of the day, the Frenchman Georges Cuvier, whose pioneering work in the field of paleontology was the first to recognize life forms in the fossil record as species that had gone extinct. With no access to specimen materials from right whales and never having seen one himself, Cuvier deferred to the views of William Scoresby Jr., an English whaling captain and explorer who had provided an excellent, authoritative account of bowhead whales based on his voyages to Greenland.[21]

According to Scoresby, the "Greenland" or "common" whale, as he referred to the bowhead, was a widespread whale that occurred "most abundantly in the frozen seas of Greenland." He wrote, "It is never met with in the German Ocean, and rarely within 200 leagues of the British coasts: but along the coasts of Africa and South America it is met with, periodically, in considerable numbers."[22]

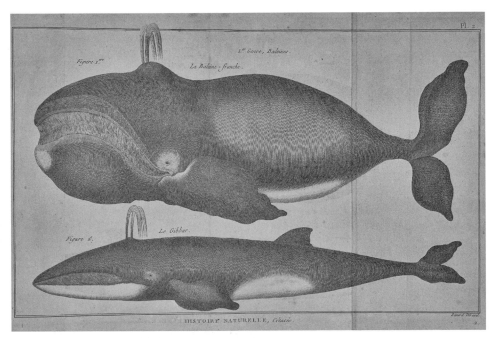

Figure 2.4. Joseph Pierre Bonnaterre gave the Nordkaper the scientific name *Balaena franche* in his 1789 *Tableau encyclopedique*. Although Nordkaper is an eighteenth-century name used by whalers to identify what is now believed to have been the right whale, Bonnaterre's illustration is clearly that of a bowhead, indicating his confusion between the two species. (Image courtesy of the Smithsonian Institution Libraries, Washington, DC.)

Scoresby's firsthand knowledge and detailed description of the Greenland whale had earned him great respect as a reliable authority on whales. Thus, his opinion that there was no difference between the whales in the Arctic and those at lower latitudes carried great weight.[23] Interestingly, his description of this composite bowhead–right whale also appears to be the first to note the many advantages now cited as the basis for the common name of "right whale," even though he did not use the term and was clearly referring to bowheads: "This valuable and interesting animal, generally called *The Whale* by way of eminence, is the object of our most important commerce to the Polar Seas—is productive of more oil than any other of the Cetacea, and, being less active, slower in its motion, and more timid than any other of the kind, of similar or nearly similar magnitude, is more easily captured."[24]

Cuvier's concurrence with Scoresby was widely accepted until the mid-1800s, when Danish zoologist Daniel Frederik Eschricht received a lithograph from a colleague showing a small whale taken at the ancient Basque whaling port of San Sebastian, Spain, in 1854. By then the landing of any right whale in the eastern North Atlantic was a rare event. Its skeleton was therefore shipped to a museum in Pamplona, Spain, for preservation. In a *eureka* moment, Eschricht recognized the whale in the lithograph as a Nordkaper and saw that it was clearly different from the Greenland whale. Moreover, he recognized that the whale bore a closer resemblance to the right whales then being hunted in the Southern Hemisphere. Acting on his flash of insight, he hastily arranged a trip to Pamplona, acquired the skeleton, and had it shipped to his museum in Copenhagen. In 1860 Eschricht published a paper presenting his new ideas on the species' taxonomy.[25] Although he did not fully describe the skeleton, he proposed that all right whales along the European coast should be recognized as a single species called *Balaena biscayensis*, or "whale of the Bay of Biscay," distinct from the Greenland whale (the bowhead).

Eschricht died in 1863 before publishing a detailed description of the San Sebastian whale's skeleton, but in an 1864 review of cetaceans in the British Isles, John Edward Gray used the name *Balaena biscayensis* for the Nordkaper, attributing it to Eschricht.[26] Gray also proposed dividing right whales into two genera—*Balaena*, including the Greenland or bowhead whale and Eschricht's *B. biscayensis*; and a new genus, *Eubalaena* (*eu* meaning "true," and *balaena* meaning "whale"), which comprised all other right whales. Later that year, William H. Flower, another respected English zoologist who would eventually head the British Natural History Museum, published a paper agreeing with Eschricht's conclusions.[27] Flower's 1864 paper, however, took Eschricht's notion that *B. biscayensis* was distinct from Greenland whales a step further. Flower noted that Southern Hemisphere right whales not only seemed more similar to *B. biscayensis* than Greenland whales but merited their own species status under the name *E. australis*. He also argued that Eschricht's whale should also be moved to the new *Eubalaena* genus. Thus, he renamed Eschricht's whale *Eubalaena biscayensis*, again attributing it to Eschricht even though Eschricht never used the genus name *Eubalaena*.

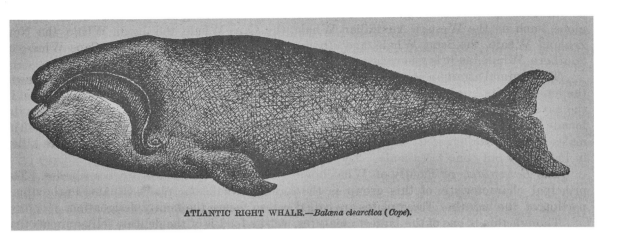

ATLANTIC RIGHT WHALE.—*Balæna cisarctica* (*Cope*).

A year after the papers by Gray and Flower, American scientist Edward D. Cope entered the fray. Cope, destined to become one of the world's foremost scholars on fishes and dinosaurs, in 1864 described a small right whale caught on the opposite side of the Atlantic, in Delaware Bay.[28] He gave it yet another name, *Balaena cisarctica*; this was Latin for the "whale on this side of the Arctic," meaning south of the Arctic (figure 2.5). Some suggest Cope may have recognized the Delaware whale as the same species described by Eschricht but gave it this new name to protest the name *biscayensis*, because it suggested a geographic range that was too narrow.

The apparent march toward recognizing right whales throughout the North Atlantic as a species distinct from bowheads soon lapsed into another round of confusion. In 1877 a small right whale was killed in the Gulf of Taranto, nestled in the heel of the Italian boot. It was the first right whale ever recorded in the Mediterranean Sea, and a paper published that same year by the eminent Italian scientist Giovanni Capellini gave the whale yet another name—*Balaena tarentina*, after the gulf in which it was caught.[29]

Suspecting the Taranto whale to be the same species as the whale killed 25 years earlier in San Sebastian, a Spanish zoologist named Francisco Gasco traveled to Copenhagen to examine the skeleton Eschricht had acquired. He then published a more complete description of its characteristics.[30] Comparing it to the whale caught in Italy, he concluded that the specimens belonged to the same species. Gasco also related that Cope, who by then had also traveled to Copenhagen to compare his Delaware whale to the San Sebastian whale, told him that he too had concluded the Delaware Bay whale was identical to Eschricht's whale. In 1885, however, in a curious about-face from his own earlier proposal in 1864, Flower published a list of cetacean specimens in the British Museum that now lumped all right whales, including bowheads, under a single genus, *Balaena*, and all "right whales" south of the Arctic, including those in both the Northern and Southern Hemispheres, under a single species name—*australis*.[31]

With few right whale skeletons in museum collections, remains of those killed on either side of the North Atlantic in the late 1800s were eagerly sought by

Figure 2.5. American naturalist Edward Cope assigned yet another scientific name—*Balaena cistarctica*—to the Atlantic right whale; this was based on his description of a whale caught off Delaware Bay in 1864. (Image courtesy of the Smithsonian Institution Libraries, Washington, DC.)

27

museums and described in the scientific literature. Right whale bones left by earlier whalers also were collected and analyzed. In 1893, for example, G. A. Guldburg examined bones left by early whalers in Finland and Norway and concluded that Greenland right whales, *Balaena mysticetus*, were indeed distinct from right whales in the genus *Eubalaena*.[32] In a separate paper he then proposed four distinct species of *Eubalaena*: *E. biscayensis* in the North Atlantic, *E. australis* in the South Atlantic, *E. japonica* in the North Pacific, and *E. antipodarum* in the South Pacific.[33]

Debate over distinctions between bowheads and right whales was finally laid to rest in the early 1900s. In 1904, Fredrick W. True, first curator of marine mammals at the Smithsonian Institution in Washington, DC, completed an exhaustive review of right whale bones in museums around the world.[34] The results dispelled any remaining doubts that right whales throughout the North Atlantic were one and the same species, distinct from Greenland right whales (or bowhead whales). Believing that Bonnaterre's publication in 1789 was the first to correctly recognize North Atlantic right whales as a separate species, True concluded that the name he mistakenly thought Bonnaterre had used—*Balaena glacialis*—should be accepted as the scientific name, notwithstanding its errant implication that they inhabited ice-covered seas. Bonnaterre, however, had actually used a different name, *Balaena franche*, for the species.[35]

A few years later, in 1908, J. A. Allen of the Smithsonian completed another review of North Atlantic right whales.[36] He agreed with True's meticulous analyses supporting separate species status under the name *glacialis*, but also concluded that Gray was correct when he split right whales and Greenland whales (bowheads) into two genera in 1864. He therefore restored the genus name *Eubalaena* for all right whales, leaving bowheads with the genus name *Balaena*. Allen was less certain about the number of right whale species. Although there was some evidence of slight physical differences between right whales in different ocean basins, Allen found the supporting data weak. Given limited specimen materials from many parts of the world, he cautioned that species differences within the genus *Eubalaena* should not be considered a settled matter.

Nevertheless, drawing on prevailing wisdom of the day that held right whales were cold-water species unable to cross warm equatorial waters,[37] Allan agreed there was compelling justification for recognizing Northern and Southern Hemisphere whales as separate species. As explained by Lieutenant Matthew Fontaine Maury, the respected nineteenth-century oceanographer and first superintendent of the US Naval Observatory, the "torrid tropics" were a "forbidden zone" for right whales. Maury argued, it "is physically as impossible for him [a right whale] to cross the equator as it would be to cross a sea of flame."[38] Allan was less confident about possible species differences between whales in the Atlantic and Pacific Oceans. Because of open, cold-water connections between the South Atlantic, Indian, and South Pacific Oceans, and possibly between the North Atlantic and North Pacific Oceans across the Russian or Canadian Arctic, Allen reasoned that species distinctions on either side of the American continents remained an open question.

For most of the 1900s, taxonomists were of two views about the number of right whale species in the genus *Eubalaena*. Some argued that right whales in the North Atlantic, North Pacific, and Southern Hemisphere were each separate species. Most, however, agreed there were just two species: the northern right whale, *E. glacialis* in the North Atlantic and North Pacific, and the southern right whale, *E. australis*, which was distributed throughout oceans south of the equator. This was the prevailing view for most of the nineteenth century, and it left the small family of balaenid whales with only three living species: the Arctic bowhead whale, *B. mysticetus*, and two right whale species.[39]

Confusion Leaves Its Stamp

In the late 1900s, the exploding field of genetics shed new light on old squabbles over splitting or lumping animals under various species names. The ability of genetic analyses to isolate and compare distinctive gene sequences in individuals across widely separated geographic regions provided a powerful taxonomic tool. With a level of precision previously impossible, scientists found that analyses of genetic code could document reproductive isolation and detect progress down separate evolutionary paths. When those techniques were trained on the question of right whale taxonomy in 2000, the results backed earlier taxonomic splitters. That is, scientific consensus shifted toward dividing right whales in the genus *Eubalaena* into three species instead of two: the North Atlantic right whale, *E. glacialis*; the North Pacific right whale, *E. japonica*; and the southern right whale, *E. australis*.[40] A decade later, a more in-depth analysis comparing physical characteristics of the skulls, flipper bones, baleen, and vertebrae of all living balaenids reached the same conclusion.[41]

Thus, for now, the small family of balaenid whales, including both bowheads and right whales, has grown from three species to four. And in a twist of fate laced with irony, North Atlantic right whales ended up with the common name first used for bowheads, and a scientific name far more befitting that Arctic cousin. In both cases the names eventually bestowed on right whales more accurately reflect the centuries of befuddlement over the relationship between bowheads and right whales than a true understanding of the species themselves.[42] That is, the common name "right whale," originally used for bowheads because *they* produced the longest baleen and greatest quantities of oil, now applies to the temperate-latitude species that actually produced somewhat shorter baleen and slightly less oil and was already rare in the North Atlantic by the time the name finally came into common use.

Similarly, under the unforgiving hand of inflexible taxonomic protocols that were devised to perpetuate Linnaeus's naming conventions, more weight was given to when a species was first named rather than to the intention that names reflect characteristic features. Thus, North Atlantic right whales received the scientific designation of *E. glacialis*, meaning "true whale of the ice," even though the species rarely ranges north of the Arctic circle and likely never encounters so much as an ice cube. And instead of an apt descriptor derived from insightful understanding of the species' biology as Linnaeus would have had it, right whales now bear a scientific name far better suited to bowhead whales. This name was assigned by one of the first disciples of Linnaean taxonomy, who in all likelihood never saw a right whale alive or dead.

NOTES

1. Allen, GM. 1916. *The Whalebone Whales of New England*. Memoirs of the Boston Society of Natural History, Vol. 8, no. 2, Boston, MA, US, pp. 114–175, 114–116.

2. Dudley, P. 1725. An essay upon the natural history of whales, with particular account of the ambergris found in the sperma ceti whale. Philosophical Transactions of the Royal Society of London 33:256–269, p. 257.

3. Dyer, MP. 2014. Why black whales are called right whales. *In* JE Ringstad (ed.), *Whaling and History IV: Papers Presented at a Symposium in Sandefjord on the 20th and 21st of June 2013*. Vestfoldsmuseene IKS, avdeling Hvalfanstmuseet. Sandefjord, Norway, pp. 165–177.

4. Mead, JG, and ED Mitchell. 1984. Atlantic gray whale. *In* ML Jones, SL Swartz, and S Leatherwood (eds.), *The Gray Whale (Eschrichtius robustus)*. Academic Press. New York, NY, pp. 33–53. Bryant, PJ. 1995. Dating remains of gray whales from the eastern North Atlantic. Journal of Mammalogy 76(3):857–861.

5. Eschricht, DF, and J Reinhardt. 1866. On the Greenland right whale (*Balaena mysticetus*). *In* WH Flower (ed. and trans.), *Recent Memoirs of the Cetacea*. Published for the Ray Society by Robert Hardwicke. London, UK, pp. 1–150.

6. Dyer, Why black whales (n. 3).

7. Gesner, C. 1558. *Historiae animalium, qui est de piscium et aquatilium animantium natura*, Vol. 4. Christoph Froschauer. Zurich, Switzerland.

8. Ray, J. 1713. *Synopsis methodica avium & piscium: opus posthumum*. William Innys. London, UK, Pt. 2, pp. 6–17.

9. Current conventions used by taxonomists to name species use the tenth edition of Linnaeus's *Systema naturae*, rather than his first edition published in 1735, to mark the formal beginning of his classification system. Thus, any scientific names ascribed to species before 1758 are considered invalid for purposes of designating current names. By the same token, naturalists who proposed scientific names before 1758 receive no recognition for their work, even if their name was subsequently used after 1758 by other scientists. Linnaeus, C. 1758–1759. *Systema naturae per regna tria naturæ, secundum classes, ordines, genera, species, cum characteribus, differentiis, synonymis, locis*. 10th ed., revised. 2 vols. Laurentii Salvii. Stockholm.

10. Knight, C. 1866. *Natural History or Second Division of "the English Encyclopedia."* Bradbury, Evans. London, Vol. 1, p. 870 (quotation).

11. Committee on Taxonomy. 2015. List of marine mammal species and subspecies. The Society for Marine Mammalogy. http://www.marinemammalscience.org, accessed on December 12, 2015.

12. Gesner, K. 1551–1558, 1587. *Historiae animalium*. 5 vols. Christoph Froschauer. Zurich, Switzerland (for vol. 4, see n. 7). Gesner, K. 1560. *Nomenclator aquatilium animantium icones animalium quadrupedum uiuiparorum et ouiparorum*. Christoph Froschauer. Zurich, pp. 160–185 (includes figures 2.1, 2.2, and 2.3 [bottom]).

13. Allen, JA. 1908. The North Atlantic right whale and its near allies. Bulletin of the American Museum of Natural History 24:277–329, 287. Kline, JT. 1741. *Historiae piscium naturalis*. Letteris Schreiberianis. Gdansk, Poland, p. 12.

14. Hershkovitz, P. 1961. On the nomenclature of certain whales. Fieldiana-Zoology 39(2):547–565. Hershkovitz. 1966. Catalogue of living whales. Bulletin of the United States National Museum 246:1–259, pp. 190–191.

15. Brisson, MJ. 1756. *Le regne animal divise en IX classes, ou Methode contenant la division generale des animaux en IX classes*. Chez Cl. Jean-Baptiste Bauche. Paris, France, pp. 350–351.

16. Linneaus, Carl von. 1788. *Systema naturae per regna tria naturae, secundum classes, ordines, genera, species, cum characteribus, differentiis, synonymis, locis*. Vol. 1. 13th ed., revised, JF Gmelin (ed.). Beer. Leipzig, Germany, pp. [1–12], cited in Hershkovitz, Catalogue of living whales (n. 14), p. 190.

17. Linneaus, *Systema naturae* (n. 16), p. 223.

18. Müller, OF. 1776. *Zoologiae Danicae prodromus*. Impensis Auctoris. Typis Hallageeriis. Copenhagen, Denmark, p. 7.

19. Bonnaterre, PJ. 1789. *Tableau encyclopedique et methodique des trois regnes de la nature: Cetologie*. Chez Panckoucke. Paris, France.

20. La Cépède, BG. 1804. *Histoire naturelle des cétacées*. Chez Plassan. Paris, France, pp. 103–110, plates 2, 3. JA Allen, The North Atlantic right whale (n. 13), p. 288.

21. Scoresby, W., Jr. 1820. *An Account of the Arctic Regions with a History and Description of the Northern Whale Fishery*. Vol. 1. Archibald Constable. Edinburgh.

22. The rarity of these whales off northern Europe by the 1800s may have reflected their elimination by earlier whalers. Ibid., Vol. 1, p. 473 (quotation).

23. Allen, JA, The North Atlantic right whale (n. 13), pp. 287–290.

24. Scoresby Jr., *An Account* (n. 21), pp. 449–450.

25. Eschricht, DF. 1860. Rev. Mag. Zool., Paris. (2), 12: 229, as cited in Hershkovitz, Catalogue of living whales (n. 14), p. 190.

26. Gray, JE. 1864. On the Cetacea which have been observed in the seas surrounding the British islands. Proceedings of the Zoological Society of London. 1864:195–284.

27. Flower, WH. 1864. Notes on the skeletons of whales in the principal museums of Holland and Belgium with descriptions of two species apparently new to science. Proceedings of the Zoological Society of London 1864:384–420.

28. Cope, ED. 1865. Note on a species of whale occurring on the coasts of the United States. Proceedings of the Academy of Natural Sciences of Philadelphia 17:168–169.

29. Capellini, G. 1877. Della balaena di Taranto, confrontata con quella della Nuove Zelanda e con talune fossill del Belgio e della Toscana. Memorie della Reale Accademia delle Scienze dell'Istituto di Bologna, series 3, 7:1–34.

30. Graso, F. 1879. Il balenotto catturato nel 1854 a San Sebastiano (Spagna) (*Balaena biscayensis*, Eschricht) per la prima volta descritto. Annali del Museo Civico di Storia naturale di Genova 14:573–608.

31. Flower, WH. 1885. *List of the Specimens of Cetacea in the Zoological Department of the British Museum*. Taylor and Francis. London, pp. 1–5.

32. Guldberg, JA. 1884. The North Cape whale. Nature 30:148–149.

33. Guldburg, JA. 1893. Zur Kenntniss des Nordkapers (*Eubalaena biscayensis* Eschr.). Zoologische Jahrbücher. Abteilung für Systematik, Geographie und Biologie der Tiere 7:1–22.

34. True, FW. 1904. The North Atlantic right whale, *Balaena glacialis* (Bonnaterre). In *The Whalebone Whales of the Western North Atlantic Compared with Those Occurring in European Waters, with Some Observations on the Species of the North Pacific*. Smithsonian Contributions to Knowledge, Vol. 33, pp. 244–268. Smithsonian Institution. Washington DC.

35. A later review of the species by Philip Hershkovitz of the Smithsonian Institution in 1966 concluded that Otto Friderico Müller's 1776 publication was the first to recognize Nordkapers as

a separate species after the 1758 publication of Linnaeus's tenth edition of *Systema naturae*, and therefore his name should be recognized despite its errant reference to their Arctic range. Thus, Müller is now officially credited as the source of the name, rather than Bonnaterre, whom True had recognized; see Hershkovitz, Catalogue of living whales (n. 14), p. 190.

36. Allen, JA, The North Atlantic right whale (n. 13), pp. 306–312.

37. The strength of this argument is now weaker than it once seemed. Female North Atlantic right whales have been documented rounding Florida's tropical southern tip, at least occasionally, to enter the Gulf of Mexico.

38. Maury, MF. 1858. *Sailing Directions*, 7th Edition. P. 253, as cited in JA Allen, The North Atlantic right whale (n. 13), p. 308.

39. Rice, DW. 1998. *Marine Mammals of the World: Systematic and Distribution*. Society of Marine Mammalogy. Special publication 4. Allen Press. Lawrence, KS, US, pp. 61–65.

40. Rosenbaum, HC, RL Brownell Jr., MW Brown, C Schaeff, V Portway, BN White, S Malik, et al. 2000. World-wide genetic differentiation of *Eubalaena*; questioning the number of right whale species. Molecular Ecology 9:1793–1802.

41. Churchill, M, A Berta, and T Deméré. 2012. The systematics of right whales (Mysticeti: Balaenidae). Marine Mammal Science 28:497–521.

42. Dyer, Why black whales (n. 3).

3

FORAGING
WITH A SMILE

ONE OF THE MOST amazing facts about right whales is that for all their great size, they subsist on some of the sea's smallest creatures. There are 14 species of baleen whales divided into 4 taxonomic families that make up the cetacean sub-order Mysticeti. Although all of them filter prey from their watery world, right whales and bowhead whales consume the smallest organisms using a feeding method distinctly different from other baleen species. To put their foraging strategy in context, it's helpful to first consider the basic ways baleen whales catch their food. As a general rule, each of the four families of baleen whales rely mostly, but not necessarily exclusively, on one of three basic methods: suction feeding, lunge feeding, and continuous ram feeding.[1]

North Pacific gray whales, the sole surviving member of the family Eschrichti-idae that now parades up and down the west coast of North America on annual migrations between calving lagoons in Mexico and feeding grounds off Alaska and Russia, are the only whales known to use suction feeding. They dine principally on a menu of amphipods, clams, and other large invertebrates living in mud and sand bottoms covering the continental shelf. The suction used to pull prey into their mouth is created by their tongue, which acts like a piston forcefully drawing water in and then pushing it back out. When gray whales feed, they dive to the bottom, roll to one side—curiously, almost always the right side—and press the side of their head into the sediment.[2] Opening the side of the mouth that is in contact with the bottom while keeping the opposite side closed, prey-laden sediment is sucked into the mouth by retracting and perhaps simultaneously depressing the tongue to create a negative pressure inside the oral cavity. When the mouth is full, the head is lifted off the bottom, leaving an elliptical depression up to 2 feet deep and 6 feet long on the bottom. The muddy slurry is then expelled through baleen plates lining each side of the mouth by pushing the tongue forward or elevating it to leave a plume of sediment in the whale's wake.

The nine species of "rorqual" whales in the family Balaenopteridae, which includes blue, fin, sei, and humpback whales, are lunge-feeding gulpers. All species in this family have expandable throat pleats visible as parallel grooves on the underside of the head and belly; these run some three-quarters of the way down their length from the chin. The term "rorqual," from the Norwegian word *röyrkval,* meaning "furrow whale," is a reference to these corduroy-like pleated grooves. The pleats themselves are actually long flexible slats of inelastic fat, each joined by highly elastic fibers.

When rorquals feed they charge into a dense school of small fish, squid, or krill and pop open their mouths to an angle of 80 to 90 degrees. As soon as the mouth opens, the throat pleats billow out like a pelican's pouch as the gaping maw engulfs—or perhaps more accurately, surrounds—a massive slug of water one enormous mouthful at a time. As soon as the pouch fills, the open mouth acts like a sea anchor bringing the lunging whale to an abrupt halt. Whereas a fully grown humpback whale can engulf an estimated 15,000 gallons of seawater in a single gulp,[3] adult fin whales may take in some 18,500 gallons,[4] an amount nearly equal to the weight of the whale itself and a volume comparable to the size of a school bus (2,500 cubic feet). After each lunge, the mouth closes to a slight gape and the throat pleats are contracted, forcing water through the baleen over a span of about a minute to strain out their prey.

North Atlantic right whales and the other three species in the family Balaenidae rely on a third strategy—continuous ram feeding.[5] They simply open their mouths and power ahead at a slow, steady clip (usually 2 to 3 miles per hour) as water continuously flows in one direction through the mouth and dense patches of minute zooplankton are filtered by the baleen plates.[6] Until recently, the accepted understanding of how this process worked began with the prey-laden water first entering through a scoop-shaped cleft at the front of the mouth between a gap in the racks of baleen plates lining either side of the oral cavity (see figure 15.1). A huge dome-shaped tongue on the floor of the mouth then redirects the water to either side of the mouth, forcing it through the baleen curtains to trap the prey. The filtered water then passes into a gutter-like channel, or "sulcus," between the baleen plates and deep lower jowl running down the side of the head to an opening at the corner of the mouth near the whale's eye (figure 3.1).

Recently, however, Alexander Werth, a professor and expert in functional morphology at Hampden-Sydney College in Virginia, and his colleague Jean Potvin at Saint Louis University developed a new theory on how this filter process works.[7] Based on new analyses, they concluded that the baleen does not serve as a sieve to catch prey as water flows through its bristles, but rather it is part of an elegant dewatering mechanism that pulls water out of the prey-laden broth as it moves along the inside of the baleen racks from the front to the back of the mouth. Thus, by the time the food reaches the rear of the mouth at the esophagus, where it is swallowed, it has become an even more concentrated slurry. We will return to this later.

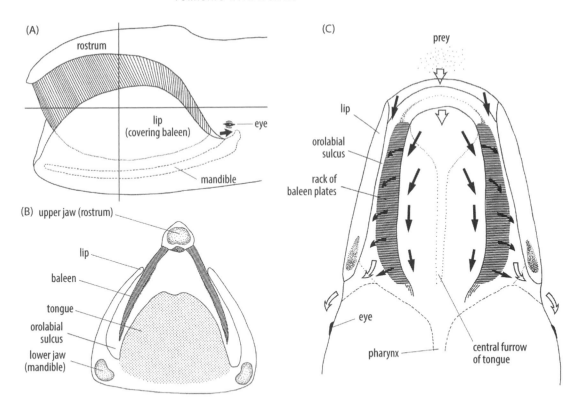

(A)
rostrum
lip
(covering baleen)
eye
mandible

(B) upper jaw (rostrum)
lip
baleen
tongue
orolabial
sulcus
lower jaw
(mandible)

(C)
prey
lip
orolabial
sulcus
rack of
baleen plates
eye
pharynx
central furrow
of tongue

Figure 3.1. Feeding structures and direction of water flow in a right whale's mouth; (A) lateral view; (B) view of cross section along the vertical plane in panel A; (C) view from above, along the horizontal plane in panel A. Arrows indicate direction of water flow. (Figure modified from Alexander Werth, 2004, Models of hydrodynamic flow the in the bowhead whale filter feeding apparatus. *Journal of Experimental Biology* 207:3569–3580.)

To force this continuous passage of water and food through their mouth with jaws agape, right whales have developed extraordinarily powerful trunk muscles. These massive muscles, along with an unusually thick covering of creamy white blubber up to half a foot or more deep, give balaenid whales their portly girth. Along with the lack of throat pleats, this rotund profile helps distinguish them from the sleeker rorquals.

A great deal has been written in the scientific literature about how baleen whales feed, yet knowledge of the details of all three feeding strategies are based to a surprising degree on good old-fashioned guesswork. Studying normal feeding behavior with captive animals has been impossible. Aside from a minke whale held briefly in a Japanese aquarium and a handful of gray whale calves kept for varying periods of time over their first year of life at SeaWorld in San Diego, California, no baleen whales have been held in captivity. Information on how the captive minke whale was fed has not been published, but the gray whales were fed either a fat-rich artificial milk formula through a tube, an unnatural diet of frozen squid dropped in blocks to the floor of their tank,[8] or an assortment of fish and squid dumped directly into their open mouths from a bucket.[9]

Opportunities to study feeding at sea are only slightly better. Most feeding occurs at depth, and therefore out of sight. Until recently, when new technological devices were brought to bear on the question, understanding was largely limited to visual observations dependent on a healthy dose of luck. The best observations from actively feeding whales involve balaenid whales seen skim feeding at the surface. More recently, feeding methods have been revealed by underwater film footage of rorquals engulfing schools of fish in clear tropical waters and by whales tracked with various types of tags affixed to their back with suction cups. The latter help identify where and when whales feed but still offer limited details about how they manipulate water and prey while feeding. For that, descriptions have been drawn largely from whales seen skim feeding at the surface, examinations of the feeding structures and stomach contents of dead whales, and various physical and mathematical models designed to simulate feeding.

Right Whales and Flamingos: Convergent Evolution on a Grand Scale?

Interestingly, a completely unrelated group of animals with a number of remarkable similarities to right whales might offer some surprising insights into the mammal's feeding. That group is the family called the Phoenicopteridae, which includes five species of flamingos. Cetacean biologists and ornithologists have long noted the curious physical resemblance between the heads of these disparate groups (figure 3.2);[10] however, few of these scientists offered more than passing thoughts as to reasons underlying the striking similarities.

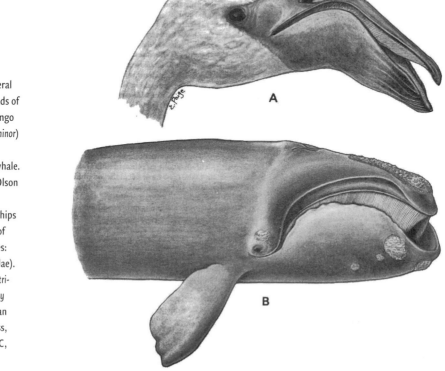

Figure 3.2. Lateral view of the heads of (A) lesser flamingo (*Phoeniconaias minor*) and (B) North Atlantic right whale. (Images from Olson and Feduccia, 1980, Relationships and Evolution of Flamingos (Aves: Phoenicopteridae). *Smithsonian Contributions to Zoology* 316. Smithsonian Institution Press, Washington, DC, p. 66.)

Yet if Charles Darwin's theory of natural selection is to be believed, this odd parallel is more a result of adaptation than superficial coincidence. Indeed, the similarity may be one of nature's grandest examples of convergent evolution (the independent development of similar anatomical structures serving similar functions in unrelated species). To appreciate this wonder of nature, and perhaps take from it a sharpened understanding of how right whales feed, bear with me in a little armchair science as we take a closer look at the heads of right whales and flamingos and come to know both of these fascinating creatures a little more intimately.

A quarter of a right whale's great length (and a third of the bowhead's) is taken up by an enormous head and a truly remarkable feeding apparatus. The extensive external landscape has several curious features that immediately catch the observer's eye. The first feature likely to grab one's attention are the callosities—light-colored incrustations scattered about the head, but most prominent on the top of the upper jaw and leading edge of the lower. Callosities are found only on right whales. They develop around the sparse hairs several inches long scattered about the head, particularly on the chin. Callosities have the feel of hard rubber and are pockmarked with ridges and depressions like miniature meteorites. To the uninformed, they could easily be mistaken for scars or wounds from an epic battle with a fellow titan, or perhaps for an alarming skin disorder. Actually, they are simply thick, callous-like patches of roughened skin.

The callosities themselves are a dark to light gray color, but at a distance they take on a brighter white or pale yellow hue. This color is a result of barnacles and tens of thousands of cyamids, or "whale lice"—small crablike amphipods of the crustacean order—that use the callosities as toeholds to cling to their mammoth host. Immediately after birth and before acquiring their lifelong community of whale lice, a calf's callosities are undeveloped, bumpy skin patches pure black in color. Within a few months the patches erupt into a thick, hardened surface. As this occurs, a founding colony of whale lice is passed from mother to offspring as the calf nuzzles and rubs its mother to nurse or when the calf gets a ride on the mother's back as she helps it keep up with her travels. In older animals, whale lice pass from whale to whale when they rub against one another during courtship or social interactions.

Cyamids range in size from a few tenths to about three-quarters of an inch long, depending on the species. Each kind of baleen whale typically hosts two or three unique species; few species of cyamids occur on more than one species of whale. Whale lice are thought to make their living feeding on sloughed whale skin or on algae and other organisms that settle on the whale's skin. Whether whales derive any benefit from their hitchhiking hoard, or are bothered by the sharp claws at the tips of the legs that cyamids use to grasp the whale's skin, is unknown.

Some callosities also occur along the edge of the massive lower lip as a string of isolated patches. The largest patches perch like fog lights on either side of the protruding chin and the tip of the upper jaw where it bends sharply downward (figure 3.3). The callosities most visible to observers confined to the surface, however, are

Figure 3.3. Head-on view of a North Atlantic right whale showing callosities, the downward-bent tip of the upper rostrum covering the deep cleft at the front of the mouth, and the high sideboard lips. (Image by Yan Guilbault, courtesy of the New England Aquarium, Boston, MA; taken under permit from the Department of Fisheries and Oceans Canada.)

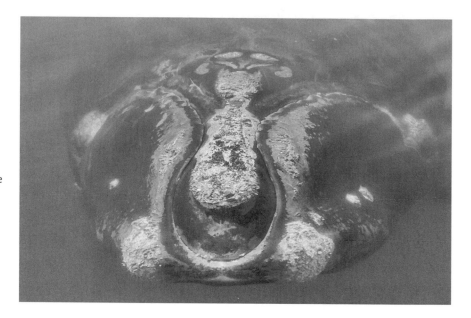

those on the top of the narrow elongated rostrum that forms the upper jaw and top of the whale's head. Whalers called these crusty-looking crests the "crown" or "combing"; others called them the "bonnet," the term used most often today. The bonnet is a meandering constellation of callosities running from the tip of the upper jaw back to the pair of *f*-shaped blowholes at the rear of the head.

The bonnet's shape and pattern is unique to each animal. Its configuration changes over the first year or two of life, but in older animals it becomes fixed, allowing biologists to use it like a fingerprint to identify individual whales.[11] Tracking individuals based on callosity patterns has revealed much of what is now known about the species' movements, habitat-use patterns, and rates of reproduction, survival, and growth. The cyamids that so effectively highlight callosity patterns also have been useful for other purposes. When a right whale's health is compromised by injury, cyamid numbers explode, spreading over abrasions, lacerations, and eventually to other areas. This casts a sickly yellow-orange pallor over much of the animal's head and body, providing a barometer for their general health. As will be seen in the following chapter, genetic analyses of cyamids have also provided tantalizing clues to the evolutionary history of right whales.

A second external feature no less arresting to even casual observers is the whale's oddly shaped mouth line. A gull's-eye view of a right whale swimming at the surface reveals a narrow, straight upper jaw that is rounded and bent downward at the tip and then flares out laterally at the rear of the head. From the surface, this great jaw covered with callosities looks like a slender, rocky reef plowing through the sea. Herman Melville, who in his early twenties shipped out in 1841 as a seaman on a whaling voyage, devoted an entire chapter of his masterpiece, *Moby-Dick*, to a description of the right whale's head. He had it pegged just about right when he observed, "if you stand on its summit [its back] and look at these two *f*-shaped spout holes, you would take the whole head for an enormous bass viola, [with the narrow upper jaw the instrument's neck and scroll] and these spiracles, the apertures in its sounding-board."[12]

The long peninsula-like upper jaw nestles snuggly in the broad embrace of a pair of massive bow-shaped lower jawbones (see figure 14.7, *bottom*). Viewed head-on with the mouth closed, the downwardly bent forward tip of the upper jaw covers a U-shaped cleft in the lower jaw. But with the mouth open, the lower jaw forms a giant scoop that bulges outward at the sides, with the upper jaw serving as a lid (figure 3.4). As noted earlier, the center of this scoop leads to the mouth's interior, but at its flanks it provides openings that lead to a pair of gutter-like channels running down either side of the head between the baleen and the cheek.

In profile, the mouth line forms an enormous arch rising upward at the snout before easing into a broad, semicircular curve midway along the head, where it plunges downward to a short S-shaped curve that cradles the low-placed eye at the back of the head. This twisted configuration—appearing to us as a half scowl, half sneer—traces a 20-foot arch, 5 to 6 feet high on either side of the head, and exudes what a casual observer might consider to be a sulking, grumpy countenance as befitting a creature with a long history of abuse. Turned upside down,

THE WHALEBONE-WHALE FEEDING UPON JELLY-FISHES.

Figure 3.4.
A skim-feeding right whale pictured in an 1883 illustration. Right whales do not feed on jellyfish and the outward spread of the sideboard lips is exaggerated, but otherwise this illustration is a reasonably accurate depiction. (Image courtesy of the Picture Collection, The New York Public Library, Astor, Lenox and Tilden Foundations.)

however, this mammoth pout becomes what appears to be a sheepish grin that seems equally and perhaps even more plausible. Indeed, some early illustrators asked to depict this whale without ever having seen it, portrayed the right whale's head upside down assuming (or perhaps simply preferring) a more cheerful disposition (plate 6).

Now switch your focus to the outward appearance of the whale's unlikely partner for comparison—the flamingo. At a casual glance, its large head features a disproportionately large beak with an arching mouth line and low-placed eye, all bearing an uncanny resemblance to the right whale. Like a right whale, the flamingo's mouth houses a filtering mechanism that, for all its great difference in size and composition, sifts food of a virtually identical size to that on which right whales dine. These are minute organisms typically ranging in size from the period at the end of this sentence to a grain of rice (about 0.5 to 4 millimeters). Resting like a lid atop the trough-shaped, boxy lower beak is a long, narrow upper bill bent down at the tip just like a right whale's upper jaw.

In profile, much like a right whale, the flamingo's strongly bent mouth line also rises upward in the front before plunging sharply downward midway along the beak and terminating at the rear near its low-placed eye. The forward half of its upper bill from tip to inflection point forms a relatively straight length; this is where the functional part of its filter is housed. Some have suggested this sharp bend evolved to allow the entire length of the beak to open with minimal jaw movement.[13] Others have offered that it allows the tips of the upper and lower beak to meet so the bird can pick out individual food items or manipulate nesting material while also allowing a constant width in the gap running along the sides of the beak when filtering prey.[14] As we will see shortly, however, another explanation is also possible.

This offers what seems to be the same haughty, disgruntled demeanor in flamingos as presented by a right whale's profile. Yet again, when turned upside down, a far more engaging grin emerges. Lining the outer edge of the upper bill along its frontward two-thirds is a series of 40 to 50 hooked "outer marginals." This rim of conical toothlike projections prevents the upper and lower bills from closing tightly and leaves a narrow gap through which flamingos draw in water and food when feeding.[15] The edge of the lower beak has a similar, but smaller, row of submarginals meshing neatly between the outer marginals of the upper bill. Together these form a toothy grill that keeps out debris too large for the filtering mechanism.

This oddly contorted beak, coupled with the bird's strange habit of lowering its head into the water in an upside-down posture when feeding, has long fascinated naturalists puzzling over the evolution of its odd form. Even before Darwin's monumental work on the origin of species in 1859,[16] some naturalists believed similarities in anatomical structures used for similar purposes in unrelated species were compelling evidence of what we now call evolution. Indeed, in the decades before Darwin's book, the major debate among naturalists was not so much *whether* such changes occurred as it was over what *caused* them to occur.[17]

In the early nineteenth century, debate over the cause of evolution had to be cloaked in a guise that would not provoke the ire of powerful religious authorities. Views on this question were divided into two opposing camps.[18] On one side were the "structuralists." They argued that changes in structural form surely must occur before an organism could effectively benefit from a modification—particularly one as bizarre and specialized as a flamingo's beak—to exploit some new ecological niche or life style. In the opposing camp were the "functionalists." They countered that organisms surely must somehow first alter an aspect of their way of life to bring about change in a structure's form. In this debate it was the functionalist Jean-Baptiste Lamarck who correctly concluded in 1809, "It is not the shape of either the body or its parts, which gives rise to the habits of animals and their mode of life; but it is, on the contrary, the habits, mode of life, and all the other influences of the environment, which have in the course of time built up the shape of the body and of the parts of animals."[19]

In simplest terms Lamarck argued that form follows function rather than vice versa. Darwin's great contribution was not in postulating the existence of biological evolution, although he is widely but mistakenly credited for doing so. Rather, his contribution was amassing a compelling body of evidence for what was then called transmutation, and also identifying a *mechanism*—natural selection—that made the process all the more plausible. What Darwin proposed was that changes can occur gradually over time only if they improve an individual's chance of surviving or reproducing, thereby allowing new advantageous traits to be passed on to future generations.[20] In most cases successful structural changes improved access or use of new or better food supplies, prevented predation, or enhanced breeding success.

Stephen J. Gould, in one of his best-known essays, "The Flamingo's Smile," explored the implications of these evolutionary insights for the form of the flamingo's odd beak and peculiar mode of feeding.[21] Flamingos feed in shallow hypersaline salt flats on the few but phenomenally abundant life forms able to survive in those harsh conditions—brine shrimp, small mollusks, salt fly larvae, seeds of salt-tolerant plants, and various algae, diatoms, and nematodes.[22] To do so, they lower their heads upside down into the shallow briny soup placing the straight, forward half of the upper beak flush against the bottom directly in front of their feet. Then, in a symphony of motion, they gracefully amble forward at a leisurely pace, tilting the beak from side-to-side as they swing their heads left and right across the surface of the organic ooze. All the while they open and close their beaks, pumping water in and out of the mouth, with muscular contractions of the tongue pulsating at rates of up to 13 cycles per second.[23] With their inverted feeding posture, flamingos are able to exploit a food source concentrated directly at the water-mud interface. The horny beak provides a tough outer surface resistant to abrasion as it sweeps back and forth across the sediment.

Gould reasoned that if it was the flamingo's strange behavior of reversing the orientation of its head while feeding that eventually gave rise to the beak's odd shape, then the functional role of the upper and lower bills should be reversed from that of other birds. That is, the structural form of the upper and lower bills of a flamingo should have evolved in ways exactly opposite to those of birds feeding in the conventional, upright posture. In this regard, the tried-and-true form of beaks used by most birds feeding in the upright posture features an upper bill that is larger than the lower bill, with the lower bill closing upward to fit into the embrace of the larger, relatively stationary, upper bill.

This standard arrangement is indeed just the reverse in flamingos. As described above, the flamingo's upper bill, which becomes its functional lower bill when the head is inverted to feed, is smaller than the lower bill. Moreover, when the mouth closes, the small upper bill fits snugly into its larger lower bill. Finally, when an ingenious study first captured high-speed video of captive flamingos feeding with their heads submerged, it revealed that the smaller upper bill had a larger vertical amplitude of movement than the larger lower mandible when the mouth opens and closes. That is, unlike bills used in upright feeding, the smaller upper bill of flamingos works against the larger, more stationary, lower bill.[24] Thus, the unusual form of the flamingo's bill seems to include the evolutionary modifications one might expect in a bird that feeds upside down.

No one has seen a right whale feeding at depth firsthand, but could it be that its contorted head, like that of a flamingo, evolved in response to an inverted feeding posture adopted to exploit prey concentrating along a solid interface? Let's consider that possibility a bit further by looking at some of the internal parts of these two disparate species.

Inward Similarities

The scaffolding for a right whale's cavernous maw is provided by the strongly arched upper jaw, which is composed of two maxillaries fused together to form a single, elongated, narrow rostrum, and a pair of bowed lower jawbones, or mandibles, held together at the tip. When opened for feeding, the lodgepole rostrum formed by the upper jaw and the two outwardly bowed, lower jawbones create an enormous mouth with an A-frame cross-section. According to Melville, early Dutch whalers likened the right whale's head to "a gigantic galliot-toed boot . . . (in which) the old woman in the nursery rhyme could have lodged her whole swarming brood."[25] With the open mouth reaching a height up to 12 feet and a length extending more than 14 feet, the oral cavity of an adult right whale might well have sheltered such a family. To support this yawning chasm, the pair of lower mandibles is massive. The lower jawbones of a 53-foot right whale killed in a 2003 ship collision off Cape Cod, Massachusetts, were nearly 14 feet long, with each weighing nearly 500 pounds.[26] Their width tapered from 12 inches at the base to about 7 inches at the forward tip. Unlike the lower jaws of most mammals, which are composed of a single bone, the lower jaws of right whales are composed of two bones bound together by a flexible connection of fibrous tissue, allowing a moderate degree of twisting.

Also unlike other baleen whales and most mammals, the base of a right whale's mandible lacks a coronoid process. In other species this bony knob, where the jaws are hinged to the skull, serves as a point of attachment for the temporalis muscle.[27] With one end of this muscle attached to the cranium, it is used to close the mouth in most animals by pulling the lower jaw up against the more stationary upper jaw. Absence of the coronoid process in right whales may reflect a tendency to open and close the mouth by lifting and lowering the narrow upper jaw, much like flamingos. Perhaps by relying less on pulling the massive lower jaw up, and more on raising and lowering the smaller upper jaw, the coronoid process became expendable for right whales, and the temporalis muscle instead grew more important for rotating the bowed mandibles outward and inward rather than up and down for feeding.

Aerial photographs of skim-feeding right whales (figure 3.5) clearly show that the width of the head is far broader when feeding than when the mouth is closed. Rotating the curved mandibles outward could serve to adjust the width of the lower mouth and position the deep, muscular, sideboard lower lips to regulate water flow through the gutter-like discharge channel when feeding.[28] Adjusting the width of that channel might even fine-tune filtering efficiency by tweaking the strength of suction pulling water through the baleen.

The most extraordinary feature of a right whale's mouth, however, is the arching system of baleen plates that hang from the upper jaw over either side of the tongue like a palm frond roof on a South Sea cabana. These plates, numbering between 250 and 350 on either side of the mouth, were known to whalers as whiskers, hog's bristles, fins, or blinds, but most commonly as "whalebone" or simply "bone." Each plate is a long, narrow shaft originating in the gum of the upper jaw where teeth normally occur in other mammals.[29] Less than a quarter of an inch separates one plate from its neighbor at the gum. Prized for their flexibility, light weight, and durability, these bendable baleen shafts were valued commodities before modern manufacturing processes produced plastic or comparably flexible

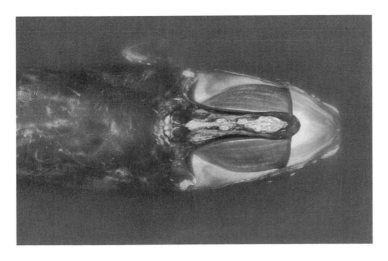

Figure 3.5. Aerial view of a skim-feeding right whale showing the narrow upper jaw (top of rostrum) encrusted with callosities, dark-colored baleen plates billowing to either side of the upper jaw, and light-colored tongue used to deflect water flow to the baleen. (Image courtesy of NOAA/NEFSC, taken by Cynthia Christman under MMPA research permit number 775-1875.)

materials. Indeed, the value of "whalebone" (at times equal to or even greater than that of a whale's oil) was an important source of profit for whalers, particularly when whale numbers and amounts of oil were declining.

Baleen plates are the signature feature of all mysticete whales. Indeed, they are the source of the name mysticete, which is derived from the Latin word roots *myst-*, meaning mustache, and *cet-*, meaning whale. Early naturalists seeking a term to distinguish baleen whales from all other whales apparently concluded that baleen shafts and their hair-like fibers lining the upper lip would remind scientists of a man's mustache.

Baleen plates are composed of keratin, the same material forming the hooves of horses and our own fingernails. Each plate is just a few millimeters thick, several centimeters wide, and composed of a matrix of embedded hair-like filaments running lengthwise along the shaft. This matrix makes them easy to split lengthwise, but requires a saw to cut across the grain. Like fingernails, they grow continuously from the base throughout life. Whereas the outer edge facing the lip comes to a clean, rounded edge, the inside margin gives rise to a frayed mat of exposed fibers. As old fibers wear away, new fibers emerge from the shaft's matrix to take their place; this renewal is probably a result of rubbing against the tongue. In cross-section, the plates show a slight S curve.[30] When water flows between them, this curvature creates a small swirl of eddies that may help pull water through the baleen.[31] When right whales open their mouths, the overlapping baleen shafts drop downward to some extent from the upper jaw and billow outward, forming a bowed fan of stiff curtains on either side of the mouth.[32]

As noted earlier, most water enters the mouth of feeding whales through the cleft between baleen racks at the forward tip (see figure 3.1). Some water, however, also enters the forward end of the gutter-like troughs along both sides of the head, between the baleen and arching lower lip.[33] This passes unfiltered down each channel (sulcus). Because the channels narrow midway along their length, the flow rate accelerates part way down. This design creates a mild suction that pulls water into the mouth's interior and through the baleen plates into the sulcus.[34] According to the new theory offered by Werth and Potvin, this movement of water through the baleen and into the sulcus is enhanced by flow rates inside the mouth between the tongue and the baleen that decrease in velocity as the water moves from the front to the back.[35] With water moving down the sulcus outside the baleen plates and its flow rate increasing, the opposing pressure differences on either side of the plates effectively pull water out of the mouth without actually pulling prey into the baleen. The result is a thick concentrate of prey inside the mouth that is ready to be swallowed by the time it reaches the esophagus.

By the same token, this flow regulation mechanism that pulls water into the mouth through the baleen plates also prevents, or at least minimizes, formation of a bow wave in front of the open mouth that would otherwise push intended prey aside. This could explain the curious observation of an Icelandic monk, who upon kayaking near a feeding bowhead whale in 1780, reported that small crustaceans apparently "found pleasure" in the whale's baleen because they seemed to go into the mouth "from all directions more or less by themselves."[36]

In any case, either by pressure from water forced into the mouth as the whale moves ahead or by suction created by water flow through lateral exit channels, the flexible baleen shafts bow outward on either side of the mouth, and the shafts thus separate until only the fibers on the inside edge of each plate overlap (see figure 3.5). To prevent baleen plates from flapping uncontrollably outward, the high arching cheeks hold them in place at their tips. When the mouth closes after a feeding bout, the shafts straighten, gaps between the shafts narrow, and (to some extent) the plates fold back and up into the mouth.

Differences in the number, width, and length of baleen plates and the thickness of the hair-like filaments are related to prey preferences of different whale species. The plates of right whales and bowhead whales are longer and straighter than all other baleen whales and have the finest filaments to enhance filtering of their minute prey.[37] In a fully grown right whale, the shafts at the front of the mouth are just a few feet long, but increase to 7 or even 9 feet in the middle third of each rack.[38] Only the baleen shafts of bowhead whales are longer, reaching 13 to 14 feet in length. The plates of both species are up to 9 inches wide at the gum and taper to a few inches at the tip.[39]

Right whales off the United States and Canada feed almost entirely on copepods just a few millimeters in length (figure 3.6).[40] Elsewhere, krill up to a half-inch long may be preferred. Krill was reported in stomachs of some of the last right whales killed off northern Europe in the early 1900s.[41] Larger fish and squid are unlikely to be caught given their greater mobility and the slow speed of feeding right whales and bowheads.

Although baleen plates are the most remarkable feature in a right whale's mouth, the tongue is also a marvel—and for a long time, it was also a puzzling one. The tongues of right whales and bowheads are said to be the largest muscular organs in the animal kingdom.[42] Their enormous bulk rests on a sling of skin and fatty tissue stretched between the lower jawbones. The tongue makes up 4 to 6 percent of the whale's total mass and weighs 2,000 to 3,000 pounds in a 50-ton

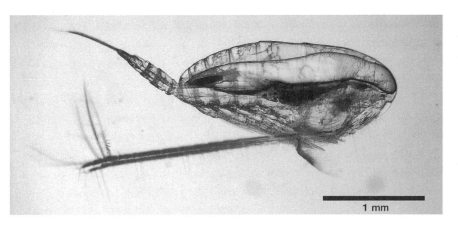

1 mm

Figure 3.6. Super-abundant blooms of the minute copepod *Calanus finmarchicus* serve as the primary food source for North Atlantic right whales. Late stage (older) copepods such as this one have nutrient-rich oil sacks that can be seen filling much of their body cavity through its trans-lucent outer shell. (Image by Dag Altin, courtesy of BioTrix.)

adult. Slightly taller than it is wide, it rises like a dome four- to five-feet-high from the floor of the mouth, and has a shallow furrow running down its center to the esophagus. Covered with a thin layer of tough keratinized cells that hold its shape even after the whale's death,[43] the tongue serves as a fixed baffle redirecting incoming water to the baleen curtains (see figure 15.1). Given the new theory discussed above about how a right whale's filtering mechanism works, the tongue's massive arch also may play crucial roles in regulating both the width of the space through which prey-laden water passes inside the mouth and its rate of flow.

It has also been suggested that the tongue is involved in detaching prey trapped in the baleen plates and moving it to the gullet.[44] However, if prey is not actually trapped in the baleen and is instead concentrated at the back of the mouth by the water flow, such a role would be unnecessary. In that case, the tongue may nevertheless play a role in maintaining the baleen's efficiency and function. This would occur when the mouth is closed and the baleen plates collapse to their unbowed, stored position. Then the broad solid shafts could become a stiff backing against which the tongue could squeeze or comb prey from the hair-like fringe, thereby grooming the filaments and removing any lodged prey or debris that could otherwise interfere with the role baleen plates play in regulating water flow when feeding. Indeed, periodic scrubbing by the tongue would seem necessary to expose new fibers embedded in the plates' matrix and would explain the regular occurrence of baleen fragments in whale feces.[45]

The tongue's interior is soft fatty tissue laced with swirls of muscle. These qualities once made it a highly prized part of the right whale. According to one New England whaler, the tongue of a large right whale killed in the 1700s yielded 20 barrels of oil, or nearly one-sixth of the total yield of 130 barrels.[46] In medieval Europe, the tongue was considered the best part of the whale for eating and was often donated as a tribute to the church.

Although not part of the skull, one final feature possibly related to right whale feeding behavior is the fusion of all seven cervical vertebrae into a solid mass of bone where the head connects to the spinal column. This mass of bone, a rare feature in most whales, creates a sturdy, inflexible neck that may help the powerful trunk muscles push the enormous head forward in a stable, fixed position. It also might function as a fulcrum or brace for raising and lowering the narrow upper jaw and cranium when opening the mouth to feed. Interestingly, the only other whale with all cervical vertebrae fused is the sperm whale, which also has a disproportionately large head.

If we now compare this internal morphology to that in a flamingo's beak, we again see several striking similarities. Notwithstanding the obvious difference in size, the skulls of both species leave a remarkably similar impression (figure 3.7). Indeed, the untrained eye might be hard pressed to tell the drawing of a right whale skull from that of the flamingo. Both have the same narrow, deeply curved upper jaw. Flamingos also have a similar trough-shaped lower jaw that provides a long, deep, box-like housing for a massively large tongue—at least by bird standards.

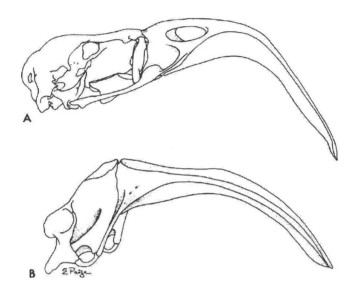

Figure 3.7. Profile view of the skulls of a lesser flamingo (A) and a right whale (B), showing their similar strongly arched, telescoped upper jaws. (Images from Olson and Feduccia, 1980, Relationships and Evolution of Flamingos (Aves: Phoenicopteridae). *Smithsonian Contributions to Zoology* 316. Smithsonian Institution Press, Washington, DC, p. 67.)

Interestingly, the flamingo's tongue was also once a highly prized food source. In ancient Rome, emperors and nobles periodically hosted week-long feasts regaling guests with exotic delicacies from far ends of the Roman Empire, including heaping platters of meaty flamingo tongues. To satisfy this epicurean excess, ships returning to Rome from northern Africa, Gaul, and western Asia brought back untold thousands of flamingos destined for slaughter, expressly for decadent gustatory displays.

The flamingo's filtering mechanism, however, is fundamentally different from a right whale's.[47] The only similarity is its location at either side of the mouth. In place of baleen plates, the flamingo's filter is a band of file-like "lamellae" inside the rim of the upper and lower beaks. Instead of the one-way, flow-through design of right whales, flamingos pump water in and out, adjusting the effective filter size by fine-tuning the height of the gap between the upper and lower bills.[48]

The only studies offering more than a passing comment on resemblances between right whales and flamingos are by ornithologists. In an excellent 1980 review of the evolution of flamingos, Storrs L. Olson and Alan Feduccia concluded, "the similarity of the heads of flamingos and whales is probably a good indication that the bent structure of the flamingo bill arose in accordance with the constraints of filter feeding alone, the 'upside down' feeding posture having evolved secondarily."[49] Commenting that this must be one of the more outstanding examples of convergent evolution, they proposed that the beak's sharp bend increased the length of the filtering mechanism in the middle portion of the rostrum.

However, two other ornithologists, S. Laurie Sanderson and Richard Wassersug, concluded that similarities between right whales and flamingos couldn't possibly be examples of convergent evolution given that their filtering mechanisms differed in both structural form and method of operation.[50] Neither team of investigators, however, considered that the similarities might be related not to the filtering mechanism, but to the advantages of a strongly bent upper jaw used for scooping up dense clusters of minute prey concentrated along a fixed aquatic interface. In the case of balaenid whales, that interface may be the sea floor or perhaps even the underside of ice sheets in polar seas, as will be discussed below.

Dining from a Different Angle
at the Right Whale Buffet

If we now add a final dash of ecology to our jambalaya of morphological detail and evolutionary mechanisms, the curious resemblance between flamingos and right whales begins to take on a more compelling mixture of rhyme and reason. By extension, the inverted feeding posture, which Gould proposed as the behavioral trigger for the anatomical changes creating the flamingo's specialized beak, might be an equally plausible explanation for some of the similarly peculiar features in right whales. By inverting their heads, flamingos found an effective way to exploit a rich, if minute, food source concentrated on the bottom of shallow hypersaline lagoons. Could right whales—or more accurately, distant ancestors of right whales and bowhead whales—have learned the same trick?

Today's right whales, like bowhead whales, often dive to the bottom when feeding;[51] this conclusion is based on observations of whales frequently surfacing with mud caked on the tops and sides of their heads. Explanations for this strange occurrence have long puzzled right whale biologists. Could it be the result of errant underwater navigation in dark murky depths? Or an intentional effort to scrape off pesky cyamids? Perhaps an inverted feeding posture used to skim swarms of prey hugging the bottom? Right whales are too well adapted for navigating dark ocean depths to erroneously and repeatedly crash into the bottom, and there is no behavioral evidence that whales find cyamids annoying. However, inverted feeding at the bottom seems more plausible and would be an exact mirror image of skim-feeding behavior often seen at the surface, where prey concentrate at the air-water boundary. As described earlier, skim-feeding right whales move along the surface with rostrums raised a foot or more out of the water (figure 3.8). If this same strategy is used in reverse along the ocean bottom, it could explain some of the more convincing correlations with inverted feeding.

First, the mud seen atop right whale heads has been reported only at feeding grounds and on feeding whales.[52] These observations could support speculation by G. M. Allen in his exhaustive 1916 review of whalebone whales in New England that callosities on a right whale's head might "serve as a bumper."[53] In the same way that a flamingo's horny beak protects against abrasion as it sweeps across the sediment, callosities may protect a right whale's head from chafing shells and rocks while feeding at the bottom. Second, film footage of the bottom in right whale feeding grounds in the Bay of Fundy, Canada, clearly shows long narrow groves etched in soft mud. These troughs are consistent with marks that might be left by

the upper jaws of inverted right whales plowing across the mud.[54] Third, electronic "D-tags" have been attached to the backs of feeding right whales with suction cups to record their orientation on dives and reveal that they often rotate onto their sides and back at midwater depths.[55] Unfortunately, because whales in those studies did not dive all the way down, their orientation when feeding at the bottom was not documented.

Both skim feeding at the surface and upside-down feeding on the bottom would be especially important for right whales and bowhead whales, because neither species is capable of chasing down prey or using tricks, such as the bubble-net feeding of humpback whales, to concentrate their minute targets. Instead, both right whales and bowheads must rely entirely on their prey's behavior and on environmental conditions to concentrate the food into patches dense enough to trigger feeding activity.[56] In this regard, both copepods, such as *Calanus finmarchicus*, which is perhaps the most important food source for North Atlantic right whales, and krill, the main food source for bowheads, undergo dramatic physical transformations and vertical migrations that make skim feeding along ocean interfaces at both the surface and bottom particularly advantageous. Indeed, right whales seem to feed only when prey densities reach a certain threshold, although that threshold may vary by type and size of prey. For example, Stormy Mayo and colleagues have found that right whales do not begin to feed in Cape Cod Bay until they can find patches of copepods at densities of 106 to 114 organisms per cubic foot.[57]

In the case of *C. finmarchicus*, eggs and early life stages, when their size is still too small for right whales to exploit, remain on or in the bottom through winter; but in spring and summer the zooplankton emerge in enormous swarms and then

Figure 3.8. View of a skim-feeding right whale with its arched upper jaw raised above the water surface, exposing the upper portion of its baleen plates. Views such as this may have inspired the name Mysticeti, or mustached whales, for the family of baleen whales. (Image by Monica Zani, courtesy of the New England Aquarium, Boston, MA, taken under NOAA/NMFS permit #15415.)

mature through a series of molts into late-stage forms with large nutrient-rich oil sacs filling most of their small bodies. These swarms migrate up and down the water column in response to daily changes in light levels. In the process, the copepods periodically concentrate in extraordinary numbers at the bottom, at the surface, at midwater depths where sharp changes in physical conditions occur (e.g., abrupt salinity or temperature transition zones), and along current margins. Relying solely on dense concentrations forming at midwater depths and the surface could severely limit the ability of right whales to exploit their food source. However, if they invert their heads like flamingos when feeding, their slender, strongly arched upper jaw could allow them to scoop up prey closer to the bottom where it often concentrates in the highest densities; this ability would greatly expand their opportunities to capture prey. The curved rostrum also might minimize injuries to the back of the head and blowholes that could otherwise be scraped and abraded when skimming upside down across the bottom while feeding. At the same time, the whales' low-placed eyes at the corner of their mouths would be raised farther off the bottom when feeding in an inverted posture and perhaps give them a clearer view of prey concentrations.

A common ancestor of right whales and bowhead whales may have found this feeding strategy and the resulting narrow, hooked jaw equally effective for skimming swarms of krill lurking just beneath the ice in frozen polar seas. Indeed, the even more pronounced arch of the upper jaw might indicate it first evolved in whales inhabiting ice-covered seas, and would thus be even more important for bowheads. Alternatively, the bowhead's ancestors may have found that an even stronger curvature in their arching upper jaw provided an additional advantage when moving into Arctic seas to exploit krill. That is, an even stronger arch might have enabled whales to use the top-forward part of their head as a battering ram to break through the ice without damaging their blowholes. Early whalers chasing bowheads along the edge of the northern pack ice were well aware of the bowheads' close affiliation with heavy ice cover in summer. This also holds true in winter.

A recent acoustic study using hydrophones mounted on the ocean floor off northeast Greenland recorded a nearly continuous cacophony of bowhead whale songs from October 2008 to April 2009 in an area where ice covered 90 to 100 percent of the ocean surface within an 18.5-mile perimeter.[58] Could those whales be using hooked rostrums to feed on swarms of krill just beneath the pack ice and on swarms hovering close to the bottom? In the northern Bering Sea between Alaska and Russia, where bowhead whales also overwinter beneath ice covering virtually the entire surface, the answer seems to be yes. Studies with tags designed to record their diving behavior just south of the Bering Strait in winter has revealed repeated cycles of diving to the bottom, presumably to feed.[59]

But if right whales and bowhead whales rub along the sea floor or ice when feeding, wouldn't cyamids clinging to their heads be rubbed off? Although some undoubtedly are, enough remain attached to maintain their colonies. For right whales, cratered callosities probably offer protected nooks to help them stay attached. However, inverted feeding may serve as a check on the size of cyamid populations on individual whales. As noted above, heavy cyamid populations on right whales are a diagnostic indicator of poor health that is often accompanied by evident losses in body fat. The correlation between heavy cyamid loads and thin body condition could reflect a disruption of normal inverted feeding behavior at the bottom that would otherwise keep cyamid populations in check and the whales in a fatter, healthier condition.

When all these factors are considered together, one is left with an intriguing, perhaps even a compelling, evolutionary hypothesis. Many of the prominent features on a right whale's head may have evolved in response to a new feeding behavior adopted to exploit prey that periodically masses in a narrow band just a foot or two above the sea floor—or perhaps a comparable distance beneath sea ice. In both cases, it could explain the strongly arched mouth line and prominent curvature of the upper jaw, callosities on the head, and the lack of a coronoid process on the lower jaw. Interestingly, the curved rostrum in today's right whales and bowhead whales is more pronounced than in fossil balaenid whales.[60] This could indicate that the strongly arched upper jaw is a recent development. Of course bowheads and right whales also feed at midwater depths where these adaptations are not critical. However, if all these morphological changes are related to a new feeding strategy that improves exploitation of prey concentrations near the bottom or ocean-ice interface, the strategy must have been particularly important over the course of balaenid evolutionary history and probably remains vital for today's whales.

Although the proposed evolution of features on a right whale's head may be speculative, the adoption of a new feeding strategy, as discussed in the next chapter, has also been proposed as the possible trigger that gave rise to evolution of the entire cetacean line. In addition, as discussed in later chapters, feeding upside down also could shed light on the risks of right whales becoming entangled in commercial fishing gear, as well as the value of possible mitigation strategies.

NOTES

1. Werth, AJ. 2001. How do mysticetes remove prey trapped in baleen? Bulletin of the Museum of Comparative Zoology 156:189–203.

2. Nerini, M. 1984. A review of gray whale feeding ecology. *In* ML Jones, SL Swartz, and S Leatherwood (eds.), *The Gray Whale: Eschrichtius robustus*. Academic Press. Orlando, FL, US, pp. 423–485.

3. Zackowitz, M. 2000. Doing a whale of a research job. National Geographic 197:140. Pivorunas, A. 1977. Fibrocartilage skeleton and related structures of ventral pouch of balaenopterid whales. Journal of Morphology 151:299–313. Pivorunas, A. 1979. The feeding mechanisms of baleen whales. American Scientist 67(4):432–440.

4. Goldbogen, JA. 2010. The ultimate mouthful: lunge feeding in rorqual whales. American Scientist 98(2):124–130.

5. Sanderson, SL, and R Wassersug. 1993. Convergent and alternative designs for vertebrate suspension feeding. *In* J Hanken and BK Hall (eds.), *The Skull: Functional and Evolutionary Mechanisms*. University of Chicago Press. Chicago, IL, US, Vol. 3, pp. 37–112.

6. Werth, AJ. 2004. Models of hydrodynamic flow in the bowhead whale filter feeding apparatus. Journal of Experimental Biology 207:3569–3580. See also Baumgartner, MF, CA Mayo, and RD Mayo. 2007. Enormous carnivores, microscopic food, and a restaurant that's hard to find. *In* SD Kraus and RM Rolland (eds.), *The Urban Whale: North Atlantic Right Whales at the Crossroads*. Harvard University Press. Cambridge, MA, US, pp. 138–171.

7. Werth, AJ, and J Potvin. 2016. Baleen hydrodynamics and morphology of cross-flow filtration in balaenid whale suspension feeding. PLoS ONE 11(2):e0150106, doi:10.1371/journal.pone.0150106.

8. Sumich, J. 2013. Gray whales in captivity. *In* J Sumich (ed.), Gray Whale: From Devilfish to Gentle Giant. Whalewatcher 41(1):5–6.

9. Stewart, BS. 2001. Introduction and background to the collected papers on the rescue, rehabilitation, and scientific studies of JJ, an orphaned California gray whale calf. Aquatic Mammals 27(3):203–208.

10. Stejneger, L. 1855. Birds. *In* JS Kingsley (ed.), *The Standard Natural History*. S. E. Cassino. Boston, MA, US, Vol. 4, p. 155.

11. Hamilton, PK, AR Knowlton, and MK Marx. 2007. Right whales tell their own stories: the photo-identification catalogue. *In* Kraus and Rolland, *The Urban Whale* (n. 6), pp. 75–104.

12. Melville, H. (1851) 1943. *Moby-Dick; or, the Whale*. Limited Editions Club. The Heritage Press. New York, NY.

13. Jenkins, JP. 1957. The filter feeding and food of flamingoes (Phoenicopteri). Philosophical Transactions of the Royal Society of London B, Biological Sciences 240(674):401–493.

14. Rooth, J. 1965. Flamingos on Bonaire (Netherlands Antilles): habitat, diet, and reproduction of *Phoenicopterus ruber ruber*. Natuurwetenschappeljke Studiekring voor Suriname en de Nederlandse Antillen 41:1–151.

15. Jenkins, The filter feeding and food (n. 13).

16. Darwin, C. 1859. *On the Origin of Species by Means of Natural Selection or the Preservation of Favored Races in the Struggle for Life*. John Murray. London, UK.

17. See, for example, Darwin, E. 1794. *Zoonomia; or, The Laws of Organic Life*. J. Johnson. London, UK, Vol. 1, pp. 478–533. Lyell, C. 1832. *The Principles of Geology, Being an Attempt to Explain the Former Changes of the Earth's Surface, By Reference to Causes Now in Operation*. Vol. 2. John Murray. London, UK.

18. Gould, SJ. 1985. *The Flamingo's Smile: Reflections in Natural History*. Norton. New York, NY.

19. Lamarck, JB. (1809) 1984. *Zoological Philosophy*. Chicago: University of Chicago Press, as quoted and cited in Gould, *The Flamingo's Smile* (n. 18).

20. Darwin, *On the Origin of Species* (n. 16).

21. Gould, SJ. 1985. The flamingo's smile. *In* Gould, *The Flamingo's Smile* (n. 18), pp. 23–39.

22. Allen, RP. 1956. *The Flamingos: Their Life History and Survival*. Research Report 5. National Audubon Society. New York, NY.

23. The rates at which water is pumped in and out differ among flamingo species. While greater and Caribbean flamingos pump water at a rate of 13 cycles per second, lesser flamingos apparently do so at about 4 cycles per second.

24. Zweers, G, F De Jong, H Berkhaudt, and JC Vanden Berge. 1995. Filter feeding in flamingos (*Phoenicopterus ruber*). The Condor 97(2):297–314.

25. Melville, *Moby-Dick* (n. 12), p. 357.

26. Campbell-Malone, R. 2007. Biomechanics of North Atlantic Right Whale Bone: Mandibular Fracture as a Fatal Endpoint for Blunt Vessel-Whale Collision Modeling. PhD diss. Massachusetts Institute of Technology and Woods Hole Institution Department of Biology.

27. Bannister, JL. 2009. Baleen whales, *Mysticetes*. *In* WF Perrin, B Würsig, and JGM Thewissen (eds.), *Encyclopedia of Marine Mammals*, 2nd ed. Academic Press. New York, NY, pp. 80–89.

28. Lambertson, RH, RJ Hintz, WC Lancaster, A Hirons, KJ Kreiton, and C Moore. 1989. *Characterization of the Functional Morphology of the Mouth of the Bowhead Whale*, Balaena mysticetus, *with Special Emphasis on the Feeding and Filtration Mechanisms*. Report to the Alaska Department of Wildlife Management, North Slope Borough. Barrow, AK, US, p. 134.

29. Rice, DW. 2009. Baleen. *In* Perrin et al., *Encyclopedia* (n. 27), pp. 61–62.

30. Lambertson, RH, KJ Rasmussen, WC Lancaster, and RJ Hintz. 2005. Functional morphology of the bowhead whale and its implications for conservation. Journal of Mammalogy 86(2):342–352.

31. Ibid.

32. Werth, How do mysticetes (n. 1).

33. Baumgartner et al., Enormous carnivores (n. 6), pp. 151–153.

34. Werth, Models of hydrodynamic flow (n. 6).

35. Werth and Potvin, Baleen hydrodynamics (n. 7).

36. Quoted in Lambertson et al., Functional morphology of the bowhead whale (n. 30), p. 343.

37. Werth, AJ. 2013. Flow-dependent porosity and other biomechanical properties of mysticete baleen. The Journal of Experimental Biology 216:1152–1159.

38. Omura, H, S Ohsumi, T Nemoto, K Nasu, and T Kasuya. 1969. *Black Right Whales in the North Pacific*. Whales Research Institute. Tokyo, Japan.

39. Folkens, P, RR Reeves, BS Stewart, PJ Clapham, and JA Powell. 2002. *National Audubon Society Guide to Marine Mammals of the World*. Alfred K. Knopf. New York, NY.

40. Baumgartner et al., Enormous carnivores (n. 6), pp. 143–146.

41. Collet, R. 1909. A few notes on the whale *Balaena glacialis* and its capture in recent years in the North Atlantic by Norwegian whalers. Proceedings of the Zoological Society of London 7:91–98.

42. Omura, H. 1958. North Pacific right whale. Scientific Reports of the Whales Research Institute 13:1–52.

43. Lambertson et al., *Characterization of the Functional Morphology* (n. 28).

44. Werth, How do mysticetes (n. 1), pp. 196–198.

45. Ibid.

46. Dudley, P. 1725. An essay upon the natural history of whales, with particular account of the ambergris found in the sperma ceti whale. Philosophical Transactions of the Royal Society of London 33:256–269, p. 257.

47. Jenkin, The filter feeding and food (n. 13).

48. Zweers et al., Filter feeding in flamingos (n. 24), p. 305.

49. Olson, SL, and A Feduccia. 1980. *Relationships and Evolution of Flamingos (Aves: Phoenicopteridae)*. Smithsonian Contributions to Zoology. No. 316. Smithsonian Institution Press. Washington, DC, p. 66.

50. Sanderson and Wassersug, Convergent and alternative designs (n. 5).

51. Citta, JJ, LT Quakenbush, SR Okkonen, ML Drukenmiller, W Maslowski, et al. 2015. Ecological characteristics of core-use areas by Bering-Chukchi-Beaufort (BCB) bowhead whales, 2006–2012. Progress in Oceanography 136:201–222.

52. Clapham, PJ (ed.). 2004. *Improving Right Whale Management and Conservation through Ecological Research*. Report of the Working Group Meeting, 16th of April 2004. Northeast Fisheries Science Center, National Marine Fisheries Service. Woods Hole, MA, US, p. 7.

53. Allen, GM. 1916. *The Whalebone Whales of New England*. Memoirs of the Boston Society of Natural History, Vol. 8, no. 2, p. 116 (quotation).

54. Moira Brown (New England Aquarium, Boston). Personal communication with author, November 2015.

55. Baumgartner, MF, and BR Mate. 2003. Summertime foraging ecology of North Atlantic right whales. Marine Ecology Progress Series 264:123–135.

56. Baumgartner et al., Enormous carnivores (n. 6), pp. 156–161.

57. Mayo, C, and L Goldman. 1992. Right whale foraging and the plankton resources in Cape Cod and Massachusetts Bays. *In* J Hain (ed.), *The Right Whale in the Western North Atlantic: A Science and Management Workshop.* April 14–15, 1992, Silver Spring, MD. Northeast Fisheries Science Center Reference Document 92-05. Northeast Fisheries Science Center, Woods Hole, MA, US, pp. 43–44.

58. Staffort, KM, SE Moore, CL Berchok, Ø Wiig, C Lydersen, E Hanson, D Kalmbach, and KM Kovacs. 2012. Spitsbergen's endangered bowhead whales sing through the polar night. Endangered Species Research 18:95–103.

59. Citta et al., Ecological characteristics (n. 51), p. 211.

60. Bisconti, M. 2005. Skull morphology and phylogenetic relationships of a new diminutive balaenid from the lower Pliocene of Belgium. Paleontology 48:793–816; and Kimura, T, K Narita, T Fujita, and Y Hasegawa. 2007. A new species of *Eubalaena* (Cetacea: Mysticeti: Balaenidae) from the Gonda formation (latest Miocene-early Pliocene) of Japan. Bulletin of Gunma Museum of Natural History 11:15–27.

4

EVOLUTION

IF THE EVOLUTION of features on the right whale's head as proposed in the previous chapter seems intriguing, it surely pales in comparison to the astonishing transformation of right whales and other large whales from small terrestrial mammals into the marine giants of today. Along with the remodeling of marine fishes into land animals 400 to 350 million years ago (Ma) and the transformation of feathered dinosaurs into birds 160 Ma,[1] reconfiguration of the mammalian body plan from one designed for the rigors of terrestrial life into one equally at home in the ocean is one of the most remarkable examples of the infinite possibilities for life's adaptability. Although seals, sea turtles, and penguins also evolved from terrestrial ancestors, only whales and sirenians (manatees and dugongs) have so completely severed their ties to land that they never return to shore—not even for brief periods to molt, rest, or lay eggs. Fascination with intermediate forms that must have existed during this incredible transition,[2] as well our own shared mammalian heritage, have no doubt fueled the special attraction whales hold for scientists and nonscientists alike.

Among the approaches scientists use to infer the course of evolution and relationships between species, two have been particularly important: comparing physical similarities between skeletal structures of species alive today with those in the fossil record, and new analyses (that is since the late 1900s) of genetic code locked in DNA samples of living and extinct animals. The traditional method relies on skeletal traits or "characteristics" shared by living and fossil specimens. Among the handful of characteristics used to distinguish cetaceans from all other mammals are the density and thickness of bone in parts of the inner ear,[3] the position of ear bones relative to surrounding bones in the skull, the shape of the molars, and the structure of limb bones. Using those characteristics, scientists have traced the cetacean line back to the early Eocene epoch some 54 to 52 Ma.[4]

Ever since Linnaeus first placed cetaceans as a subgroup of mammals in the 1758 edition of his zoological classification scheme, debate over which land mammals are their closest living relatives has been a subject of great debate. In 1883, W. H. Flower proposed that today's ungulates (hoofed mammals) are most similar to whales.[5] Recent analyses support his conclusion and have narrowed the list of candidates to the taxonomic order called Artiodactyla, which includes even-toed ungulates such as deer, hippos, and pigs. Within that order, most of today's taxonomists lean toward hippos as the closest living land relatives to whales.[6]

Tracing the lineage of whales back through the fossil record has been a challenging but vibrant field of study. Indeed, the evolution of whales is now held up as one of the best-documented transformation sequences of any major taxonomic group, thanks largely to discoveries and analyses since the 1980s.[7] Notwithstanding such accolades, any outline of evolutionary descent can be filled with more holes than a shredded fishing net. Thus, any discussion of evolutionary pathways must always be prefaced with caution and recognition that significant revisions are possible as new fossils are unearthed—or simply rediscovered in long-forgotten museum lockers—and new genetic analyses are undertaken. With that warning, let's take a brief look at the currently understood evolutionary trail leading to today's right whales.

The Ancient Whales

The earliest fossils with the requisite complement of cetacean characteristics are collectively called "archaeocetes," or ancient whales. Although the oldest of these date to the early Eocene some 54 Ma, it was not until 48 to 42 Ma in the middle Eocene that they first began to diversify broadly and invade all the world's oceans (figure 4.1). Such periodic flowering often seems to be associated with major changes in world climate or geography.[8] The initial diversification of the cetacean line in the Eocene was no exception. It coincided with a steep rise in global temperatures that created a world far warmer than today—indeed, far warmer than any time since the Eocene.

At the start of the Eocene (56 Ma), water temperatures at the ocean surface began increasing to a point that reached about 10°F to 16°F (4°C to 9°C) above current levels, while air temperatures over the continents rose to approximately 10°F to 14°F (4°C to 7°C) higher than present.[9] These "hothouse" conditions eliminated year-round ice almost everywhere on earth, including the poles.[10] Temperature gradients between the equator and poles were also far smaller than today. This allowed alligators and palm trees to stretch their range northward above the Arctic Circle, and cool-temperate flora, such as beech trees and ferns, to colonize Antarctica.[11]

World geography was also quite different at that time. The Indian subcontinent, which broke off from the supercontinent Gondwana 125 Ma, was still moving north by means of the forces of plate tectonics toward its eventual collision with Asia; the Atlantic Ocean was far narrower than it is now; the Central American

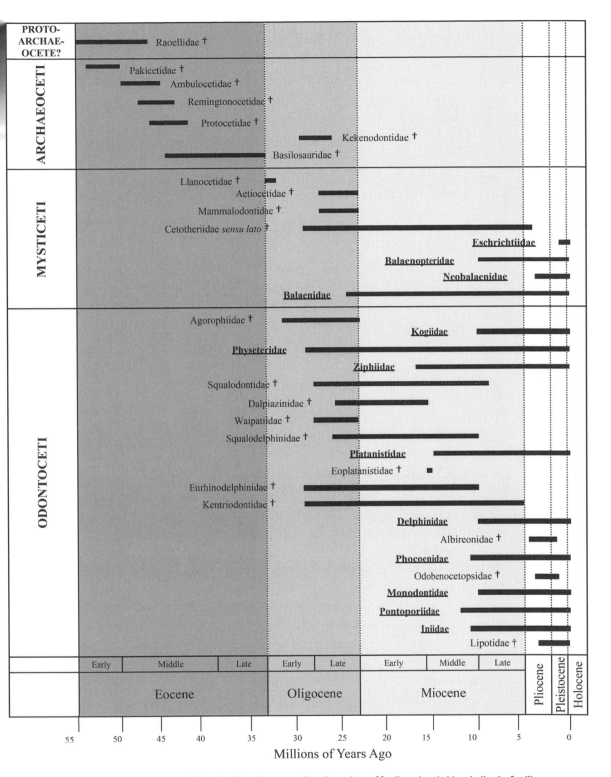

Figure 4.1. Chronology of extinct and living families of cetaceans († = all members of family extinct; bold underlined = families with living representatives). (Modified from A Berta, JL Sumich, and KM Kovacs, 2006. *Marine Mammals: Evolutionary Biology.* Academic Press, New York, NY, p. 52, courtesy of Steven Thornton.)

land bridge separating the North Pacific and Atlantic Oceans was yet to form; and a wide warm-water body called the Tethys Sea covered much of southern Europe and southeastern Asia, separating them from Africa and India. At the bottom of the world, Antarctica still maintained a tenuous link to northern lands through a pair of archipelagos and submerged ridges. One was connected to Tierra de Fuego at the southern tip of South America, and the other to Tasmania in southeastern Australia. Those connections effectively impeded the circum-Antarctic current that now sweeps uninterrupted around the continent. Today, this current blocks warm surface currents to the north from reaching Antarctica.[12]

Because the oldest and greatest diversity of Eocene archaeocetes are found in India and Pakistan, lands that once bordered the eastern Tethys Sea, it is widely assumed that this area was the cradle of cetacean evolution. New support for this supposition came to light in 2007 when Hans Thewissen, a specialist in anatomy and paleontology at Northeastern Ohio Medical University, analyzed fossil remains collected decades earlier by the Indian geologist Anne Ranga Rao. Originally removed in large blocks of rock from a middle Eocene deposit (47 million years old) in the state of Kashmir in northwest India,[13] Ranga Rao never prepared or analyzed the fossils before his death. When Thewissen prevailed upon Ranga Rao's widow, Friedlinde Obergfell, to let him study some of the unprocessed blocks, he brought to light previously unknown details about a small mammal named *Indohyus* that was no larger than a house cat.[14] First identified from a few teeth and skeletal fragments by Ranga Roa in the 1970s, the species was placed in an extinct artiodactyl family of mammals named Raoellidae (after Ranga Rao) in the same suborder (Cetruminantia) containing today's hippos.[15] By fortuitous accident during fossil preparation, one of the skull's ears broke cleanly in half. When Thewissen examined the damage, he was astounded to find that the thickness of the ear bone possessed all the hallmarks of a cetacean ear, even though the rest of the skeleton resembled a cross between a raccoon and a small deer with a hunched back.

Further analyses revealed that *Indohyus* limb bones were osteosclerotic (lined with a heavy, dense layer of bone). Such bones are typical of many aquatic mammals and help counteract buoyancy. The finding suggested that *Indohyus* spent much of its time in the water. The animals also had molars and ankle bones characteristic of early cetaceans. Putting these pieces together, Thewissen proposed that *Indohyus* represented the basal artiodactyl line that gave rise to cetaceans. However, being younger than the oldest known archaeocetes, *Indohyus* must have been a descendent of the common ancestor giving rise to whales, rather than the common ancestor itself.

The skeletal remains of *Indohyus* also held clues to its lifestyle. Wear on the molars and ratios of oxygen and carbon isotopes in the teeth—all clues to diet and whether they fed on land or in water—suggested they ate aquatic plants in freshwater streams or lakes. Taking those findings a step further, Thewissen proposed that the seminal event giving rise to cetaceans was a gradual shift in diet from terrestrial to aquatic plants. If true, today's cetaceans evolved from a line of small terrestrial herbivores or omnivores that waded into shallow tropical water bodies to graze on aquatic plants; this finding differed from the speculation of many that they had instead evolved from carnivores.

At the time of these revelations, the fossil record had already revealed remains of at least five families of archaeocetes, the last of which petered out 24 Ma in the early Miocene. Those five families included the Pakicetidae, Ambulocetidae, Remingtonocetidae, Protocetidae, and Basilosauridae (see plate 1). Collectively their fossils reveal a compelling sequence of intermediate forms leading to the cetaceans we know today. Pakicetids or "Pakistan whales," oldest of the five families, date from the early to mid-Eocene (56 to 40 Ma). Presumed decedents of a humble *Indohyus*-like ancestor, they were four-legged beasts about the size of a fox or wolf with a slender foxlike snout. Although first discovered by Richard Dehm, a German paleontologist who collected fragments of the species in northern Pakistan shortly before and after World War II, they were not recognized as early whales until the early 1980s.[16] Known only from Pakistan and India, these animals presented no outward resemblance to today's whales. Nevertheless, they shared key cetacean characteristics, including osteosclerotic limb bones and thick-walled bone in the inner ear. Although pakicetids apparently spent much of their time in freshwater pools or lakes, they were poor swimmers. Like crocodiles, they probably lay in wait for unwary fish or animals attracted to their freshwater ponds and river habitats; however, they rested and bore their young on land.[17]

Not long after the pakicetids came the ambulocetids and remingtonocetids. Again found only in Pakistan and India, their oldest members first appeared about 49 Ma.[18] Ambulocetids or "walking whales" are best known from a single, relatively complete skeleton found by Muhammad Arif and Hans Thewissen in 1991, just a few miles from the first pakicetid discoveries.[19] They were squat, crocodile-like animals somewhat larger than pakicetids with short powerful legs, enormous feet, and a thick muscular tail capable of undulating up and down for swimming. They too were fully capable of crawling about on land. Their head featured a long snout with nasal openings at the tip and a pair of narrow-set eyes. Perched high atop the head, this eye placement would have worked well for peering stealthily above the water as crocodiles and hippos do today. Ambulocetids also apparently lived in lower parts of rivers, lying in wait to ambush unsuspecting prey like pakicetids.

Remingtonocetids are also poorly known. Only a skull and parts of a few species have been discovered to date, all from India and Pakistan. They were first recognized as early whales by a pair of pioneering Indian paleontologists, Ashook Sahini and V. P. Mishra, who first collected teeth and bone fragments in the early 1970s from the Kutch District along India's coast, near the Pakistan border.[20] Remingtonocetids were named in honor of Remington Kellogg, one of the first experts on fossil whales, who served as director of the United States National Museum of Natural History from 1948 to 1962. The archaeocete's march to the sea was further represented by remingtonocetids, which occupied coastal estuarine habitats.

They were more streamlined, otter-like animals with short legs, a long tail, large ears, and a long slender snout with sharp front teeth likely used to grab fish. Interestingly, their lower jaws featured unusually large ducts (mandibular foramen), which provided a pathway for soft tissues. In modern whales, these ducts house nerves, blood vessels, and a large pad of fat through which clicks and whistles are transmitted. If the large size was designed in part to accommodate this fat, the jaws of remingtonocetids could represent an important early step toward the advanced echolocation capabilities of today's whales.

The first archaeocetes to strike out beyond the warm confines of the eastern Tethys Sea and cross wide ocean stretches to other coastal environments were the Protocetidae. Their fossil record spans the middle Eocene from 49 to 40 Ma and includes more than a dozen genera discovered to date. Featuring broad paddle-like feet and a long, substantial tail, they must have been powerful swimmers. Their remains have been found in coastal sediments from Asia, North Africa, Europe, and North America. They too were squat four-legged animals similar to ambulocetids and remingtonocetids, but beefier, and with several other important differences. Instead of close-set eyes high atop the head, their eyes had migrated to either side of the head, which possibly provided a better focus for capturing aquatic prey. In addition, their nasal openings were set farther back from the tip of the snout, closer to the position of blowholes near the rear of the skull in modern whales. At least some protocetids also had fluked tails.

At some point before the protocetids disappeared some 41 Ma, they gave rise to the first fully aquatic whales with distinctly cetacean-like forms. These were the Basilosauridae, or "king lizards." Named in 1832 by naturalist Richard Harlan who mistakenly took their remains to be some sort of sea serpent, they were first discovered along a Louisiana river bank. This family of toothed archaeocetes—and more specifically a subfamily called Dorudontinae—is believed to be the source of the Mysticeti and Odontoceti suborders that include today's whales, dolphins, and porpoises. Basilosaurids still had obvious vestiges of their terrestrial ancestry in the form of much reduced hind limbs, but they marked a clear shift from paddle-powered propulsion with feet and limbs to a more efficient swimming system reliant on an undulating tail.[21] Although some basilosaurids were long eel-like animals reaching lengths of 80 feet, others (the dorudontinae subfamily) were smaller dolphin-like animals with fluked tails. Basilosaurids extended the range of cetaceans from the tropics to subtropics in all the world's oceans.

A focal point for archaeocete evolution was the change in their means of propulsion from four-legged paddling to vertical strokes of a fluked tail. Because tail flukes are made of soft tissue, they rarely fossilize. However, a structural change in the caudal vertebrae (i.e., an increase in the ratio of height to width of each vertebra's centrum) where the body narrows just in front of the tail is a reliable indicator of the presence of flukes. The changes in propulsion made possible by a fluked tail were likely driven by incremental improvements in the ability to catch prey or avoid predators. The progression is thought to have passed through four stages:[22]

(1) four-legged paddling as seen in the some protocetids; (2) vertical undulation of the lower back and tail with hind limbs serving as hydrofoils to stabilize and steer, as is typical of ambulocetids and most protocetids; (3) hydrofoil functions (steering and stability) shared by the hind limbs and fluked tail, as seen in some protocetids; and (4) total dependence on the fluked tail for both propulsion and hydrodynamic stability, with a near loss of hind limbs seen in basilosaurids.

Recent analyses of a late protocetid from the middle Eocene named *Georgiacetus vogtlensis* have revealed important clues to the later stages in this hypothesized transformation.[23] Found in the southeastern United States along the Savannah River during construction of a nuclear power plant in 1983, *Georgiacetus* was the first of several protocetids discovered in North America. Apparently capable of transoceanic travel, *Georgiacetus* neatly fills a gap between protocetids that still relied on hind limbs and feet for propulsion and hydrofoils, and later basilosaurids with much reduced hind limbs and a fully aquatic way of life.[24] Although *Georgiacetus* still had large hind legs, its pelvis was no longer attached to the spine. This would have greatly limited its ability to move on land, and it probably paved the way for eventual loss of hind limbs and complete reliance on the tail for propulsion and stability.

Georgiacetus featured other interesting structural changes from earlier forms. It continued a trend toward increasing spine length behind the diaphragm; thus, the tail lengthened, and its propulsive power without paddling legs was enhanced. At the same time, the number and length of lumbar vertebrae increased, and the thoracic vertebrae (those supporting the ribs), shifted further forward. The caudal vertebrae at the end of the tail also developed a slightly square profile not seen in earlier protocetids, but common in later basilosaurids with flukes; this suggested *Georgiacetus* was one of the first whales with fluked tails.[25] Its ribs also show signs of osteosclerosis to counteract buoyancy.

Bridging the Gap between Ancient and Modern

Between the very end of the Eocene and the early Oligocene (36 to 34 Ma), cetaceans underwent another momentous change.[26] During this time, the basilosaurids, last of the archaeocetes, disappear from the fossil record, with representatives of modern mysticetes (baleen whales) and odontocetes (toothed whales) taking their place. Over this two- to three-million-year period, a mere blink of the eye in evolutionary time, virtually all of the diverse sizes, body plans, and diets of today's cetaceans appeared for the first time.[27] This abrupt diversification coincided with another profound change in oceanographic and climatic conditions. The slender ridges linking Antarctica to Australia and South America finally began to split as the forces of plate tectonics resulted in the two northern continents moving farther north, thereby opening an uninterrupted path for ocean currents to sweep around Antarctica. This further isolated Antarctica and its surrounding seas from the moderating influences of warm equatorial currents, which previously had kept the nearshore polar seas and land mass ice free in summer. At the same time, the high levels of atmospheric carbon dioxide that kept the Eocene so warm plummeted from over 1,000 parts per million by volume to about 560.[28]

These changes plunged Earth into "icehouse" conditions that persisted through most of the Oligocene epoch of 34 to 24 Ma. In as little as 200,000 years, this abrupt freeze covered Antarctica beneath a glacial ice sheet.[29] It also brought about a major shift in plankton communities surrounding Antarctica, which in turn may have fueled a burst of evolutionary changes not only in whales, but also in penguins.[30] In the Northern Hemisphere, winter temperatures also plunged. Over Greenland they fell from an average of 41°F to 28.4°F (5°C to -2°C), creating year-round ice at higher elevations, locking shorelines in a blanket of winter ice, and producing a more pronounced seasonal temperature cycle.[31]

At some point during the start of this cold phase, the cetacean line split in two, with one branch leading to today's toothed whales (odontocetes) and the other to baleen whales (mysticetes). Unlike the baleen whales of today, the first mysticetes (including members of the families Mammalodontidae and Aetiocetidae) had a full complement of widely spaced teeth on the upper and lower jaws. (All of today's mysticetes have tooth buds during embryonic stages and some also have small teeth on the lower jaw that disappear during the first year of life.) To distinguish toothed mysticetes from the odontocetes, paleontologists rely on diagnostic skull features other than teeth that are characteristic of mysticetes but lacking in odontocetes. Particularly helpful in this regard is a thin plate of bone forming a flange along the margin of the upper jaw, a broad bone plate located beneath the eye orbit, and outwardly bowed lower jaws.[32] Gaps between the teeth of these early mysticetes may have been used to filter small schooling fish or krill, much like the widely spaced teeth of crabeater seals occasionally do today.

The oldest toothed mysticete found so far is *Llanocetus denticrenatus*, a large whale from the late Eocene (35 Ma); this was shortly before dramatic cooling ushered in the Oligocene.[33] Found on the Antarctic Peninsula, *L. denticrenatus* is currently known only from a fragmentary toothed jaw and a nearly complete 6-foot-long skull. Its jaw had widely spread teeth separated by what appears to be protobaleen—bone laced with channels for blood vessels, as found in the jaws of modern baleen whales. It is uncertain if *Llanocetus* also had baleen, because the material rarely fossilizes. Other toothed mysticetes from later Oligocene deposits (34–23 Ma) have been found scattered around the North Pacific (Japan and the western United States)[34] and the South Pacific (Australia and New Zealand).[35] One of the earliest examples (*Fucaia buelli*) dating to the early Oligocene (33.2–31 Ma) is also one of the smallest—comparable in size to today's harbor porpoise— and may have been a suction-assisted feeder.[36] Toothed mysticetes apparently survived into the early Miocene (17–19 Ma); this conclusion is based on the recent discovery of a yet-to-be described toothed baleen whale found while widening a roadcut in California.[37]

The first baleen whales clearly lacking functional teeth (family Eomysticetidae) began to appear near the end of the Oligocene (27–25 Ma) when global warming restored temperatures to levels not seen since the late Eocene.[38] Indeed, by the end of the Oligocene all four of today's mysticete families (Balaenidae, Balaenopteridae, Eschrichtiidae, and Neobalaenidae) had made their appearance, albeit with representatives now extinct. The first such representative of these surviving families of modern mysticetes was a balaenid whale named *Morenocetus parvus*, which was discovered in a New Zealand rock formation dating to 28 Ma.[39] Its relationship to living balaenids is uncertain, but a similarity to bowhead whales suggests that the line leading to today's bowheads predates that for right whales and has its roots in the Antarctic, reaching back to the late Oligocene.[40]

A long gap in the fossil record follows *Morenocetus*, which possibly suggests that balaenids did not diversify broadly until the late Miocene or early Pliocene, some 6 to 3 Ma.[41] One hypothesis suggests this gap was related to prey availability around Antarctica. An enormous proliferation of diatoms (microscopic plants or phytoplankton) in sediment cores pulled from the muddy sea floor surrounding Antarctica suggests there was a major change in the Antarctic circumpolar current about 10 Ma. The resulting profusion of diatoms may have supported superabundant swarms of krill and other zooplankton, and these would have offered new foraging opportunities for baleen whales.[42]

In any case, the balaenid fossil record does not pick up again until the late Miocene and Pliocene epochs 5 to 2 Ma, when several new species appear. Like *Morenocetus*, some of these species seem more similar to bowheads in today's *Balaena*. These include the small whale *Balaenella brachyrhynu*, found in 1974 in Belgium;[43] the large whale *Balaena montalionis*, found in 1873 in Italy;[44] and the large whale *Balaena ricei*, discovered in 1960 on the US Atlantic coastal plain in Virginia.[45] Other species, however, seem to be more closely related to right whales in the genus *Eubalaena*. These include small whales assigned to an extinct genus *Balaenula*, discovered in Pliocene deposits in Japan, Italy, and Belgium. Although numerous studies have sought to sort out relationships between these extinct balaenids and today's living species,[46] matters are far from settled. But the small size of most extinct species in both the right whale and bowhead lines may indicate that early balaenids tended to be far smaller than the giants we know today.[47]

Emerging from the Past

In 1938 an important right whale fossil was discovered in Japan's Nagano Prefecture, west of Tokyo. Unfortunately, its remains were not collected until the 1970s, after most of its postcranial parts had been lost to erosion. A description of the specimen, eventually published in 2007, gave it the name *Eubalaena shinshuensis*,[48] thus making it the oldest known representative of today's right whale genus. Recovered from a formation deposited 4 to 6 Ma at the Miocene-Pliocene boundary, its age suggests the split between the *Balaena* and *Eubalaena* genera reaches back into the Miocene. Moreover, because living right whales and bowheads look so much alike, the specimen suggests that the rates at which new balaenid species evolve are exceptionally slow.[49]

Factors affecting the rate at which species branch off into new lines are poorly understood. However, slower rates appear to be correlated with large body size, low reproductive rates, and long life spans,[50] all of which describe right whales to a tee. Like most large animals, right whales have a slow metabolism that helps them live to a ripe old age. Because it takes them eight to nine years to reach sexual maturity and they produce just a single calf every three to five years, the opportunity for mutations that give rise to new species occurs at a rate far slower than in animals with short life spans and faster reproduction rates.[51] Indeed, right whales are among the longest living and slowest breeding of all animals. Photo-identification records reveal that North Atlantic right whales can live at least 60 to 70 years,[52] but much older ages are possible.

Their close relatives, bowhead whales, can live to at least 100 years of age. Evidence of such longevity comes from the recovery of stone harpoons last used by Eskimo whalers late in the nineteenth century from the blubber of several whales killed by subsistence hunters between 1993 and 2007 (figure 4.2). Yet even older ages are possible. Using a technique called aspartic racemization to estimate age, scientists analyzed lens tissues in 48 eyeballs collected from bowheads landed by Eskimo whalers in the late 1990s.[53] As expected, given that commercial whalers nearly eliminated bowhead whales off Alaska in the early 1900s, most whales were estimated to be between 20 and 60 years old. A few, however, were far older. One was estimated to be 135 years old, two were thought to be 159 and 172, and the oldest was estimated to be 211 years old. With an error range for this aging technique estimated at plus or minus 16 percent, the oldest animal might have been between 177 and 245 years old. If those ages are accurate, Native whalers could be hunting some of the very same whales hunted by their ancestors in the late 1700s before their first contact with Europeans!

Nonbiological factors, including climate change and plate tectonics, can also influence the formation of new species.[54] As indicated above, bursts of cetacean diversification often seem to coincide with dramatic changes in climate. Unfortunately, imprecise estimates of fossil ages and dates of temperature change and geological events make such correlations difficult to pin down with certainty. The opening and closing of connections between adjoining seas over the past 50 million years also could have disrupted genetic exchange leading to the formation of new species. Some of the best documented examples of such effects, possibly including right whales, are associated with the relatively recent formation of the Central American land bridge.

The geology of southern Central America was created by a confusing jumble of plate tectonics, volcanic uplift, geologic faulting and folding, and rising and falling sea levels. As a result, exact dates for formation of the Central American land

Figure 4.2. These stone and iron harpoon tips were found in the blubber of bowhead whales killed by Eskimo whalers between 1981 and 1997. Believed to date from the eighteenth or early nineteenth centuries, they offer compelling evidence that bowhead whales can live to at least 100 years of age. (Image courtesy of Craig George and Robert Suydam, Alaska Eskimo Whaling Commission, North Slope Borough, Alaska.)

bridge to South America are uncertain. What is not in dispute is that fossils of rhinoceroses, peccaries, horses, bears, and other land animals found in both South and Central America attest to a continuous arc of land that first joined the two land masses in the Miocene, between 19 and 6 Ma.[55] From about 6 to 2.6 Ma, a Central American Seaway reclaimed the isthmus temporarily, reconnecting the Pacific and Caribbean basins. This opening began to close a second time, about 4.5 to 4 Ma.[56] When it closed the second time, it left an enduring mark on many marine species. Not only did it isolate populations of species (including monk seals) that have since evolved into a distinct species, but it changed ocean habitats and climatic conditions worldwide.

With exchange of waters between the two oceans blocked, warm surface water blown west across the tropical Atlantic was shunted north through the Caribbean; this invigorated the Gulf Stream passing through the Florida Strait and then bending north along the east coast of North America.[57] The heat carried north by this invigorated current had several profound effects: evaporation and precipitation in northern latitudes increased; the Atlantic became saltier; the range of tropical and temperate fauna expanded northward; seasonal temperature extremes at midlatitudes moderated; and by some accounts, winter sea surface temperatures as far north as the North Sea off Europe increased by as much 18°F (10°C).[58] Similarly, the North Pacific had its own cascade of oceanographic changes that affected the depth of ocean thermoclines, the strength of coastal upwelling, and faunal distribution and abundance.

For both physical and physiological reasons, these changes could have disrupted movements of, and gene flow between, right whales in different ocean basins. Before the seaway closed, when the Gulf Stream was weaker and the tropical zone was narrower, at least a few right whales may have migrated far enough south to move between the Atlantic and Pacific Oceans. Even with today's strong Gulf Stream and a relatively wide tropical zone, right whales occasionally round the southern tip of Florida into the Gulf of Mexico and move as far west as Texas.[59] However, as this low-latitude Atlantic-Pacific connection gradually became shallower and eventually closed between about 4.5 and 2.6 Ma, the movements of large whales through the strait would have become impossible. At about the same time, between about 3.2 and 2.7 Ma, global cooling created glacial conditions in the arctic regions of North America and Eurasia that blocked whale movements along the northern shores of those continents. As a result, right whales in the North Atlantic and North Pacific likely become isolated, and each group embarked on different evolutionary pathways.

Formation of the Central American land bridge also could have eliminated right whale movements between the Northern and Southern Hemispheres. With more tropical water moving toward the poles, right whales migrating to subtropical calving grounds would have encountered these increased temperatures and curtailed their travel southward. By the same token, the wider tropical expanse separating northern and southern right whales could have ended previous movements by the odd right whale that had once maintained genetic continuity between the two hemispheres.

Today, right whales in the North Atlantic, North Pacific, and Southern Hemisphere are recognized as separate species largely due to genetic differences.[60] Those differences also suggest that North Pacific right whales (*E. japonica*) are more closely related to southern right whales (*E. australis*) than to North Atlantic right whales (*E. glacialis*). Nevertheless, with virtually no physical differences between whales in any of the three regions, it would seem they could actually be incipient species still very closely related to one another and likely able to interbreed with each other if they ever had opportunities to do so.

Further support linking closure of the Central American Seaway to the origin of the three existing right whale species has come from an unlikely source—the cyamids, or "whale lice," that cling to callosities and outer crevasses on their mammoth hosts (figure 4.3). As it happens, each of today's right whale species carries a related yet distinct set of three cyamid species. Because whale lice have no

free-swimming larval stage, these crustaceans spend their entire lives clinging to their host (unless they jump between hosts when whales come into direct physical contact with one another). Thus, whale lice and their hosts always travel together and share identical ecological histories.

Because cyamids are far more numerous, reproduce far more rapidly (perhaps twice a year), and produce far more offspring per individual than the whales, their genes mutate and mix far faster. Over time, this has magnified genetic differences and species distinctions among the cyamids on each of the three right whale species more than the differences among the three right whale species. Scientists therefore examined the genetic diversity of whale lice on each right whale species as a proxy for speciation in right whales. Using a mutation rate previously derived for a similar marine crustacean (snapping shrimp), they determined that each of the three cyamid species on each species of right whale became isolated from those on the other right whale species between 3.6 and 9.9 Ma, with a best estimate centered at about 6 Ma.[61] This indicates that when the seaway closed, a single globally distributed right whale species was split into three separate populations that eventually became the three right whales we see today in different ocean basins. Although the range of years marking this separation is imprecise, the lower end of the range falls within the same time frame that the Central American Seaway last closed (between 4.5 and 2.6 Ma). This matches the date that Caribbean and Hawaiian monk seals are believed to have split into separate species, which is also attributed to isolation caused by the seaway's closure.[62]

Finally, this genetic study revealed a mutation in one of the corresponding pairs of whale lice species found on Southern Hemisphere and North Pacific right whales, but not in the related species on North Atlantic right whales. Investigators interpreted this surprising finding as an indication that at least one southern right whale, but not more than a few, must have crossed the equator to mix with North Pacific right whales about 1 to 2 Ma. That mixing could explain genetic discoveries that suggest North Pacific right whales are more closely related to southern right whales than to North Atlantic right whales.

Figure 4.3. *Cyamus ovalis,* a species of whale lice or "cyamid" found only on right whales: *left,* female and male specimens; *right,* a live specimen shown walking. (Images courtesy of V. Rowntree, University of Utah.)

This brings us to the Pleistocene, a geologic period lasting a mere 2 million years and ending just 10,000 to 12,000 years ago when a series of glacial ice ages gave way to our current warm interglacial phase—the Holocene. During the Pleistocene, global climates cycled back and forth between frigid ice ages (each lasting roughly 100,000 years) and mild interglacial periods lasting a few thousand years each. The latter were marked by warm temperatures much like those we enjoy today. This succession of warm–cold periods had significant effects on the distribution of bowhead whales[63] and probably also affected right whale distribution.

During glacial periods, shorelines advanced seaward toward the outer edge of submerged continental shelves as ice and snow accumulated on land. How right whales responded to those conditions is uncertain, but for right whales that spend most of their time in cool coastal waters over the continental shelf, this would have restricted shallow feeding and calving habitat. It perhaps forced them to spend more time in deep water along margins of the continental shelf. During these glacial periods, their migratory patterns also may have shifted to lower latitudes, with winter calving grounds closer to the equator than they are today. One can only speculate about such changes, but because glacial conditions predominated throughout most of the Pleistocene, right whales—and indeed all coastal whales—may be better adapted to ice age conditions than to today's warm interglacial climates. In any case, the current interglacial phase with its ascendency of human influence (hunting, collisions, and entanglements) has tested the species' resilience to a degree unlike any interglacial period experienced in its 4 to 6 million years of existence.

Over the past two chapters we have traveled through geologic time, viewing right whale evolution from the proverbial 30,000-foot level where only broad outlines of their existence can be discerned. With the onset of the Holocene and the dawn of human civilization, we will apply the brakes in the following chapter, and begin thinking in time scales measured in the thousands and hundreds of years that are more suitable to the consideration of human history.

NOTES

1. Lee, MSY, A Cau, D Naish, and GJ Dyke. 2014. Sustained miniaturization and anatomical innovation in the dinosaurian ancestors of birds. Science 345:562–566.

2. Gingrich, PD. 2011. Evolution of whales from land to sea. Proceedings of the American Philosophical Society 156(3):309–323.

3. The size of ear bones in all whales today, great and small, are roughly the same size.

4. Berta, AJ, L Sumich, and KM Kovacs. 2006. *Marine Mammals: Evolutionary Biology*. Academic Press. New York, NY. Thewissen, JGM. 2014. *The Walking Whale: From Land to Water in Eight Million Years*. University of California Press. Oakland, CA, US.

5. Flower, WH. 1883. On whales, past and present and their probable origin: a discourse. Proceedings of the Research Institute of Great Britain 10:360–376.

6. Gingerich, PD, M ul Haq, IS Zalmout, I Hussain Khan, and M Sadiq Malkani. 2001. Origin of whales from early Artiodactyla: hands and feet of Eocene Protocetidae from Pakistan. Science

293(5538):2239–2242. Gatesy, J, and MA O'Leary. 2001. Deciphering whale origins with molecules and fossils. Trends in Ecology and Evolution 16:562–570.

7. Thewissen, JGM, LN Cooper, JC George, and S Bajpai. 2009. From land to water: the origin of whales, dolphins and porpoises. Evolution Education Outreach 2:272–288.

8. Marx, FG, and MD Uhen. 2010. Climate, critters and cetaceans: Cenozoic drivers of the evolution of modern whales. Science 327:993–996.

9. Ross, S, PD Gingerich, KC Lohmann, and KG MacLeod. 2010. Continental warming preceding the Paleocene-Eocene thermal maximum. Nature 467:955–958.

10. Pross, J, L Contreras, PK Bijl, DR Greenwood, SM Bohaty, JA Bendle, U Röhl, et al. 2012. Persistent near-tropical warmth on the Antarctic continent during the early Eocene epoch. Nature 448(7409):73–77.

11. Prothero, D. 1994. *The Eocene-Oligocene Transition: Paradise Lost*. Columbia University Press. Chichester, UK, pp. 17–18.

12. Berta et al., *Marine Mammals* (n. 4), p 117; and Fordyce, RE. 1980. Whale evolution and Oligocene southern ocean environments. Palaeogeography, Palaeoclimatology, and Palaeoecology 31:319–336.

13. Thewissen, JGM, LN Cooper, MT Clementz, S Bajpai, and BN Tiwari. 2007. Whales originated from aquatic artiodactyls in the Eocene epoch of India. Nature 450:1190–1194.

14. Thewissen, JGM. 2014. *The Walking Whales: From Land to Water in Eight Million Years*. University of California Press. Oakland, CA, US, pp. 191–198.

15. Ibid., pp. 198–206.

16. Gingerich, PD, NA Wells, DE Russell, and SM Ibrahim Shah. 1983. Origin of whales in epicontinental remnant seas: new evidence from the early Eocene of Pakistan. Science 220:403–406; and Uhen, M. 2010. The origins of whales. Annual Review of Earth and Planetary Sciences 38:198–216, doi:10.1146/annurev-earth-040809-152453.

17. Gingerich, PD. 2012. Evolution of whales from land to sea. *Proceedings of the American Philosophical Society* 156:309–323, pp. 312–314.

18. Thewissen et al., From land to water (n. 7), pp. 58–66, 102–116.

19. Thewissen, *The Walking Whales* (n. 4), pp. 38–50, 58.

20. Sahni, A, and VP Mishra. 1975. Lower Tertiary vertebrates from western India. Monographs of the Paleontological Society of India 3, cited in Thewissen, *The Walking Whales* (n. 4), p 109.

21. Gingerich, Evolution of whales (n. 17), pp. 319–322.

22. Gingerich et al., Origin of whales (n. 6). Fish, FE. 1996. Transitions from drag-based to lift-based propulsion in mammalian swimming. American Zoology 36:628–641. Thewissen, *The Walking Whales* (n. 4), pp. 53–58.

23. Hulbert, RC. 1998. Postcranial osteology of the North American Eocene protocetid *Georgiacetus*. *In* JGM Thewissen (ed.), *The Emergence of Whales: Evolutionary Patterns in the Origin of Whales*. Plenum Press. New York, NY, pp. 235–267.

24. Bucholtz, EA. 1998. Implications of vertebral morphology for locomotor evolution in early cetaceans. *In* Thewissen, *The Emergence* (n. 23), pp. 325–357.

25. Uhen, MD. 2008. New protocetid whales from Alabama and Mississippi and a new cetacean clade, Pelagiceti. Journal of Vertebrate Paleontology 28(3):589–593.

26. Steeman, MF, MB Hebsgaard, RE Fordyce, SYW Ho, DL Rabosky, R Nielsen, C Rahbek, et al. 2009. Radiation of extant cetaceans driven by restructuring of the oceans. Systematic Biology 58(6):573–585.

27. Slater, GJ, SA Price, F Santini, and ME Alfaro. 2010. Diversity versus disparity and the radiation of modern cetaceans. Proceedings of the Royal Society of London B, Biological Sciences 277(3):97–104, doi:1098/rsbp.2009.1650.

28. Person, PN, GL Foster, and BS Wade. 2009. Atmospheric carbon dioxide through the

Eocene-Oligocene climate transition. Nature 461(22):1110–1113.

29. DeCanto, RM, and D Pollard. 2003. Rapid Cenozoic glaciation of Antarctica induced by declining atmospheric CO2. Nature 421:245–249.

30. Houben, AJP, PK Bijl, J Pross, SM Bohaty, S Passchier, CE Stickley, U Röhl, et al. 2013. Reorganization of the Southern Ocean plankton ecosystem at the onset of Antarctic glaciation. Science 340 (3160):341–344.

31. Eldrett, JS, DR Greenword, IC Harding, and M Huber. 2009. Increased seasonality through the Eocene to Oligocene transition in northern high latitudes. Nature 459:969–973.

32. Berta et al., *Marine Mammals* (n. 4), pp. 61–62.

33. Mitchell, ED. 1989. A new cetacean from the late Eocene La Meseta formation, Seymore Island, Antarctic Peninsula. Canadian Journal of Fisheries and Aquatic Science 46:2219–2235. Fordyce, RE. 2009. Cetacean fossil record. *In* WF Perrin, B Würsig, and JGM Thewissen (eds.), *Encyclopedia of Marine Mammals*, 2nd ed. Academic Press. New York, NY, pp. 207–215.

34. Barnes, IG, M Kimura, H Furusawa, and H Sawamura. 1994. Classification and distribution of Oligocene Aetiocetidea (Mammalia; Cetacea; Mysticeti). Island Arc 3:392–431. Marx, FG, C-H Tsai, and RE Fordyce. 2015. A new Oligocene toothed whale (Mysticiti: Aetiocetidae) from western North America; one of the oldest and smallest. Royal Society Open Science, doi:10.1098/rsos.150476.

35. Fitzgerald, E. 2006. A bizarre new toothed mysticete (Cetacea) from Australia and the early evolution of baleen whales. Proceedings of the Royal Society of London B, Biological Sciences 273(1604):2955–2963. Fordyce, Whale evolution (n. 12). Fordyce, Cetacean fossil record (n. 33), pp. 211–212.

36. Marx et al., A new Oligocene toothed whale (n. 34), p. 31.

37. Rivin, M. 2013. Before they were giants: the fossil record of toothed baleen whales. Abstract. Paper presented at the Annual Meeting of the American Association for the Advancement of Science, Boston, MA, US, February 14–18, https://aaas.confex.com/aaas/2013/webprogram/Session5818.html.

38. Bonssenecker, RW, and RE Fordyce. 2015. Anatomy, feeding ecology, and ontogeny of a transitional baleen whale: a new genus and species of Eomysticetidae (Mammalia: Cetacea) from the Oligocene of New Zealand. PeerJ 3:e1129, doi:10.7717/peerj.1129; and Berta et al., *Marine Mammals* (n. 4), p. 63.

39. Fordyce, RE. 2002. Oligocene origins of skim-feeding right whales: a small archaic balaenid from New Zealand. Journal of Vertebrate Paleontology 22:54A.

40. Bisconti, M. 2005. Skull morphology and phylogenetic relationships of a new diminutive balaenid from the lower Pliocene of Belgium. Paleontology 49(4):793–816.

41. Steeman et al., Radiation of extant cetaceans (n. 26).

42. Kimura, T. 2009. Review of the fossil balaenids from Japan with a redescription of *Eubalaena shinshuensis* (Mammalia, Cetacea, Mysticiti). Quaderni del Museo di Storia Naturale di Livorno 22:3–21.

43. Bisconti, Skull morphology (n 40).

44. Capellini, G. 1904. Balene fossili Toscane II. *Balaena montalionis*. Memori della Reale. Accademia delle Scienze dell' Instituto di Bologna 1:1–11.

45. Westgate, JW, and FC Whitmore Jr. 2016. Balaena ricei, a new species of bowhead whale from the Yorktown formation (Pliocene) of Hampton, Virginia. Smithsonian Contributions to Paleobiology (93):295–312.

46. Churchill, M, A Berta, and T Démére. 2012. The systematics of right whales (Mysticeti: Balaenidae). Marine Mammal Science 28(3):497–521.

47. Bisconti, Skull morphology (n. 40), p. 809.

48. Kimura, T, K Narita, T Fujita, and Y Hasegawa. 2007. A new species of *Eubalaena* (Cetacea:

Mysticeti: Balaenidae) from the Gonda Formation (latest Miocene-Early Pliocene) of Japan. Bulletin of the Gunma Museum of Natural History 11:15–27. Kimura, Review of the fossil balaenids from Japan (n. 42).

49. Ho, SYW, U Saarma, R Barnett, J Haile, and B Shapiro. 2008. The effect of inappropriate calibration: three case studies in molecular ecology. PLoS ONE 3(2):e1615, doi:10.1371/journal.pone.0001615.

50. Martin, AP, and SR Palumbi. 1993. Body size, metabolic rate, generation time, and the molecular clock. Proceedings of the National Academy of Sciences 90:4087–4091.

51. Ibid.

52. Kraus, SD, and RM Rolland. 2007. *The Urban Whale: North Atlantic Right Whales at the Crossroads*. Harvard University Press. Cambridge, MA, US, pp. 1–3, 22.

53. George, JC, J Bada, J Zeh, L Scott, SE Brown, T O'Hara, and R Suydam. 1999. Age and growth estimates of bowhead whales (*Balaena mysticetus*) via aspartic racemization. Canadian Journal of Zoology 77(4):571–580.

54. Steeman et al., Radiation of extant cetaceans (n. 26), pp. 573–574.

55. Kirby, MX, DS Jones, BJ McFadden. 2008. Lower Miocene stratigraphy along the Panama Canal and its bearing on the Central American Peninsula. PLoS ONE 3(7):e279, doi:10.1371/journal.pone.000279.

56. Bartoli, G, M Sarnthein, M Weinelt, H Erienkeuser, D Garb-Schönberg, and DW Lea. 2005. Final closure of Panama and the onset of northern hemisphere glaciation. Earth and Planetary Science Letters 237:33–44.

57. Lunt, DJ, PJ Valdes, A Haywood, and IC Rutt. 2008. Closure of the Panama Seaway during the Pliocene: implications for climate and Northern Hemisphere glaciation. Climate Dynamics 30:1–18, doi:10.1007/s00382-007-0265-6.

58. Williams, M, AM Haywood, EM Harper, AL Johnson, T Knowles, MJ Leng, DJ Lunt, et al. 2009. Pliocene climate and seasonality in North Atlantic seas. Philosophical Transactions of the Royal Society of London A, Mathematical, Physical and Engineering Sciences 367(1886):85–108.

59. Ward-Geiger, L, AR Knowlton, AF Amos, TD Pitchford, B Mase-Gutherie, and BJ Zoodsman. 2011. Recent sightings of the North Atlantic right whale in the Gulf of Mexico, North. Gulf of Mexico Science 29(1):74–78.

60. Rosenbaum HC, RL Brownell Jr, MW Brown, C Schaeff, V Portway, BN White, S Malik, et al. 2000. Worldwide genetic differentiation of Eubalaena: questioning the number of right whale species. Molecular Ecology 9:1793–1802.

61. Kaliszewska, ZA, J Seger, VJ Rowntree, SG Barco, R Benegas, PB Best, MW Brown, et al. 2005. Population histories of right whales (Cetacea: Eubalaena) inferred from mitochondrial sequence diversities and divergences of their whale lice (Amphipoda: Cyamus). Molecular Ecology 14:3439–3456.

62. Scheel D-M, GJ Slater, S-O Kolokotronis, CW Potter, DS Rotstein, K Tsangaras, AD Greenwood, and KL Helgren. 2014. Biogeography and taxonomy of extinct and endangered monk seals illuminated by ancient DNA and skull morphology. ZooKeys 409:1–33.

63. Foote, AD, K Kaschner, SE Schultze, C Garilao, SYW Ho, K Post, TF Higham, et al. 2013. Ancient DNA reveals that bowhead whale lineages survived Late Pleistocene climate change and habitat shifts. Nature Communications 4:1677, doi:10:1038/ncoms2714.

5

THE ORIGIN
OF WHALING

The frigid waters of the Bering Sea are one of the most biologically productive habitats on earth. Opening like a fan between Alaska and Russia, the sea's vast expanse covers 880,000 square miles—roughly the size of Alaska and Texas combined (figure 5.1). At its northern point, just below the Arctic Circle, lies the Bering Strait, a shallow 58-mile-wide gap that separates Asia from North America between the Chukotka and Seward Peninsulas. When winter ice begins to break up in the spring, bowhead whales, walruses, ice seals, and other marine mammals stream north between the strait's opposing capes on annual migrations to summer feeding grounds in Arctic seas. At the Bering Sea's southern extreme lie the Aleutian Islands, a looping 1,700-mile-long crescent of craggy volcanic peaks stretching from the Alaska Peninsula in the east to Russia in the west. The Aleutian Islands and their passes (channels between islands) serve as both a geographic boundary and an ecological gateway, with the vast North Pacific Ocean to the south. Between the islands, tens of thousands of right whales, gray whales, and other large whales once traveled at least twice a year on seasonal migrations. Somewhere along the shores of the Bering Sea, the technology and techniques for killing large whales were almost certainly mastered for the first time.

Based on evidence as skimpy as it is tantalizing, some archeologists argue that people first reached the Americas by crossing the Bering Land Bridge 40,000 years "before present" (BP).[1] If they did, they left only faint traces of their presence over the following 25,000 years, which suggests they were confined to small local groups at best. Far more certain is that a later wave of nomadic hunters ventured eastward across that same land bridge 16,500 to 13,500 years ago. They carried the seeds of humanity that would flourish into a rich diversity of Native cultures throughout the Americas.[2] Their travels from the cold, glacier-free Siberian steppe were aided by an extended warming trend that marked the beginning of the end of

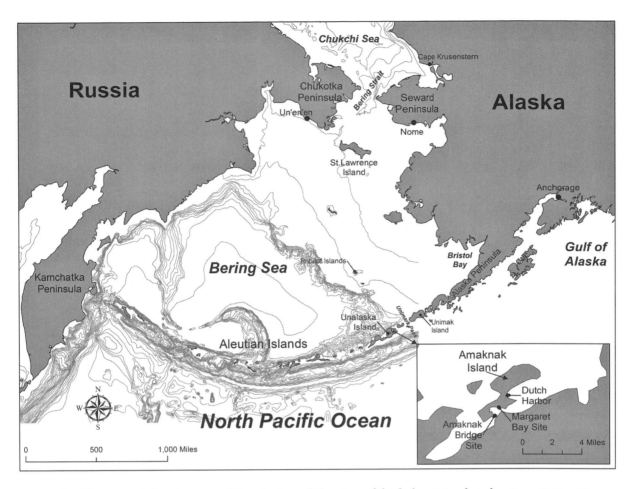

Figure 5.1. Map of the
Bering Sea region.
(Map courtesy of
Linda White.)

the Pleistocene's last ice age and foreshadowed the start of the balmy interglacial epoch enjoyed today—the Holocene.

The Holocene's first several millennia—from about 11,000 to 5,000 BP—were also its warmest. Scientists refer to this warm phase as the "thermal maximum" or "hypsithermal" period. As early as 11,000 to 9,000 BP, the average maximum temperatures in Alaska were 1.8°F to 3.6°F (1°C to 2°C) warmer than those at the close of the twentieth century.[3] Warming in the Canadian Arctic, however, lagged behind the Bering Sea region by some 4,000 years. This was due to cooling effects of the massive Laurentide Ice Sheet that once covered the interior of North America as far south as present-day Ohio, Indiana, and Pennsylvania. As melting ice sheets in North America and elsewhere retreated, the Bering Land Bridge was slowly inundated. It was completely covered by 11,000 BP, but sea levels continued to rise for another two to three millennia. Today, most of this former land bridge lies beneath 150 to 300 feet of water and forms a continental shelf extending as much as 300 miles off the coast of Alaska.

Although much of the evidence of early immigrants who crossed over to North America now rests underwater on the continental shelf, several land-based archeological sites firmly fix an early human presence in various parts of Alaska, including the Aleutian Islands, as far back as 13,700 to 11,900 BP.[4] Those early people moved from campsite to campsite following seasonal migrations of caribou and salmon. There is no evidence of whaling and little evidence of sealing at those early dates, but the people of that time left behind an assortment of stone tools, including projectile points, small stone blades, burins, and scrapers indicative of an industrious way of life and impressive hunting skills.

Between 3,600 and 2,500 BP, if not earlier, the remains of fish, shellfish, and seals in the waste pits, or "middens," of Arctic coastal habitation sites mark the development of a maritime tradition that was dependent on food from the sea.[5] Eventually that tradition gave rise to the Thule (pronounced "Too-lee") and modern Eskimo cultures that would rely heavily on sealing and whaling and spread eastward across the Arctic from Russia to Greenland. Research suggests that whaling became an integral part of that lifestyle by at least 1,500 BP, and possibly even earlier than 3,000 BP when ancient Egyptians were building the world's first step pyramids. The people of the Arctic left behind no such stone monuments, but their ability to survive and prosper in such harsh, unforgiving Arctic and sub-Arctic environments is itself a lasting monument to their ingenuity, endurance, and resourcefulness.

Probably the most complete record yet discovered of the development of Arctic maritime culture over time is from Cape Krusenstern, on the shores of Norton Sound just north of Alaska's Seward Peninsula. There, in the 1950s and 1960s, an eminent archeologist named John Lewis Giddings, along with his colleagues, began excavating cultural remains at a series of neighboring sites that spanned the entire Holocene. On inland hills they found Paleolithic artifacts ranging from 10,000 to 7,000 BP, while along the coast they discovered remains of settlements dating from about 4,000 BP through historic times. The coastal settlements were conveniently located on a succession of old dune ridges that were left behind as sea levels receded during a brief mid-Holocene neoglacial phase. Among remains dating from about 2,900 to 2,700 BP,[6] Giddings found semi-subterranean houses, a large lance head, long-bladed butchering tools, and a litter of whale bones.[7] Given the size of some of the tools and their similarity to later whaling implements, he concluded that he had found the earliest evidence of whaling yet discovered; he dubbed the cultural horizon containing them the "Old Whaling Culture." Curiously, characteristic whaling artifacts disappear from the site's cultural record as abruptly as they had appeared. This limited time frame of whaling artifacts and their sudden disappearance have led many to caution that the whale bones and implements could have been merely coincidental, or reflected efforts to scavenge stranded whales rather than active whaling.[8]

More recent excavations at two sites on Amaknak Island (pronounced "ə-mæk-næk") in the eastern Aleutians have yielded further evidence of the development of an early maritime tradition that included whaling. Located just a few hundred yards off the north shore of Unalaska Island near the Aleutian Islands' largest present-day community, Dutch Harbor, the two sites are just 70 miles west of Unimak

Pass, the easternmost connection between the Bering Sea and North Pacific Ocean. The older site, called Margaret Bay, was occupied from about 4,000 to 3,000 BP during a neoglacial phase lasting from 4,700 to 2,500 BP, when temperatures plunged to their lowest point since the end of the last ice age.[9]

Then, as now, the Aleutians were a mecca for fish, seabirds, and marine mammals. Yet even with its abundance of food, the region must have been a grim place to make a living. Under neoglacial conditions, the islands' rocky, treeless slopes would have been buffeted by biting winds and fierce storms sweeping in off the cold Bering Sea and North Pacific. In winter its shores were locked in pack ice. In at least some years during this mini–ice age, the Bering Strait may even have remained choked in ice year-round, blocking spring and summer migrations of marine mammals to the Arctic Ocean.[10]

To survive those harsh conditions, Paleo-Eskimos at Margaret Bay literally had to dig in. Excavations of the semi-subterranean houses that made year-round life possible reveal a succession of improvements. By 3,500 BP, the circular belowground portions of the dwellings were roughly 12 feet across and 3 feet deep. Belowground walls were lined with cobbles and small boulders, while above the ground they were presumably made of sod. Each house featured a hearth with connections to subfloor heating, ventilation ducts, and subfloor storage pits.[11] Cultural ties with upland residents are evident from tools found at the site that bear a strong similarity to a widespread Arctic Small Tool tradition. Kitchen middens at Margaret Bay were dominated by remains of shellfish and seals, particularly fur seals, indicating they relied on the ocean for food. Although the middens included no large whale remains, an abundance of juvenile and adult Dall's porpoise bones indicate that skin boats must have been used for at-sea hunting. A plentitude of bones from the pups of ring seals and bearded seals—both born on the ice edge in early spring—attest to the presence of sea ice in the southeast Bering Sea throughout the spring.[12] Today such spring conditions extend south only to the northern parts of the Bering Sea.

The second location is barely a kilometer away from the Margaret Bay site, across a narrow cove. Positioned in the path of a bridge project scheduled to connect Dutch Harbor on the small island of Amaknak to the much larger island of Unalaska, the site was named Amaknak Bridge. Looming construction deadlines required excavation in far more haste than archeologists would have preferred. Nevertheless, between 2000 and 2006 a great deal was learned before the site was bulldozed and paved. Radiocarbon dates fix its occupancy at 3,300 to 2,700 BP, which is slightly younger than, but briefly overlapping, the older Margaret Bay site. Compared to Margaret Bay, artifacts and zoological remains at Amaknak Bridge reveal distinct cultural differences and innovations, including whaling. In fact, its tools differ so markedly from the Margaret Bay site that they suggest the arrival of an entirely new group of people with their own cultural traditions and ideas. Earlier similarities to the Arctic Small Tool tradition were not apparent, while new tools not found at Margaret Bay emerged. Among the new tools were asymmetrical knives; elaborately barbed, bone harpoon points; and toggling harpoon heads.[13]

Invention of the toggling harpoon was a landmark event for subsistence hunters around the Bering Sea. Although first used for sealing, it was perhaps the last piece of equipment that Paleo-Eskimos needed to add whaling to their subsistence repertoire. Other important prerequisites were already in place, including a cohesive social structure for cooperative hunting; well-developed techniques for constructing skin boats from walrus hides; an ability to craft strong line from animal sinew and hide; and a compound harpoon with a bone foreshaft and a separate socket piece, which allowed harpoon heads to detach from the shaft once embedded in the victim. The missing ingredient was an ability to hold fast to a whale once it was struck. The toggling harpoon elegantly solved that problem.

The toggling harpoon is a deceptively simple yet ingenious invention (figure 5.2). Essentially it is an ivory, bone, or stone projectile point with a hole drilled through its core, and an offset barb (a barb on only one side of the base or longer than its mate on the opposite side). A strong line tied through the hole runs back along the harpoon shaft where it is held taut by the hunter's fingers. After a seal or whale is struck, the harpoon head detaches from the shaft and remains lodged in the animal's flesh. When the animal attempts to flee and the line held by the hunter becomes taut, the offset barb prevents the head from being pulled straight back out of the entry wound. Instead, the head rotates 90 degrees until its long axis is perpendicular to the entry wound. This provides a far stronger attachment to secure the animal. Over time, multiple offset barbs were added to further enhance its pivoting action beneath the skin.

The first toggling harpoons were relatively small, allowing hunters to hang on to and retrieve struck seals, dolphins, and perhaps even small whales. Larger harpoon heads worked well for large whales, but animals of their size were too powerful to haul in by hand. The toggling harpoons were therefore used to tether a float or "drogue" that served at least three functions. First and foremost was safety. The pulling force of a whale could easily shred a hunter's hands if they attempted to hold the line, and the sudden surge of a struck whale could capsize the skin boats, or *umiaks*. The drogue avoided the need for whalers to hang onto the line once the whale was struck. Second, the drogue created a constant drag that could tire even the largest whales. Third, the floats provided a surface marker that hunters could use to track the whale when it submerged. When the whale tired and slowed, trailing hunters could then move in and dispatch the animal with repeated thrusts of a lance whose sharp point (lacking barbs) could penetrate deep into internal organs, yet be pulled out easily for a series of quick lethal stabs.[14]

Toggling harpoon heads have been found throughout the Arctic at archeological sites where sealing and whaling was practiced, and those at Amaknak Bridge are among the earliest known in the Bering Sea region. This innovation appears to have developed independently in different parts of the world. The oldest example comes from a 7,500-year-old burial mound at L'Anse Amour, along the Strait of

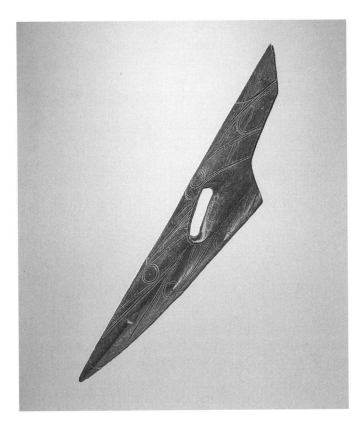

Figure 5.2. An exceptional example of a large toggling harpoon point made of walrus ivory from the Old Bering Sea culture, circa 0–500 AD. Stained a dark color by centuries of burial and beautifully inscribed with a finely etched geometric design, its large size (nine inches long) indicates it was made to catch whales, rather than seals or walruses. (Image courtesy of Pinchas Mendelson and the Princeton University Art Museum, Princeton, NJ.)

Belle Isle in Labrador, Canada, where seals, walruses, and perhaps small whales, but not large whales, were taken.[15] Despite the toggling harpoon's early date at L'Anse Amour, almost no others have been found anywhere prior to 3,500 BP.

Skeletal remains of large whales at Amaknak Bridge also indicate whaling was undertaken. Although only a small proportion of the site's middens were examined, genetic analyses revealed bones from at least two humpback whales and two right whales.[16] The dates ascribed to the whaling tools and whale bones at Amaknak Bridge make them roughly contemporary to Giddings's "Old Whaling Culture" at Cape Krusenstern and one of the oldest sites with evidence of successful whaling. The findings also represent the oldest evidence of whaling for right whales, which in this case would have been North Pacific right whales.

The bones of large whales were also incorporated as architectural features into the semi-subterranean houses at Amaknak Bridge. Almost every excavated dwelling had a pile of large whale vertebrae in the center that were likely stacked atop one another to create pillars for roof support. Even more suggestive of whaling being practiced, however, was the discovery of smaller bones from tails, flippers, and hyoid (tongue) structures of large whales in kitchen middens. Those parts were prized for food. Historical accounts tell us that early Eskimo whaling captains distributed them to crew members—as they still do today—as a form of payment and respect to those who took part in the hunt.[17] Although whale bones amounted to less than 1 percent of the bones identified to species in Amaknak

Bridge middens, they indicate that the residents of this community dined at least occasionally on fresh meat from whales they killed.

Doubts that residents of Amaknak Bridge actively hunted whales, rather than simply scavenged carcasses cast ashore, could be eliminated by an astonishing find in 2007, some 800 miles north, on the shores of Russia's Chukotka Peninsula. There, at the northern extreme of the Bering Sea, a team of Russian and US archeologists excavated another ancient Paleo-Eskimo settlement called Un'en'en. This site was perched on the crest of a steep slope overlooking a protected bay. During the first year of work the archeologists unearthed an array of whale bones and tools characteristic of whaling, but on the final day of the short digging season they uncovered a 20-inch-long thin slab of etched walrus ivory from the floor of a semi-subterranean house.[18] It was announced (perhaps prematurely) as being one of the most amazing discoveries ever made in the field of Arctic archeology.

Inscribed on either side of the slab are figures and scenes representing a veritable handbook of early marine mammal subsistence hunting (figure 5.3). Indeed, the artifact may have been just that—a teaching aid to indoctrinate young hunters into the community's subsistence lifestyle. Its meticulously etched figures depict a lone hunter in a kayak with a harpoon resting at his side, a polar bear, a seal, teams of seven people hauling a rope possibly pulling a large animal ashore or onto the ice, and two scenes of men in large skin boats hunting whales. The ivory slab was found beneath a collapsed wooden roof support that was radiocarbon dated to 3,000 BP.[19]

The whaling scenes illustrate a fully developed whaling tradition virtually identical to the first historical accounts of Inuit and Thule whaling in the nineteenth century. The whales in the two scenes are etched with an attention to detail that seems to identify characteristics of two different kinds. One appears to be a bowhead whale attached by a line to an umiak carrying five people. Close inspection of this carefully etched whale reveals what appears to be a distinctively arched jaw, a large broad flipper, and a smooth, finless back that is crosshatched to impart a heavy dark color. All of these features describe the bowhead whale still prized by Native subsistence whalers. Halfway along the line between the whale and umiak is what appears to be a large sealskin float with a wooden or bone crosspiece. Historical accounts tell us that Eskimo whalers used such floats as drogues to tire and track whales.

The second whaling scene etched on the opposite side of the slab shows what appears to be an umiak with a four- or five-man crew attached to a different kind of whale. This whale lacks the high arching jaw line and broad flipper of the whale on the other side. It also includes an exaggerated dorsal fin and a more pointed head, which suggest it could be a large porpoise. As in the other whaling scene, a float lies along the line linking the animal to the boat. Unlike the other, however, it includes what seems to be a net attached to the stern. Perhaps this "net" is unrelated to the whale and instead represents a fishing technique simply making use of the same umiak illustration. If it is part of the whaling scene, it might represent some early form of a sea anchor used to increase drag and speed the animal's exhaustion.

Figure 5.3. Details of subsistence whaling scenes on opposite sides of an etched slab of walrus ivory found at Un'en'en on the Chukotka Peninsula, Russia, in 2007. The *top* panel shows a whale with no fin and arching jaw that may be a bowhead whale; the *bottom* shows a different animal with a fin; this may be a porpoise. The artifact was initially estimated to be 3,000 years old, based on the date of an adjacent wood fragment. If the preliminary age estimate is confirmed, whaling in the Bering Sea may date even further back into antiquity. (Images courtesy of Sergey Gusev, Un'en'en Expedition.)

However, this would have been useful only if the whalers held on to struck whales, which could indicate that this animal is a smaller porpoise.

Floats were also essential components of whaling equipment no less important than the boats, toggling harpoons, or line. In the 1800s, they were made from the skin of a single large seal; often a ringed seal was cleaned from the inside out after removing the skull.[20] Where the head once was, flaps of skin were wrapped around a wooden or bone crosspiece, stitched tight with a lashing of walrus rawhide, and baked in ashes to form a rigid, airtight buoy. Removing all internal organs without puncturing the skin required exceptional skill, especially with the crude tools they had at their disposal. The crosspiece used to close the float's opening also provided a functional cleat to tie on the harpoon line and a handle to throw the float overboard quickly once a whale was struck. Apparently because of their perishable nature, no early skin floats have been found at archeological sites, and this leaves it uncertain when they were developed.

A full analysis of the inscribed slab has yet to be published. Some archeologists are skeptical that its age is as great as the associated piece of wood and believe it could be just a few hundred years old. Nevertheless, it presents many intriguing possibilities. For example, etchings on the two sides may depict hunting practices in different seasons. In any case, the fully formed nature of the technology illustrated on the artifact indicates that whaling was already a well-established tradition when it was created and must have developed long before this etching was made. If the initial date ascribed holds up, the roots of whaling could lie even centuries earlier than 3,000 BP.

Moving East

Around 4,750 BP, shortly before the Holocene's neoglacial conditions warmed to a thermal optimum, a wave of Paleo-Eskimos began moving east from Siberia across the Bering Sea and eventually reached Newfoundland, Labrador, and western Greenland. These emigrants provided the roots of the Saqqaq (pronounced "saːkək") culture that flourished on the shores of Davis Strait between Canada and Greenland until about 2,500 BP. Genetic samples from a 4,000-year-old tuft of human hair recovered at a Saqqaq settlement on the southwest coast of Greenland indicates they were descendants of people living in Siberia about 5,500 BP.[21] Middens at their settlements document a strong maritime tradition dependent on seals, birds, fish, and shellfish, much like the people of the Aleutian Islands. Although some investigators have suggested they hunted bowhead whales,[22] this conclusion seems to be based more on speculation than hard evidence. Saqqaq tools were similar to the Arctic Small Tool tradition and were not sizeable enough to kill large whales. If the Saqqaq were indeed descendants of migrants from Siberia and did not practice whaling, it might indicate that their ancestors had not yet developed a whaling tradition when they began moving east 4,500 to 5,500 years ago.

What is more certain is that whaling with sealskin drogues was an integral part of the lives of later Arctic aboriginals who moved across the Canadian Arctic from the Bering Sea in another wave of emigration between 1,200 to 1,000 BP (800 to 1000 AD). Those people were the Thule, who recolonized coastal lands abandoned by the Saqqaqs in western Greenland and northeastern Canada. The Thule were probably the first people to successfully hunt large whales on seas bordering the North Atlantic Ocean. Their eastward migration coincided with yet another warm phase called the Medieval Optimum that left the Canadian Arctic largely ice free in summer. This opened waters in northern Canada to a variety of marine mammals, including bowhead whales. The Thule were expert hunters who relied almost entirely on seals and whales, particularly bowhead whales, for food, clothing, and other resources needed for everyday life. They were a nomadic people, following seasonal wildlife migrations that were in turn dictated by the advance and retreat of seasonal pack ice.

Zooarcheologist Susan J. Crockford at the University of Victoria in British Columbia has proposed that the Thule's nomadic lifestyle and eastward migration may have developed as the neoglacial period gave way to the Medieval Optimum, between 4,700 and 2,500 BP. By following seasonal wildlife migrations that reached farther and farther north as temperatures warmed, the Thule may have gradually extended their own migrations north and east until eventually finding a clear path all the way to Greenland.[23] Their nomadic existence depended on dogsleds for transportation and apparently changed little between about 1,200 AD, when they first arrived on the shores of eastern Canada and western Greenland, and the early 1800s, when the first historic descriptions of the Thule were recorded by Europeans.[24] Because they preferred ice-edge hunting in the far north, it seems unlikely that Thule hunters encountered right whales, although something of their whaling expertise may have been transferred to later Norse hunters.

Although most Thule lived in western Greenland and eastern Canada, archeological evidence reveals at least one outpost in northeastern Greenland as recently as the 1500s.[25] In the early 1600s, Europeans whalers off eastern Greenland found stone harpoons embedded in the blubber of some bowheads.[26] Unaware of the Thule presence along eastern Greenland, European whalers took the recovered implements as evidence that bowheads must have been struck by Thule hunters in the Davis Strait, on the west side of Greenland, and migrated around the island's southern tip into the Greenland Sea. Although such movements are possible, it may be that the stone harpoons found by European whalers were relics of unsuccessful Thule whale hunts off Greenland's eastern coast.

Up to about 800 AD, whaling remained a skill that was the sole domain of Arctic Natives. Although by then the practice had moved east from the Bering Sea to the Arctic seas just north of the North Atlantic Ocean, its target focused on bowhead whales. It is likely that very few, if any, right whales were killed up to that time. Indeed, given the sparse populations of Arctic Natives and their subsistence economy, it is probably safe to assume that even bowhead whale populations were still virtually unscathed by hunters. Nevertheless, Arctic Natives had developed effective whaling methods fundamentally identical to those that would be used by Europeans in later centuries. Perhaps through contact with Europeans, the Thule passed along some of their whaling knowledge. If details of their whaling techniques and know-how were not transferred during contact with Norse voyagers before 1000 AD, they may have at least planted the idea that even the largest of whales could be killed to meet many human needs. Precisely what knowledge may have been shared is uncertain, but the Norse began hunting large whales during the latter half of that millennium. And with that, the insulation from significant human pursuit that right whales had enjoyed up to that time would soon end.

NOTES

1. Fagan, BM. 1995. *Ancient North America: The Archaeology of a Continent*. Thames & Hudson. New York, NY, p. 78.

2. Curry, A. 2012. Coming to America. Nature 485:30–32.

3. Kaufman, DS, TA Ager, NJ Anderson, PM Anderson, JT Andrews, et al. 2004. Holocene thermal maximum in the western Arctic (0–180°W). Quaternary Science Reviews 23(5–6):529–560.

4. Jensen, AM. 2014. The archaeology of north Alaska: Point Hope in context. *In* CE Hilton, BM Auerbach, and LW Cowgill (eds.), *The Foragers of Point Hope: The Biology and Archaeology of Humans on the Edge of the Alaskan Arctic*. Cambridge University Press. Cambridge, UK, pp. 11–34, 19.

5. Ibid., p. 21.

6. Darwent, J, and C Darwent. 2006. Occupational history of the old whaling site at Cape Krusenstern, Alaska. Alaska Journal of Anthropology 3(2):135–154.

7. Giddings, JL. 1967. *Ancient Man in the Arctic*. Alfred A. Knopf. New York, NY. Giddings, JL, and DD Anderson. 1986. *Beach Ridge Archeology of Cape Krusenstern: Eskimo and Pre-Eskimo Settlements around Kotzebu Sound, Alaska*. Publications in Archeology 20. US National Park Service. Washington, DC.

8. Darwent, CM. 2006. Reassessing the old whaling locality at Cape Krusenstern, Alaska. *In* J Arneborg and B Grønnow (eds.), *Dynamics of Northern Societies: Proceedings of the SILA/*

NABO Conference on Arctic and North Atlantic Archaeology, Copenhagen, May 10–14, 2004. Studies in Archaeology and History, Vol. 10. National Museum of Denmark. Copenhagen, Denmark, pp. 95–102.

9. Knecht, RA, and RS Davis. 2008. The Amaknak bridge site: cultural change and the neoglacial in the eastern Aleutian Islands. Arctic Anthropology 45(1):61–78. Crockford, SJ, and SG Frederick. 2007. Sea ice expansion in the Bering Sea during the neoglacial: evidence from archaezoology. The Holocene 17(6):699–706.

10. Crockford SJ. 2008. Be careful what you ask for: archeological evidence of mid-Holocene climate change in the Bering Sea and implications for the origins of the Arctic Thule. *In* S O'Connor, G Clark, and F Leach (eds.), *Islands of Inquiry: Colonisation, Seafaring and the Archaeology of Maritime Landscapes.* Terra Australis 29. ANU Press. Canberra, Australia, pp. 113–131, 115.

11. Knecht and Davis, The Amaknak bridge site (n. 9), pp. 65–71.

12. Crockford, Be careful (n. 10), pp. 118–119; and Crockford, SJ, and SG Frederick. 2011. Neoglacial sea ice and life history flexibility of ringed seals and fur seals. *In* TJ Braje and TC Rick (eds.), *Human Impacts on Seals, Sea Lions, and Sea Otters: Integrating Archaeology and Ecology in the Northwest Pacific.* University of California Press. Berkley, CA, US, pp. 65–91, 76–78.

13. Knecht and Davis, The Amaknak bridge site (n. 9), pp. 73–75.

14. Mason, OT. 1902. *Aboriginal American Harpoons: A Study in Ethnic Distribution and Invention.* US Government Printing Office. Washington, DC, US.

15. Tuck, JA. 1976. Paleoeskimo cultures of northern Labrador. *In* MS Maxwell (ed.), *Eastern Arctic Prehistory: Paleoeskimo Problems.* Memoirs of the Society of American Archaeology, No. 31. Washington, DC, pp. 89–102. Tuck, JA, and R McGhee. 1975. Archaic cultures in the Strait of Belle Isle Region, Labrador. Arctic Anthropology 12(2):76–91.

16. Crockford, Be careful (n. 10), p. 120; and Frey, A, SJ Crockford, M Myer, and G O'Corry-Crow. 2005. Genetic analysis of prehistoric marine mammal bones from an ancient Aleut village in the southeastern Bering Sea. Abstracts of the 16th Biennial Conference on the Biology of Marine Mammals, San Diego, CA, 12–16 December, https://www.marinemammalscience.org/wp-content/uploads/2014/09/Abstracts-SMM-Biennial-San-Diego-2005.pdf, p. 98.

17. Savelle, JM, and AP McCartney. 2003. Prehistoric bowhead whaling in the Bering Strait and Chukchi Sea regions of Alaska: a zooarchaeological assessment. *In* AP McCartney (ed.), *Indigenous Ways to the Present: Native Whaling in the Western Arctic.* Canadian Circumpolar Institute Press. Edmonton, Alberta, Canada, pp. 167–184.

18. Witze, A. 2008. Whaling scene found in 3,000-year-old picture. Nature, doi:10.1038/news.2008.714.

19. Pringle, H. 2008. Signs of the first whale hunters. Science 320(5873):175.

20. Roswell Schafer (Nome, Alaska). Personal communication with author, 2009.

21. Rasmussen, M, Y Li, S Lindgreen, JS Pedersen, A Albrechtsen, I Moltke, M Metspalu, et al. 2010. Ancient human genome sequence of an extinct Palaeo-Eskimo. Nature 463:757–762.

22. Mobjerg T. 1999. New adaptive strategies in the Saqqaq culture of Greenland, c. 1600–1400 BC. *In* P Rowley-Conway (ed.), World Archaeology. Special issue. Arctic Archaeology 30(3):452–465.

23. Crockford, Be careful (n. 10), p. 114.

24. Savelle, JM, and AP McCartney. 1999. Thule Eskimo bowhead whale interception strategies. Arctic Archaeology 30(3):437–451.

25. Knuth, E. 1967. Archaeology of the musk ox way. Contrib. cent. Etude Arc. Finn. - Scand. (Paris) 5:1–70, as cited in Reeves, RR. 1980. Spitsbergen bowhead stock: a short review. Marine Fisheries Review. Sept–Oct., pp. 65–69.

26. Gray, R. 1935. Greenland whales and Eskimos of N.E. Greenland. Scott. Nat. 211:75–77, as cited in Reeves, Spitsbergen bowhead stock (n. 25).

6

MEDIEVAL WHALING IN NORTHERN EUROPE

IN 650 AD northern Europe was on the cusp of major change. Its human population—a scant 2 million people in what are now Scandinavia and Germany, and perhaps 500,000 more in the British Isles—was scattered among small coastal villages and farms wherever tracks of arable land could be found.[1] As descendants of fierce Germanic tribes, these people were no strangers to armed conflict. They had already successfully fended off the Roman Empire at the height of its powers, thereby preserving their pagan traditions; this was accomplished even as the Romans were spreading Christianity across their own realm from central and southern Europe to the southern half of Great Britain behind Hadrian's Wall.

Throughout human history climate change has left its stamp, and the dawn of the European Middle Ages was no exception. Temperatures in the middle of the first millennium were considerably cooler than they are now, but were about to cycle through a thousand-year roller coaster ride. In the ninth century, a warm phase called the Medieval Optimum or the Medieval Warm Period raised temperatures across northern Europe; they were then an average of 1.8°F (1°C) above those at the close of the twentieth century. By the early 1300s, they would again drop into a 500-year cold period called the Little Ice Age, with temperatures some 3.7°F to 10.8°F (2°C to 6°C) below those of the twentieth century. The Medieval Optimum brought longer growing seasons and an increase in the frequency and intensity of storms sweeping in off the Atlantic. This was a boon for crops and livestock, and by the time the Little Ice Age set in, the human population around the Baltic ballooned to 11 million, with another 5 million in the British Isles.

It's likely no coincidence that the start of the Medieval Optimum marked the beginning of the age of expansion for the Vikings—or more properly, the Norse.[2] With versatile vessels of their own design called longboats and a somewhat broader version called *knarrers*—both powered by oars, a large square-rigged sail, and a crew of some two dozen—the Norse mastered the art of ocean voyaging across open stretches of the storm-tossed Atlantic. Contemporary illustrations of these state-of-the-art craft are depicted on the remarkable 225-foot-long Bayeux Tapestry, completed in the 1070s, and the Överhogdal tapestries, possibly dating to the ninth century. At that time, Norse seafarers were establishing new colonies in the northern British Isles, Faroe Islands, and Iceland (figure 6.1). Exercising their maritime prowess in a relatively balmy climate, they mounted raids on settlements as far south as Iberia and as far east as Constantinople in the Mediterranean. As Norse influence spread along the shores of Europe, the Iron Age was drawing to a close and the first written records—the marker separating prehistoric and historic eras—were replacing oral traditions. It was probably during this period, but possibly earlier, that the Norse became the first Europeans to take up whaling.

The Distribution and Movements of Right Whales

Before describing the roots of whaling in the North Atlantic, a brief digression would be useful to explain the geographic distribution, movements, and population structure of what would become the principal target of whalers—North Atlantic right whales. From information on where and when right whales were killed over the past millennium and what is known of their movements and ecology, scientific consensus now holds that right whales were once divided into two relatively discrete groups, or "populations," living on opposite sides of the Atlantic.[3] At various times in the 1900s two other distribution patterns were thought possible—a single ocean-wide range, and a three-population model with two occurring on either side of the Atlantic and a third occupying the deep mid-Atlantic waters between the continental shelves of Europe and North America. The single stock proposal is now out of favor and the notion of a third group in the central Atlantic has been discounted given the species' current fidelity to continental shelf habitats. Under the presently accepted two-population paradigm, at least part of each population, principally pregnant females and some juveniles, presumably undertook annual migrations as they do today along the North American coast, over the continental shelf between northern feeding grounds and southern calving areas in subtropical latitudes.

As new information develops, however, an open mind is always advisable when considering population structure and movements of large whales. Recent studies of gray whales in the North Pacific provide a case in point. Scientists have assumed that gray whales in the Pacific consisted of two coastal populations living on opposite sides of the ocean basin. The western gray whales along the coast of Asia were all but eliminated by nineteenth-century whalers; perhaps 150 individuals now survive. Eastern gray whales off the west coast of North America were also devastated by whalers but have made one of the most remarkable recoveries of any large whale. Now numbering over 15,000, the eastern population may be at or

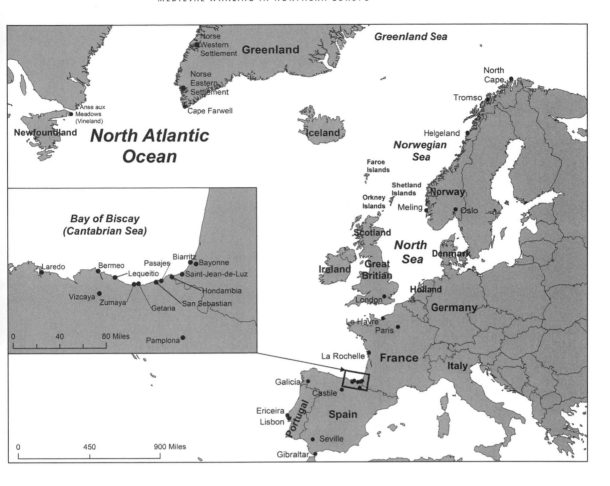

Figure 6.1. The medieval European whaling region. (Map prepared by Linda White.)

approaching their pre-exploitation abundance. Whereas western gray whales were thought to migrate between feeding grounds off Russia and a yet-to-be discovered calving ground off southeast Asia, the well-studied eastern population undertakes one of the longest migrations of any mammal on earth—an annual roundtrip of some 10,000 to 12,000 miles between calving lagoons off Baja California, Mexico, and feeding grounds in the Bering and Chukchi Seas between Alaska and Russia.[4]

Recently, however, scientists were dumbfounded when Bruce Mate at Oregon State University, a pioneer in tracking whales with satellite telemetry, reported on the movements of a western North Pacific gray whale after he successfully tagged it while it was feeding off Sakhalin Island, north of Japan.[5] Two months after being tagged in October 2010, instead of turning south down the Asian coast as all expected, the whale moved steadily northeast into the Bering Sea. Over the next two months, it traveled east across the Bering Sea to the Pribilof Islands in Alaska, and then southeast through the Aleutian Islands at Unimak Pass, across the open Gulf of Alaska, and down to the coasts of Washington and Oregon where its tag transmissions stopped, presumably because the tag fell off. It seemed clear that this "western" gray whale was headed for the same calving grounds used by the eastern gray whale population in Baja, Mexico.

This startling implication was confirmed the following year when Bruce Mate tagged another western gray whale off Sakhalin Island. That whale was tracked as it traveled nearly 14,000 miles in 172 days to Baja California and back to Sakhalin Island. This is the longest known annual migration ever documented for any mammal.[6] Subsequent comparisons between photo-identification catalogues and genetic samples of western and eastern gray whales revealed at least twelve matches between animals in the "western" and "eastern" populations, including the two tracked whales.[7] Yet genetic studies also reveal that western and eastern gray whales have distinct differences that indicate a significant degree of reproductive isolation.

Implications of those revelations are still being digested. Although some western gray whales are known to move south along the coast of Asia (a conclusion based on strandings in Japan and Korea), these new findings will undoubtedly spark a spirited debate over whether western and eastern gray whales—and perhaps other large whales occupying opposite sides of the same ocean—are in fact discrete populations or parts of a single population. For now, however, scientists still hold that North Atlantic right whales once consisted of separate eastern and western North Atlantic populations.

In winters during the Middle Ages, eastern North Atlantic right whales presumably calved off northwest Africa and southern Europe, and perhaps even in the Mediterranean Sea. In spring they turned north and followed the coastline around and through the British Isles to summer feeding grounds off Scotland, Iceland, the southern tip of Greenland, and Norway. Some may even have continued west to feeding grounds off North America, where they would have intermingled with right whales in the western population. Whereas adult males and nonpregnant females may have remained in northern waters year-round, pregnant females and at least some juveniles would have retraced their path in the fall to southern calving grounds in order to complete their migratory cycle.

The western population off North America likely followed the same pattern they do today. Presumably, they wintered on calving grounds off the southeastern coast of North America and perhaps occasionally in the Gulf of Mexico and spent the summer months feeding off New England, Nova Scotia, New Brunswick, and perhaps Quebec, Newfoundland, and southern Labrador. At least some animals migrating north probably continued traveling northeast to feeding grounds off the southern tip of Greenland, Iceland, and Norway, where they would have mixed with eastern North Atlantic right whales. Occasional movements to European waters are still evident today. Several right whales photographed over the past several decades along the eastern United States have been matched to photographs of rare right whale sightings off the southern tip of Greenland,[8] North Cape in northern Norway,[9] Iceland,[10] and the Azores.[11]

Migration routes appear to become ingrained during the first year of life as calves accompany their mothers. Those routes are then followed by most whales throughout their lives. As a result, even though some right whales from each population might cross the Atlantic to summer feeding grounds on the opposite side of the ocean, they almost always return to their natal calving grounds to give birth. This maintains two relatively discrete populations. Before commercial whalers devastated both right whale and bowhead whale populations, the ranges of the two species may have overlapped in places such as the Strait of Belle Isle between Labrador and Newfoundland, Norway, and Iceland. In those areas, bowhead whales arriving in early winter at the southern limit of their calving grounds may have encountered northern right whales lingering on or arriving early at the northern limit of their feeding range; that overlap could also have occurred when bowheads were leaving for the north in the late spring.

Norse Whaling

With arable land in short supply in high latitudes, the Norse had to rely on fish and shellfish to stock their larders. Whaling was a natural outgrowth of this dependence on the sea. As would happen elsewhere, it undoubtedly began by scavenging stranded whale carcasses, a practice that continued opportunistically even as whaling methods were being mastered. Although not written down until the thirteenth century, the sagas of early Norse settlers in Iceland described disputes over stranded whale carcasses from at least the ninth and tenth centuries. They were often settled by force despite unwritten codes intended to prevent such altercations.[12]

The *Saga of Grettir* describes one such brawl that erupted during a time of famine between two groups contending for a stranded whale. It washed ashore in an autumn storm on the Island of Rifsker, off eastern Iceland.[13] The battle was fought with the axes, cleavers, and knives intended for butchering the animal, as well as parts of the whale itself. One claimant was said to have been butchering a whale while "standing in a foot-hold cut into the whale just behind its head" (suggesting it was a large whale), when he was set upon from behind and beheaded with an axe.[14] In the melee that followed, at least one other man was killed, and those lacking weapons resorted to wielding whale bones and any other readily available whale parts. A verse commemorating the incident painted an unbecoming picture that would be amusing were it not for its deadly outcome:

> *Hard were the blows which were dealt at Rifsker;*
> *no weapons they had but stakes of the whale.*
> *They belaboured each other with rotten blubber.*
> *Unseemly methinks is such warfare for men.*[15]

What may be the earliest evidence of at-sea whaling anywhere in Europe are petroglyphs depicting whaling scenes; these are found at various locations on the Scandinavian Peninsula. The age of rock art is difficult to estimate, but most Norwegian petroglyphs are thought to be at least 1,000 years old and possibly much older. According to E. J. Slijper, author of a definitive book on whales and whaling, one example occurs near the port of Meling, Norway, 200 miles east of Oslo, and illustrates a whale surrounded by four boats, one of which seems to have been overturned by the thrashing tail.[16] Another example has been found on islands in the Ångermanälven River off the northern Baltic Sea. Thought to be at least 3,200 years old, the petroglyph depicts men in boats hunting what may be a beluga whale.[17]

Other early signs of whaling in northern Europe point to dates between at least 650 and 1000 AD. Northern Europe's Iron Age was then coming to a close and the Norse were expanding their realm with settlements in Scotland, the Shetland Islands, and the Orkney Islands around 800 AD; the Faroe Islands close to 825 AD; Iceland in 870 AD; and Greenland in 986 AD. By then the stone projectile points of regional hunters had been replaced with iron spear points and arrow tips.[18] Examples of these points have been recovered from the graves of Norse hunters where they were placed as a tribute or sacrificial gift. The offerings were often accompanied by bones of the animals they presumably hunted. Ocean hunters first used these spear points to kill seals, porpoises, and dolphins, but they were also apparently used to kill large whales. Supporting evidence includes an early ninth-century grave in Hundhom on Norway's north coast that contained a large spear point and the hyoid bone of a large whale, possibly a right whale.[19]

In the latter part of the first millennium, written records of whaling begin to appear. The earliest apparent reference to whaling is in a work credited to King Alfred the Great, the Anglo-Saxon ruler of Wessex in southern England between 880 and 901 AD. King Alfred had a passion for geography and avidly sought information on lands throughout northern Europe. The fruits of his obsession were recorded in a volume entitled *History of Orosius*, which was based on the writings of Paulus Orosius, a widely traveled Christian priest in the Roman province of Galacecia (the northwest coast of Iberia). Orosius's fifth-century description of northern Europe was still the most authoritative geographic reference in Alfred's day. In his own text, King Alfred added new details gathered from firsthand accounts by Norse navigators. One of his sources was a wealthy merchant and explorer named Ohthere (Ottar in some texts), from northern Norway.

Alfred wrote that Ohthere "told his lord, King Alfred that he dwelt northmost of all Northmen [and] . . . that, at a certain time, he wished to find out how far the land lay right north." Ohthere sailed north for three days until "he was as far north as Whale-hunters ever go."[20] Fridtjof Nansen, the renowned Norwegian Arctic explorer of the early 1900s, analyzed Alfred's text and concluded that Ohthere's voyage must have taken place between 880 and 890 AD and that it originated from a location near present-day Malangen, about 100 miles north of the Arctic Circle and 20 miles southwest of Tromsø, Norway. Nansen concluded that Ohthere probably sailed north to Norway's North Cape and then east along the coast of what is now Finnmark, Noway's northernmost county, and then westernmost Russia.[21]

King Alfred offered no description of whaling methods, but he did say that Ohthere hunted both large whales and walruses, the latter called "horse-whales" in the original 1852 English translation. Alfred related Ohthere's description of horse-whales as animals "not longer than seven ells, with very good bone in their teeth [tusks]," which were hunted by the Finns, east of North Cape.[22] He added that Ohthere reported the best whale hunting was in his own country, with the whales "eight-and-forty ells long, and the largest fifty ells long." King Alfred wrote, "of these he said that he [was] one of six [who] killed sixty in two days."[23]

The contrast in size between horse-whales and the much larger whales Ohthere was said to hunt suggests that his target included animals the size of right whales or rorquals. Yet something is plainly amiss in this account. Nansen pointed out that it stretched the limits of credibility to believe that six Norse ships of the era could have killed sixty large whales in two days. As he put it, "given that the sobriety of Ottar's narrative makes it very improbable that he made boasts of the sort, King Alfred evidently did not grasp the essential differences between walruses and whales."[24] Nansen added that even if it was true that 60 whales were killed in two days, "Ottar and his men could not deal with that quantity of blubber and flesh in two weeks, to say nothing of two days."

Perhaps King Alfred misunderstood the length of time during which the catch was made, or maybe Ohthere told the king that he simply saw 60 whales in two days. Even more likely, the syntax of Ohthere's account was misunderstood and the catch total referred to 60 walruses. These and other interpretations of this oft-quoted passage illustrate the difficulties inherent in deciphering ancient texts. The passage nonetheless offers tantalizing evidence that large whales were hunted in King Alfred's day, and at least some historians have suggested that the whales Ohthere hunted must have been right whales.[25]

Norwegian and Icelandic laws leave no doubt about the importance of whales to the Norse. Norway's oldest surviving legal code, the Gulaþing, was written in the mid-eleventh century. Thought to reflect unwritten practices dating back to late Paleolithic centuries, the code's provisions served as a model for laws adopted elsewhere in the Norse realm, with modifications to meet local customs and conditions. Community leaders in Iceland, for example, adopted the Grágás code, while Faroe Islanders relied on the Seðabrævid code.[26] The Gulaþing code confirms that the Norse not only salvaged whales dying and stranding naturally but also actively hunted whales on the open sea. One section echoed in later codes declared the following: "A man may hunt/catch whales wherever he can. If a man hunts a whale and it is wounded on the deep, the whale is his whether it be smaller or larger. If a man shoots and hits [a whale] and [it] runs ashore, then he who hunts [it] owns one-half and one-half [goes to] him who owns the land."[27]

Even more elaborate provisions were added to Norwegian and Icelandic laws in the mid-thirteenth century. Their detail stands as strong testimony to the importance of whales to coastal residents. Indeed, whereas the text of Icelandic laws on hunting in 1280 included 3 paragraphs on elk, 4 on deer, and 7 each on wolves and bears, it devoted 19 paragraphs to whales and whaling.[28] Among other things, it describes how stranded whales, including those dying naturally as well as those killed by whalers, were to be divided between hunters and land owners. This was based on what part of the shore profile a whale stranded (e.g., a beach vs. the intertidal zone), who first struck the whale, whether the whale was killed at sea or after it stranded, how large it was, and who owned the land on which it beached. To prove who struck a whale, this and other laws, such as Magnús Lagabætir's national code of Norway written in 1274, called on hunters to mark their weapons with a distinctive brand registered with local officials explicitly for the purpose of distinguishing the harpoons of individual whalers found in stranded whales.[29]

Most whaling historians writing before the late 1900s assumed that Norse whalers used harpoons with attached lines to manage and kill whales at sea in a manner similar to earlier Arctic Natives and later Basques. However, hard evidence supporting this assumption is rare. A few large, barbed slate points have been recovered by archeologists from waters off the coast of Norway, but their age and the animals they were designed to kill are unknown.[30] Moreover, no examples of medieval harpoons capable of attaching lines to whales have survived. Thus, instead of whaling in the manner of Arctic Natives and Basques, with drogues or by tow whaling—when whalers hung on to the harpoon line after a successful strike—some historians believe the early Norse relied on a method called spear whaling to kill the larger animals.[31]

Spear whaling involved weapons without attached lines. Hunters hurled the spears with small barbed or unbarbed iron points into the backs of any whales they could approach; if they could get close enough, they stabbed them. After doing their best, they then hoped the animal would die and either drift ashore of its own accord or be recovered later drifting at sea. In some cases, spear tips may have been covered with rotting whale meat or some kind of poison intended to cause infections that would kill whales or cause them to become disoriented and strand.[32] The small size of Norse spear points, however, strongly suggests they were not adequate to either kill a large whale outright or secure an attachment sufficient for towing a drogue or whaling boat.

When a struck whale stranded, brands etched on spear shafts or points left in the whale were used to identify who deserved credit for the kill. Hunters who could prove ownership of the spear were entitled to the "shooter's share" or "whaler's share"—usually a third to a half of the carcass, depending on local custom.[33] Spear whaling probably began as an opportunistic venture when fishermen happened across whales at sea. Over time the practice became more organized, with multiple vessels working as a team. Whaling was a risky business for early Norse hunters—and not just because of the physical danger inherent in attacking a large animal at close quarters in small boats.[34] Whalers risked not only their lives and equipment, but could also be fined if their battle with a whale interfered with fishermen or their fishing gear; and they could be ordered to compensate those who had offered to join the hunt for a share if a whale was lost. For example, Iceland's

Grágás code stipulated, "if [a whaler should] refuse assistance of other men, and the whale is lost, they are responsible for [profits lost] to those who would have helped."[35]

Spear whaling must have been terribly inefficient. Most whales probably escaped with minor injuries. If a whale was mortally wounded, the lack of ropes to manage it probably meant it would swim off unclaimed, only to die elsewhere with little chance for the whalers to recover the carcass. Even if it died quickly and whalers managed to stay close, if it was not a right whale, there was a good chance it would sink. But dead whales that sink often resurface a few days later as decomposition gases bloat the carcass until it refloats. However, unless whalers happened to be waiting when it resurfaced, winds and currents could just as easily carry it out to sea as drive it ashore. Nevertheless, the bounty from those that were recovered must have made the effort worthwhile, given the lengths to which Norse legal codes detailed a whaler's rights and how struck whales were to be apportioned among claimants.

The Norse were keen observers of the whales they hunted. This is abundantly clear from another ancient text written by an anonymous author in the original Old Norse language, and entitled *Konungs Skuggsjá* (see plate 2). Written as an educational text during the reign of Norwegian king Häkon Häkonsson, circa 1250 AD, the manuscript is also called *Speculum regula* in a later Latin version, and the *King's Mirror* in the English translation based on the Latin edition.[36] The text is in the form of a colloquy between the king and his son; the king answers a series of questions from his future heir about the wonders and ways of the Norse realm.

An entire section of the tutorial entitled "Marvels of the Icelandic Seas: Whales and the Kraken" provides a remarkably lucid description of both small and large whales hunted by the Norse in Iceland during the thirteenth century. Indeed, the description of whales is so detailed that it is one of the few early texts in which it is possible to make a reasonably educated guess as to species being discussed. More than a dozen different kinds of whales are mentioned, most of which were said to be good to eat. In the original English version, published in 1917, translator Laurence Marcellus Larson gave the name "right whale" to one, clarifying in a footnote that this animal was likely "*Balaena mysticetus*, also called [the] Greenland or bowhead whale."[37] Yet, in another section of the text, he described a different animal that he translated as the "Greenland whale." At the time of Larson's translation, the distinction between right whales and bowhead whales was still a subject of confusion, even for experts. It's uncertain therefore whether the two sections refer to two different species or one, and if one, which one.

This ancient text's initial passage on "right whales" is brief and only noted the following: "this [whale] has no fins along the spine and is about as large as the sort we mentioned last [the sperm whale described as 30 to 40 ells long]. Sea-faring men fear it very much for it is by nature disposed to sport with ships."[38]

The description of a large whale with no fin on its back could be either a right whale or a bowhead whale. However, the description of its feisty nature "disposed to sport with ships" is more consistent with a right whale. Comparisons between right whales and bowhead whales by later whalers experienced with hunting both almost always described right whales as being far more dangerous and unpredictable.[39]

Following the brief paragraph on "right whales" and several more paragraphs on other kinds of whales, a passage describes what Larson translated as the "Greenland whale." This whale was described as a "fish [that] grows to a length of eighty or even ninety ells and is as large around as it is long; for a rope that is stretched the length of one will just reach around it where it is bulkiest." The description continued as follows: "Its head is so large that it comprises fully a third of the entire bulk.... This whale rarely gives trouble to ships. It has no teeth and is fat and good to eat."[40] Even though the name of this whale was different, Larson commented in a footnote that it was probably the same kind that was mentioned in the earlier paragraph on "right whales." A whale as large around as it is long with a head one-third its length describes the bowhead whale to a tee. A right whale's head is somewhat smaller, taking up only about a quarter of its length. The milder disposition and larger size also suggests this whale was a bowhead. Considering the separate descriptions and large gap between the two passages, Larson may have been incorrect to suggest that they referred to the same kind of whale.

In the 1300s, before bowhead and right whale numbers were decimated by commercial whalers, the range of the two species may have overlapped off Iceland as mentioned earlier, although not necessarily at the same time of year.[41] Based on current migratory patterns, bowheads overwintering in Iceland likely would have migrated north in early spring, following the retreating pack ice to summer feeding grounds in the high Arctic. Right whales, on the other hand, probably would have fed off Iceland in summer and moved south in the fall to more temperate latitudes in winter. Unfortunately, for all its details about the whales of Iceland, the *King's Mirror* includes no information on the seasonality of the whales discussed.

If Larson were correct and the two separate passages do refer to the same species, North Atlantic right whales are probably the more likely candidate given what is now known of the species' distribution. Shore whalers in the early 1900s reported catches of right whales, but not bowhead whales, off Iceland even though both species were rare by then. In addition, a few right whales, but again no bowheads, have been seen off Iceland and the tip of Greenland since the late 1980s.[42] And with Iceland and northern Europe still under the influence of the Medieval Warm Period when the *King's Mirror* was written, the warmer water temperatures around Iceland would have favored the presence of right whales over bowheads.[43] Indeed, they even may have enabled right whales to stay around Iceland year-round.

The description of the "Greenland whale" in the *King's Mirror* also reported the following: "This whale is very cleanly in choice [clearly in search?] of food; for people say it subsists wholly on mist and rain and whatever falls into the sea from the air above. When one is caught and its entrails are examined, nothing is found in its abdomen like what is found in other fishes that take food, for the abdomen is empty and clean. It cannot readily open and close its mouth for the whalebone in it will rise and stand upright in the mouth when it is opened wide; and consequently whales of this type often perish because of their inability to close the mouth."[44] A whale with no prey in its stomach might fit the description of a bowhead whale having fasted through the winter before heading north to summer feeding grounds. However, it also could describe a female right whale arriving at Iceland in early spring after fasting through the winter on southern calving grounds. In either case and despite the errant assessment of its feeding habits, this passage may be the first written record of an attempt to determine what a balaenid whale ate by examining its stomach contents.

Still other sections of the *King's Mirror* hint of Norse contact with the Thule in Greenland, leaving a faint possibility that Arctic Natives may have influenced the start of Norse hunting of large whales. A description of Norse merchants tells us, "the man who is to be the trader will have to brave many perils, sometimes at sea and sometimes in heathen lands." Larson speculated that the reference to "heathen lands" might indicate trade in Arctic areas inhabited by the Finns or possibly the "Esquimaux."[45] In another section that discussed an ill-fated colony near the southern tip of Greenland (the Norse Eastern Settlement in figure 6.1), which lasted from 986 AD to the late 1300s AD, we are told that colonists subsisted partly on a diet of whale meat[46] and had at least occasional contact with Native residents living farther north. However, given the absence of any reference to active whaling by the colonists, it would seem that their whale meat must have come from scavenging stranded carcasses. Indeed, considering the desperate struggle to secure food in the years before the colony's collapse, what may be more remarkable is how little the colonists learned from Thule hunters about capturing whales and seals and surviving in the cold, harsh environment. Perhaps inflexible adherence to familiar Norse ways or an aggressive attitude toward strangers prevented a more amicable, productive relationship with their Thule neighbors.[47]

While the *King's Mirror* offers no description of whaling methods, the spear-whaling techniques implied by Norse legal codes seem to be corroborated by a remarkable series of coincidences recounted by a wealthy nobleman named Björn Einarsson, who lived in northwest Iceland in the late 1300s. According to a journal he kept that apparently survived into the 1600s but has since been lost,[48] late in the summer of 1385, Einarsson was returning to Iceland from a trading trip in Norway when a storm pushed him off course and forced an emergency landing at the colony on southern Greenland.

Einarsson could not have arrived at a worse time. Reaching the colony too late in the season to double back for Iceland, he and his crew resigned themselves to a hard winter layover. The struggling Greenland colony was also in its final stages of collapse as crops and livestock shriveled under the onset of the Little Ice Age,[49] and shifting cultural priorities stymied local adaptability.[50] Advised that colony food stores were dangerously low with nothing to spare, Einarsson and his crew were left to lay in their own supply of food for the winter. With winter rapidly closing in and virtually no options to hunt or forage on the barren landscape, they seemed to face an impossible task. Their fortunes turned, however, when Einarsson came across a large whale called a *reyör* that had beached while still alive. From the description of this kind of whale in the *King's Mirror*, it was likely a large rorqual, possibly a blue or fin whale that Icelanders considered to have the best meat of all whales. According to the *King's Mirror* "because of its quiet peaceful nature, it often falls a prey to the whale fishers."[51]

While butchering the whale, Einarsson discovered a whaler's spear with an inscribed brand he recognized as belonging to Ōlafur Isfirðingur,[52] a whaler from his own district in northwest Iceland. The whale had apparently died of wounds inflicted by Einarsson's neighbor. It proved to be their salvation. Upon returning to Iceland in 1387, Einarsson would have been obligated to reimburse Isfirðingur for his shooter's share, which under the circumstances, probably would have been a payment gladly made. His account seems to confirm that Icelanders not only hunted large rorqual whales, but also relied on chance strandings of struck victims to reap their reward.

Whaling by the Normans and Flemish

Whereas the Norse were likely the first Europeans to hunt large whales in the North Atlantic, by the end of the first millennium, their methods and techniques were being copied and improved upon by the Normans and Flemish. King Alfred's reference to Norse whaling in his *History of Orosius* may have been silent on whaling methods, but it provides undeniable proof that the exchange of cultural knowledge and practices was already underway between the Norse and other people in northern Europe at the close of the of the ninth century. That exchange almost certainly led the Normans and Flemish living along the English Channel and North Sea to take up the trade perhaps as early as the last centuries of the first millennium.[53] Although no details were provided, an 875 AD text entitled *Translation and Miracles of Saint Vaast* (*de la translation et des miricals des Saint Waast*) mentioned a whale fishery along the French coast.[54]

Another more substantive reference appears in a fictitious dialogue written circa 990 AD by Ælfric Bata, a prolific writer and Benedictine abbot living at the Abby of Cerne in Dorset, England. Similar to the *King's Mirror* written 200 years later, Ælfric's *Colloquy* is a dialogue between a teacher and his student that explains the ways of various tradesmen and laborers. The colloquy was written in Latin as a lesson plan to teach students the language, but in an effort to make his homework more interesting, Ælfric provided a description of the daily lives of ordinary citizens engaged in different trades. Today his colloquy offers a rare glimpse into everyday life in late tenth-century England. In response to questions posed about the life of fishermen and catching whales, the instructor has the following reply: "It is a risky business catching a whale. It's safer for me to go on the river with my boat, than to go hunting whales with many boats . . . I prefer to catch a fish that I can kill, rather than a fish that can sink or kill not only me but also my companions with a single blow . . . Nevertheless, many catch whales and escape danger, and make a great profit at it."[55] Ælfric's reference to "hunting whales with many boats" could describe the "drive" fishing methods used to herd pods of small whales ashore. However, the observation that their quarry was capable of killing the fisherman and all his companions with a single blow suggests he was referring to large whales, rather than the small porpoises or pilot whales targeted by drive fishermen.

Another early reference to whaling in the British Isles comes by way of the Spanish geographer al-'Udhrī during the period of Muslim occupation of Iberia, circa 1058. His account, which was repeated in a later work by another Spanish writer al-Qazwīnī around 1283, described coastal whaling by "Hiberno-Norse-men" (people in northern Great Britain) using multiple ships in the months of October to January.[56] He wrote that whalers targeted calves of large whales using iron-tipped spears with large rings attached for the purpose of tying on a rope. Once killed, whales were towed ashore, cut up, and the meat was then preserved with salt. Some elements of al-'Udhrī's description seem unrealistic and hard to interpret, including a claim that whalers enticed calves to their ships by clapping and shouting and then soothed them by briskly rubbing their heads before driving their blades into the skull with the blows of a mallet. The latter could be a garbled description of the drive whaling in which small whales were herded ashore and killed by stabbing. However, the description of iron-tipped harpoons with ropes attached to kill calves is consistent with well-documented practices employed by later whalers. This is also the first reference to the use of harpoons with attached lines for managing struck whales and suggests that this important advance in whaling methods occurred by at least the eleventh century, and perhaps was an invention of the Norse.

By the twelfth century, references to the sale of whale meat in markets on both sides of the English Channel become common. In Normandy, whales were caught by associations of "walmanni" (a Norse term literally meaning "whale men") who worked together in close-knit teams using multiple boats.[57] These associations may have been formed for the purpose of conducting drive fisheries for porpoise and small whales, but apparently handheld spears were also used to kill large whales, as with the Norse. One account tells of a miraculous hunt in 1116, when Flemish fishermen wounded a whale with arrows and lances while the animal blew sprays of water into the air. It then leaped into the air, and after disappearing briefly, reappeared and attacked the hunters so vehemently that they made an oath to Saint Arnulf, promising him a piece of the meat from this whale.[58] To strike such fear in the hunters, this could have been a large whale rather than a porpoise.

Perhaps the most thorough description of whaling in the middle ages comes from an account by Albertus Magnus (1193?–1280). Also known as Albert the Great, Albertus was a Dominican friar and bishop with the Catholic Church in Cologne, France. He possessed one of the most gifted scientific intellects of his day. In a monumental legacy of medieval literature, Albertus compiled a 38-volume compendium of knowledge on everything from botany to zoology, geography to astronomy, and chemistry to physiology. Five of those volumes are known collectively as *De animalibus* (22 through 26) and describe many of the world's animals, including whales and other ocean creatures. Many of his marine animals are farfetched beasts that were based on fanciful descriptions by people far less discriminating than he. However, mixed among the specious accounts from others, are descriptions of whaling based on Albertus's own observations. Because his more clear-eyed accounts are interspersed among descriptions of impossible creatures, his works have largely been overlooked by whaling historians. Recently, however, Vicki E. Szabo of Western Carolina University sifted out the more credible passages from Albertus's work in an excellent book on whales and whaling in the Middle Ages.[59]

Albertus provided a description of medieval whaling that was based at least in part on his own knowledge of whaling in Friesland, a province along the coast of Holland. According to his text, several small boats, each with a crew of three men, worked together when hunting a large whale. Each time the whale surfaced, they hurled harpoons attached to long ropes into the animal's back. If the whale dove and did not take out all of their line, the rope was used to pull back to the whale when it resurfaced. They then resumed the attack. This was repeated as many times as necessary to kill their quarry. If a diving whale was still taking out line near the end of its length, the rope would be cut and both the whale and their equipment would be lost. When a whale was killed, rope was tied around its tail and it was towed ashore for butchering. A variation of this approach apparently involved firing a spear into the whale with a "powerful ballista," or oversized crossbow.[60] Albertus relayed the following:

> The harpooner's weapon consists of a shaft fashioned from pine wood to make it lighter; at the end of the shaft near the place where the harpooner grasps it in his hand is a hole through which the end of the rope is knotted; this stout rope is extremely long, and prior to the hunt is carefully coiled in a circular fashion at the bottom of the boat, so that when the harpoon is unleashed, the attached rope will follow the trajectory of the weapon without a hitch. The point of the harpoon has below it a triangle projection like a barbed arrow, whose lateral angle is very sharp and honed for easy penetration into the whale's skin. The two edges of the triangular barb come to a point like the sharpness of a dagger . . . At the center of the triangular barb's base, opposite the point, is welded a cylindrical piece of steel, perpendicular to the base of the triangle and measuring about a cubit in length or slightly more.[61]

This account, and the one by al-'Udhrī, are the only contemporary descriptions of medieval whaling methods and implements. Whether the era's whalers used true harpoons designed to hold fast to a whale has been questioned by some historians; they maintain that handheld whaling weapons in the Middle Ages were essentially spears and lances incapable of towing a drogue or allowing hunters to maintain contact with struck whales.[62] The above references, however, suggest otherwise and point to the possibility that this method may have first been developed by the Normans or Flemish as early as the eleventh century.

Albertus also described the shore-based butchering operation: "We once witnessed the bulk of a whale which was so large that the body, after being cut into manageable sections of flesh and bone, filled three hundred wagons. Such large whales are not commonly captured, but our contemporaries often capture specimens that require from one hundred and two hundred and fifty wagons, more or less, for haulage."[63] The need for hundreds of wagons to transport meat and blubber from a single whale indicates that the one Albertus watched being butchered must have been very large. James J. Scanlan, who translated the original text from Latin in 1987, noted that a four-wheeled cart in the Middle Ages was pulled by a pair of draught horses capable of hauling loads of 500 to 1,000 pounds. From that he guessed that the whale Albertus described must have weighed between 75 and 150 tons and was likely a "Biscay right whale."[64] The comment by Albertus that hunters towed large whales ashore after they were killed also means that they stayed afloat, which also might suggest they were right whales.

With no catch records kept by Norse or other northern European whalers before 1500, their effect on right whales or any other large whales is necessarily speculative. Because Norse, Flemish, English, and Icelandic whalers apparently exploited a range of whale species using inefficient spear-whaling methods, and considering that the use of harpoons with lines attached would still have been in its infancy and unperfected, it seems unlikely that northern Europeans had a significant effect on whale populations, even though they probably killed far more whales than they successfully landed. Their effect, however, cannot be dismissed entirely, because they were not Europe's only medieval whalers. As described in the following chapter, the Basques, and to a lesser extent the Portuguese, also began hunting right whales in the Middle Ages in southern Europe. And unlike northern Europeans, they focused primarily on right whales, and took whaling to a new level.

NOTES

1. Russell, JC. 1972. Populations in Europe. *In* Cipolla, CM (ed.), *The Fontana Economic History of Europe, Vol. I: The Middle Ages*. Collins/Fontana. Glasgow, UK, pp. 25–71. Medieval sourcebook: tables on population in medieval Europe. *In* P. Halsall (ed.), Internet History Sourcebooks Project. 1996. http://legacy.fordham.edu/halsall/source/pop-in-eur.asp.

2. Szabo, VE. 2008. *Monstrous Fishes and the Mead-Dark Sea: Whaling in the Medieval North Atlantic*. Koninkhjke Brill. Leiden, Netherlands, p. 81.

3. Reeves, RR. 2001. Overview of catch history, historic abundance and distribution of right whales in the western North Atlantic and Cintra Bay, West Africa. *In* PB Best, JL Bannister, RL Brownell Jr., and GP Donovan (eds.), Right Whales: Worldwide Status, Special issue 2, Journal of Cetacean Research and Management, pp. 87–192.

4. Jones, ML, and SL Swartz. 2009. The gray whale, *Eschrichtus robustus*. *In* WF Perrin, B Würsig, and JMG Thewissen (eds.), *Encyclopedia of Marine Mammals*, 2nd ed. Academic Press. New York, NY, pp. 503–511.

5. Mate, BR, VY Ilyashenko, AL Bradford, VV Vertyankin, GA Tsidulko, VV Rozhnov, and LM Irvine. 2015. Critically endangered western gray whales migrate to the eastern North Pacific. Biology Letters 11(4), doi:org/10.1098/rsbl.2015.0071.

6. Ibid.

7. Ibid. Weller, DW, A Klimek, AL Bradford, J Calambokidis, AR Lang, B Gisborne, AM Burdin, et al. 2012. Movements of gray whales between the western and eastern North Pacific. Endangered Species Research 18:193–199. Lang, A. 2012. Insight into the behavior of gray whales using genetics. *In* J Sumich (guest ed.), Grey Whales: from Devilfish to Gentle Giant. Whale-watcher 41(1):36–41.

8. Hamilton, PK, AR Knowlton, and MK Marx. 2007. Right whales tell their own story: the photo-identification catalogue. *In* SD Kraus and RM Rolland (eds.), *The Urban Whale: North Atlantic Right Whales at the Crossroads*. Harvard University Press. Cambridge, MA, US, pp. 75–104, 94.

9. Jacobsen, K-O, M Marx, and N Øiel. 2004. Two-way trans-Atlantic migration of a North Atlantic right whale (*Eubalaena glacialis*). Marine Mammal Science 20(1):161–166.

10. Hamilton et al., Right whales tell their own story (n. 8), p. 94.

11. Hamilton, PK, RD Kenney and TVN Cole. 2009. Right whale sightings in unusual places. Right Whale News 17(1):9–10. Silva, MA, L Steiner, I Cascão, MJ Cruz, R Prieto, T Cole, PK Hamilton, and M Baumgartner. 2012. Winter sighting of a known western North Atlantic right whale in the Azores. Journal of Cetacean Research and Management 12(1):65–69.

12. Schnall, U. 1993. Medieval Scandinavian laws as sources for the history of whaling. *In* B Basberg, JE Ringstad, and E Wexelsen (eds.), *Whaling and History: Perspectives on the Evolution of the Industry*. Sandefjordmuseene. Sandefjord, Norway, pp. 11–15.

13. Hight, GH (trans.). 1914. *The Saga of Grettir the Strong (Grettir's Saga)*. EP Dutton. New York, NY, Vol. 2, pp. 21–24.

14. Fox, D, and H Pálsson (trans.). 1994. *Grettir's Sag*. University of Toronto Press. Toronto, Canada, pp. 19–20 (quotation), as cited in Szabo, *Monstrous Fishes* (n. 2), pp. 147–148.

15. Clark, JGD. 1968. *Prehistoric Europe: The Economic Basis*. Reprint of 1952 edition. Stanford University Press. Stanford, CA, US, p. 65.

16. Slijper, EJ. 1979. *Whales*, 2nd ed., trans. AJ Pomeras. Cornell University Press. Ithaca, NY, US, pp. 11–12, fig. 2.

17. Sapwell, M, and L Janik. 2015. Making community: rock art and the creative acts of communication. *In* H Stebergløkken, R Berge, E Lindgaard, and HV Stuedal. *Ritual Landscapes and Borders within Rock Art Research: Papers in Honor of Professor Kalle Sognnes*. Archaeopress Archaeology. Oxford, UK, pp. 47–58, 48.

18. Lindquist, O. 1995. Whaling by fishermen in Norway, Orkney, Shetland, Faeroe Islands, Iceland and Norse Greenland: medieval and early modern whaling methods of inshore legal regimes. *In* Basberg et al., *Whaling and History* (n. 12), pp. 17–54.

19. Lindquist, O. 1994. Whales, Dolphins, and Porpoises in the Economy and Culture of Peasant Fishermen in Norway, Orkney, Shetland, Faroe Islands, and Iceland ca. 900–1900 AD and Norse Greenland ca. 1000–1500 AD. PhD diss., University of St. Andrews, Vol. 1, p. 402.

20. Alfred Committee. 1852. *The Whole Works of Alfred the Great: With Preliminary Essays Illustrative of the History, Arts and Manners of the Ninth Century*, Jubilee ed., JF Smith. Oxford, UK, Vol. 2, p. 40 (quotation).

21. Nansen, F. 1911. *Northern Mists: Arctic Exploration in Early Times,* trans. AG Chater. Frederick A. Stokes. New York, NY, Vol. 1, pp. 170–177.

22. An "ell" is an old Germanic measurement of length equivalent to one cubit, the length of a man's forearm, or about 18 inches.

23. Alfred Committee, *The Whole Works* (n. 20), pp. 39–43 (quotations, with bracketed text in original translation).

24. Nansen, *Northern Mists* (n. 21), p. 172, n. 1.

25. Carwadine, M. 1995. *Whales, Dolphins, and Porpoises.* DK Publishing. New York, NY, p. 44.

26. Szabo, *Monstrous Fishes* (n. 2), pp. 243–275.

27. Lindquist, Whales, Dolphins, and Porpoises (n. 19), Vol. 2, p. 606 (quotation).

28. Schnall, Medieval Scandinavian laws (n. 12).

29. Lindquist, Whales, Dolphins, and Porpoises (n. 19), Vol. 2, p. 613.

30. Ibid., Vol. 1, pp. 399–400.

31. Ibid., Vol. 1, pp. 399–414. Reeves, RR, TD Smith, and EA Josephson. 2007. Near-annihilation of a species: right whaling in the North Atlantic. *In* Kraus and Rolland, *The Urban Whale* (n. 8), pp. 39–74.

32. Reeves, RR, and TD Smith. 2006. A taxonomy of world whaling operations and eras. *In* JA Estes, DP DeMaster, DF Doak, TM Williams, and RL Brownell (eds.), *Whales, Whaling, and Ocean Ecosystems.* University of California Press. Berkeley, CA, US, pp. 82–101.

33. Lindquist, Whales, Dolphins, and Porpoises (n. 19), Vol. 1, pp. 405–409; and Szabo, *Monstrous Fishes* (n. 2), pp. 248–249.

34. Szabo, *Monstrous Fishes* (n. 2), pp. 257–258.

35. Laws of early Iceland: *Grágás,* 217 (quotation), as cited in Szabo, *Monstrous Fishes* (n. 2), p. 258.

36. Larson, LM (trans.). 1917. *The King's Mirror (Speculum regale-Konungs skuggsjá) Translated from the Old Norwegian.* Scandinavian Monographs 3. American-Scandinavian Foundation. Humphrey Millford. London. Magnús Lagabætir, or "Magnus the law-mender" was the son of King Håkon Håkonsson who was asking the questions in the *King's Mirror.*

37. Ibid., p. 121 (quotation).

38. Ibid.

39. Larson also noted that the name "nor-caper (*Balaena glacialis*)" had been ascribed to yet another kind of whale in *The King's Mirror.* That whale was also called a "fish driver." The text describes this whale as particularly useful because it chased herring close to shore where fishermen could catch them. For that reason, it was not to be killed. This description strongly suggests it was a fish-eating rorqual, possibly a fin or minke whale, rather than a right whale. See ibid., pp. 120–121.

40. Ibid., pp. 123–124 (quotation).

41. De Jong, C. 1983. The hunt of the Greenland whale: a short history and statistical sources. *In* MF Tillman and GP Donovan (eds.), Historical Whaling Records, Special issue 5. Journal of Cetacean Research and Management, pp. 83–106.

42. Hamilton et al., Right whales tell their own story (n. 8), p. 94.

43. Grove, JM. 2001. The initiation of the "Little Ice Age" in regions around the North Atlantic. Climatic Change 48(1):53–82. Water temperatures around Iceland in the 1200s were perhaps 1°C warmer than they are today and 4°C higher than they would become during the Little Ice Age.

44. Larson, *The King's Mirror* (n. 36), pp. 123–124 (quotation).

45. Ibid., p. 79 (quotation).

46. Ibid., p. 145.

47. Diamond, J. 2005. *Collapse: How Societies Choose to Fail or Succeed.* Viking Penguin Group. New York, NY.

48. Lindquist, Whales, Dolphins, and Porpoises (n. 19), Vol. 2, pp. 720–721.

49. Buckland, PC, T Amorosi, LK Barlow, AJ Dugmore, PA Mayewski, TH McGovern, AEJ

Ogilvie, et al. 1996. Bioarchaeological and climatological evidence for the fate of Norse farmers in medieval Greenland. Antiquity 70:88–96.

50. Dugmore, AJ, TH McGovern, O Vesteinsson, J Arneborg, R Streeter, and C Keller. 2012. Cultural adaptation, compounding variables, and conjunctures in Norse Greenland. Proceedings of the National Academy of Sciences of the United States of America 109(10):3658–3663.

51. Larson, *The King's Mirror* (n. 36), p. 124 (quotation).

52. Szabo, *Monstrous Fishes* (n. 2), p. 256.

53. De Smet, WMA. 1981. Evidence of whaling in the North Sea and English Channel during the middle ages. *In* FAO Advisory Committee on Marine Resources Research Working Party on Marine Mammals, *General Papers and Large Cetaceans*, Vol. 3 of *Mammals of the Seas*. United Nations Food and Agricultural Organization. Rome, pp. 301–309.

54. Jenkins, JT. 1921. *A History of the Whale Fisheries*. H. F. & G. Witherby. London, p. 60.

55. Ælfric's Colloquy, ed. G. N. Garmonsway (London: Methuen and Co., Ltd., 1939). "Ælfric's Colloquy." In M. Swanson (ed.), *Anglo-Saxon Prose*. (London: J. M. Dent, 1993). Pp. 169–177. Both as cited in Szabo, *Monstrous Fish* (n. 2), pp. 56–59, 57 (quotation).

56. Dunlap, DM. 1957. The British Isles according to medieval Arabic authors. Islamic Quarterly 4:11–28. As cited in Szabo, *Monstrous Fishes* (n. 2), pp. 191–192.

57. De Smet, Evidence of whaling (n. 53), pp. 304–305; and Lindquist, Whales, Dolphins, and Porpoises (n. 19).

58. De Smet, Evidence of whaling (n. 53), p. 304; and Scoresby, W, Jr. 1820. *An Account of the Arctic Regions, with A History and Description of the Northern Whale Fishery*. Archibald Constable. Edinburgh, UK, Vol. 2, pp. 11–12.

59. Scanlan, JM (tr.). 1987. *Albert the Great, Man and the Beasts: De Animalibus (Books 22–26)*. Medieval Renaissance Texts and Studies Series #47. Binghamton, NY, 1–5, as cited in Szabo, *Monstrous Fishes* (n. 2), pp. 61–65.

60. *Albert the Great*, xxiv, xvii. As cited in Szabo, *Monstrous Fishes* (n. 2), p. 64.

61. *Albert the Great* xxiv, ixx. As translated from Latin in Szabo, *Monstrous Fishes* (n. 2), pp. 64–65 (quotation).

62. Lindquist, Whales, Dolphins, and Porpoises (n. 19), pp. 18–19.

63. *Albert the Great*, xxiv, 16, 337. As cited in Szabo, *Monstrous Fishes* (n. 2), p. 63 (quotation).

64. Scanlan, JM (tr.). 1987. Footnote 16.2. As cited in Szabo, *Monstrous Fishes* (n. 2), p. 337.

7

GHOST WHALERS

IN THE NINTH CENTURY, if not earlier, a new whaling force emerged along the southwestern coast of Europe. This one was carried out by an industrious yet enigmatic people straddling the border between France and Spain. They were the last descendants of Paleolithic tribes that had occupied Europe for thousands of years, and then survived invasions by the Romans from the west in the third century BC, the Visigoths from the north in the fifth century AD, Muslims and Moors from Africa in the eighth century, and Vikings from the north in the ninth century. These people were the Basques.

The same qualities that sustained the Basques' cohesive, distinctive culture, while surrounding people were eliminated or absorbed by invading armies, helped them create the commercial whaling industry. That notable accomplishment can rightfully be credited to the Basques. They were eminently practical, resourceful, and self-reliant, with a proclivity for trade. Content to live within their realm, they harbored no ambitions for foreign conquest or empire. Their perseverance in the face of a succession of invading armies had taught them to be wary of outsiders and to avoid provoking strangers. The Basques learned to bend with political winds, while taking advantage of new ideas and exploiting opportunities for the commerce that invaders invariably brought with them. Although not always friendly, they were always polite. Yet if any interlopers attempted to take Basque lands or pillage their villages, they banded together to defend their native soil with a ferocity that earned respect from all.

Aiding their longevity and autonomy was an oral tradition rooted in an ancient language so alien from any other that it was virtually indecipherable to outsiders. Indeed, the first book written entirely in Euskera, the ancient Basque language, was not published until 1554.[1] When it came to sharing knowledge, the Basques were tight lipped with any information that might be used against them to compete with their livelihoods. Confronted with such an impenetrable façade, many people have branded the Basques as a secretive sect. Yet those who know them best now are quick to denounce the label and note their long history of coexistence and trade with foreigners both within and beyond their homeland.

Armed with those traits and against all odds, the Basques quietly maintained a cohesive tradition whose endurance has been matched by few others anywhere in the world. Admiration for this feat was noted by none other than John Adams, who spent time in Paris in the late 1770s while negotiating an alliance with France during the American war for independence. In 1794, Adams wrote of the Basques that "while their neighbors have long since resigned all their pretensions into the hands of Kings and priests, this extraordinary people have preserved their ancient language, genius, laws, government and manners without innovation, longer than any other nation of Europe."[2]

Over the course of their whaling history, the Basque homeland was split into regions subject to English, French, and Spanish rule. Yet their shared heritage and business interests bound the Basques together as much or more than geopolitical ties. In times of peace between France and Spain, Basque whalers on both sides of the border maintained collegial, if competitive, relationships. In times of war, demands by their respective monarchs might have stretched, but could never break, their cultural bonds. Their discretion with outsiders, flare for business, and unique language enabled the Basques to keep their whaling methods and locations a secret; this in turn allowed them to retain control over the whaling industry they had created into the 1600s. Unfortunately, those same traits now conceal their whaling achievements and activities, particularly those from the Middle Ages. Indeed, details are so sparse that the Basques could justly be called "ghost whalers."

Most of what is known of their early whaling activities comes not from the typical sources, such as whaling logs or journals, but from archeological remains, accounts by mariners and coastal residents of other nations, and unlikely sources such as old legal disputes and even baptismal records. As a result, substantial gaps and inconsistencies exist. How they developed and carried out whaling operations, the dates of major technological advances, and even where they caught whales are all cloaked in uncertainty. What is clear, however, is that the Basques exploited North Atlantic right whales throughout most of their range and transformed whaling from a parochial subsistence endeavor into an international industry. More than any other people, the Basques laid the foundation for the modern whaling industry.

Some suggest that the Basques learned their whaling skills from the Norse when the latter swept south along the European coast between 800 and 1000 AD. This notion is largely based on unsubstantiated conjecture by early historians. Another possibility is that Norman or Flemish whalers introduced new whaling methods to the Basques in the tenth or eleventh centuries.[3] In either case, any suggestion that the Basques simply adopted methods learned from northern Europeans, but applied them on a grander scale, would be an injustice to the ingenuity of their whalers and the influence they had on the development of whaling.

When the Basques began whaling is uncertain. It could have been as early as

670 AD when a French Basque supplier in Bayonne signed a contract to deliver 40 *moyos*, or casks, of whale oil to light an abbey in Jumieges, near the French city of Le Havre on the English Channel.[4] Whether they were killing whales at that early date or merely salvaging stranded animals is unknown. Yet, this contract implies that the Basques were already trading in whale oil, had the technical skills to render oil from whale blubber, and had a marketing network that extended beyond their native homeland. To that foundation the Basques applied their shrewd instincts with superior shipbuilding and navigational skills to become Europe's preeminent whalers and fishermen. Their reputation was built largely on a single-minded pursuit of just a few select species. For whalers this was right whales, and later bowhead whales; for fishermen it was cod and hake.

Over the course of their whaling history, the Basques employed at least three different whaling strategies: (1) shore whaling; (2) offshore or "pelagic" whaling, where whales were butchered at sea but the rendering of oil still relied on onshore-based facilities; and (3) pelagic whaling, with onboard furnaces or "tryworks" to render oil.

Shore Whaling

The Basques' first strategy was shore whaling. This is the only method known with certainty to have been used by them during the medieval era. Shore whaling was carried out with dories deployed from shore to kill passing whales; the whales were then towed back to land for processing. These dories were first launched from homeland villages along an open arm of the Atlantic now called the Bay of Biscay, but known to the Spanish Basques as the Cantabrian Sea. Initially the whales they chased were spotted from limestone bluffs rising along the oak- and chestnut-forested coast. Before long, perhaps as early as the seventh century,[5] coastal villagers began erecting stone watchtowers, or *atalaya*, explicitly for spotting whales. Most of those towers have long since crumbled, but remains of the tower that lorded over the storied Spanish Basque whaling port of San Sebastian can still be seen today (figure 7.1). Throughout the winter whaling season, these towers were manned by paid lookouts. When a spout was sighted, smoke from a signal fire lit in the watchtower or the pounding of a drum alerted the town below,[6] and all available whalers rushed to their *txalupa*—open dories called *chalupas* in Spanish and "shallops" in English—that were moored in coastal coves or drawn up onto ocean beaches.

Like Eskimo umiaks, chalupas were functional works of grace and art. They were sturdy, multipurpose craft typically 20- to 26-feet long that were built for both fishing and whaling. Pointed at both ends, with a shallow, rounded keel and low sides, they could be maneuvered with equal ease forward and backward; they could pivot on a dime, and go from full speed to full stop in a single boat length.[7] Above the waterline, they were "lapstrake" or "clinker-built" (with overlapping

planks) for strength. Below the waterline they were constructed caravel style (edge-to-edge planking) to decrease drag and increase speed.[8] The boats were ideal for chasing whales. Indeed, the same basic design would be used by whalers into the twentieth century.

A chalupa's whaling crew numbered from 5 to 10. Included were a helmsman at the steering oar in the stern, three to eight oarsmen, and a harpooner in the bow who doubled as an oarsman. To muffle sounds that could spook their quarry, iron oarlocks, oar handles, and occasionally even oar blades were sometimes wrapped tightly with rope coils or animal hide. These versatile craft could also be fitted with one or two masts; one stood near the bow and carried a small jib-like lug sail, and the other was placed amidships and fitted with a larger square-rigged sail. Sails were common when fishing for finfish, but for whaling they were used only on boats that followed whalers to assist in retrieving carcasses.

To catch whales, the Basques used a harping iron, or *arpoi*, from which the English word "harpoon" is derived. Although the Basques may have appropriated some design concepts from the Norse, Normans, or Flemish, the English origin of the word suggests it was a design they invented or significantly improved, perhaps through modifications of the point or the line attachments. The arpoi had a large V-shaped head with a deeply notched base for holding fast to large whales. The large point was forged to the end of a long iron shank, with an iron loop welded to the opposite end for tying on a stout rope (figure 7.2). Alternatively, a 6.5-foot-long beech shaft with a trailing line was lashed to the iron shank.[9] By the 1500s, if not earlier, Basque harpoons were modified to incorporate molded sockets on the end of short iron shanks into which wooden harpoon shafts could be inserted. At some point—likely in the 1600s or later—the Basques also developed their own version of a toggling harpoon similar in concept to that used by Arctic Natives (see chapter 5).[10]

Like the toggling harpoons used by Arctic Natives, the arpoi was not intended to kill whales but rather to secure a drogue that would cause them to tire and allow whalers to track their movements. Instead of the sealskin drogues used by aboriginals, the Basques used square slabs of wood cut roughly 3 feet across. Chalupas had no bollards or loggerheads to tie off and manage harpoon lines,[11] yet tow whaling was practiced at least occasionally by the 1500s. Evidence for the method comes from a pair of sketches found in baptismal records circa 1526–1586 in the archives of Zumaya, a port on the Spanish Basque coast of Gipuzkoa (figure 7.3; and see plate 5). The crude drawings are of chalupas with crews of five to seven men; plate 6 includes what appears to be a line tub (a container holding the coiled line) in the stern of a boat being towed by a whale.[12]

Once launched from shore, a contingent of chalupas would set a course to intercept the whale. When nearing the quarry, whalers used all possible stealth. The harpooner would swivel from his position on the lead oar to stand and face the target, and direct the helmsmen to a spot where experience told him the whale would surface next. Upon reaching that spot, the crew would ship oars and wait silently with the harpooner standing in the bow, harpoon in one hand and a coil of rope in the other. If they were lucky and the whale surfaced within a few yards, the harpooner would hurl his arpoi into the whale as its back arched above the waves to take a breath. The wooden drogue tied to the end of the harpoon with a separate line rested nearby in the boat and was immediately tossed overboard once a whale was struck.

The whale's reaction was a swift and powerful stroke of its flukes that propelled it down to the safety of the depths. In shallow coastal waters, the depth of a whale's dive was limited and the drogue's buoyancy soon brought it back to the surface. At that point, the chase was on. The chalupas would set off following the bobbing drogue that marked the whale's course. When the whale tired, boats converged to set additional harpoons or attempt a kill with repeated thrusts of

Balæna cum fuscina,qua confixa capitur,&c.

Figure 7.2. Illustration of a whale and harpoon at the beginning of Konrad Gesner's review of cetaceans in his 1560 *Icones nomenclator*. (Image courtesy of the Biodiversity Heritage Library, Museum of Natural History, Smithsonian Institution, Washington, DC.)

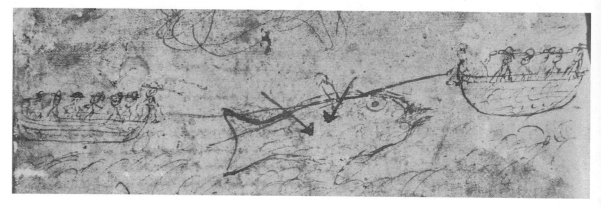

Figure 7.3. A sketch of Basque whaling drawn in the margins of a 1526–86 baptismal and marriage registry from the Basque port of Zumarraga, in northern Spain. The registry is now preserved at the Diocesan Archive of San Sebastian, Gipuzkoa, Spain. (Image courtesy of Untzi Museoa--Museo Naval, Diputación Floral de Gipuzkoa, San Sebastián, Spain.)

a long-bladed iron lance that could pierce the lungs or heart. This was the most dangerous time for whalers. They were within feet of a now-panicked animal that could be more than twice the length of their boats. If the lines held fast and the whale didn't upend their craft, newly planted harpoons and lance blows would send the whale on another run, with the boats again in hot pursuit. This process might be repeated several times over a span of hours, depending on the whale's elusiveness and wounds. Its imminent demise was usually signaled when the white mist of its spout turned crimson; this indicated a lung had been pierced by a lance and was filling with blood.

If whalers attempted to hang on to the harpoon line after a successful strike, the line had to be long enough to reach the bottom of the whale's initial dive. To be sure the line ran out smoothly after a strike, it was coiled carefully in a tub on the floor of the boat near the helmsman. The line would then have run through a guide on the bow made of a pair of stout wooden pegs; this would prevent it from swinging wide and taking crewmen with it. Even then, however, the crew needed to exercise extreme care to avoid legs or arms being caught in a loop or kink of uncoiling line as it raced out. A careless or unlucky whaler could lose a limb or be yanked overboard and pulled underwater in an instant to certain death. A whale's initial run could take out line so quickly that water would have to be poured over the bow guide to prevent the friction from igniting a fire.

When the whale reached bottom and turned back to the surface, the line might slow or go slack for a moment, giving whalers a chance to grab hold. Then bracing themselves against the seat in front them, or the gunnels, they would hang on as the boat bolted forward in what Yankee whalers would later call a "Nantucket sleigh ride." When the victim's strength was finally spent, whalers used the line to pull up close and deliver a series of killing blows with their lances.

The thick coat of blubber up to a foot deep on the back of a large right whale provided assurance, although not certainty, that the whale would remain afloat once killed. If a whale was prone to sinking due to fat reserves having been depleted by nursing, a long winter's fast, or poor health, it could still be kept afloat by looping lines underneath to form a sling between a pair of chalupas positioned on either side of the carcass. A line was then lashed around the tail stock for the laborious task of towing the animal to a processing station on shore. If they had the right tools with them, the whalers might have lopped off the large pectoral fins to ease the towing chore. Depending on winds and currents and how far their chase had taken them, trips back to shore stations could take four or five whaleboats several hours. If they were lucky and had a favorable wind, a trailing chalupa with sails set might have assisted the retrieval. If it was too late in the day to make it back to shore, whales could be anchored overnight and recovered the next day. Because dead whales had to be towed ashore, the range of shore whalers was perhaps 5 to 10 miles up or down the coast from their base.

The central feature of shore processing stations was a series of two to five brick or stone ovens built on a leveled terrace perhaps 100 feet back from the water's edge. The ovens were essentially enclosed fireboxes three to four feet high with circular openings on top. Fifty-gallon copper cauldrons set into the openings were used to melt the whale's blubber into oil. Over time, facilities for "trying out," or rendering, blubber into liquid oil came to be called "tryworks." The ovens were protected from rain by an open-sided pavilion covered with red terracotta tiles; this structure provided ample ventilation for smoke from the fires and steaming cauldrons. Initial butchering took place in shallows as close to the ovens as possible. A winch or "crab" might have been installed to tow whales or whale parts partially out of the water, and a wood or stone ramp leading up to the ovens provided a path for shuttling the blubber to the ovens.[13]

The butchering process began by climbing onto the whale's back and slicing off the strips of blubber (figure 7.4). Axe-like tools that evolved over time into specialized long-handled knives were used to score the skin and cut into the soft underlying fat down to the muscle in a process called flensing. Slabs of blubber several feet long and two or more feet wide were then peeled from the carcass like rind from a mammoth orange. If the fins had not been lopped off for towing, they were removed with the tail at the shore station so the carcass could be rolled to reach blubber on the whale's underside. Stripped slabs were then cut into small blocks on shore and carted up the ramp to the ovens. There they were diced into even smaller cubes and tossed into the copper kettles. As the fat melted, oily skin and impurities sank to the bottom where they were periodically fished out and thrown into the fire in a self-sustaining process that essentially rendered the whale's oil using its own fat for fuel. The crinkled appearance of the fried skin led some to call it "fritters," while others called it "crackling" for the snapping sound it made when thrown into the fire box.

Once melted, hot oil was ladled into barrels or a trough of water next to the ovens to cool. In some cases an old chalupa filled with water was used for this purpose. Cooled oil floating on top of the water was then transferred to large barrels called *barricas* for storage and transport to market. Most barrels held about 56 to 58 gallons,[14] but the capacity of some may have ranged from 52 to 64 gallons, about the size of today's oil drum. Adult right whales could yield 40 to 45 barricas of oil. The baleen plates or "whalebone" and tongue were also removed. By some

Cetus ingens, quem incolæ Faræ insulæ ichthyophagi, tempestatibus appulsum, unco comprehensum ferreo, securibus dissecant, & partiuntur inter se.

Nautæ

accounts, the tongue was considered a delicacy and often donated to the church.[15] Finally, the meat was cut from the bones and salted, and any other internal organs considered useful were removed. The bones were left to bleach in the sun and seawater until they too were salvaged for fence posts, steps, or other uses. A large whale could take several days to process.

Whale oil, also called *traen,* or "train" oil, was a translucent brown color and sold for use in the tanning industry, the manufacture of broadcloth, as fuel for lamps, lubricant for machinery, an ingredient for soaps, cosmetics, and paint, medicinal ointments, or an alternative to other types of cooking oil.[16] Whalebone (baleen) would not become a highly prized material until the late 1500s, when its value soared. Its many uses included plumes for military helmets after being shredded and dyed, corset stays, hoops for skirts, ribs for umbrellas and parasols, springs for chair seats, riding crops and buggy whips, gun ramrods, knife handles, fishing rods, brushes, and fibers woven into other fabrics to increase their durability and resilience.[17] The meat was salted and sold in local markets or shipped off with the oil and baleen to distant markets in Spain and France.

Municipal and church records trace Basque shore whaling back to at least 1059 AD, when a royal decree was issued for marketing whale products in the French Basque town of Bayonne. By 1180 similar marketing privileges and warehouse duties were established in San Sebastian by King Sancho the Wise of Navarre.[18] Figuring prominently on the list of merchandise subject to warehouse duties was whale oil. A few years later, in 1197, King John of England, whose realm then included southern France, imposed what may have been the first tax on whales; he directed that, as an annual fee, part of the proceeds from the first two whales taken at the coastal village of Biarritz, near Bayonne, be paid to a local official named Vital de Biole and his heirs.[19] By 1261 a tithe was levied on all whales taken near Bayonne.[20]

Taxes were also levied on whales taken on the Spanish side of the border. In 1237 King Ferdinand III ordered that "in accordance with custom, the King should have a slice of each whale, along the backbone, from the head to the tail."[21] This early reference to past "custom" has been taken by some as evidence that Basque whaling had roots reaching far back into antiquity.[22] Another document from 1381 in archives from the Spanish Basque whaling town of Lequeitio, farther west, ordered that whalebone from landed whales be divided equally into three parts: two for repairing the boat harbor, and one for making fabric for the church.[23]

By the end of the fourteenth century, shore whaling had spread west through the Spanish provinces of Gipuzkoa and Vizcaya, to Galicia, north of the border with Portugal. Whaling to at least some extent had been taken up by nearly 50 Basque towns dotting the 400-mile stretch of coast along the northern shore of the Iberian Peninsula.[24] Although all of those villages may not have been active every year or to the same degree, by the late thirteenth century, several towns, including Fuenteffabia in the province of Gipuzkoa and Bermeo in the province of Vizcaya, had adopted official town seals that featured whaling (figure 7.5).[25]

Figure 7.4 (*facing page*). A whale being flensed, as illustrated in Konrad Gesner's 1560 *Icones nomenclator* (top) and in Andre Thevet's 1574 *Cosmographie universelle* (bottom). (Images courtesy of the Smithsonian Institution's Biodiversity Heritage Library, Washington, DC, and the New Bedford Whaling Museum, New Bedford, MA, respectively.)

Figure 7.5. A town crest for the Basque port of Fuenterrabia (now Hondarribia), adopted circa 1297. The crest implies that as early as the thirteenth century, whaling was of great importance to the town's economy and culture. (Image courtesy of Untzi Museoa—Museo Naval, Diputación Floral de Gipuzkoa, San Sebastián, Spain.)

This early recognition of whaling as the lifeblood of at least some communities is further evidence that shore whaling must have been flourishing by the 1200s.

Some early historians point to the number of documents referencing Basque whaling as evidence that the practice in the medieval era might have reached a peak as early as the twelfth or thirteenth centuries,[26] and that the Spanish Basques had assumed the lead in whaling as French Basque whaling efforts flagged in the thirteenth century,[27] perhaps in deference to the pursuit of cod. By the fourteenth and fifteenth centuries, the importance of whales as a major source of income is more apparent, and whaling methods had been honed to an efficient, well-run operation. Shore whaling, however, was strictly a seasonal activity limited to October through March, when adult female right whales moved south from northern feeding grounds to give birth and nurse their young in the region's nearshore waters.[28]

Offshore/Pelagic Whaling

A second strategy attributed with less certainty to the medieval Basque is offshore whaling beyond the range of shore whalers.[29] This method was essentially the same as shore whaling, except that chalupas were launched from ships, and the blubber and whalebone were stripped at sea. When lookouts posted on the stern-castle or atop a mast spotted a whale, chalupas were lowered over the side. After the whale was killed, it was made fast to the side or stern of the mother ship with a rope sling. A chalupa then sidled up to the secured whale and flensers clambered onto the floating carcass to strip the blubber and whalebone just as they would have done at shore stations. To remove the whalebone, a hook was lowered from the rigging or high poop deck on the sterncastle to lift the mouth open, allowing a man to crawl inside and hack out the baleen and tongue with an axe.[30] After they were removed, the fins and tail were lopped off so the carcass could be rolled to strip blubber from the sides and belly.

The raw blubber was minced, stowed in barrels, and returned to a shore-based tryworks for boiling. With four tons of blubber producing three tons of oil,[31] even relatively small vessels might carry the fat from one or two whales. Lopped-off fins were often hauled aboard and used as chopping blocks when dicing the blubber; this saved wear and tear on the flensing blades and deck.[32] Because the meat was perishable and of modest value compared to oil, it was left on the carcass and cast adrift for the sharks once the blubber and whalebone were removed. This still left whalers with a handsome profit. Offshore whaling was likely first practiced by small ships patrolling coastal waters just beyond the range of shore whalers, with blubber returned for processing within a day or two. Once stripped and stored in barrels under cool conditions aboard ship, raw blubber might have held up for processing for perhaps one to two weeks.[33] Assuming Basque ships were capable of sailing 100 to 150 miles a day, whaling operations conceivably could have ranged well over 1,000 miles from home processing ports; this meant that distant-water pelagic whaling could have occurred as far away as the British Isles or perhaps even Iceland.

While medieval Basque whalers might have constructed seasonal processing stations on whaling grounds in northern Europe as they would later do, there are no tangible remains of facilities built before the 1600s. Alternatively, they may have pulled into secluded bays or coves to process blubber using a portable tryworks consisting of little more than a tripod from which to hang a large caldron over a pit fire.[34] In either case, fees demanded for whales taken or processed on foreign shores, or even worse, the risk of hostilities with local residents who perceived Basque whalers as a threat, would have been compelling reasons to avoid processing oil far from home. Thus, pelagic whaling in distant waters with fat returned to home-based processing stations may have been an appealing and profitable strategy, despite the inconvenience of long voyages.

If the Basques did develop offshore and pelagic whaling methods before the 1500s, it would have been fully consistent with the far-ranging pursuit of cod by Basque fishermen, and their advanced skills in shipbuilding and navigation. Indeed, because of the scant information on so many aspects of Basque whaling, many historians have turned to a thorough examination of their shipbuilding and seafaring practices to shed light on their possible whaling capabilities.[35] On both counts, the Basques had a reputation for excellence that makes offshore and pelagic whaling by at least the late medieval period all the more plausible. They had two broad categories of vessels that could have met the needs for distant-water whaling: a class of large ships called *naos*, and a smaller class called *pinazas*, or pinnaces. Either could have carried chalupas lashed to their deck.

By the end of the fifteenth century, the Basques were building some of the finest seagoing vessels of the day. Their proficiency was based on centuries of experience and experimentation, a homeland blessed with high-quality old-growth timber husbanded specifically for that purpose, and foundries producing some of Europe's finest iron products. Several ships used by Columbus on his various voyages to the Americas were built by the Basques, possibly including the *Santa Maria*, the flagship of his first voyage in 1492. At least one (the *Victoria*) of the five vessels taken by Magellan on his 1519 landmark expedition around the world was also Basque built. As it turned out, the *Victoria* would be the only vessel to complete the expedition, making it the first ship to circumnavigate the globe.

Basque navigators and seamen were no less renowned. They helped fill out crews accompanying many of the era's most famous explorers, including both Columbus and Magellan. After Magellan's death in a battle with Philippine Natives three-quarters of the way into the expedition, it was a Spanish Basque commander named Juan Sebastian de Elcano from Guetaria who led the *Victoria* and its 18 bedraggled survivors (out of an initial company complement of 237 men) back to their starting point of Seville.[36]

By the 1500s, the largest class of Basque ships (*naos*) ranged in size from 150 to 750 *toneladas*—a Basque measure of displacement equal to about 225 to 980 metric tons. Naos were a type of galleon equipped with three masts, square-rigged sails, and three to four decks.[37] Like chalupas, they were multipurpose vessels. They could be used for fishing one year, whaling the next, hauling cargo another year, and as corsairs when requisitioned in the name of the king for military campaigns.[38] Larger vessels tended to be used for whaling at remote shore stations because of their larger capacity, an important asset when returning large quantities of oil from distant whaling grounds. Ships at the smaller end of the range and pinnaces were preferred for cod fishing. At times, however, the same ship was used for both fishing and whaling on the same voyage.

Although the Basques are widely credited with developing offshore or pelagic whaling methods by the fifteenth century if not earlier,[39] no contemporary Basque accounts have been found describing those strategies, or noting when or where they started. Most contemporary accounts postdate the early 1600s and do not come from Basques sources, but from seafarers of other nationalities. The few that predate 1600 involve reports of Basque ships off Iceland and Great Britain that were assumed to be engaged in whaling, but could have been fishing for cod. Thus, like much of the shadowy outline of medieval Basque whaling, the use of offshore and pelagic whaling methods rests on shaky ground, although whaling historians continue to ascribe its use to Basque whalers well before 1600.[40]

If some form of at-sea whaling was practiced, the lack of contemporary Basque accounts would be entirely consistent with the rarity of *any* written records describing their whaling activities. Taxation also may have played a role in the absence of documentation. During the middle ages, offshore and pelagic whaling may have been exempt from the taxes levied on whales caught by home-based shore whalers, thereby eliminating a major reason for recordkeeping.[41] Taxes were as unpopular in medieval times as they are today. King Alfonso XI of Castile acknowledged complaints from shore whalers at the port of Lequeitio in 1334 when he reduced duties from one-fifteenth to one-eighteenth of each whale for a five-year period.[42]

What little evidence there is for distant-water pelagic whaling suggests it might have begun as early as the eleventh century, if not earlier. Basque ships were reported around the Faeroe Islands, a right whale feeding ground north of the British Isles, as early as 875 AD.[43] However, no proof has been offered to confirm the ships were whaling, rather than fishing for cod or hake. The description of whaling in the British Isles by the Spanish geographer al-'Udhrī, circa 1058, also has been attributed to Basque whalers by some historians.[44]

In the early fourteenth century, Basque use of the English Channel and waters off Scotland is far clearer.[45] Indeed, trips were so frequent that the Basque port of Bermeo negotiated a treaty with King Edward III of England in 1351 to secure safe use of those waters. The pact was perhaps the first international agreement to incorporate principals of freedom of navigation on the high seas.[46] Whether the agreement was prompted solely by fishing interests or also included whaling concerns is unknown. Records from Basque ports in that period document landings and sales of cod from those waters, but none has been found that listed landings of whale oil or baleen.

Use of waters even more distant also has been suggested. In 1412 as many as 20 Basque ships, claimed by some to have been whaling, were reported off Iceland.[47] Again there is no proof the Basques were whaling, rather than cod fishing. Several historians have also suggested pelagic whaling developed in response to a growing scarcity of whales off home ports.[48] This is certainly possible, but the lure of profit even before whales became scarce could have provided all the incentive Basque entrepreneurs needed to pursue whales far beyond their local shores—if they knew where to find them.

As noted earlier, the Basques were guarded in dealings with foreigners, but routinely made arrangements among themselves to organize and carry out whaling ventures. Sightings by Basque cod fishermen working on right whale feeding grounds off the British Isles or elsewhere could have provided all the intelligence and incentive needed to begin experimenting with pelagic whaling expeditions. Initially, cod fishermen may have fitted out their boats for both fishing and whaling. Distant-water cod fishing clearly dates back to before 1000 AD, when Basques perfected a new technique for preserving their catch by splaying and salting the fish before they were dried.[49] This created a superior food product that remained edible for months. In the age before refrigeration, many Basques were quick to capitalize on this momentous advance. They developed markets for salted cod throughout Spain and France between the tenth and fourteenth centuries to supply a growing demand for "cold foods" that grew with the spread of Catholicism. Their nutritious, long-lasting qualities also made possible the extended voyages of exploration, trade, and whaling that helped lift medieval Europe out of the Dark Ages. The Basque marketing network for cod almost certainly would have provided ready-made connections for selling whale products.[50]

Considering their flare for trade, superior shipbuilding, and seamanship, it would be unwise to dismiss the possibility that the Basques exploited whales almost anywhere from northern Europe to North Africa that was within a one- to two-week sail of shore-based oil rendering stations in their home ports. Unfortunately, because of the meager written records and dim prospects for finding archeological evidence of offshore or pelagic whaling, it may never be known when, where, and how extensively the Basques engaged in distant-water whaling before the 1500s.

Pelagic Whaling with Onboard Tryworks

The third whaling strategy attributed to the Basques is a variation of pelagic whaling in which whalers not only killed and stripped whales at sea but also rendered their blubber into oil with onboard tryworks.[51] Although no convincing accounts of the practice prior to the seventeenth century have been found, by the 1670s the French Basques were definitely experimenting with the technique.[52] Some authorities have speculated that the Basques dabbled with it off northern Europe as early as the fifteenth or sixteenth centuries,[53] but this seems doubtful. At those early dates, fire aboard wooden ships was a frightful prospect, and fitting galleons with heavy brick or stone ovens without creating stability problems would have been a notable feat. Onboard tryworks did not become standard equipment for whaling vessels until the eighteenth century (figure 7.6). Given scant evidence for its use before the 1600s and the long delay in time before it became common practice, it seems likely that any flirtations the Basques may have had with the method in the sixteenth century, much less the Middle Ages, were short lived and long forgotten by the time onboard tryworks were developed and widely adopted in the mid- to late 1700s.

Figure 7.6. This 1794 print illustrates the cross-section of a Basque whaling vessel equipped with onboard tryworks to render blubber into oil at sea. Some historians credit Basque whalers with developing this approach as early as the fifteenth century, but evidence of its use before the eighteenth century is weak. (Image courtesy of New Bedford Whaling Museum, New Bedford, MA.)

Portuguese Whaling

The Basques were doubtlessly the preeminent medieval whalers based in southern Europe, but they were not alone. The Portuguese also developed an interest in whaling, apparently independently of the Basques and possibly as early as the twelfth century.[54] Whether they were hunting live whales at that early date or simply salvaging stranded animals is unclear, but a document ascribed to King Afonso II and dated March 4, 1206, lists whale meat and whale fat among the taxed commodities at the seaport of Atouguia de Baliera, some ten miles north of Lisbon. The document describes the trade as a "customary practice," suggesting it even predated the thirteenth century. The seeds of this pursuit may have been sown by northern Europeans with whaling experience, rather than the Basques. The evidence is equivocal, but shortly before the above-noted tax on whale products was established, a group of *franci* (Frank) crusaders from Normandy, Flanders, or possibly the southern parts of Great Britain was granted settlement privileges in Atouguia by King Afonso I (Afonso II's grandfather) in return for their assistance in a victorious assault on Lisbon in 1147.[55]

By 1229, the town of Ericeira, just a few miles farther north, imposed a tax of one-twentieth on each *baleia negra*, or "black whale," and also prohibited foreigners from whaling around their community.[56] The prohibition on foreigners could be a faint hint of pelagic whaling at this early date. Unfortunately, details regarding whaling methods, catch levels, and how long whales had been caught in the region are unknown, but the number of Portuguese documents referencing whales and whaling before 1800 (38 references) peaks between 1200 and 1400.[57] Skeletal remains of no less than three right whales with distinct linear cuts from flensing tools were excavated in 2001 and 2002 at Peniche, a small Portuguese fishing village adjacent to Atouguia. Carbon dating pins the age of those bones at between the mid-fifteenth and late sixteenth centuries.[58] This suggests local whaling for right whales continued into the 1500s.

Effects of Medieval Basque Whaling

Catch records for shore whaling along the Bay of Biscay are frustratingly thin. Indeed, the only totals yet to be found are from the Spanish Basque port of Lequeitio, in Gipuzkoa, some 400 to 500 years after Basque whaling is thought to have begun. Over a 145-year span from 1517 to 1662, those records report one or more landings in just 28 years for a grand total of 68 whales.[59] This equates to an average of one whale every two years over the entire 145-year period, or 2.5 whales per year for the 28 years with records. The highest one-year total is six whales in both 1536 and 1538, but there is no reason to believe the records are complete, even for years with reported whales.

From those meager records, Alex Aguilar of the University of Barcelona concluded in 1986 that shore whaling along the Basque coast probably peaked in the

1500s or 1600s. He also estimated that the combined annual catch by the nearly 50 Basque villages engaged in whaling to some degree totaled perhaps a few dozen whales in most years and as many as 100 in the best years.[60] Whether such catches alone could have brought about a population decline is unclear, but Aguilar noted that the practice of selectively targeting mother-calf pairs would have magnified the potential for detrimental effects. Experienced whalers deliberately harpooned calves first and held them at bay, knowing that maternal bonds would keep mothers close to the besieged offspring and thereby increase the chances of taking the mothers as well.[61] This was so important that harpooners and crews striking a calf first earned an extra share if the mother was also taken. Given the importance of adult females and calves for replenishing population numbers, centuries of this practice could have reduced the eastern Atlantic right whale population.

Aguilar considered the possibility that Basque shore whaling peaked earlier, but ruled it out. After sifting through 224 documents referencing shore whaling before 1800, he found no evidence for higher levels of effort before 1500 other than a peak in written references during the thirteenth century; this peak was in the relatively small French Basque area where their shore whaling is thought to have originated.[62] No further information has been found since 1986 to shed light on catch levels before 1500. Yet, given the right whale population's apparent decline during the 1500s, the possibility of higher catch levels in the 1400s or earlier cannot be ruled out.

By the 1500s, Basque whalers had been perfecting their craft for over four centuries. Investments in the construction of watchtowers and adoption of town crests featuring whaling by the thirteenth century imply that whaling was already an economic mainstay of several coastal villages well before the 1500s. There certainly was a strong economic incentive for whaling. By one analysis, a single barrel of whale oil in the 1500s would have returned the equivalent of $5,000 in 2007 dollars;[63] this would make a single large right whale worth more than $200,000 today. Notwithstanding such exorbitant riches realized from a single whale, it seems questionable that individual Basque communities would have built watchtowers and adopted town whaling crests at such an early date had catch levels not exceeded the one to two whales per year recorded at Lequeitio after 1517.

Total whale mortality also must account for animals that died after being struck and lost, something not reflected in raw catch totals. Struck-loss estimates from the 1600s in other areas ranged between about 20 and 80 percent.[64] In the medieval period, relatively crude methods and technology probably produced struck-loss rates at the high end of that scale if not higher. Considering available catch records, the number of engaged whaling villages, the economic incentives, and whales struck but lost, as many as 200 right whales might have been killed in some years by the fifteenth century, if not earlier. In any event, with no data on catch levels before 1517, it seems unwise to dismiss the possibility that shore whalers in the 1400s caught as many or more right whales than they did in the 1500s.

Offshore or pelagic whaling also could have increased the potential Basque catch, but with no information on the extent to which it was practiced in the Middle Ages, this is unknown. However, annual catch levels in the low hundreds would have been even more likely if offshore or distant-water pelagic whaling was a common practice. Although direct evidence of the practice is weak during the Middle Ages, local ordinances, such as ones adopted in the Portuguese town of Ericeira in 1229 and in Galicia in 1531, that prohibited foreigners from catching whales off their shores,[65] imply a pelagic whaling capability. They also could imply that local communities were by then already concerned about the availability of whales. For such reasons, numerous historians have speculated that medieval whalers had brought about a decline in right whale abundance by the 1500s, if not earlier.[66]

Despite uncertainty about whaling methods and the numbers of whales killed, virtually all scholars agree that the Basques were the most proficient and productive medieval whalers and were well on their way to transforming whaling into a commercial endeavor before the 1500s. When whaling by northern Europeans and the Portuguese is added, the possibility of a decline in the eastern North Atlantic right whale population off Europe during the Middle Ages seems all the more likely. Moreover, assuming that at least a few right whales crossed the Atlantic to forage off Greenland, Iceland, and Norway as they do today, medieval Basque and northern European whalers could have been the first to kill western North Atlantic right whales. Whether they did or did not, the Basques would soon expand their efforts to even more distant whaling grounds, demonstrating that their capacity to catch whales exceeded local opportunities to do so by the early 1500s. At that time, the first well-documented distant-water whaling by the Basques began in what is now Labrador, Canada.

NOTES

1. Kurlansky, M. 1999. *The Basque History of the World.* Penguin Books. New York, NY, p. 24.

2. Ibid., p. 5 (quotation).

3. Slijper, EJ. 1979. *Whales,* trans. AJ Pornerans. Cornell University Press. Ithaca, NY, US, p. 17. Proulx, J-P. 2007a. Basque whaling in Labrador: an historical overview. *In* R Grenier, M-R Bernier, and W Stevens (eds.), *The Underwater Archaeology of Red Bay: Basque Shipbuilding and Whaling in the 16th Century,* 5 vols. Parks Canada. Ottawa, Canada, Vol. 1, pp. 25–41. Scoresby, W., Jr. 1820. *An Account of the Arctic Regions, with A History and Description of the Northern Whale-Fishery.* Archibald Constable. Edinburgh, UK, Vol. 2, p. 11.

4. Proulx, Basque whaling (n. 3), p. 26.

5. Kurlansky, *The Basque History* (n. 1), pp. 49–50.

6. Sanderson, IT. 1956. *Follow the Whale.* Cassel. London, UK.

7. Harris, R, and B Loewen. 2007. A Basque whaleboat: chalupa no. 1. *In* Grenier et al., *Underwater Archaeology of Red Bay* (n. 3), Vol. 4, pp. 315–369.

8. Tuck, JA, and R Grenier. 1989. *Red Bay, Labrador: World Whaling Capitol A.D. 1550–1600.* Atlantic Archaeology. St. Johns, Newfoundland, Canada, p. 37.

9. Proulx, J-P. 2007b. Basque whaling methods, technology, and organization in the 16th

century. *In* Grenier et al., *Underwater Archaeology of Red Bay* (n. 3), Vol. 1, pp. 42–96, 55.

10. Markham, CR. 1881. On the whale-bone fishery of the Basque Provinces of Spain. Proceedings of the Zoological Society of London 63:969–976.

11. Tuck and Grenier, *Red Bay, Labrador* (n. 8), p. 37.

12. Proulx, Basque whaling methods (n. 9), p. 53; and Harris and Loewen, A Basque whaleboat (n. 7).

13. Proulx, Basque whaling methods (n. 9), pp. 70–72.

14. Ibid., p. 73. Loewen, B. 2007. Casks from the 24m wreck. *In* Grenier et al., *Underwater Archaeology of Red Bay* (n. 3), Vol. 2, pp. 5–55, 40–41.

15. Jenkins, JT. 1921. *A History of the Whale Fisheries*. H. F. & G. Witherby. London, UK, p. 61.

16. Proulx, Basque whaling methods (n. 9), p. 77.

17. Ross, WG. 1993. Commercial whaling in the North Atlantic sector. *In* JJ Burns, JJ Montague, and CJ Cowles (eds.), *The Bowhead Whale*. Special publication no. 2, Society for Marine Mammalogy. Allen Press. Lawrence, KS, US, pp. 511–561.

18. Markham, On the whale-bone fishery (n. 10), pp. 970–971. The year of this edict is often cited incorrectly as 1150 due to an early typographical error perpetuated by a succession of writers; see R Collins. (1986) 1990. *The Basques: The People of Europe*, 2nd ed. Basil Blackwell. Oxford, England, UK.

19. Jenkins, *A History* (n. 15), p. 61.

20. Perrin, WF, B Würsig, and GM Thewissen. 2009. *The Encyclopedia of Marine Mammals*, 2nd ed. Academic Press. New York, NY, p. 1232.

21. Markham, On the whale-bone fishery (n. 10), p. 971 (quotation).

22. Ibid., p. 971.

23. Ibid.

24. Aguilar, A. 1981. The black right whale, *Eubalaena glacialis* in the Cantabrian Sea. Reports of the International Whaling Commission 31:454–459.

25. Kurlansky, *The Basque History* (n. 1), p. 51.

26. Jenkins, *A History* (n. 15), p. 61.

27. Du Pasquier, T. 1986. Catch history of French right whaling mainly in the South Atlantic. *In* RL Brownell, PB Best, and JH Prescott (eds.), Right Whales: Past and Present Status, Special issue 10, Reports of the International Whaling Commission, pp. 269–274.

28. Aguilar, A. 1986. A review of old Basque whaling and its effect on right whales, *Eubalaena glacialis*, in the North Atlantic. *In* Brownell et al., *Right Whales* (n. 27), pp. 191–199.

29. Reeves, RR, TD Smith, and EA Josephson. 2007. Near-annihilation of a species: right whaling in the North Atlantic. *In* SD Kraus and RM Rolland (eds.), *The Urban Whale: North Atlantic Right Whales at the Crossroads*. Harvard University Press. Cambridge MA, US, pp. 39–74.

30. Sanderson, *Follow the Whale* (n. 6).

31. Scoresby, *An Account of the Arctic Regions* (n. 3), Vol. 1, p. 461.

32. Cumbaa, SL, MW Brown, and BN White. 2002. Zooarchaeological and molecular perspectives on Basque whaling in 16th century Labrador. Canadian Zooarchaeology 20:6–5, 9.

33. Frances Gulland (Director of Veterinary Science and Senior Scientist, The Marine Mammal Center, Sausalito, CA, US). Personal communication with author, June 2013.

34. Hacquebord, L. 1996. Whaling stations as bridgeheads for exploration of the Arctic regions in the sixteenth and seventeenth century. *In* J Everaert and J Parmentier (eds.), *Proceedings of the International Conference on Shipping, Factories and Colonization, Brussels 24–26 November 1994*. Koninklijke Academie voor Wetenschappen, Lettern en Schone Kunsten Van Belgie. Brussels, Belgium, pp. 289–297.

35. Sanderson, *Follow the Whale* (n. 6); Barkham, MM. 2007. Report on 16th century Spanish Basque shipbuilding, ca. 1550 to ca. 1600. Loewen, B. 2007. The Basque shipbuilding industry: ship types and units of measure. Both *in* Grenier et al., *Underwater Archaeology of Red Bay* (n. 3),

Vol. 5, pp. 1–12, and Vol. 3, pp. 13–24, respectively.

36. Kurlansky, *The Basque History* (n. 1), pp. 61–73.

37. Barkham, Report (n. 35), pp. 1–2.

38. Proulx, Basque whaling methods (n. 9), p. 43.

39. Proulx, J-P. 1993. *Basque Whaling in Labrador in the 16th Century*. Parks Service, Environment Canada. Ottawa, p. 13.

40. Kurlansky, *The Basque History* (n. 1), p. 58. Reeves et al., Near-annihilation of a species (n. 29), p. 46.

41. Ciriquiain-Gaiztarro, M. 1961. *Los Vascos en la pesca de la ballena*. Biblioteca Vascongada de los Amigos del Pais. San Sebastian, Spain.

42. Kurlansky, *The Basque History* (n. 1), p. 54.

43. Ibid., p. 58.

44. Lindquist, O. 1994. Whales, Dolphins, and Porpoises in the Economy and Culture of Peasant Fishermen in Norway, Orkney, Shetland, Faroe Islands, and Iceland ca. 900–1900 AD and Norse Greenland ca. 1000–1500 AD. PhD diss., University of St Andrews, Vol. 1, p. 427.

45. Bailac, JB. 1827. *Nouvelle croniqaue de la ville de Bayone*. Duhart-Fauvet. Bayonne, France, p. 57. Ruspoli, Mario. 1955. *A la recherché du cachalot*. Editions de Paris. Paris, France, p. 66. Both cited in Proulx, Basque whaling in Labrador (n. 3), pp. 26–27.

46. Kurlansky, *The Basque History* (n. 1), p. 56.

47. Ruspoli, *A la recherché* (n. 45). Cited in J.-P. Proulx, Basque whaling in Labrador (n. 3), p. 27.

48. Slijper, *Whales* (n. 3), p. 17. Proulx, Basque whaling in Labrador (n. 3), p. 30.

49. Richards, JF. 2003. Cod and the new-world fisheries. In *Unending Frontier: An Environmental History of the Early Modern World*. University of California Press. Berkeley, CA, US, pp. 547–573.

50. Kurlansky, M. 1998. *Cod: A Biography of the Fish That Changed the World*. Penguin Books. New York, NY.

51. Sanderson, *Follow the Whale* (n. 6).

52. Martens, F. 1675. *Spitzbergische oder Groenlandische Reise Beschreibung: gethan im Jahr 1671*. Hamberg. Germany. As cited in Reeves et al., Near-annihilation of a species (n. 29), p. 51.

53. Reeves et al., Near-annihilation of a species (n. 29).

54. Brito, C. 2011. Medieval and early modern whaling in Portugal. Anthrozoö 24(3):287–300.

55. Teixiera, A, R Venâncio, and C Brito. 2014. Archaeological remains accounting for the presence and exploitation of the North Atlantic right whale *Eubalaena glacialis* on the Portuguese coast (Peniche, West Iberia), 16th to 17th century. PLoS ONE:9(2):e85921, doi:10.1371/journal.pone.0085971.

56. Castro, A. 1966. A evolução económica de Portugal nos séculos XII a XV. Lisboa: Portuguália. As cited in Brito, Medieval and early modern whaling (n. 54), pp. 296–300, 296.

57. Brito, Medieval and early modern whaling (n. 54).

58. Teixeira et al., Archaeological remains (n. 55), table 3.

59. Markham, On the whale-bone fishery (n. 10). Aguilar, The black right whale (n. 24).

60. Aguilar, A review of old Basque whaling (n. 28), p. 194.

61. Ciriquiain-Gaiztarro, *Los Vascos* (n. 41), p. 30.

62. Aguilar, A review of old Basque whaling (n. 28), p. 193.

63. Cumbaa, SL. 2007. Reflections in the harbour: inferences on Basque whaling and whale processing. *In* Grenier et al., *Underwater Archaeology of Red Bay* (n. 3), Vol. 3, pp. 168–180.

64. Reeves, RR, and E Mitchell. 1986. The Long Island, New York, right whale fishery: 1650–1924. *In* Brownell et al., *Right Whales* (n. 27), pp. 201–214.

65. Aguilar, A review of old Basque whaling (n. 28), p. 194.

66. Slijper, *Whales* (n. 3), p. 17. Proulx, Basque whaling in Labrador (n. 3), p. 30.

8

BASQUE WHALING
IN TERRANOVA

IF THE BASQUES were not already whaling on a commercial scale in the 1400s or earlier, they certainly were in the 1500s. During that century their whaling abilities surely reached a zenith when they caught thousands and likely tens of thousands of whales. Whereas medieval Basque whaling is buried in a fog of obscurity, recent studies have revealed amazingly rich detail on their sixteenth-century activities. By then they were clearly hunting not only right whales but also bowhead whales on a new whaling ground they called Gran Baya (Grand Bay) in Terranova (the New Land). Today, Grand Bay is known as the Gulf of St. Lawrence in southeastern Canada (figure 8.1). Most of the Basque whaling outposts were nestled into coves lining the mainland side of the Strait of Belle Isle, an arm of the gulf jutting northeast between Newfoundland and Labrador. This was an ideal place to whale. Its narrow passage, 20 to 30 miles wide and 45 miles long, formed a natural funnel forcing whales close to shore as they moved between the Gulf of St. Lawrence and the open Atlantic southwest of Greenland.

The era of Basque Terranova whaling coincided with the heart of the Little Ice Age, a cold phase lasting from 1300 to 1850. With winters distinctly colder than they are today, the strait's coves and bays, as well as vast stretches of open water in both the Strait and Gulf of St. Lawrence, were locked in ice from December through May and even into early June in some years. The icy conditions made Terranova whaling a seasonal activity limited to open-water months from late spring to early winter.

The Basques' Terranova era was a pivotal chapter in whaling history. It was the first time a major new whaling ground had been discovered, and thereby instigated what would become a global search for productive new whaling areas. Details of the era have come to light only in the last 40 years, thanks to a large contingent of dedicated scientists who scoured dusty archives in Basque ports

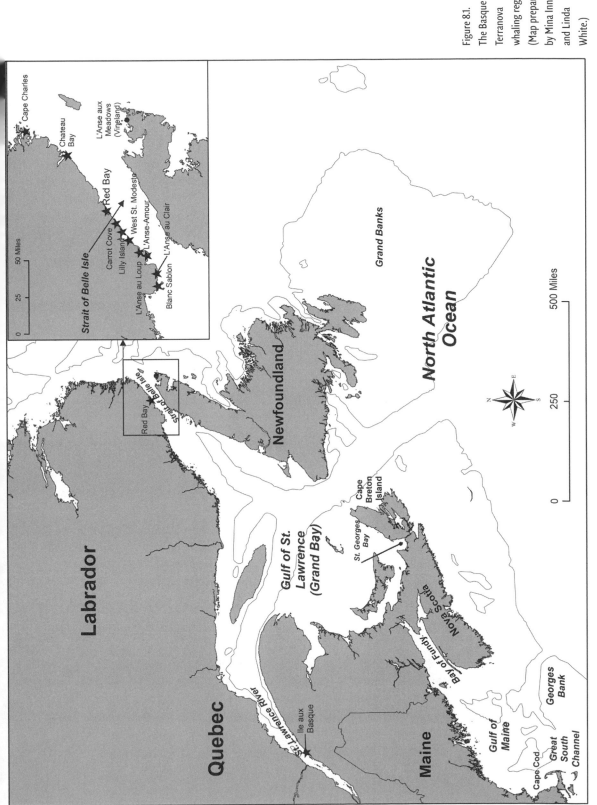

Figure 8.1.
The Basque
Terranova
whaling region.
(Map prepared
by Mina Innes
and Linda
White.)

in Spain and France, and gingerly pried archeological relics from the shores and frigid waters of Canada. One investigator in particular merits special recognition—a geographic historian named Selma Huxley Barkham. Great grandniece of Thomas Henry Huxley, who was a leading proponent of Darwin's theory of evolution in the years following its publication in 1859, Barkham had a keen sense for science and a nose for history.

Spending countless hours examining thousands of civic and church records squirreled away in various archives in Spain, she uncovered old wills, maps, contracts for building and provisioning whaling ships, and testimony from old legal disputes that revealed long-forgotten accounts of sixteenth-century Terranova whaling.[1] The records pointed to a Basque whaling enterprise in eastern Canada. When she followed those clues to the Strait of Belle Isle to look for physical evidence, her findings led to a cascade of extraordinary archeological discoveries illuminating one of the first successful commercial ventures by Europeans in the New World. The discoveries included sixteenth-century whaling ships,[2] stone cookeries for rendering whale blubber into oil, graves of Basque whalers, and a cornucopia of whaling artifacts.

Although barely recognizable after the ravages of centuries of wind and weather, many whaling stations were found virtually untouched since they were abandoned nearly 400 years ago. Barkham's first clue was the discovery of hundreds of shards of red terracotta fragments lying openly on the ground. From inventories of provisions requisitioned for Basque whaling ships she had examined in Spain, she immediately recognized them as fragments of roofing tiles ordered for the purpose of covering the pavilions whalers built to shelter their ovens (figure 8.2).[3] The rest, as they say, is history.

Decades of work would be required to completely excavate all of the whaling stations and wrecks found so far, but much has already been revealed from what was likely the hub of Basque whaling activity. This was a cove that the French Basques christened Havre de Buttes, or "port surrounded by flat-topped hills," and the Spanish Basques called Buytes, Buitres, and Butus for the same reason.[4] Today the cove is known as Red Bay, home to a small, bucolic fishing village of some 300 hearty souls. A multivolume summary of the first 30 years of work was published in 2007 by Parks Canada, which sponsored most of the research.[5] It weaves together findings from both archival and archeological studies, including the excavation of a whaling ship thought to be the *San Juan*, which foundered in Red Bay in 1565 at the height of the Basques' activities.

Although the Basques were clearly enterprising opportunists, they were neither colonists bent on creating an empire nor obsessed seekers of gold and silver. This set them apart from most of contemporary Spanish, English, Dutch, French, and Portuguese voyagers who traveled to the New World. True to their cultural roots, Basque interest in the New World centered on the familiar fishing and whaling pursuits they knew best. Although earning them a handsome profit in whale oil, whalebone, and cod, their early presence in North America has largely been

Figure 8.2. Basque whalers stripping blubber from a whale carcass at a shore processing station in Red Bay, Canada, in the mid-1500s. The shelter closest to the water illustrates a covered pavilion where ovens rendered blubber into oil; the shelter farther up the hillside is the cooperage, where barrels were fabricated. (Painting by Robert Schlicht, courtesy of National Geographic Creative; appeared in *National Geographic Magazine*, July 1985.)

ignored by historians focused on the more flamboyant drama that unfolded over the same era to the south. There, early explorers mounted conquests of Native civilizations, sent back galleons laden with gold and plunder, and skirmished with other Europeans scrambling to stake their claims to the new lands.

In their own quiet way, however, the Basques set a precedent that later whalers would repeat many times over: following close on the heels of famous explorers, whalers would exploit and exhaust new whaling grounds, and in the process fill in a more complete picture of the nature and geography of new lands. The famous Renaissance explorers, which school children are tasked with memorizing, rightly deserve great credit for advancing broad outlines of world geography; yet it was often anonymous whalers and fishermen who made first contact with Native inhabitants and filled in gaps on early maps with the coves, capes, rivers, and headlands we recognize today.

Grand Bay Whaling

How and when the Basques came to know of Grand Bay has been a topic of considerable debate. Some early historians speculated that Basque cod fishermen reached the Canadian Maritime Provinces even before Columbus left on his first voyage of discovery.[6] A line of support for this now discredited hypothesis cites the seemingly endless supply of salted cod Basque vendors sold throughout much of medieval Europe. Because Basques rarely revealed the locations of their fishing grounds and were not often reported in areas used by other European fishermen, early historians argued that they must have discovered the rich cod stocks off Newfoundland in the 1400s or even earlier.

Another theory held that Basque whalers followed a trail of whales to the new world. This claim drew a sharp rebuke from Selma Barkham: "Contrary to the spurious claims of writers on the history of whaling who have based their findings on secondary evidence, the Basques never, at any point, chased whales further and further out into the Atlantic until they collided with North America. This ridiculous legend must be laid to rest at once and for all."[7] It has.

Still others suggest the Vikings somehow passed along knowledge of the region's geography gained from their settlements on Greenland and even farther west on the shores of North America, when they swept south along the coast of Europe between 800 and 1000 AD. As it happens, remains of the Norse colony of Vinland, whose brief existence circa 1000 AD marked the first European settlement in North America, were discovered in 1960 at L'Anse aux Meadows on Newfoundland's northern tip, just 35 miles across the Strait of Belle Isle from Red Bay. Such are the coincidences that perpetuate conspiracy theories. However, the overwhelming weight of evidence makes the possibility of any connection between the Basque discovery of Terranova and Norse exploration vanishingly small. Today, no reputable historians subscribe to those theories.

A more plausible explanation of how the Basques came to know of Terranova whaling and fishing grounds is that an indiscrete Breton cod fisherman passed along information about a phenomenally rich cod fishing ground found in the far western seas at a place now called the Grand Banks off Newfoundland. The Bretons are widely credited with this discovery, and coastal place names of Breton origin throughout the Canadian Maritimes still bear silent witness to their early presence.[8] Although it remains unclear precisely when the Bretons discovered those storied fishing grounds, it was probably in the early 1500s, a few years after John Cabot, the Italian navigator sailing under an English commission in 1497, made his voyage of discovery to Canada; or more accurately, it was his rediscovery of Canada given the Norse presence in Newfoundland 500 years earlier. In any case, the name of the Basque captain who first wet his lines for cod off Terranova and the circumstances that led him there have been lost in the mist of time.

What is known is that the Basques sent great numbers of ships to the region and outcompeted the Bretons for the region's bounty of cod. The first documented evidence of a Basque vessel in Terranova comes from archives in Bordeaux documenting a 1511 exploratory expedition by Jean de Agramonte de Lireda to "New Found Land."[9] He was soon followed by a Basque cod fisherman from Bayonne in 1517.[10] By the 1520s the French Basque were making annual trips to the Grand

Banks and surrounding fishing grounds.[11] Leaving home in early spring, they fished through early summer and returned by August or September with holds full of fish and stories of more whales than they had ever seen before. Fishermen sometimes filled their holds with cod so quickly they made two or even three voyages a year. By the 1530s they outnumbered the Bretons.

Acting on reports of whales brought back by cod fishermen, the French Basques began outfitting some vessels to catch both cod and whales. By the 1530s at the latest, many were sailing exclusively for whales.[12] News of the plentiful fish and whales soon spread to the Spanish Basques, who began sending ships of their own in the 1540s. Over the next several decades the Spanish Basques took the lead in Terranova whaling. Along with a few French Basques, they had the whaling grounds all to themselves. English mariners seeking cod and seals off Newfoundland and Nova Scotia cast an envious eye toward the Basque whaling enterprise, but they lacked the skills to catch whales and were unprepared to make any inroads in the whaling trade.

At the peak of Basque whaling, between the 1550s and 1570s, some 20 to 30 ships, mostly Spanish Basque, were making annual voyages to Grand Bay and returning with an average of 20,000 barrels of oil per year.[13] During the first half of the century, whaling captains timed their departure to arrive at the Strait of Belle Isle as soon as the ice pack broke up in late May or June. They filled their holds by midsummer and were back in Europe by August or September to deliver their oil to market. Because of the time-consuming effort required to set up whaling camps and catch and process whales, they rarely made more than one trip per year.

In the 1550s whaling vessels began returning to Europe in December and January, rather than late summer. The lengthened stay likely reflected a decline in the availability of whales in summer and a need to remain until late November or early December; then they could take advantage of a late fall and early winter influx of bowhead whales returning to the strait for the winter calving season after feeding in the high Arctic through the summer.[14] Soon, late fall through early winter became their most important season,[15] and at least some whalers bound for Terranova delayed their departures from April to early June to concentrate on the later whaling season.[16] With that change, Basque whaling ships continued returning home with full holds for at least a few more years.

By the 1570s and 1580s, however, some ships began returning half empty as whales apparently became scarce in both seasons.[17] The availability of whales continued to decline in the final decades of the century, and by 1610 the supply was almost exhausted. A few whalers continued venturing to the area in the following decades, but most shifted operations south and diversified their focus to include cod, seals, other whales, including beluga whales, and trade with the local Native communities. Between 1585 and 1635, for example, the Basques maintained a station with whaling ovens on Île aux Basques on the St. Lawrence River.[18] By the time it was established, it may already have been apparent that the days of Terranova whaling were numbered. Nevertheless, marine mammal expert

Alex Aguilar has estimated that over the entire 80-year period between 1530 and 1610, the Basques caught an average of 300 to 500 whales per year, with a total of 25,000 to 40,000.[19]

The Basque Economic Engine

For the Basques, Terranova whaling and cod fishing fueled an economic engine on a par with the US automobile or steel industries in the twentieth century.[20] In addition to providing livelihoods for whalers and fishermen, it supported a host of tradespeople, shipowners, provisioners, money lenders, and a long chain of merchants and middlemen.[21] In the mid- to late 1500s, more than a dozen Basque shipbuilding ports produced some 20 vessels per year, ranging in size from 100 to 750 toneladas.[22] Most were destined for the Terranova whaling and cod fleets. Basque shipbuilders in southern France faced shallower ports, so they tended to produce smaller vessels.

Whereas cod-fishing vessels were generally less than 150 toneladas with crews of 14 to 30 men,[23] the whaling vessels that established remote whaling stations often exceeded 300 toneladas and occasionally reached 650 to 750 toneladas. These larger vessels carried 150 men and were preferred because their greater capacity meant fewer trips to transport valuable cargo and less chance of interception by pirates and privateers. With an estimated capacity of 3.06 barricas, or barrels per tonelada,[24] a 650-tonelada vessel could carry nearly 2,000 barricas. Basque vessels had an expected lifespan of about 15 years, but at the peak of Terranova whaling, owners often sold their ships in Seville after just two or three trips, relegating them to end their days hauling cargo and booty between Spain and the West Indies.[25]

The need for distant-water whaling and fishing ships meant steady employment not only for carpenters, caulkers, and shipbuilders but also for foresters, loggers, operators of lumber mills, sailmakers, iron ore miners, and blacksmiths. Basque mines and forges produced some of Europe's highest quality iron, much of which went into the production of nails and fittings for new ships. The cost of provisioning a whaling ship in the mid-1500s was often more than 2,000 ducats,[26] which easily exceeded the cost of the ship itself. Ship provisioning provided employment for still other tradespeople. In addition to supplying basic whaling gear, such as chalupas and pinnaces, harpoons, and flensing tools, provisioners had to equip their ships with navigation instruments; cordage and sails; medicines and ointments; copper kettles for boiling oil; roofing tiles, nails, and lumber for pavilions sheltering the ovens; staves, hoops, and carpentry tools for fabricating barrels; axes for cutting firewood; heavy artillery, versos (small deck-mounted swivel guns), harquebuses, crossbows, pikes, and lances to fend off pirates and hostile Natives; lanterns, candles, and oil lamps for light; and food and drink for the crew. The chalupas and pinnaces, like barrels, were shipped in pieces and had to be assembled after arriving in Terranova. As many as 10 chalupas were stocked on large vessels.

The size of whaling crews varied with the size of ships. The 200-tonelada whaling ship *Nuestro Senora de la Guadalupe* carried 55 men, whereas the *Conçepçion*, a ship of 450 toneladas, carried a crew of 120.[27] Larders were stocked with bacon, salt pork, salt cod, biscuits, sardines, flour, cheese, vegetables (peas and broad beans), cider, wine, salt, olive oil, garlic, and spices. Despite the considerable costs of building and provisioning ships, successful voyages were so lucrative that vessel owners and outfitters had a good chance of recouping costs for both building and provisioning a ship in a single trip, and still be left with a handsome profit. With whale oil selling at 6 ducats per barrel in the 1560s, a moderately sized vessel of 250 toneladas carrying 1,000 barrels could realize 6,000 ducats in oil alone; a 650-tonelada ship with 2,000 barrels could earn twice that.[28]

To finance expeditions, the Basques had their own venture capital system.[29] If shipowners or outfitters were unable to cover all costs themselves, *armadores* (those who chartered ships and organized fishing or whaling voyages) could arrange loans from a network of wealthy investors, nobles, and friends living in both French and Spanish Basque communities. Because Terranova operations involved high risk—particularly with the loss of vessels or cargo due to storms and privateers—interest rates for loans backing Terranova whaling and fishing trips were high, typically between 25 and 35 percent. This was particularly true in the late 1500s, when the Spanish economy was wracked by inflation from debts incurred to finance a succession of military campaigns.

To earn such high rates, loans were often made *ala gruesa venture*, meaning that if a ship was lost, lenders lost both capital and interest.[30] By the 1550s, investors had the option of hedging their investments by purchasing insurance against the loss of a ship on all or part of its voyage (e.g., the outbound trip, the return, or both). At the end of a voyage, proceeds were distributed with one-third going to the crew, a quarter to the vessel owner, and the remainder to the outfitter.[31] With separate loans often arranged for different aspects of a trip, disputes over profits and loan payments routinely ended in court and dragged on for years.

The Whaling Cycle

The straight-line distance from the Bay of Biscay to Grand Bay is just over 2,000 miles. On outbound voyages, ships bucking the easterly flow of the Gulf Stream took two months to reach their whaling stations. Conditions aboard ship were Spartan to say the least. The crew was divided into three groups.[32] *Oficiales*, the highest paid members, received a full share of the crew's portion of the catch. The oficiales included not only the captain, pilot, boatswain, mate, and steward but also vital tradespeople, including carpenters, caulkers, coopers, surgeons, and gunners. The second group included normal seamen who received something less than a full share. The smallest shares went to apprentices and pages whose duties included cleaning the ship and serving the crew. They were often boys just 11 or 12 years old who were sons or nephews of oficiales and seamen.

When underway the crew ate twice a day. The morning meal, eaten standing up, might have included biscuits, sardines or cheese, a few cloves of garlic, and some cider or wine. The main meal was served by pages in late afternoon on a table set up on deck. The fare was similar to breakfast but with the addition of vegetables and either a meat or fish course. Because foodstuffs were often delivered to ships in poor condition at the outset, by the time they were consumed they were often rotten, infested with bugs, and foul tasting. By the end of a voyage, remaining scraps were palatable only when heavily dosed with salt and spices to camouflage the horrid taste and smell. On larger vessels, the captain might have had a small cabin in the sterncastle, but otherwise all crew regardless of rank slept together on straw mats in the "focsal" at the ship's bow. They slept in their work clothes, and the spaces below deck, particularly the sleeping quarters, quickly soured to a dank, foul-smelling enclosure. With water far too precious for anything but drinking, the crew neither bathed nor washed and personal hygiene was ignored. Toilets were nonexistent. When sailors needed to relieve themselves, they did so over the side or off the rear of the poop deck at the ship's stern and used tarred rope for toilet paper.

Upon arrival at Grand Bay, ships pulled into one of a dozen sheltered coves on the mainland side of the Strait of Belle Isle, along its narrowest point. The Basques preferred coves with an island at the center or entrance to provide a secure base of operations. Mainland sites, however, were also common when coves lacked islands or island anchorages were already occupied. To date, remains of over 40 shore-based cookeries have been found scattered along the strait, and at least a few more were located farther south on the northeast shore of the Gulf of St. Lawrence in the eastern panhandle of present-day Quebec.[33] Once anchored, the rigging of at least some large ships was removed to protect them from storm damage, and for the duration of the season, ships served as a floating headquarters for whaling operations.[34] They provided sleeping quarters for most crew, warehouses for storing processed oil, supply centers for dispensing equipment, and at times, flensing stations for removing blubber and whalebone from whales.

The first order of business was to prepare for the whaling season. The ship's carpenter headed ashore to build or repair the cooper's cabin, erect or refurbish the tiled roofs sheltering the ovens, and assemble the chalupas. In some cases, pinnaces were also assembled for coastwise sailing, presumably to search for whales beyond the range of shore whalers or else to catch fish to feed the crew. Crew members responsible for boiling blubber also went ashore to tend to the construction or repair of ovens and install the copper cauldrons. Other crew members either assisted with those chores or were assigned jobs, such as collecting firewood, hunting and fishing to stock the larder, unpacking and preparing whaling equipment, or repairing damage to the ship incurred during the voyage. The cooper, carpenter, and men in charge of the ovens were the only crew members who lived ashore; they slept in the cooperage.[35] With an abundance of cod and salmon, the occasional deer or wild duck, and stores of cider and wine, Terranova whalers likely ate better in the New World than most of their counterparts did at home.[36]

In summer, at the peak of whaling operations in the 1560s and 1570s, 10 or more whaling ships and over 1,000 men worked side-by-side in Red Bay alone (see plate 3). An equal or even greater number of ships dispersed to other coves, including Chateau Bay, Carrot Cove, East and West St. Modeste, Forteau, and a few stations farther west on the northeast shore of the Gulf of St. Lawrence.[37] In all, as many as 20 to 30 ships and more than 2,000 men might have been engaged in whaling operations in late summer on the Labrador coast.[38] Still other vessels engaged in cod fishing would share anchorages for parts of the summer in order to dry and process fish.[39] The archives at French ports indicate the number of fishing vessels at work was also substantial. For example, the records at La Rochelle note 65 trips to fish for cod off Terranova between 1523 and 1550, while those at Bordeaux note 100 more between 1517 and 1550.[40]

Of course, lands around whaling stations were already occupied by Native communities, but few references describe their relationships with the Basques. Missionaries along the banks of the St. Lawrence River and in Nova Scotia in the early 1600s encountered Natives speaking a language that was half Basque and half their own, indicating earlier contact and trade with Basques.[41] Further insight is provided from a 1542 inquiry into Jacques Cartier's 1535 voyage to southeastern Canada. Clemente de Odelica, who sailed with Cartier, testified that he knew of Grand Bay and understood local Natives to be on very friendly terms with Basque fishermen.[42]

Whaling methods used by the Basques in Terranova are poorly documented, but certainly included shore-whaling methods honed to perfection along their Bay of Biscay homeland in earlier centuries. Lookouts were posted on coastal headlands to spot passing whales, and at every opportunity chalupas were launched to kill and retrieve their quarry. However, pinnaces and perhaps even some larger ships almost certainly struck out on their own to patrol for whales farther offshore or elsewhere along the coast.

One difference in shore whaling at Grand Bay was that locally caught whales were often towed to moored whaling vessels for butchering, rather than taken directly to the shore station. Perhaps butchering whales off moored ships kept rotting carcasses from cluttering or stinking up ramps leading up to the ovens. Evidence of this alternative practice includes the specification of slings to secure whales to ships on supply lists for Terranova whale ships.[43] Other evidence comes from a veteran harpooner Doming de Segura, who testified in a lawsuit regarding a whale that was allegedly stolen in 1575 from the crew of a ship under the command of Captain Juan Lopez de Reçc by the crew of another ship led by Captain Nicolas de la Torre. Segura claimed that shortly after he had killed the whale and tied it up in a cove a short distance south of Red Bay to prevent its loss due to strong winds and tides, he "had found the said whale alongside the ship of the said Nicolas de la Torre who had it in the slings and it had started to be flensed and cut up in order to render it down."[44] The slings and methods used to strip whales from moored ships would have been identical to those needed for pelagic whaling.

Once secured to the ship, slabs of blubber were peeled from whales in chunks roughly a yard square. After punching holes through the skin, the blubber's buoyancy allowed whalers to string slabs together and tow them behind a chalupa to onshore ovens for boiling and barreling. A description of this process was provided by Robert Fotherby and Thomas Edge, both of whom participated in the first English whaling voyages to Spitsbergen;[45] these voyages hired Basque whalers between 1613 and the early 1620s. Fotherby and Edge's description of what happened after a whale was killed is likely also applicable to Basque whaling methods used just a few decades earlier in Terranova. According to them, once a whale was killed, the crew would "fasten a rope to his tayle, and with the Shallops, one made fast to another, ... towe him towards the ship with his tayle foremost." Their description continued:

> Then doe they lay him crosse the sterne of the ship, where he is cut up in this manner; two or three men in a Boat or Shallop come close to the side of the Whale, and hold the Boat fast there with a Boat-hooke; and another standing either in the Boat, or most commonly upon the Whale, cutteth the fat (which we call Blubber) in square pieces with a cutting Knife, three or foure foot long. Then to race it from the flesh, there is a Crane or Capsten placed purposely upon the poope of the ship, from whence there decendeth a rope with a hooke on it; this hooke is made to take hold of a piece of Blubber; and as the men winde the Capsten so the cutter with his long knife looseth the fat from the flesh, even as if the lard of a Swine were to be cut off from the lean. When a piece is in order cut off, then they lower the Crane, and let downe the Blubber to flote upon the water, and make a hole in some part of it, putting a rope thorow it; and so they proceed to cut off more, fastning ten or twentie pieces together to bee towed a shoare at one time, being made fast to the stern of a Boat or Shallop. These pieces being thus brought unto the shoare side, they are drawne by one and one upon the shoare with a high Crane, or carried up by two men on a Barrow unto a Stage, there to be cut into small pieces about a foot long, but thin.[46]

The Basques were not immune to squabbles, or worse, heated arguments between one another. As noted above, disputes often generated lawsuits over whaling rights, salvaging lost cargos, and the ownership of whales. The most severe conflicts may have been between French and Spanish Basque crews when their respective monarchs declared war with each other. In 1554, when the two nations were at war, four Spanish Basque whaling ships at the Labrador anchorage of Los Hornos were seized by a contingent of French Basque ships and taken to Red Bay.[47] Eventually, the detainees were crowded aboard a vessel and allowed to sail home.

At other times, relationships between whaling crews were not only civil, but cooperative. Testimony in another legal dispute describes joint efforts by chalupas from different whaling ships to tow dead whales to flensing stations for processing.[48] Sometimes vessel captains took cooperation a step further, forming seasonal alliances to share equipment, whaling tasks, and even whales themselves. One such pact is mentioned in a lawsuit contesting arrangements struck in 1589 by four captains working in the same cove.[49] According to the testimony, they agreed to pool all oil from whales producing 40 barrels or less by dividing it among the four ships in proportion to the size of their crews. For whales yielding more than that, however, lots were drawn to determine the order in which all oil and whalebone from a whale would be allotted to ships on a rotating basis. Once the order of allocation was set, the first large whale was awarded to the first ship on the list, the next whale went to the second ship, and so on. Given that a large right whale produced about 40 barrels, whereas a large bowhead could yield nearly twice that amount (70 barrels or more[50]), this arrangement may have reflected an agreement to split the oil from right whales caught elsewhere by pelagic whalers and returned to shore stations packed in barrels, but to award the yields from more valuable bowheads caught locally on a rotating basis.

For captains of ships with less than a full hold, the fall arrival of bowheads must have been a seductive incentive to push the seasonal weather window. Extending stays in Grand Bay into early winter, however, was a risky proposition. An early freeze that locked a ship in ice could force a crew to overwinter through bitter cold, with a questionable chance of survival. Sometimes these wagers were lost. In 1566 the *Conçepçion* escaped an early freeze only by a quick departure, leaving behind both whaling gear and several unprocessed whales.[51] In the winter of 1574–75, the *Magdeleine*, a 200-tonelada ship with 450 barrels of oil in its hold, was less fortunate. Trapped by a sudden freeze in Red Bay, the crew had to abandon their icebound ship and escape aboard another vessel.[52] After salvaging some cargo and equipment the following spring, disputes over remaining valuables ended up in a court battle.

In what must have been the greatest disaster in the history of Basque Terranova whaling, several vessels were trapped by an early freeze in Red Bay in the winter of 1576–77. A statement recorded in 1619 claimed that over 500 whalers perished in the frigid cold before relief could arrive the following spring.[53] Many historians now think that number was exaggerated. While 500 whalers may have been marooned that winter, most may have been rescued the following spring. One who was not, however, was Juan Martinez de Larrume. On his deathbed along the shores of Red Bay on June 22, 1577, he dictated a last will and testament now believed to be one of the oldest surviving documents written by a European in North America north of Florida.[54]

Yet another ship lost in Red Bay was the 250-tonelada ship *San Juan*, a three-masted galleon built in the Spanish Basque port of Pasajes and outfitted in San Sebastian. Testimony in a lawsuit over salvage rights tells us that in 1565 the *San*

Juan was moored close to shore and fully loaded with nearly 1,000 barrels of oil when it was lost late in the season. As it was preparing to leave, a strong gust of wind from a late fall gale snapped its bow mooring and drove it ashore.[55] The crew was able to salvage some cargo and gear and return home aboard another vessel, but the ship was a total loss. The vessel was almost certainly the same ship found in 1978 as a result of Selma Barkham's research. Located in shallow water just a few tens of yards off Saddle Island in the mouth of Red Bay, the ship was meticulously excavated in one of the most thorough underwater archeological studies ever undertaken up to that time. Its keel measured 50 feet, but its overall length with spars and rigging was probably twice that. Recovered relics included a nearly complete chalupa (figure 8.3) and hundreds of barrels, all yielding a wealth of information about Basque shipbuilding and whaling technology.

Remains of the casks are particularly significant to historians interested in estimating the numbers of whales killed by the quantities of oil produced. When pieced together, the casks provide a direct measure of their capacity. Some casks were small barrels called *quintals*, used to store supplies or oil in odd spaces left after larger barrels were stacked and loaded. The larger barricas varied in size by some 25 percent from about 50 to 63.4 gallons.[56] When full, each large barrel would have averaged about 510 pounds, including oil and wood. Interestingly, the range of sizes conflicts with some assertions that the capacity of barrels was rigorously controlled to avoid fraud.[57] Perhaps a more important consideration was adjusting barrel sizes for the best fit aboard ship.

The oak and beech barrel staves and alder hoops apparently came from suppliers dispersed throughout Europe, including Brittany and the Loire valley of France. Such a broad distribution of suppliers is indicative of the Basques' widespread trad-

Figure 8.3. A Basque whaling boat or "shallop" (*chalupa* in Spanish) circa 1565, recovered by archeologists in Red Bay, Labrador, Canada. (Image by George Vanervlugt, courtesy of Parks Canada.)

ing network and suggests they went to great lengths to preserve their own coastal forests for shipbuilding timber and production of charcoal for their ironworks. Barrels were recycled whenever possible. Once emptied by retailers, serviceable barrels were disassembled and returned to whalers for reuse on later voyages.

Return trips to Europe were considerably faster than outbound voyages. With the Gulf Stream now at their back, crossings back to Europe could take as little as 30 days. Most ships sailed directly home to sell their oil at central markets. For the Spanish Basques, Bilbao and San Sebastian were major off-loading ports and distribution centers. For the French Basques, some of their oil was landed in Bayonne and St. Jean de Luz, but because those ports were too shallow for large heavily laden vessels, much of their product also went to the deeper Spanish Basque ports. From there some of it was sold in local markets or shipped south to Spain and Portugal, but most was exported to northern Europe.[58] Some vessels sailed directly to northern European ports, such as Bristol, Southampton, London, Flanders, Nantes, and Le Havre, where whalers sold their oil and took on cargos of salt, grain, or other commodities not produced locally in the Basque homeland.[59]

Return voyages were fraught with danger, and not just because of unpredictable seas and early winter storms. Pirates and privateers were an ever present threat.[60] Because Basque whaling ships tended to be large and built for versatility, including military service, they were by no means defenseless. Large galleons of 600 to 700 toneladas could carry a dozen cannons, and even small galleons of 250 toneladas carried at least one or two cannons and small artillery (versos). The value of versos, also called "murderers," was apparently far less assuring than their ominous nickname suggests. Apparently the moniker was earned not from their effectiveness in battle, but because they often exploded when fired, thereby killing the gunner.[61] Details of pirate attacks are lacking, but at least some ships apparently fell victim, particularly after the 1550s when Spain was at war with England, Holland, and France, and corsairs of those countries were encouraged to interdict any Spanish ships wherever possible.[62]

End of the Terranova Whaling Era

By the 1580s, when some Terranova whaling ships began returning half empty,[63] profit margins still may have been maintained due to a doubling of the price of whale oil (between the 1560s and early 1600s) and a growing demand for whalebone, whose value increased so much it rivaled that of oil. By 1610, however, bowhead whales and right whales were all but gone from Grand Bay and adjacent areas. At the same time, whales along the Basque homeland were also fast disappearing.[64] With failing catches in both areas, a door opened for others to step in and fill oil shortfalls by hunting at other whaling grounds. By the second decade of the 1600s, the Basque whaling monopoly would be broken as other European nations developed their own whaling skills.

Declining whale stocks, however, were just one of a cascade of calamities undermining the Basques' long dominance over whaling. In the late 1500s, when Spain was at the height of its influence and power, it was ruled by a manipulative monarch and devout Catholic, Philip II. Philip had taken it upon himself to serve as Catholicism's principal defender. His arrogant, self-righteous demeanor kept Spain embroiled in political intrigue and shifting struggles over religious dominance throughout Europe. Always ready to use military force to defend his interests and beliefs, Philip kept Spain in a nearly constant state of war with one country or another and ran up staggering debts, despite the flood of bullion flowing in from the New World. By the late 1500s, Philip's military adventures escalated to the point of devastating Basque whaling and fishing fleets.[65] Their ships were routinely requisitioned for military service and denied permission to leave port in order to ensure their availability to the king. Moreover, it was increasingly difficult to find able-bodied crewmen due to the conscription of sailors for Royal service. Even when those constraints were overcome, ships had to contend with the growing scourge of privateers acting in the interests of nations at war with Spain.

In the 1580s the situation became intolerable. With few ships available for Terranova operations and so many seamen impressed for military service, merchants began to fear that experienced whalers would seek to escape their military duties by offering their services to foreign whaling interests and reveal confidential whaling methods to potential competitors. As a result, local ordinances, such as one in 1584 in the Spanish Province of Guipúzcoa, threatened imprisonment of any Basque whaler sailing on a foreign vessel.[66] Such ordinances, however, did nothing to ease the underlying problem. Fearing economic ruin, the province of Guipúzcoa wrote to Philip II in 1586, "The damage that has been done in this province by the general embargo of all ships and boats has impeded the navigation of *Terranova* and Andalúcia . . . and if the embargo is not soon lifted the whale fishery will be lost just as the cod fishery has been lost and the damages will total more than 200,000 ducats."[67]

The situation, however, only worsened. In 1588 Philip II commissioned a massive armada for the invasion of England. Whaling merchants were forced to suspend most of their expeditions. In May of that year, more than 150 vessels, including more than 100 merchant ships and 25,000 sailors departed Lisbon (then under Spanish rule) on an ill-fated campaign that culminated in disastrous defeat. The following fall, after being outmaneuvered and routed by a much smaller English fleet in the English Channel, remains of the armada attempted to flee for home against an unfavorable wind that forced them to circumnavigate the British Isles. In the process, they fell victim to a cascade of misfortunes brought on by deadly navigation mistakes and a wretched series of gale-force storms that slammed a monstrous swell into the western flanks of the Irish coast. Many ships were driven to ruin on its rocky shore. By the time remains of the armada limped back to Spain, half its ships and more than half its men had been lost. The disaster left Basque maritime interests, including the whaling industry, mired in economic difficulties that only worsened as the war with England dragged on until 1604. The unrelenting demand for ships and men for military service created a dire situation that was expressed bluntly in a 1590 plea for relief from the governing council of San Sebastian to Philip II: "Both the principle and common people are dying and being finished off and they are losing and consuming ships, estates, and patrimonies. Finally, everything and the few people left are being laid to waste and destroyed."[68]

What was a disaster for the Spanish Basques, however, was a boon for French Basques. As the Spanish Basques dropped out of the Terranova whale fishery, the French Basques filled the void and began taking over the market for whale oil. With Spanish ships and men in short supply, at least some French Basque merchants drew upon their connections with Spanish Basque moneylenders to finance the former's whaling ships. With Spanish Basques having few options to invest in their own Terranova ventures, many apparently backed French initiatives. Because French ports were too shallow to accommodate large, heavily laden whaling vessels, the whalers arranged to land their oil in the deeper Spanish Basque ports. Although the oil was to be exported once landed to avoid competition with the few remaining Spanish Basque merchants, much of it was nevertheless shipped inland to Spanish markets.

Spanish Basques managed to continue sending a few ships to Terranova after 1600, but they never regained the dominant position they had held before the 1570s. The French Basques, on the other hand, expanded their efforts in the late 1500s and maintained a substantial whaling effort into the early 1600s. Nevertheless, with whale stocks in Terranova nearly exhausted and vessels returning with far less than full loads, many fortunes made earlier in the century on both sides of the Spanish-French border were lost in the final years of the 1500s and early 1600s.[69] The disarray left many experienced whalers with no opportunities to ply their trade. In better times the Basques had successfully guarded their whaling knowledge and prevented inroads by envious entrepreneurs in other countries, but with whalers facing economic ruin, their wall of confidentiality began to crumble.

In the second decade of the 1600s, Basque outfitters with few options struck agreements with English, Dutch, and German merchants eager to plant a foothold in the lucrative whaling business. The northern Europeans had no whaling expertise of their own, but had discovered new stocks of whales in Arctic waters north of Europe. To capitalize on the discovery, they needed the know-how to catch and process whales. Perhaps naively hoping to not divulge secrets of their whaling skills while learning where these new stocks of whales could be found, some Basque outfitters agreed to supply experienced harpooners, flensers, oil cookers, and all necessary equipment to carry out whaling aboard northern European vessels. Once they did, it was only a matter of time before all the Basques' whaling secrets were revealed and their whaling monopoly was shattered.

Effects of Basque Whaling

Although much has been learned about Basque whaling in Terranova, knowledge about the species caught has been something of a moving target. In the 1980s, when Barkham and Aguilar first estimated that Basque whalers caught perhaps 25,000 to 40,000 whales between 1530 and 1610,[70] it was assumed the entire catch was made up of right whales. In addition, their estimates were derived from assumptions about yields of oil per whale and sizes of barrels that have since been shown to be incorrect. For example, Barkham's catch estimate that assumed each whale produced 50 barrels of oil was based on an unidentified source. Aguilar's estimates relied on two references from the later Spitsbergen whaling era. One from 1625 reported typical yields ran between 70 and 140 barrels per whale, with each barrel containing 57.5 gallons; this came to 3,328–6,763 gallons/whale.[71] The other was an early 1600s source that reported 80 to 100 barrels of oil per whale, with each barrel holding 37 gallons; this one came to 2,959–3,698 gallons/whale.[72] Remains of casks recovered from Red Bay, however, subsequently revealed that Basque barrels used in Terranova held an average of 57 gallons, about 20 to 50 percent more than the latter reference cited by Aguilar. In addition, both studies concluded that a fully grown bowhead could produce 70 to 90 barrels of oil, but that an adult right whale yielded only 42 to 45 barrels per whale.[73] As a result, the estimates of barrel sizes and amount of oil produced per whale that Barkham and Aguilar used to calculate the numbers of whales caught need to be revisited.

More important than these faulty estimates, however, is new information indicating that bowhead whales made up a significant part of the Basques' catch in Terranova. The first clue to the importance of bowheads came to light in the mid-1980s, when 21 whale bones were recovered during archeological excavations of a vessel that sank in Red Bay in 1565. After comparing those bones to reference collections for each species, Steven L. Cumbaa, a zooarcheologist with the Canadian Museum of Nature in Ottawa, concluded that while half of the bones were from right whales, half were from bowheads.[74] However, close similarities between skeletal features of the two species made it challenging to distinguish one species from the other based on physical similarities alone, and this prompted scientists to carry out further studies.

In 2004, scientists reexamined those same 21 bones using new genetic techniques that could distinguish the two species more conclusively. Their results were totally unexpected. Only one bone—a flipper bone—was found to come from a right whale and all the rest were from bowheads.[75] To further investigate those startling findings, scientists visited the Strait of Belle Isle to collect additional whale bones strewn about the stations. Using the same genetic techniques, they analyzed 218 bones, including the initial 21 bones excavated at Red Bay. Their findings, published in 2008, again found only one right whale bone—the same flipper bone identified in the 2004 study.[76] Except for a few bones from rorquals (blue, fin, and humpback whales), all of the rest were from bowheads. Because none of the bones was radiocarbon dated, the researchers were unable to confirm whether the bones they collected were left by whalers or were from more recent natural strandings. Nevertheless, from these results the investigators asserted with great confidence that bowhead whales, not right whales, were the primary target of the region's sixteenth- and seventeenth-century Basque whalers.

Whether the Basques had previously encountered bowheads on whaling voyages to northern European seas is unknown. By the early 1600s, when the English first caught bowheads in Arctic waters north of Europe using hired Basque whalers, the animals were called "Grand Bay whales." This name suggests that Grand Bay was the first place Basques caught bowheads in large numbers, even if it was not the first place they took them. If there was any difference in the quality, use, or value of oil from bowheads compared to right whales, it must have been negligible, because no records seem to have been kept to distinguish one kind from the other, either in terms of numbers killed or prices charged in markets.

The researchers' assertion that bowheads alone were taken by Basque whalers in the Strait of Belle Isle, however, is at odds with their current distribution. Today, other than a rare straggler, neither bowheads nor right whales occur in the strait during spring or summer. Yet the Basques recognized both a summer and a fall whaling season and made a point of arriving at the strait in late spring or early summer during the early 1500s.[77] To explain the remains of so many bowhead bones, but only one right whale bone, and the presence of Basque whalers in the early spring and summer, the researchers suggested that conditions at those seasons in the Little Ice Age must have been cold enough to keep right whales well south of the strait, while allowing bowheads to stay and feed.

Results of their studies offer compelling evidence that virtually all whales killed in the Strait of Belle Isle were bowheads. The conclusion that Basque whalers must have caught bowheads in the spring and summer, however, may be premature. Because the extent to which the Basques killed right whales is central to understanding their role in the species' collapse, this conclusion merits a closer look.

Migratory patterns of bowhead whales today are tightly bound to the advance and retreat of seasonal pack ice. It seems doubtful that such ingrained migratory patterns were abandoned in the Little Ice Age. In spring, virtually all bowheads now move north through leads and heavy ice (when there is up to 90 percent ice cover or more) as soon as the pack ice begins to break up. While colder winters in the sixteenth century may well have expanded winter sea ice cover, and thus pushed the winter range of bowheads farther south into the Gulf of St. Lawrence, their deep-rooted migratory patterns still would have had overwintering bowheads leaving the area for feeding grounds in the high Arctic before Basque whalers could safely reach their whaling stations through the melting ice. Over the past century, the only evidence of bowhead whales south of the Davis Strait in spring or summer is from a dead, stranded bowhead found in Newfoundland in April 2005[78] and a lone whale seen in Cape Cod Bay and the Gulf of Maine in both 2012 and 2013 (the same individual each year).

The possibility that bowheads lingered in the strait into the summer is also at odds with later accounts by Dutch and English whalers who hunted bowheads west of Greenland in the Davis Strait in the 1700s and early 1800s under similar Little Ice Age conditions. Very little whaling by the Dutch and English west of Greenland in those centuries took place south of the Davis Strait,[79] and virtually none occurred in the Strait of Belle Isle.[80] This discrepancy with Basque whaling was noted in 1861 by the acclaimed Danish whale expert Daniel Frederick Eschricht. After describing whaling records that indicated "Greenland right whales" (*Balaena mysticetus,* or bowhead whales) migrated from wintering grounds in the Davis Strait to high Arctic latitudes in summer, he puzzled over the species of whale that Basque whalers chased in the summer even farther south in the Gulf of St. Lawrence in the 1500s. While not ruling out the possibility that they might have been bowheads, he cautioned his readers:

> We should at least hesitate before we conclude that the whale tumbling about in the Gulf of St. Lawrence and along the shores of Newfoundland and Acadia in the midst of summer is the same species as the Greenland whale. We have seen that the latter [bowheads], in the Davis Strait, leaves winter stations even much more northerly before this season, in order to go still farther northwards; and it would be quite contrary to all that we know about its nature and manner of living if it regularly and constantly stayed in the latitudes so southerly, at a season when the temperature both of the air and the sea is much higher than that in which it is accustomed to live.[81]

Yet, if the Basques, who focused almost entirely on bowheads and right whales, made a point of arriving at the strait in late spring or early summer when neither bowheads nor right whales were present, what species were they after and where did they catch them? Investigators studying the whale bones recovered from the strait did not consider the possibility that Basque whalers used pelagic whaling methods to chase whales elsewhere in summer. As noted in the previous chapter, the Basques may have been whaling in the open sea off Europe as early as the thirteenth century. If so, they likely put this practice into use in Terranova in the sixteenth century as well. Because blubber stripped from whales at sea and packed in barrels might have kept for one to two weeks,[82] Basque whalers camped on the Strait of Belle Isle could have hunted right whales hundreds of miles away from their cookeries in summer and returned with only unprocessed blubber and whalebone. This would have left few bones at whaling stations to document their take. Interestingly, the one right whale bone that was recovered at Red Bay was a flipper bone. Pelagic whalers often severed flippers from whale carcasses and put them to use as chopping blocks while they diced whale blubber into small chunks for storage in barrels. Thus, a flipper bone might be one of the few right whale bones pelagic whalers brought back aboard ship to their whaling stations.

To capitalize on their late spring arrival in Terranova, the Basques might have continued whaling through the summer, far beyond the strait's confines, by using some of their larger ships or the small pinnaces that had been assembled once they arrived.[83] As noted earlier, slings to secure whales alongside ships for flensing were on requisition lists for Terranova whaling ventures and could have been used on pelagic whaling ships just as easily as ships moored at whaling stations. This availability of slings would have enabled Basque whalers to target right whales on summer feeding grounds in the Gulf of St. Lawrence or perhaps even as far south as the Gulf of Maine. In late fall, however, when bowheads returned to the strait from Arctic feeding grounds, whalers could have sailed back to the Strait of Belle Isle to devote full attention to this new target. Even then, however, pelagic whaling was probably necessary. It is difficult to believe that shore-whaling crews alone could have caught enough whales to fill the holds of 10 or more ships concentrated in Red Bay, given their limited range of perhaps a dozen miles from camp.

Pelagic whaling also might explain the terms of the above-noted agreements between Basque whaling captains to share oil from whales producing less than 40 barrels (right whales), while allocating oil from whales producing more (bowheads), on a rotating basis. Perhaps those agreements were struck to form pelagic whaling crews in summer months, while keeping at least some crew at their shore stations to process blubber, catch food, and maintain a secure base for their operations.

Proof that the Basques sought right whales south of the strait is faint at best. Eschricht claimed that the whalebone brought back to Europe by Basque whalers in the first 20 to 30 years of their operations in the Gulf of St. Lawrence was 8 feet long. This is roughly the maximum length of right whale baleen, but far shorter than that from bowheads, which can reach lengths of 14 feet. Eschricht therefore concluded that a separate species of "short-boned whale" was likely present in the gulf;[84] however, with no other whales having baleen of this length, it must have been the right whale.

Whaling south of the Strait of Belle Isle could also explain a few early English observations. In 1594, the English barque *Grace,* under the command of M. Rice Jones, recorded what may be the earliest firsthand account of Basque whaling in North America by a non-Basque source. Jones had hoped to hunt walruses off Newfoundland but arrived too late in the spring to catch them ashore, so he pushed on south to Nova Scotia. There at the base of a bluff on St. George Bay—the western opening of a narrow strait separating Cape Breton from the Nova Scotia mainland (see figure 8.1)—he reported finding the remains of two Basque whaling ships half buried beneath a mudslide from the cliffs above. His account notes the following discovery: "the wrackes [wrecks?] of 2 great Biskaine ships, which had bene cast away three [there?][85] yeeres before: where we had some seuen or eight hundred Whale finnes" as well empty casks from which "al their traine [whale oil] was beaten out with the weather."[86]

Farther south in the Gulf of Maine, evidence of the Basques is even more speculative. Journals from a 1602 voyage by the English explorer Bartholomew Gosnold reported the presence of a "Biscay shallop" in the possession of Native residents near Penobscot Bay. The account suggests that either Basque whalers or cod fishermen plied the coast of Maine before 1602.[87] If the shallop came from a Basque whaler, it could explain a curious subsequent account of Native whaling from a 1605 voyage under the command of Captain George Weymouth (see chapter 10 on Weymouth's 1605 trip).[88]

If Basque whaling for right whales did occur in the Gulf of St. Lawrence or farther south, a logical question is why the Basques didn't build shore stations in those areas. At least some stations, such as the one on Île aux Basques along the St. Lawrence River, did exist south of the Strait of Belle Isle, but they apparently were few in number and built after the stations in the strait. However, there are several possible explanations for this rarity. Perhaps shore stations were initially built along the strait to be close to Basque cod fishermen with the recognition that there was safety in numbers. Or perhaps there was a relatively discrete group of feeding right whales that occupied the strait early in the 1500s, but was quickly exhausted; their bones may have subsequently been buried deep in sediments and are yet to be unearthed. If subsequent pelagic whaling farther south kept the strait's cookeries busy through summer, the labor and expense of building new shore facilities may have seemed unnecessary, particularly after the discovery of early winter bowheads passing their camps.

Another explanation for the rarity of whaling stations south of the strait is the possibility that the Basques used portable tryworks to render oil from whales caught elsewhere and returned processed oil rather than blubber to their base camps. There is no evidence of this strategy in Terranova, but as will be seen in the next chapter, Basque whalers hired by English and Dutch merchants used portable tryworks of their own design just a few decades later in Spitsbergen; this suggests the strategy must have already been part of the Basque whaling repertoire in the 1500s. Yet another possibility lies in the whalers' relationships with local Natives. Basque whalers may have decided their security in Terranova was best assured by concentrating camps along the strait. Perhaps they were able to establish good relations with the local people along the Strait of Belle Isle, but those farther south were more hostile and more numerous.

Of course, Basque ships also could have been sent north in summer to catch bowheads in the high Arctic. Evidence for this is even more speculative. No records confirm the hunting of whales amid ice floes or along the ice edge in the 1500s, and this could have been a far less appealing proposition than chasing available right whales in milder climates farther south. Moreover, if they had discovered bowheads in the Davis Strait and high Arctic west of Greenland, where later whalers would successfully exploit them by the thousands in the eighteenth century, the Terranova whaling era would not have collapsed when it did.

The only hint of Arctic whaling lies in the aggressive behavior of what appear to have been the Thule living to the north. Occasionally, they apparently mounted raids on Basque whaling stations hundreds of miles south of their homeland,[89] which could suggest the Basques did something to incite their anger. Perhaps the attacks were in response to Basque whaling ships on their hunting grounds, or they may have perceived a decline in bowheads and attributed it to the Basques. In any case, for these Arctic Natives to have traveled so far south to raid Basque whaling stations is curious. In 1625, the Basque historian Lope de Isasti wrote that men "called Eskimaos . . . were inhuman, because they suddenly attack our men with their bows and arrow (with which they are very dexterous) and kill and eat them."[90] During one such raid, the local Natives tried to warn the Basques of approaching "Eskimaos." The warning went unheeded, leaving the whalers unprepared for the attack. Church records between 1575 and 1618 list the deaths of several Basque whalers by "savages," which presumably was a reference to the "Eskimaos."

How Many Whales Did They Take?

When all of the above possibilities are evaluated together, it seems plausible that Basque whalers could have used pelagic whaling methods to take significant numbers of right whales south of the Strait of Belle Isle during the summer right whale feeding season. Indeed, given the time Basque whalers spent whaling in spring and summer, it seems more likely that right whales rather than bowheads were caught and that perhaps as much as half of their overall Terranova oil production might have come from right whales hunted on feeding grounds south of the strait. If we consider that possibility along with the alternative scenario that virtually all whales caught were bowheads,[91] revised catch estimates can be calculated using new information on the size of Basque barrels and the amounts of oil produced per whale.

If bowheads made up all of the Basques' catch in Terranova between 1530 and 1610, revised estimates indicate the total catch would have been a third fewer whales than Aguilar's estimate; that is, the total would be about 14,500 to 28,100 bowheads versus 25,000 to 40,000 right whales (see also appendix table A.1). On the other hand, if equal amounts of oil were produced from right whales and bowhead whales, the total number of whales caught over the 80-year period—about 21,800 to 37,400 whales—would have been slightly lower than Aguilar's estimate, but would have been a mix of some 14,500 to 23,400 right whales and 7,300 to 14,000 bowhead whales. If only a third of the oil produced by the Basques came from right whales, the catch still could have included 9,700 to 15,600 right whales.

To estimate the total number of whales killed, it is necessary to account for whales that died after being struck but were lost before their oil could be processed. Records have not been found describing such struck-loss rates by Basque whalers in Terranova; however, information is available from 10 voyages by Basque whalers in the 1700s. Those records report that out of 294 whales struck, 70 (23 percent) were lost alive, 41 (14 percent) sank after being killed, and 182.5 (62 percent) were actually landed.[92] If half of the whales that were struck and lost subsequently escaped and died, and 14 percent sank when killed, the total number of whales killed but not landed would have been about 25 percent greater than the total successfully landed and processed. During the earlier Basque Terranova whaling era, that percentage might have been even greater.

Equipment inventories for Terranova whaling ships also may hint that struck-loss rates and associated mortality exceeded 25 percent. Lists for two Terranova whaling ships, the *Guadalupe* with a cargo capacity of 600 barrels and the *Conçepçion* with a capacity of 1,350 barrels, each included supplies of 100 harpoons (figure 8.4).[93] Considering the number of processed whales required to fill their holds (roughly between 10 and 20 whales), it seems provisioners anticipated the possibility of losing five to ten harpoons for every whale landed successfully.

If all of the whales landed by the Basques over the 80-year Terranova whaling era were bowhead whales (14,500–28,100 whales, as suggested above) and an additional 25 percent died after being struck without being recovered, the total number of bowheads killed might have ranged from 18,200 and 35,100 whales. Alternatively, if equal amounts of oil came from bowheads and right whales, the total number killed, including those struck and lost, might have ranged from 18,100 to 29,000 right whales and 13,500 to 17,400 bowhead whales. If only a third of the oil produced came from right whales, the total kill of right whales could have ranged from about 12,125 to 19,500.

Basque Whaling after 1610

Although the Basques remained a force in whaling through the 1600s, they were never again the leading producers they had been before 1610. Faced with a near complete collapse of whale stocks in the Terranova and Bay of Biscay whaling grounds, they shifted efforts to Iceland, Greenland, Norway, Jan Mayen Island, and Spitsbergen. Their presence in Iceland is still marked by the remains of an old whaling station believed to have been built in the early 1600s on the remote Westfjords Peninsula, which juts out from the island's northwestern quadrant.[94] Located at a site called Strákatangi, at the base of a small fjord named Steingrims-fjörður, are the ruins of ovens for rendering oil and a small stone building likely used as a cooperage and living quarters. Excavations in 2005 revealed ovens very similar to those built on the Strait of Belle Isle. Radiocarbon dates place the site's period of use between 1610 and 1650. If further work confirms those dates, the structure would be the oldest building erected in Iceland by people other than those of Norse decent.

Annals of the Alþing, Iceland's ruling general assembly, as well as annals from the Westfjords Peninsula, record the presence of Basque whalers as early as 1608.[95] In 1615, Jón Guðmundsson the Learned, a local resident and self-taught naturalist, wrote of several Spanish and French Basque ships moored in Steingrimsfjörður in 1613 that were said to have caught 17 whales with consent from the local sheriff. Encouraged by those results, as many as 26 ships were sent the following year.

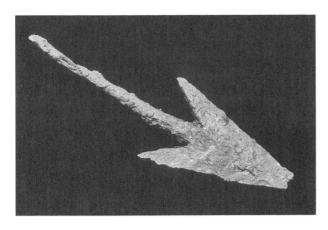

Figure 8.4. Basque whaling harpoon circa 1550 to 1600, found on Saddle Island in Red Bay, Labrador, Canada. (Image courtesy of © Government of Canada, Canadian Conservation Institute, CCI 128250-0001.)

Only 10 apparently made it to Iceland; the others were either scattered or robbed by English privateers or were forced, or chose, to move elsewhere. With the help of a local priest, relations between the Basques and local Icelanders were cordial and the Basques earned a reputation for dealing fairly in matters of trade. In 1614, however, the priest died and relations took a turn for the worse.

In the summer of 1615, 16 whaling ships arrived in Iceland, but 12 soon moved on to whaling grounds farther north.[96] Their arrival in summer suggests they were after right whales, rather than bowheads. If bowheads overwintered around Iceland, they probably would have already left for Arctic feeding grounds by summer. Given the few whaling vessels that remained in Iceland, Basque whalers may have considered opportunities to catch right whales too sparse to support a larger whaling contingent. By September, the remaining ships had taken at least 11 whales. As they were preparing to leave for home, they were caught by a sudden storm or early freeze that sank three ships and stranded 82 men.

Facing a hard winter in Iceland, the men split into two groups. One group of 32 men took their chalupas west in search of either winter quarters or a lingering Basque ship to take them home. That group split again, with one party of 14 men reportedly ransacking an abandoned trading post as they looked for supplies to get them through the winter. After settling into an unoccupied fishing station a few days later, they were attacked by an angry mob of local residents bent on revenge for damage to their trading post. All but one of the whalers was killed. At the instigation of the local sheriff, the other group of 18 whalers, who had settled in for the winter at another location and knew nothing of their comrades' activities or fate, were also set upon and killed. The remaining 50 whalers managed to survive winter at yet another remote location and escaped the following spring by commandeering an English fishing vessel.

The residents of Westfjords feared retaliation for the 1615 massacre, but none occurred. Revenge was counter to Basque instincts for avoiding confrontation and minimizing the footprints of their activities. There are no reports of whaling ships around Iceland over the following decade, but in 1626 a French Basque ship returned and caught 20 whales.[97] The following year, two English warships captured a French Basque whaler close to the Westfjords Peninsula.

The tumultuous events in Iceland were a harbinger of events that would soon embroil the Basques as well as other northern European whalers as they took up the trade in the Arctic. After being driven out of Spitsbergen by newly minted English and Dutch whalers early in the second decade of the 1600s, the Basques retreated to pelagic whaling in the Greenland Sea and Norway. They probably accounted for a modest part of the overall catch of whales in the 1600s and early 1700s, but their effect on right whales may have been considerable. With Spanish merchants still reeling from wars and other problems, most Basque whaling was carried out by the French Basques. Incomplete records from the 1600s confirm that in most years, at least 2 to 5 whaling ships were sent out by the French Basques, but in 1664, a rare year with relatively complete data, no fewer than 20 ships were dispatched. This may be a better reflection of French Basque whaling effort in the mid-1600s.[98] In the 1670s and 1680s, their fleet increased to between 35 and 40 ships before being cut back to a dozen or fewer in the last decade of the century and first decade of the next. The Spanish Basques also continued to deploy at least a few ships through the 1700s, but to what effect is unknown. By the end of the 1600s, the stature of the Basques as an important force in whaling was a fading memory.[99]

A final surge of French Basque whaling occurred after the Dutch discovered new stocks of bowhead whales in the Davis Strait, west of Greenland, in 1719. When the Basques learned of this, they sent 10 to 30 ships to whale among the hundreds of vessels from Holland and the British Isles working this new ground.[100] Bowheads dominated the catch in those Arctic waters, but the Basques also continued to roam sub-Arctic areas between Jan Mayen, Norway, and Iceland during the summer. If holds of ships returning from Arctic waters were not already full by the end of summer, at least some whalers made a point of trying to top off remaining cargo space by taking one, two, or perhaps as many as half a dozen right whales on each voyage.[101] Considering whales struck and lost, this practice might have killed 50 to 100 right whales per year throughout much of the 1600s. To the extent that whalers of other nations followed a similar routine, the catch of right whales conceivably could have been twice that amount, although the numbers still would have remained small compared to bowheads.

From the 1740s to the end of the eighteenth century, the French Basque whaling fleet dwindled to no more than a handful of ships. By the early 1800s, both bowhead and right whales were so rare in the North Atlantic and adjacent Arctic seas that almost all whalers had shifted their attention to unexploited stocks of these and other large whales in other ocean basins. At home, Basque shore whalers continued taking a few right whales in the Bay of Biscay through the 1600s. With whaling so deeply rooted in their history and culture, they were always ready to send out chalupas at every opportunity, but those occasions became few and far between. In the 1800s only four documented landings were recorded over the entire century. The last right whale taken by Basque shore whalers was in 1893 at San Sebastian.

NOTES

1. The Basque government bestows the "Lagun Onari" distinction upon Selma Huxley Barkham, researcher into Basque presence in Newfoundland and Labrador. 2014. Irekia, a website of Autonomous Government of the Basque Country, http://www.euskadi.eus/contenidos/noticia/ 2014_03_11_18235/en_18235/18235.html. Michael M. Barkham (son of Selma and Brian Barkham). Personal communication with author, April–May 2015.

2. Bernier, M-A, and R Grenier. 2007. Dating and identification of the Red Bay wrecks. *In* R Grenier, M-A. Bernier, and W Stevens (eds.), *The Underwater Archaeology of Red Bay: Basque Shipbuilding and Whaling in the 16th Century*, 5 vols. Parks Canada. Ottawa, Canada, Vol. 4, pp. 291–308.

3. Barkham, S. 1978. The Basques: filling a gap in our history between Jacques Cartier and Champlain. Canadian Geographical Journal 96(1):8–19, p. 16. Proulx, J-P. 1993. *Basque Whaling in Labrador in the 16th Century*. Parks Service, Environment Canada. Ottawa, p. 9.

4. Barkham, S, The Basques: filling a gap (n. 3), p. 13. Tuck, JA, and R Grenier. 1989. *Red Bay, Labrador: World Whaling Capital, A.D. 1550–1600*. Atlantic Archaeology. St. Johns, Newfoundland, Canada.

5. Grenier et al., *Underwater Archaeology of Red Bay* (n. 2).

6. Van Beneden, PJ. 1878. Un mot sur la péshe de la baleine et les premiéres expeditions arctiques. *In* Graells. (1889), as cited in A Aguilar. 1986. A review of old Basque whaling and its effect on the right whales (*Eubalaena glacialis*) of the North Atlantic. *In* RL Brownell, PB Best, and JH Prescott (eds.), Right Whales: Past and Present Status, Special issue 10, Reports of the International Whaling Commission, pp. 191–199. Terán, M de. 1949. La *Balaena biscayensis* y los balleneros españoles del mar Cantábrico. Est Geog. (C.S.I.C.) 37:639–668. As cited in Aguilar, A review of Basque whaling (this note). Kurlansky, M. 1999. *The Basque History of the World*. Penguin. New York, NY, pp. 56–57.

7. Barkham, SH. 1984. The Basque whaling establishments in Labrador 1536–1632—a summary. Arctic 37(4):515–519, 515 (quotation).

8. Ibid., p. 516.

9. Proulx, *Basque Whaling* (n. 3), p. 14.

10. Proulx, J-P. 2007a. Basque whaling in Labrador: an historical overview. *In* Grenier et al., *Underwater Archaeology of Red Bay* (n. 2), Vol. 1, pp. 25–41, 28.

11. Richards, JF. 2005. Cod and new world fisheries. *In* JF Richards (ed.), *The Unending Frontier: An Environmental History of the Modern World*. University of California Press. Berkeley, CA, pp. 547–573.

12. Ciriquiain, M. 1979. *Los Vascos en la pesca de la ballena*. Ed. Vascas Argitaletxea, San Sebastian. 354p. As cited in Aguilar, A review of old Basque whaling (n. 6), p. 195.

13. Barkham, S, The Basques: filling a gap (n. 3), p. 10.

14. Ross, WG. 1993. Commercial whaling in the North Atlantic sector. *In* JJ Burns, JJ Montague, and CJ Cowles (eds.), *The Bowhead Whale*. Special publication no. 2. Society for Marine Mammalogy. Allen Press. Lawrence, KS, pp. 511–561.

15. Cumbaa, SL. 2007. Reflections in the harbour: inferences on Basque whaling and whale processing. *In* Grenier et al., *Underwater Archaeology of Red Bay* (n. 2), Vol. 2, pp. 168–180, 171–172.

16. Barkham, MM. 2007. Aspects of life aboard Spanish Basque ships during the 16th century with special reference to *Terranova* whaling voyages. *In* Grenier et al., *The Underwater Archaeology of Red Bay* (n. 2), Vol. 5, pp. 45–64, 51.

17. Barkham, S, The Basques: filling a gap (n. 3), p. 10.

18. Turgeon, L. 1990. Basque–Amerindian trade in the Saint Lawrence during the sixteenth century: new documents, new perspectives. *Man in the Northeast* 40:81–87. Turgeon, L. 1998. French fishers, fur traders, and Amerindians during the 16th century. William and Mary Quarterly 55(4):585–610.

19. Aguilar, A review of old Basque whaling (n. 6), p. 196.

20. Hacquebord, L. 1996. Whaling stations as bridgeheads for exploration of the Arctic regions in the sixteenth and seventeenth century. In *Proceedings of the International Conference on Shipping, Factories and Colonization, Brussels 24–26 November 1994*. Koninkluke Academie. The Hague, Netherlands, pp. 289–297.

21. Proulx, J.-P. 2007b. Basque whaling methods, technology, and organization in the 16th century. *In* Grenier et al., *Underwater Archaeology of Red Bay* (n. 2), Vol. 1, pp. 42–96.

22. Proulx, *Basque Whaling* (n. 3), p. 26. The Basque tonelada was a measure of displacement roughly equal to 1.5 US tons.

23. Barkham, SH, The Basque whaling establishments (n. 7), pp. 517–528.

24. Proulx, Basque whaling methods (n. 21), p. 45. Ringer, J. 2007. Cargo lading and ballast. *In* Grenier et al., *Underwater Archaeology of Red Bay* (n. 2), Vol. 4, pp. 180–224, 207–213.

25. Proulx, Basque whaling methods (n. 21), p. 46.

26. Ibid., p. 52.

27. Ibid., p. 56.

28. Ibid., p. 75.

29. Barkham, MM. 1994. French Basque "New Found Land" entrepreneurs and the import of codfish oil and whale oil to northern Spain, c. 1580 to c. 1620: the case of Adam de Chibau, Burgess of St. Jean de Luz and Sieur de St. Julien. Newfoundland Studies 10(1):1–43.

30. Ibid., p. 12.

31. Proulx, Basque whaling methods (n. 21), pp. 175–177.

32. Ibid., pp. 58–64. Barkham, MM, Aspects of life (n. 16), pp. 52–53.

33. Barkham, S, The Basques: filling a gap (n. 3), p. 16.

34. Ibid., p. 12.

35. Barkham, MM, Aspects of life (n. 16).

36. Barkham, SH, The Basque whaling establishments (n. 7), p. 517.

37. Barkham, S, The Basques: filling a gap. (n. 3), pp. 10–16.

38. Ibid., p. 10.

39. Richards, Cod and new world fisheries (n. 11).

40. Proulx, Basque whaling (n. 10), p. 28.

41. Bakker, P. 1989. "The language of the coast tribes is half Basque": a Basque-American Indian pidgin in use between Europeans and Native Americans in North America, ca. 1540–1640. Anthropological Linguistics 31(3–4):117–147.

42. Proulx, Basque whaling (n. 10), p. 28.

43. Barkham, MM, Aspects of life (n. 16), p. 54.

44. Ibid.

45. Conway, WM. 1906. *No Man's Land: A History of Spitzbergen from Its Discovery in 1596 to the Beginning of the Scientific Exploration of the Country*. Cambridge University Press. Cambridge, UK, pp. 85–88. [Edge T, or R Fotherby]. (1626) 1905. The description of the severall sorts of whales, with the manner of killing them; whereto is added the description of Greenland. *In* S Purchas, *Hakluytus Posthumus, or Purchas His Pilgrimes: Contayning a History of the World in Sea Voyages and Lande Travells by Englishmen and Others*. Reprint. James Maclehose & Sons. Glasgow, UK, Vol. 13, pp. 26–34. Further research into equipment inventories mentioning slings could help clarify when pelagic whaling first began.

46. [Edge or Fotherby], The description (n. 45), p. 28 (quotation).

47. Barkham, S, The Basques: filling a gap (n. 3), p. 13. Proulx, Basque whaling methods (n. 21), p. 32.

48. Barkham, S, The Basques: filling a gap (n. 3), p. 13.

49. Proulx, Basque whaling methods (n. 21), pp. 65–66.

50. Cumbaa, Reflections in the harbour (n. 15), p. 172.

51. Proulx, Basque whaling methods (n. 21), p. 74.

52. Bernier, M-A., and R Grenier. 2007. Dating and identification of the Red Bay wrecks. *In* Grenier et al., *Underwater Archaeology of Red Bay* (n. 2), Vol. 4, pp. 291–308.

53. Proulx, Basque whaling (n. 10), p. 33.

54. Barkham, SH, The Basque whaling establishments (n. 7).

55. Tuck and Grenier, *Red Bay, Labrador* (n. 4).

56. Loewen, B. 2007. Casks from the 24 M wreck. *In* Grenier et al., *Underwater Archaeology of Red Bay* (n. 2), Vol. 2, pp. 5–46.

57. Ciriquiain, Los Vascos (n. 12).

58. Proulx, Basque whaling methods (n. 21), p. 77. Barkham, MM, Aspects of life (n. 16), pp. 74–78.

59. Barkham, SH, The Basque whaling establishments (n. 7), p. 518. Proulx, Basque whaling methods (n. 21), p. 77.

60. Barkham, SH, The Basque whaling establishments (n. 7), p. 516.

61. Sanderson, IT. 1956. *Follow the Whale*. Cassel. London. Stevens, W, D LaRoche, D Bryce, and J Ringer. 2007. Shipboard activities and vessel use. *In* Grenier et al., *Underwater Archaeology of Red Bay* (n. 2), Vol 4, pp. 158–161.

62. Barkham, SH, The Basque whaling establishments (n. 7), p. 518. Barkham, MM, French Basque (n. 29), p. 4.

63. Barkham, SH, The Basque whaling establishments (n. 7), p. 518.

64. Aguilar, A review of old Basque whaling (n. 6), p. 194.

65. Barkham, S, The Basques: filling a gap. (n. 3), pp. 17–18. Barkham, MM, French Basque (n. 29), pp. 2–5.

66. Proulx, Basque whaling (n. 10), p. 34.

67. AGM, CVP, Vol. 1, doc. 35, translation by M. Barkham. As cited in Barkham, MM, French Basque (n. 29), p. 3 (quotation).

68. AGM, CVP, Vol. 7, doc. 3, No. 3, f 22 v. f. 24, translation by M. Barkham. As cited in Barkham, MM, French Basque (n. 29), p. 4 (quotation).

69. Barkham, MM, French Basque, (n. 29).

70. Aguilar, A review of old Basque whaling (n. 6), p. 196. Barkham, SH, The Basque whaling establishments (n. 7), p. 518.

71. Martinez de Isasti, Lope de. 1625. Copendio Historical de Guipuzcoa Facs. La Gran Enciclopedia Vasca. 1972. As cited in Aguilar, A review of old Basque whaling (n. 6), p. 195.

72. Allen, JA. 1908. The North Atlantic right whale and its near allies. Bulletin of the American Museum of Natural History 24:277–329.

73. Cumbaa, Reflections in the harbour (n. 15), p. 172.

74. Cumbaa, SL. 1986. Archaeological evidence of the 16th century Basque right whale fishery in Labrador. *In* Brownell et al., *Right Whales* (n. 6), pp. 187–190.

75. Rastogi, T, MW Brown, BA McLeod, TR Frasier, R Grenier, SL Cumbaa, J Nadarajah, and BN White. 2004. Genetic analyses of 16th-century whale bones prompts a revision of the impact of Basque whaling on right and bowhead whales in the western North Atlantic. Canadian Journal of Zoology 82:1647–1654.

76. McLeod, BA, MW Brown, MJ Moore, W Stevens, SH Barkham, M Barkham, and BN White. 2008. Bowhead whales, and not right whales, were the primary target of 16th-to-17th-century Basque whalers in the western North Atlantic. Arctic 61(1):61–75.

77. Cumbaa, Reflections in the harbour (n. 15).

78. Ledwell, W, S Benjamins, J Lawson, and J Huntington. 2007. The most southerly record of a stranded bowhead whale, *Balaena mysticetus*, from the western North Atlantic Ocean. Arctic 60(1):17–25.

79. Eschricht, DF, and J Reinhardt. 1866. On the Greenland right whale (*Balaena mysticetus*,

Linn.). *In* Eschricht, Reinhardt, and W. Lilljeborg. *Recent Memoirs of the Cetacea.* WH Flower (ed. and trans.), Published for the Ray Society by Robert Hardwicke. London, England, pp. 1–150, 17–18.

80. Douglas, W. 1760. *A Summary, Historical, and Political, of the First Planting, Progressive Improvements, and Present State of the British Settlements in North America.* J. S. Dodsley. London, Vol. 1, p. 298. Ross, WG. 1993. Commercial whaling in the North Atlantic sector. *In* Burns et al., *The Bowhead Whale* (n. 14), pp. 536–537.

81. Eschricht and Reinhardt, On the Greenland right whale (n. 79), pp. 17–18 (quotation).

82. Frances Gulland (Director of Veterinary Science and Senior Scientist, The Marine Mammal Center, Sausalito, CA, US). Personal communication with author, June 2012.

83. Proulx, *Basque Whaling* (n. 3), p. 33.

84. Eschricht and Reinhardt, On the Greenland right whale (n. 79), pp. 20–21.

85. There is no explanation in the passage about how Jones could have known the two wrecks were three years old. Therefore, the word "three" may simply be a typographical error for the word "there."

86. Haklyut, R. (ca. 1600) 1889. E Goldsmid, ed. *America.* Vol 13 of *The Principle Navigations, Voyages, Traffiques, and Discoveries of the English Nation,* 16 vols. E & G Goldberg. Edinburgh, UK, Part 2, pp. 60–61 (quotation).

87. Archer, G. 1626. The relation of Captaine Gosnols voyage to the north part of Virginia, begun the sixe and twentieth of March, anno 42, Elizabethae Reginae 1602, and delivered by Gabriel Archer, a gentleman in the said voyage. *In* Purchas, *Hakluytus Posthumus* (n. 45), Vol. 18, pp. 302–313, 304.

88. Rosier, J. (1605) 1860. A true relation of the most prosperous voyage made this present year, 1605, by Captain George Waymouth, in the discovery of the land of Virginia. *In Rosier's Narrative of Waymouth's Voyage to the Coast of Maine, in 1605: Complete. With Remarks by George Prince, Showing the River Explored to Have Been the George River.* Eastern Times Press. Bath, Maine, US, pp. 15–39.

89. Barkham, SH, The Basque whaling establishments (n. 7). Proulx, Basque whaling (n. 10), p. 34.

90. Barkham, SH, The Basque whaling establishments (n. 7), p. 518 (quotation).

91. McLeod et al., Bowhead whales (n. 76), p. 71.

92. Du Pasquier, T. 1986. Catch history of French right whaling mainly in the south Atlantic. *In* Brownell et al., *Right Whales* (n. 6), pp. 269–274.

93. Proulx, Basque whaling methods (n. 21), p. 54.

94. Edvardsson, R., and M. Rafnsson. 2006. *Basque Whaling around Iceland: Archeological Investigation in Strákatangi, Steingrimsfjörður.* The Natural History Institute of Vestfirðir. Bolungarvíc, Iceland.

95. Ibid.

96. Ibid.

97. Ibid.

98. Du Pasquier, Catch history (n. 92), p. 269.

99. Aguilar, A review of old Basque whaling (n. 6), pp. 196–198.

100. Du Pasquier, Catch history (n. 92), p. 270.

101. Ibid.

Plate 1. Fossil archaeocetes or "early" whales reveal intermediate forms through which small terrestrial mammals evolved into today's whales and porpoises. Shown here from oldest (*bottom*) to youngest (*top*) are *Indohyus*, *Pakicetus*, *Ambulocetus*, *Kutchicetus* (a reminingtonocetid), and *Basilosourus*, as discussed in chapter 4. (Painting by Jacqueline Dillard, in J.G.M. "Hans" Thewissen, 2014, *The Walking Whales: From Land to Water in Eight Million Years*. University of California Press.)

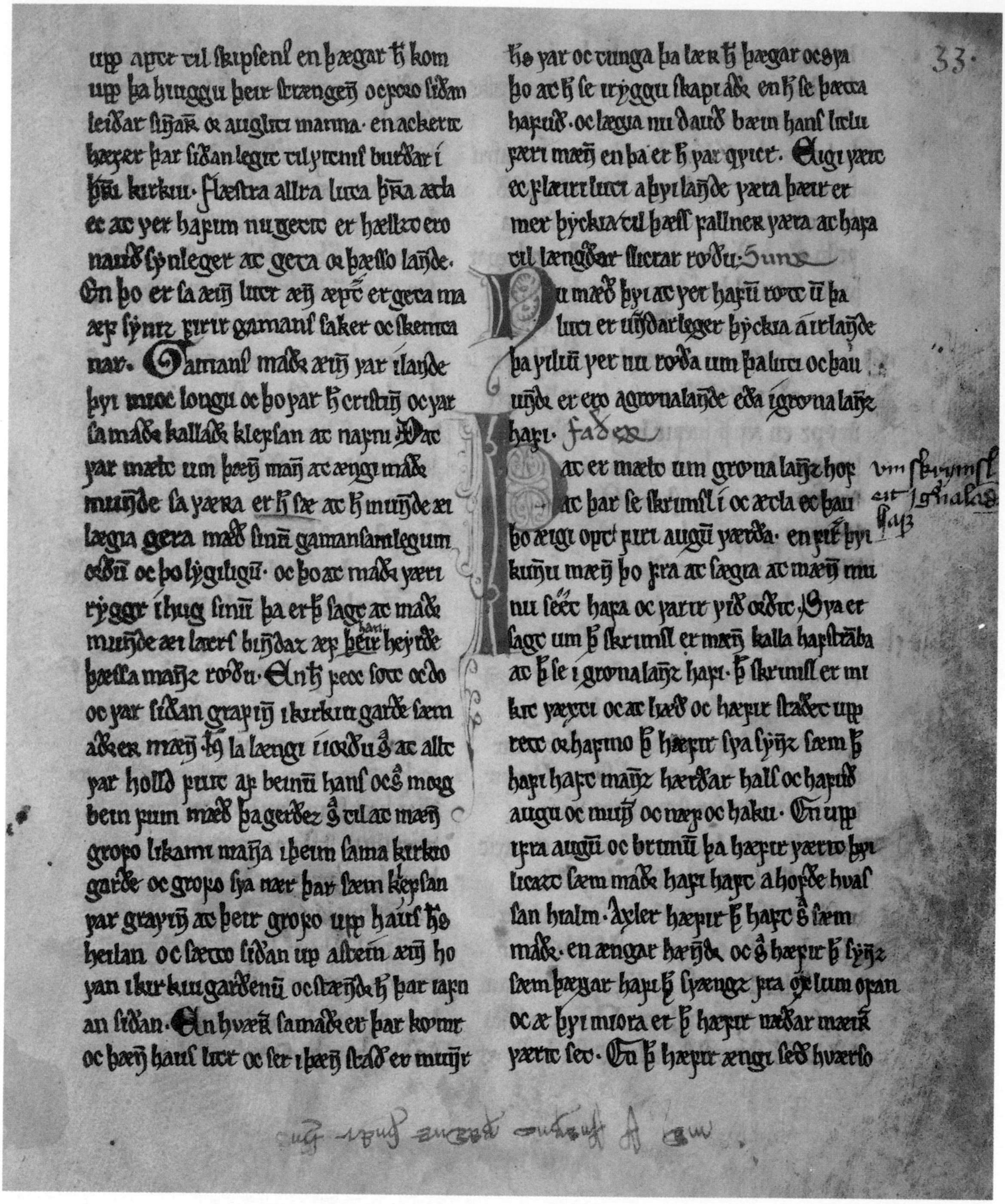

Plate 2. A page from the (circa) 1275 edition of the *Konungs Skuggsjá* (The King's Mirror), originally written around the year 1250. The text, written in Old Norse, is a colloquy between King Häkon Häkonsson and his son. On this page the son ("Sonr," in red) asks about the marvels in Greenland and the father ("Faðer," also in red) replies, "It is told that the waters about Greenland have monsters in it, though I do not believe they have been seen very frequently." (Image courtesy of the Arnamagnæan Collection, AM 243 b a fol. (Speculum regale), p. 33. Photograph: Suzanne Reitz. Used with permission of the Arnamagnæan Institute, Copenhagen, Denmark.)

Plate 3. Basque whaling galleons riding at anchor in Red Bay, Labrador, during the peak of the Terranova whaling era in the mid-1500s. Smoke rises from cauldrons rendering blubber into oil at shore stations. (Painting by Robert Schlicht, courtesy of National Geographic Creative.)

a

b

c

d

e

Plate 4. Basque whalers were hired by the Moscovy Company of London to take part in the first English whaling expeditions. These illustrations from Robert Fotherby's 1613 journal of a voyage to Spitsbergen show Basque whalers (*a*) harpooning a whale, (*b*) towing it to a flensing station, (*c*) stripping its blubber after the whale was secured to the stern of a ship, (*d*) cutting up blubber slabs at a shore station, and (*e*) rendering blubber into oil in caldrons erected ashore. (© American Antiquarian Society.)

Plate 5. Sketch of whaling in the margins of a 1526–1586 Basque baptismal and marriage registry from the port of Zumarraga in northern Spain. The drawing is undated, but its ink is similar to that used in the registry, which is now preserved at the Diocesan Archive of San Sebastian, Gipuzkoa, Spain. (Image courtesy of Untzi Museoa-Museo Naval, Diputación Floral de Gipuzkoa, San Sebastián, Spain.)

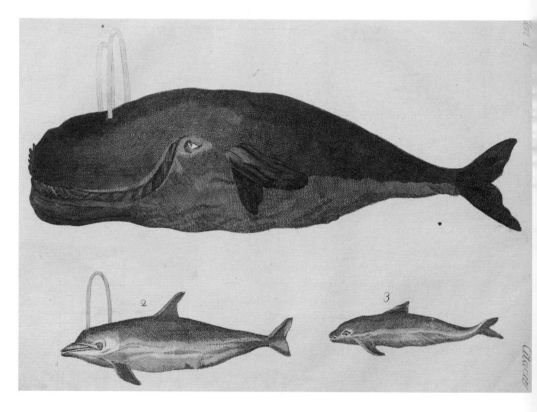

Plate 6. Copper engraving (1794) of a right whale with its mouth line mistakenly inverted. Because early illustrators rarely saw the large whales they were asked to depict, inaccurate renditions were typical before 1800. (Image courtesy of the New Bedford Whaling Museum, New Bedford, MA.)

Plate 7 (*facing page*). Rare 1853 whaling chart by Matthew Fontaine Maury entitled, "A Chart Showing the Favorite Resort of the Sperm and Right Whale." Based on early nineteenth-century whaling log books (*blue* represents areas of right whale catches; *yellow* shows areas of sperm whale catches). The records of right whale catches in the mid-Atlantic have since been found to be incorrect. (Image courtesy of Patrick J. Slattery.)

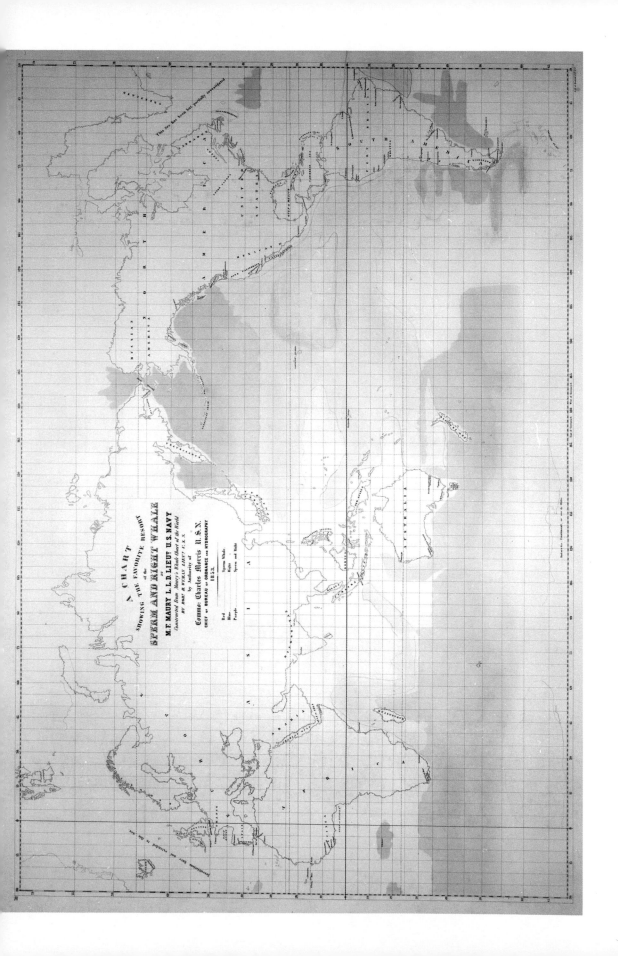

Plate 8. Shore whalers cutting up a whale landed on Long Island, New York, in the 1800s, as illustrated by W. P. Bodfish in the February 10, 1885, edition of *Harper's Young People Weekly*. (Image courtesy of the New Bedford Whaling Museum, New Bedford, MA.)

THE S.S CONTE DI SAVOIA AND THE WHALE

Plate 9. A colorized cigarette card (a trading card used to stiffen cigarette packages) issued in 1934 by the Carreras Tobacco Company of London. The card was based on a 1933 Ripley's Believe it or Not© cartoon claiming the whale was carried for more than two hours on the bow of the SS *Conte di Savoia*; this was an Italian passenger liner christened in 1931 that nearly captured the Blue Riband Prize for the fastest transatlantic crossing before World War II. (Image courtesy of the Picture Collection, The New York Public Library, Astor, Lenox and Tilden Foundations.)

9
THE DAWN OF INTERNATIONAL WHALING

WHEN OTHER NATIONS finally broke the Basque whaling monopoly in the early 1600s, they did so in what must be one of the most unlikely places on earth—a remote archipelago in the high Arctic originally called Spitsbergen, a Dutch name meaning "pointed mountains" (figure 9.1). Today the archipelago is part of Norway and known as Svalbard, meaning "cold coast." Its largest island, however, still carries the original Dutch name of Spitsbergen. There is a good chance that no human eyes, including those of Arctic Natives, had ever set sight on this frigid, desolate land more than 15 years before the first whale was taken off its shores in 1611. Perched 500 miles north of Norway and just 750 miles from the North Pole, it lies closer to the pole than just about any other land above the Arctic Circle.

Today Spitsbergen is a frigid, lonely outpost occupied year-round by a few hundred adventurous researchers studying climate change and the biological adaptions that make life possible at the extreme limits of tolerance to cold. Perhaps its greatest claim to fame is an underground cold storage vault serving as a Noah's Ark to preserve seeds of endangered plants disappearing around the world. In summer, cruise ships offer ecotours to a growing clientele of hardy souls looking for a taste of the planet's most exotic, out-of-the-way places. In the early 1600s, however, it was the center of a struggle to wrest control over the lucrative trade in whale oil. Each spring as winter ice finally relinquished its grip on the bays and fjords cutting into Spitsbergen's western flank, productive foraging grounds opened and attracted hungry bowhead whales. During the first half of the 1600s they also became the base of operations for Spitsbergen's "bay whaling" period.

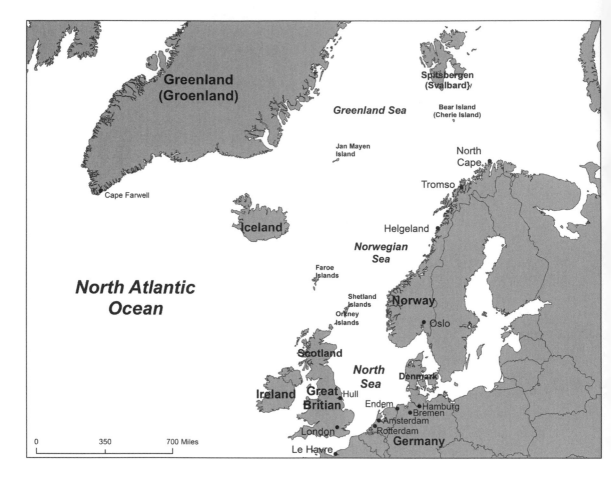

The map shows labels including: Greenland (Groenland), Spitsbergen (Svalbard), Greenland Sea, Bear Island (Cherie Island), Jan Mayen Island, North Cape, Tromso, Cape Farwell, Iceland, Helgeland, Norwegian Sea, North Atlantic Ocean, Faroe Islands, Shetland Islands, Orkney Islands, Norway, Oslo, Scotland, North Sea, Denmark, Ireland, Great Britian, Hull, Endem, Hamburg, Bremen, Amsterdam, London, Rotterdam, Germany, Le Havre. Scale bar: 0 — 350 — 700 Miles.

Figure 9.1. The Arctic and subarctic whaling region used by European whalers in the 1600s. (Map courtesy of Linda White and Mina Innes.)

Bay whaling in Spitsbergen lasted barely 50 years before evolving into a broader regional effort along the open ocean ice pack that continued into the late 1800s. This whaling era is usually told from the perspective of one species—bowhead whales. Given the tens of thousands of bowheads killed along Spitsbergen's coast and the nearby polar pack ice over this 250-year period, the focus on this species is hardly surprising. Yet competition was so fierce among Arctic whalers in the 1600s and early 1700s that some were forced south, into sub-Arctic latitudes where right whales were more likely targets. Moreover, all whalers traveling to and from the high Arctic passed through sub-Arctic waters, and most were always on the look-out for the odd right whale to fill any empty cargo space. Although whaling in sub-Arctic waters was a minor part of regional hunting, it could have delivered a devastating blow to an already depleted right whale population.

The seeds of Spitsbergen whaling were sown in the elusive quest for a new trade route to the Far East in the early seventeenth century. Instead of opening a door to China, however, this search laid the foundation for an international whaling campaign bounded roughly between Greenland, Iceland, Norway, and Spitsbergen. It began on Spitsbergen's western shore but expanded quickly to Jan Mayen Island between Greenland and northern Norway, and to a lesser extent to

Iceland, northwestern Norway, and possibly Bear Island, an isolated rocky peak rising halfway between Norway and Spitsbergen. Unforgiving winter temperatures at those latitudes restricted whaling to nonwinter months, principally May through September.

These Arctic hunting grounds, however, were more than just a new whaling arena. They were also the proving ground for a new generation of whalers and merchants never before involved in the pursuit. Their eagerness to stake a claim in the profitable business made for tense, and sometimes violent, confrontations that added yet another layer of risk to what was already a risky business, especially when conducted in such a forbidding environment. Yet, if the number of whales killed is a measure of success, the opening decades of the Spitsbergen whaling era were a triumph. By one estimate, 15,000 whales were taken by a combination of English, Dutch, Danish, and German whalers operating chiefly from bases on Spitsbergen and Jan Mayen between 1610 and 1669; another 86,644 were taken between 1670 and 1800.[1] Most were bowhead whales.

In the low Arctic to the south, however, right whales made up at least part of the mid- to late summer catch in waters around Iceland, Norway, and southern Greenland, and perhaps even Jan Mayen and Bear Islands. Warm ocean currents splitting off the northern arch of the Gulf Stream flowed north along the eastern flank of Greenland, past Jan Mayen Island to the western coast of Spitsbergen, where the last of their warmth opened a tongue of gray-green water; this allowed right whales to forage off Iceland, the northwest coast of Norway, and perhaps Jan Mayen Island during the height of summer.

There is a faint possibility that an occasional right whale even reached Spitsbergen.[2] In 1680, a surgeon named Johann Dietz served aboard the Dutch whaling ship *Hoffnung* as the crew worked off Spitsbergen.[3] Dietz wrote in his journal that the crew "killed nine whales and a Nordcaper, which was a remarkable success." He added: "It meant also a great deal of money for the ship for one whale may richly repay all the expenses, and still leave something over."[4] Separate references in the same sentence to "whales," which in the context of Spitsbergen meant bowheads, and a Nordkaper, a name now linked with right whales in recognition of their occurrence off Norway's northernmost cape, suggests that both species were taken on the same trip. Of course the Nordkaper may have been picked up on the outbound or return voyage as *Hoffnung* passed North Cape. Such a practice was common for Spitsbergen whalers returning in the 1600s with space in their hold for a few more barrels.[5] Indeed, at least a few whalers spent most of their season along the Norwegian coast. By 1614, if not earlier, the Basque plied the fjords off northwest Norway, and were soon followed by Dutch, Flemish, German, and Norwegian whalers.[6]

In the early 1600s, Jan Mayen Island was also a new whaling destination, no less important than Spitsbergen. With a narrow, 30-mile spit of land anchored at its northeast end by a tall shield volcano known as the Devil's Thumb (figure 9.2), Jan Mayen rises alone in the middle of the Greenland Sea just below the Arctic

Figure 9.2. An early eighteenth-century print by an anonymous artist of Dutch whalers off the northern tip of Jan Mayan Island and near its shield volcano, known as the Devil's Thumb. (Image courtesy of the New Bedford Whaling Museum, Bedford, MA.)

Circle, 300 miles north of Iceland, 300 miles east of Greenland, and 600 miles west of northern Norway. Under the influence of the Little Ice Age, waters around Jan Mayen were well suited as winter calving habitat for bowhead whales. Even during summer, some bowheads apparently remained around the island to feed. However, given that right whales were then being caught in summer at the same latitude 600 miles to the east off Norway and perhaps even at Bear Island,[7] both species may have occurred and been hunted around Jan Mayen during the same season in the early 1600s.

Spitsbergen

The start of bay whaling around Spitsbergen began innocently enough in 1596 when a pair of Dutch ships under the command of Jan Corneliszoon Rijp and Jacob van Heemskercke Hendrickszoon discovered the archipelago while searching for a route to China through the high Arctic. Although it was Rijp who insisted on plotting the northerly course from North Cape, Norway, which led their ships to the archipelago, it was the pilot on van Heemskerck's vessel, Willem Barentsz, who recorded the discovery in his journal and is today credited with the accomplishment.[8] When the ships returned without having found a passage to the East, the newly discovered land inspired no interest from the voyage's Dutch sponsors. This was not the case a decade later when another explorer seeking a passage east "discovered" the same archipelago, and perhaps also Jan Mayen Island.

That adventurer was Henry Hudson. He had been hired in 1607 by the Muscovy Company, a London cartel of merchants formed in 1556 to carry on trade with Russia along a sea route hugging the coast of northern Europe. Tasked with finding a new trade route to China, Hudson sought to test a theory that the Arctic's 24-hour-a-day summer sun might melt the polar ice cap and open a pathway to the Pacific across the North Pole. This may have been the same idea that compelled Rijp to chart his course north. When Hudson reached the polar ice cap, thereby proving to himself the theory was false, he turned east, hoping to find an open route around the obstruction. Just a few days later, on June 27, 1607, he found his way blocked by the middle of Spitsbergen's western shore, which ended his hopes for an eastern sea route to China.[9]

Finding an unbroken field of ice extending north from the archipelago's northern shore, Hudson assumed this frozen expanse was not ocean, but rather the arm of a vast northern continent that also included Greenland. He therefore called this new expanse "Groenland," the name then used for Greenland. With his way blocked, Hudson spent the next month exploring the same stretch of coast (western Spitsbergen) that Barentsz had sailed eleven years earlier (figure 9.3). Unlike Barentsz, however, his account recorded observations of many "whales" and "morses" (walruses). In one inlet at the north end of a frozen strait later named Foreland Sound, Hudson found whales in such abundance he called it "Whale Bay." He wrote, "in this Bay . . . we saw many whales and one of our company having a Hooke and Line over-boord to trie for Fish, a Whale came under the Keel of our ship and made her held, yet by Gods mercie we had no harme, but the losse of the hooke and three parts of the line."[10] This observation may be the first record of a whale entanglement in fishing gear, a problem that would become a vexing conservation concern, particularly for right whales, four centuries later. Summarizing his impressions of the new land, Hudson concluded, "I think this land may bee profitable to those that will adventure it."[11]

Unlike Barentsz, Hudson recorded a detailed account of the abundance of whales and walruses, which generated immediate interest from Muscovy Company officials. In response, they promptly commissioned Jonas Poole to lead another expedition to the archipelago to explore its economic potential. On April 12, 1610, Poole and his pilot Nicholas Woodcocke left Scotland, setting a course first for Norway's North Cape and then north past Cherie Island (an English name for Bear Island bestowed by Poole when he thought he had discovered it in 1604). They reached Groenland (Spitsbergen) on May 16.[12] Poole spent the next three months cruising Spitsbergen's western flank hunting walrus, polar bears, and reindeer. When he returned with the blubber from walruses and many "Whale fins" (baleen) found lying on the shore, he reported a "great store of Whales," which confirmed Hudson's account.[13] Perhaps just as important, Poole visited and named most of the major sounds along Spitsbergen's west coast that would later become major whaling centers, including Horn and Bell Sounds to the southwest, Ice Sound (now Isfjorden) midway along the coast, and both Whale Bay and

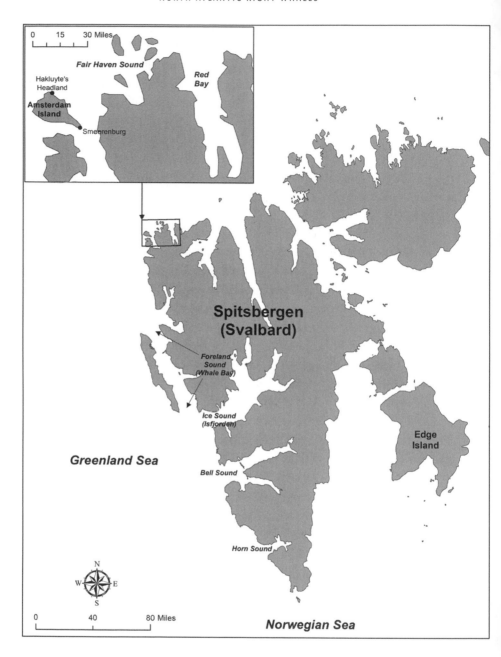

Figure 9.3. Major whaling grounds in Spitsbergen during the early 1600s. (Map courtesy of Mina Innes and Linda White.)

Foreland Sound further northwest. His account of those areas included the following: "I found the weather very warme and farrer temperater then I have found at North Cape at this time of the yeare. . . . [and] the ayre temperate in the Lands and nothing so cold as I have found at Cherie Iland in five severall voyages."[14]

Now convinced of the new land's economic potential, the Muscovy Company petitioned King James I for exclusive rights to develop its resources. After they were granted, the Company commissioned two ships for an expedition in 1611: the 150-ton *Mary Margaret* under the command of Thomas Edge, and the 50-ton bark *Elizabeth*, led by Poole.[15] While the smaller *Elizabeth* was to hunt walruses, the *Mary Margaret* was fitted out for "the killing of a whale, or two or three (and then) to discover farther north along the said land to find whether the same be an Iland or a Mayne [part of a continent]."[16] According to some historians, the company's initial interest in whaling was driven by the need for funds to subsidize their search for a trade route east. Whatever the motivation, company officials recognized that the crews of their ships were completely naïve in all matters related to whaling. Therefore, possibly following the suggestion of Nicholas Woodcocke (one of their employees), the Muscovy Company made arrangements with a French Basque outfitter in St. Jean de Luz to supply whaling equipment and six seasoned whalers.

Fully aware that their two expedition leaders knew nothing of whaling, Muscovy Company officials provided a tutorial on the whales they would be hunting and their duties as part of the commissions. They instructed Edge and Poole to treat the hired Basque with all due courtesy and to heed their advice in choosing which whales to kill. They were further instructed to watch the Basques closely and "observe and diligently put in practice the executing of that business in striking the whale."[17] Finally, they were told to render oil from the whales and walruses *before* returning to England to avoid unspecified "problems" (possibly the processing of rancid, rotting blubber) experienced when Poole returned with walrus blubber on his previous trip. Edge's commission went on to describe the characteristics of eight different kinds of whales that company officials understood they might encounter. Because those officials also had no direct knowledge in these matters themselves, this information must have come from either the Basque outfitter with whom they contracted or their hired whalers. In part, the commissions described the three most valuable kinds of whales—bowhead, right, and sperm whales—in decreasing order of importance:

> The first sort of Whales, is called the Bearded Whale [bowhead whale], which is black in color, with a smoothe skinne, and white under the chops; which Whales is the best of all the rest: and the elder it is, the more it doth yeelde. This sort of Whale doth yeelde usually foure hundred and some sometimes five hundred finnes [whalebone], and betweene one hundred and one hundred and twentie Hogsheads of Oyle. The second sort of Whale is called Sarda [right whale], of the same colour and fashion as the former, but somewhat lesse, and the finnes not above one fathom [6 feet] long, and yeeldeth in Oyle, according to his bignesse, sometimes eightie, sometimes a hundred Hogsheads. The third sort of Whale is called Trumpa [sperm whale], being as large as the first but not so thicke, of colour Grey, having but one Trunke [blowhole] in his head, whereas the former have two. He hath in his mouth teeth of a span long, as thicke as a mans wrist, but no fines, whose head is bigger then either of the two former, and in proportion farre bigger then his body. In the head of this whale is the Sperma-ceti, which you are to keepe in Caske apart from your other Oyle: you may

put the Oyle you find in the head and the Spermaceti altogether, and marke
it from the other Oyle, and at your comming home, we will separate the
Oyle from the Spermaceti. The like is to be done with the Oyle of this sort
of Whale, which is to be kept apart from the Oyle of the other Whales . . .
In this sort of Whale is likewise found the Ambergreese, lying in the entrals
and guts of the same, being the shape and colour like unto Kowes dung.
We would have you therefore your selfe to be present at the opening of this
sort of Whale, and cause the residue of the said entrals to be put into small
Caske, and bring them with you into England. . . . The Teeth likewise of this
sort of Whale we would have you cause to be reserved for a traill . . . This
Whale is said to yeelde in Oyle fortie hogshead, besides Spermaceti."[18]

As it turned out, the expedition in 1611 was a disaster. On June 12, Basque
whalers aboard the *Mary Margaret* "killed a small Whale [presumably a bow-
head], which yeelded twelve Tunnes of Oyle, being the first Oyle that was ever
made in Groenland [Spitsbergen]."[19] Shortly thereafter, however, while the crew
was ashore hunting walruses in Whale Bay, their ship broke loose from its moor-
ings and foundered, stranding the crew. After a harrowing 600-mile trip in open
shallops, Edge and some of the crew summoned help from Poole, who was hunting walruses on Bear Island. The *Elizabeth* then returned to rescue the crew that
had remained at Whale Bay and salvage what they could of their goods. When it
arrived in Whale Bay and its crew began transferring oil and walrus skins caught
by the *Mary Margaret* to the *Elizabeth*, it too was lost when its cargo suddenly
shifted and rolled the vessel onto its side, causing it to capsize. Fortunately, the
crew and part of their catch were saved by a chance encounter with yet another
English ship, the *Hopewell* from Hull, under the command of Thomas Marmaduke. Hunting walrus without permission from the Muscovy Company, Marmaduke's pilot was none other than Nicholas Woodcocke, who apparently had quit
his employment with the company in anger after being rejected as an officer for
their new Spitsbergen expedition.[20]

Despite their losses, the Muscovy Company's faith in the region's economic
potential was unshaken. In 1612 it again commissioned Edge and Poole to lead two
ships to Spitsbergen. Once more, both were equipped for whaling and provided
with a complement of Basque whalers. This time the expedition was a resounding
success. According to Poole, between June 9 and July 2 in Foreland Sound, they
killed 15 whales and struck but lost 8 others.[21]

The Basques on the voyage attempted to keep their whaling methods confidential; according to Poole, they "would not have us have any insight into their
businesse."[22] Nevertheless, the Basques were unsuccessful at keeping their secrets.
It was impossible to conceal their methods for rendering whale blubber into oil
because this was done with portable tryworks on land, which the English could
easily see. Basque attempts to conduct whaling operations out of view of the
English were also soon dashed. To their great dismay, one of their own countrymen

was persuaded to take five Englishmen to whale in a shallop for several days, while the rest of the Basques were off conducting their own hunt out of view. The English crew guided by the Basque whaler successfully killed three whales, while the all-Basque crew killed but one.[23] The journal of Robert Fotherby, who served as master's mate on English voyages to Spitsbergen in 1613 and 1614, provided a detailed description and several drawings of the whaling techniques used by their hired Basques (see plate 4). The drawings of whales being killed and processed at sea are the first solid evidence of Basque pelagic whaling and portable tryworks. In any case, such were the indiscretions that revealed Basque whaling methods and shattered their centuries-long whaling monopoly.

In 1612, the two Muscovy Company ships again found they were not alone in Spitsbergen. They encountered two English interlopers hunting walruses, three foreign vessels, including a Spanish Basque ship with the former Muscovy Company employee Nicholas Woodcocke as pilot, and two Dutch ships from Amsterdam.[24] One of the Dutch vessels was piloted by Allen Sallowes, yet another former Muscovy Company employee.[25] Like the English, the Dutch had hired Basque crews to conduct their whaling. The Basque ship from San Sebastian returned at the end of the season fully loaded and earned a great profit.

The success of Basque whalers launched a rush to exploit the region's whale populations even as the Dutch and English squabbled over rights to the region. Whereas the Dutch claimed discovery of the archipelago by Barentsz,[26] the English countered with claims based on Hudson's "discovery." To bolster the Dutch position, whaling merchants in ports or "chambers" in the Netherlands formed their own whaling cartel fashioned after England's Muscovy Company. Chartered in 1614 by Holland's States General under the name Noordsche Company (the Northern Company), the Dutch Prince of Orange granted its members exclusive rights to establish shore stations for the purpose of processing whale oil on Spitsbergen's shores.[27]

Between 1613 and 1620, both the English and Dutch attempted to solidify their claims by sending armed consorts to protect their whaling vessels.[28] Tense confrontations ensued as each side sabotaged the other's whaling stations, stole whaling equipment, confiscated ships and cargo, and even killed at least some whalers.[29] Despite these tensions, they both recognized the superior whaling skills of the Basques. As a result, the English and Dutch agreed to cooperate in ousting any Basque vessels they encountered. In this endeavor, they were successful.

Although the Basques moved aggressively to exploit the new whale stocks by sending at least 13 whaling ships to Spitsbergen in 1613, it was not in their nature to use force to defend their whaling interests. The English and Dutch, however, had no such reluctance to do so, and bending to their threats of armed conflict, the Basques grudgingly abandoned whaling at Spitsbergen. It is unclear where the banished Basques went instead, but it is assumed they turned to pelagic whaling, possibly with onboard or portable tryworks.[30] This could have been accomplished in the lower latitudes off Norway, Jan Mayen, or Iceland, where right whales likely made up most of the catch. Some of their vessels also may have returned to Grand Bay.

Despite tensions between the English and the Dutch, both sides began earning handsome profits. In 1616, English ships caught so many whales they were forced to leave some processed oil in Spitsbergen at the end of the season.[31] By one account, they killed 130 whales that year.[32] In 1617 they again had success, and killed some 150 whales and thousands of walruses by mid-August.[33] However, in 1618, aggressive efforts by the Dutch and their allies to assert their whaling rights left the English with few whales to chase.[34] In the following years, confrontations cut sharply into both Dutch and English profits. After a protracted series of uneasy negotiations between 1619 and 1621, an unwritten understanding was reached that resulted in splitting Spitsbergen's whaling grounds between the two sides. The English solidified claims to the bays and sounds along most of Spitsbergen's west coast, including Horn, Bell, and Foreland Sounds, while the Dutch were restricted to the northwestern corner of the archipelago.

By 1620, both the English and Dutch had fully adopted Basque shore-whaling methods and were catching as many as 300 to 450 whales per year.[35] Both countries continued hiring Basques to kill and process whales, but gradually they were replaced with whalers from their own countries. By the 1670s, virtually all their whalers were English, Dutch, or German.[36] To carry out whaling on their northwest corner of the archipelago, the Dutch established a station called Smeerenberg, so named because *smeer*, the Dutch word for fat, was found everywhere during its peak of operations. It would become Spitsbergen's most prominent whaling station between 1620 and 1640.[37] Although its name can make no claim as source for the contemporary expression "fat city," one might be excused for considering the link, as the blubber rendered at Smeerenberg was a source of immense wealth to all who were associated with it.

By the 1650s, whales were becoming scarce in all the bays and sounds off Spitsbergen; this forced whalers to shift to pelagic whaling. Initially, blubber flensed at sea was packed in barrels and returned to cookeries on Spitsbergen, particularly in Smeerenberg, for processing.[38] However, as whalers ranged farther and farther afield in search of whales, they soon found it just as easy to return to home ports in Europe rather than to Spitsbergen. That realization marked the beginning of the end for Spitsbergen's whaling stations. By 1660 all of its stations were abandoned, including Smeerenberg. As a result, the Noordsche Company allowed its charter for shore whaling to lapse, which ultimately dissolved the company. Pelagic whalers roamed the ice edge west of Spitsbergen to Jan Mayen, and perhaps even farther south to southern Greenland and Iceland. When their holds were filled with barrels of raw blubber, they turned for home,[39] where independent tryworks had sprung up near whaling ports. Most of those facilities had been processing fat from land animals, but now much of their business involved rendering blubber from whales and walruses.[40]

Jan Mayen Island

Shore whaling off Jan Mayen Island lagged just a few years behind the activities at Spitsbergen. Between 1616 and 1625, the Dutch sent an average of 10 vessels per year to Jan Mayen and stationed over 200 workers on its shores to operate half a dozen temporary tryworks.[41] Although Dutch ships dominated Jan Mayen whaling, the English and Basques also sent ships to the island to set up seasonal tryworks. During the peak of its shore-whaling era between 1615 and 1630, as many as 20 whaling ships and 1,000 men worked on and around Jan Mayen throughout the summers. As on Spitsbergen, early temporary tryworks evolved into semipermanent seasonal stations. In the 1620s, they became more substantial with brick furnaces, wood huts, and capstans. Gun emplacements with one or two cannons were soon added to guard at least one station, although there is no evidence they ever saw action.

For a brief period between 1616 and 1624, catch rates at Jan Mayen rivaled those of Spitsbergen. Noordsche Company whalers alone caught nearly 1,000 whales over that period.[42] The log book of a whaler named Matthijus Jansz Hoepstock recorded an amazing single-season haul of 44 whales that yielded 2,300 casks of oil in 1619.[43] Given its location just below the Arctic Circle amid currents branching off the Gulf Stream, right whales may have made up part of this summer catch at Jan Mayen; further evidence for this is that Hoepstock's average yield was just 52 casks of oil per whale, an amount well below what might be expected if all whales were bowheads. By the 1630s, however, whales around the island became scarce. In 1633, 11 Dutch ships managed to catch just 47 whales, and in 1635, 10 ships caught only 42. The Dutch abandoned Jan Mayen shore stations altogether in the 1640s as the area's whalers took up pelagic whaling.[44]

Another whaling ground more clearly occupied by right whales emerged 600 miles to the east of Jan Mayen, along the coast of Norway, and particularly around North Cape in the country's northernmost county of Finnmark. This was the same area where King Alfred's 880 text reported Norsemen hunting whales at the end of the ninth century. The development of commercial whaling in northwest Norway coincided with that at Jan Mayen. The Basques had begun whaling off Norway by at least 1614 and continued sending some ships annually for the remainder of the decade. Their presence, along with other foreign whaling vessels, prompted the Dano-Norwegian Navy to dispatch vessels to assert their claim to the region by confiscating the catch of unlicensed whalers and expelling illicit vessels from their waters. Nevertheless, by the mid-1600s, Finnmark was a well-known destination for Danish, Dutch, German, and French whalers, as well as for the Basques.

Little was recorded about the number of whales caught along the Norwegian coast, but levels comparable to those reported for Jan Mayen seem possible for at least some years. Christian Bullen, coxswain on a 1667 whaling voyage that spent time off the Norwegian coast, claimed he counted some 20 whaling ships in the Loppa Sea west of present day Tromsø, Norway.[45] Twenty years later, surgeon Johann Dietz reported 75 ships, most of which he presumed were whaling.[46] Nearly a score of shore-based processing stations built by foreign whalers in the 1600s have been identified through archival and archeological research.[47] Considering the many decades of whaling in the area, several thousand right whales might have been killed along the Norwegian coast by the close of the seventeenth century.

The End of the Beginning for International Whaling

By the mid-1600s, The Netherlands had emerged as the world's preeminent whaling nation, a position it would hold through the late 1700s. Their success was based on some of the same advantages that enabled the Basques to retain their long dominance over whaling before the 1600s: a world-class shipbuilding industry, a deeply rooted maritime tradition, an entrepreneurial ethic backed by a financial system that encouraged investment in whaling, and a healthy rivalry between regional chambers in the Noordsche Company and independent practitioners.[48]

Of particular importance, however, was the Dutch legal system. It encouraged merchants throughout Holland to enter the whaling business and unite their efforts with those based in Danish and Flemish ports. Because the Dutch adhered to the principle of *mare librum* (freedom of the sea), they were less inclined to confer exclusive privileges to a select few at the expense of fellow countrymen. As a result, the Dutch whaling fleet flourished and out-competed the English by sheer numbers. In contrast, Muscovy Company officials looked inward. Putting their connections with the crown to use, they sought to tighten control over domestic markets for whale oil by favoring ships of a few influential London merchants to the exclusion of ships from other English ports, such as Hull and Scotland. The bitter rivalries sparked by the Muscovy Company's exclusionary policies turned nonaligned English merchants in ports outside of London away from becoming potential allies who could have strengthened their country's position in Spitsbergen. Instead, those potential partners grew into resentful competitors bent on inflicting their own measure of revenge. For example, in 1626, before the Muscovy Company's fleet could reach Spitsbergen, several Hull vessels previously accosted as interlopers by company ships destroyed its shore station in Horn Sound. In the process they took eight shallops, burned casks, smashed oil cookeries, and demolished huts and gun emplacements.[49]

The transition from shore whaling to pelagic whaling between the 1630s and 1650s not only liberated whaling operations from fixed shore stations, it opened a door to further expansion of the whaling industry. In the years that followed, the number of Dutch ships equipped for whaling ballooned. By 1661, well over 100

Dutch ships were outfitted for whaling; just a few years later, in the mid-1660s, their numbers multiplied to nearly 250. Although greatly increasing the capacity of the Dutch fleet to kill whales, the number of ships is somewhat misleading because many served part time as merchant traders that took up whaling only when oil prices were high or freight rates low. Demise of the Noordsche Company in 1642 and the rise of pelagic whaling also encouraged German merchants to enter the trade. By 1669, German shipowners in Hamburg, Bremen, and Emden were sending nearly 40 whaling ships a year to work the Greenland Sea west of Spitsbergen.[50]

The spread of pelagic whaling marked the end of Spitsbergen's tumultuous bay-whaling era and the ultimate triumph of Dutch, English, Danish, and German merchants to wrest control of whaling from the Basques. Although surviving Noordsche Company records are incomplete and most of those from the Muscovy Company were lost in the Great Fire of London in 1666, an estimated 15,000 whales were killed by English, Dutch, Danish, and German whalers off Spitsbergen between 1611 and 1669.[51] An even higher total during this period is conceivable if the effects of Basque whaling are added. But true to their history, the Basques left no catch records and their contribution to the era's landings are not available. Basque whaling was undoubtedly overshadowed by the English and Dutch once they became proficient, yet the Basques continued to send out whaling ships through the late 1600s, and it would be a mistake to disregard them entirely, especially with regard to understanding the decline of North Atlantic right whales. Due to their expulsion from Spitsbergen, they spent more time than others in the 1600s scouring sub-Arctic whaling grounds where right whales would have been caught.

While the catch of right whales by northern Europeans was certainly small compared to bowhead whales, they could have also taken a significant number of the former. Unfortunately, distinctions as to which whales were caught where and when are blurred by catch statistics that referred only to "whales" and barrels of oil produced. Although most whalers worked the waters off Spitsbergen and along the edge of pack ice, some roamed North Cape, Iceland, the southern tip of Greenland, and perhaps Jan Mayen, where right whales were more likely targets. Considering their catch in those areas plus that by the Basques, thousands of right whales may well have been killed over the course of the 1600s.

This would have devastated the remaining eastern North Atlantic right whale population and perhaps also affected the western population to the extent that some migrated to summer feeding areas off Iceland, Norway, and possibly Jan Mayen Island. By one estimate, at least 1,000 to 2,000 right whales still must have survived in all the North Atlantic in the mid-1600s to account for the number of whales taken after that date.[52] But because later whaling occurred primarily along North American shores, most surviving whales must have been western North Atlantic right whales. By the end of the 1600s, the eastern population off Europe had likely dwindled to a few hundred at most.

NOTES

1. Hacquebord, L. 1999. The hunting of the Greenland right whale in Svalbard, its interactions with climate and its impact on the marine ecosystem. Polar Research 18:375–382.

2. Sanger, CW. 2005. The origins and development of shore-based commercial whaling at Spitsbergen during the 17th century: a resource utilization assessment. The Northern Mariner 15(3):39–51.

3. Richards, JF. 2005. Whales and walruses in the Northern Oceans. *In* Richards, *The Unending Frontier: An Environmental History of the Modern World*. University of California Press. Berkeley, CA, US, pp. 547–617, 600.

4. Miall, B (trans.). 1923. *Master Johann Dietz Surgeon in the Army of the Great Elector and the Barber to the Royal Court*. George Allen & Unwin. London, UK, p. 122 (quotation).

5. Smith, TD, K Barthelmess, and RR Reeves. 2006. Using historical records to relocate a long-forgotten summer feeding ground of North Atlantic right whales. Marine Mammal Science 22(3):723–734.

6. Ibid., p. 724.

7. Sanger, The origins (n. 2), pp. 43–44.

8. Barents, W. (1626) 1905. The third voyage northward to the kingdoms of Cathaia, and China in anno 1596. S Purchas, *Hakluytus Posthumus, or Purchas His Pilgrimes: Contayning a History of the World in Sea Voyages and Lande Travells by Englishmen and Others*. Reprint. James Maclehose & Sons. Glasgow, UK, Vol. 13, pp. 62–87. Van der Werf, SY. 1998. Astronomical observations during William Barents's third voyage to the north, 1596–1597. Arctic 52(2):142–154.

9. Playes, J, and H Hudson. 1626. Divers voyages and northerne discoveries of that worthy and irrecoverable discoverer Master Henry Hudson. *In* Purchas, *Hakluytus Posthumus* (n. 8), Vol. 13, pp. 294–314, 300. Conway, M. 1906. *No Man's Land: A History of Spitsbergen from Its Discovery in 1596 to the Beginning of the Scientific Exploration of the Country*. Cambridge University Press. Cambridge, UK, p. 22.

10. Playes and Hudson, Divers voyages (n. 9), p. 305 (quotation). Asher, GM. 1860. *Henry Hudson the Navigator. The Original Documents in which His Career Is Recorded, Collected, Partly Translated, and Annotated with an Introduction*. The Hakluyt Society. London, UK.

11. Playes and Hudson, Divers voyages (n. 9), p. 307 (quotation).

12. Poole, J. (1626a) 1905. Divers voyages to Cherie Iland in the yeers 1604, 1605, 1606, 1608, and 1609. *In* Purchas, *Hakluytus Posthumus* (n. 8), Vol. 13, pp. 265–293.

13. Poole, J. (1626b) 1905. Voyage set forth by the Right Worshipfull Sir Thomas Smith, and the rest of the Muscovie Company, to Cherry Iland: and for a further discoverie to be made towards the North Pole. *In* Purchas, *Hakluytus Posthumus* (n. 8), Vol. 14, pp. 1–23, p. 9 (quotation).

14. Ibid., pp. 6, 13 (quotation).

15. A commission for Jonas Poole and a commission for Thomas Edge . . . given the last day of March 1611. (1626) 1905. *In* Purchas, *Hakluytus Posthumus* (n. 8), Vol. 14, pp. 24–33.

16. Conway, *No Man's Land* (n. 9), p. 44 (quotation).

17. Ibid., p. 42 (quotation).

18. A commission (n. 15), pp. 31–32 (quotation).

19. Edge, T. (1626a) 1905. Greenland first discovered by Sir Hugh Willoughbie; the voyages of Frobisher, Pet and Jackman, Davis, the Dutch; first morse and whale killing, with further discoveries. *In* Purchas, *Haklutys Posthumus* (n. 8), Vol. 13, pp. 4–16.

20. Conway, WM. 1904a. Introduction to Hessel Gerritsz's *Histoire du Pays nommé Spitsberghe*, 1612. *In* Conway (ed.), *Early Dutch and English Voyages to Spitsbergen in the Seventeenth Century*. Hakluyt Society, London, UK, pp. 1–8.

21. Poole, J. (1626c) 1905. A relation written by Jonas Poole of a voyage to Greeneland in the yeere 1612. *In* Purchas, *Hakluytus Posthumus* (n. 8), Vol. 14, pp. 41–47.

22. Ibid., p. 46 (quotation).

23. Ibid., p. 46.

24. Edge, T. (1626b) 1905. Dutch, Spanish, Danish disturbance; also by Hull men, and by a new patent, with the succeeding successe and further discoveries till this present. *In* Purchas, *Haklutys Posthumus* (n. 8), Vol. 13, pp. 15–26.

25. Conway, Introduction (n. 20), p. 3.

26. Baffin, W. (1626) 1905. A journal of the voyage made to Greenland with five English ships and a pinnasse, in the yeere 1613. *In* Purchas, *Hakluytus Posthumus* (n. 8), Vol. 14, pp. 47–60.

27. De Jong, C. 1978. A *Short History of Old Dutch Whaling*. University of South Africa. Pretoria, South Africa.

28. Conway, WM. 1904b. Description of the new country called by the Dutch Spitsbergen. *In* Conway, *Early Dutch and English Voyages* (n. 20), pp. 13–38.

29. Gerritszoon van Assum, H. (1618) 1904. History of the country called Spitsbergen. Its discovery, its situation, its animals. BH Soulsby (trans.). Conway, WM. 1904c. Introduction to troubles at Spitsbergen. Both *in* Conway, *Early Dutch and English Voyages* (n. 20), pp. 9–65, and pp. 39–41, respectively. Conway, Description (n. 28).

30. Conway, *No Man's Land* (n. 9), p. 69.

31. Edge, Greenland (n. 19), pp. 15–16.

32. *Calendar of State Papers, Domestic, 1611–1618*, p. 392. As cited in Conway, *No Man's Land* (n. 9), p. 92.

33. Heley, W. (1625) 1905. Divers and other voyages to Greenland. *In* Purchas. *Hakluytus Posthumus* (n. 8), Vol. 14, pp. 91–108.

34. Beversham, J. (1618) 1905. Letter of James Beversham to Master Heley from Faire-Haven, the 12 of July, 1618. *In* Purchas, *Hakluytus Posthumus* (n. 8), Vol. 14, pp. 96–97.

35. De Jong, *A Short History* (n. 27).

36. Ibid.

37. Ross, WG. 1993. Commercial whaling in the North Atlantic sector. *In* JJ Burns, JJ Montague, and CJ Cowles (eds.), *The Bowhead Whale*. Special publication no. 2. Society for Marine Mammalogy. Allen Press. Lawrence, KS, US.

38. Conway, Introduction to troubles (n. 29).

39. Hacquebord, The hunting of the Greenland right whale (n. 1).

40. De Jong, *A Short History* (n. 27).

41. Hacquebord, L. 2004. The Jan Mayen whaling history: its exploitation of the Greenland right whale and its impact on the marine ecosystem. *In* S Skreslet (ed.), *Jan Mayen Island in Scientific Focus*. NATO Science Series. IV Earth and Environmental Sciences, Vol. 45. Kluwer Academic Publishers. Dordrecht, The Netherlands, pp. 229–238.

42. Ibid.

43. Ibid.

44. Ibid.

45. Barthelmess, K. 2003. Das erste gedruckte deutsche Walfangjournal. Christian Bullens einer Hamburger Fangreise nach Spitzbergen und Nordnorwegen im Jahre 1667. De Bataafsche Leeuw, Amsterdam, & Deutsches Schiffahrtsmeuseum, Bremerhaven. As cited in Smith et al., Using historical records (n. 5), pp. 224–227.

46. Miall, *Master Johann Dietz* (n. 4), p. 152.

47. Jørgensen, R. 1994. Kommersiell kvalfangst i Nord-Norge på 1600-tallet. Tromura. Tromsø Museums Rapportserie, kulturhistorie, 26. Tromsø Museum, Tromsø. As cited in Smith et al., Using historical records (n. 5), pp. 735–736, table 1.

48. De Jong, *A Short History* (n. 27).

49. Conway, WM. 1904d. Disputes between English whalers. *In* Conway, *Early Dutch and English Voyages* (n. 20), pp. 173–175.

50. Ross, Commercial whaling (n. 37).

51. Hacquebord, The Jan Mayen whaling industry (n. 41).

52. Reeves, RR, JM Breiwick, and E Mitchell. 1992. Pre-exploration abundance of right whales off the eastern United States. *In* J Hain (ed.), *The Right Whale in the Western North Atlantic: A Science and Management Workshop, 14–15 April 1992, Silver Spring, MD*. Reference Document 92-05. Northeast Fisheries Science Center, National Marine Fisheries Service. Woods Hole, MA, US, pp. 5–7.

10

A FITFUL START FOR COLONIAL WHALERS

BY THE MID- TO LATE 1600s, the only significant number of right whales left in the North Atlantic were along the eastern seaboard of what would become the United States. This was the last part of the species' range to be scoured by whalers. This is not to say that right whales inhabiting the region had not yet been hunted. As already mentioned, Basque whalers based along the Strait of Belle Isle in the 1500s and early 1600s may have killed large numbers of whales in the Gulf of St. Lawrence and perhaps the Gulf of Maine, and at least a few others were probably killed off northern Europe. Despite those losses, a large contingent of right whales greeted the first European explorers and colonists when they arrived on American shores in the sixteenth and seventeenth centuries. They were keenly aware of whales but their ability to catch them would remain elusive for decades.

Of course, the Atlantic coast of North America was already occupied by Native Americans, and some historians have asserted or assumed that they had their own whaling traditions even before first contact with Europeans. Therefore, before picking up the trail of European whalers in the New World, those claims deserve consideration.

Native American Whaling?

What may be the first description of Native whaling—and also one of the most preposterous—was recorded by a Spanish Jesuit missionary named Father Jose de Acosta. Acosta is sometimes called the "Pliny of the New World" after the first-century Roman author and naturalist Pliny the elder, whose wide-ranging work *Naturalis historia* served as the model for subsequent encyclopedias. The Spanish Jesuit Acosta was the first European to compile a detailed description of Natives and natural history of a small portion of the Americas. This influential book on the natural and moral history of the West Indies was completed in 1590. Acosta had sailed to Mexico in 1572 and spent 15 years traveling between that country and Chile before returning to Spain in 1587 to write his account.

Acosta claimed that Native Americans in what the Spanish even then called Florida used canoes or reed floats to sneak up on whales as they slept at the surface. When they were close enough, he maintained that they would jump on the whale's back, drive a wooden stake into each nostril of the blowhole, and straddle the whale until its struggles ceased and it could be towed ashore (figure 10.1).[1] According to Acosta, Florida natives hunted "a whale as bigge as a mountaine." He described their means of doing so as follows:

> The manner the Indians of Florida use (as some expert men have tolde me) to take these whales (whereof there is great store) is, they put themselves into a canoe which is like barke of a tree, and swimming approach neere the whales side; then with great dexteritie they leap to his necke, and there ride as on horsebacke, expecting his time, he thrusts a sharp and strong stake, which he carries with him, into the nosthrill, for so they call the hole or vent by which they breathe; presently he beates it in with an other stake as forcibly as hee can; in the meane space the whale dooth furiously beate the sea, and raiseth mountaines of water, running into the deepe with great violence . . . the Indian still sittes firme, and to give him full payment for his trouble, he beates another stake into the other vent or nosthrill so as he stoppeth him quite, and takes away his breathing; then hee betakes him to his canoe, which he holdes tied with a corde to the whales side, and goes to land . . . then a great number of Indians come unto the conquered beast to gather his spoils and kill him and cut his flesh to peeces, this do they drie and beate into powder, using it for meate dooth last them long."[2]

This account was clearly based on a muddled description from a second- or thirdhand source, because Acosta himself never reached Florida. It was certainly not a description of hunting large whales. If there is any credence to the tale at all, it must have been based on descriptions of either a Native drive fishery, in which porpoises or other small cetaceans were herded into the shallows of a cove or along a beach by a small fleet of canoes, efforts to salvage dead whales or porpoises found floating offshore, or a hunt for manatees.

The most compelling evidence of precontact whaling for large whales—and probably the uncited source for many claims of its occurrence—comes from the narrative of a voyage by the English Captain George Weymouth (also Waymouth) in the spring of 1605 to what is now the coast of Maine. Weymouth was probing for possible settlement sites and spent an eventful month exploring the coast a few miles south of Penobscot Bay. There he mapped the coast, traveled 40 miles up a river later named the Georges River after his sponsor Sir Ferdinando Gorges, traded for furs with the local Natives (probably Penobscots), and recorded fauna and flora including whales.

Figure 10.1. A 1597 print by Theodore Bry illustrating purported Native whaling. The artist depicted reed rafts used to approach whales and the hammering of wooden stakes into blowholes. The illustration is based on a 1587 account by the Spanish missionary Father Jose de Acosta. (Courtesy of the New Bedford Whaling Museum, New Bedford, MA.)

At the end of his stay he abducted two local Natives, a common practice of early English explorers. Said to be treated with all due respect and civility, Natives captured by the English were typically taken to England, schooled in Christianity, and taught the English language and customs. In the process, English "hosts" sought to learn all they could about languages, customs, tribal relations, and geographic features of the captives' homelands. After a few years in England, captives were usually pressed into service as guides or interpreters on return voyages, or simply repatriated with their home villages in hopes they would serve as cultural ambassadors to ease distrust in future meetings. The two men captured by Weymouth in 1605 were assigned as guides to Captains George Popham and Raleigh Gilbert in a failed attempt to establish a settlement near the mouth of the Kennebec River in 1607 under authority granted by King Henry IV. This was made by a group of English business interests that would soon form the Plymouth Company. Meanwhile, both of their guides apparently escaped at the first opportunity to rejoin their home communities.

Thomas Rosier accompanied Weymouth as a journalist hired to record notable discoveries and events, and Rosier published a narrative of the trip immediately upon returning to England in mid-July of 1605.[3] His account offers a richly detailed day-by-day description of their adventures, including numerous encounters with indigenous people. Although this daily log provided no description of whaling, a summary of major discoveries included at the end of his narrative is a short account of what they had "learned from the [two] savages" since they took them.[4] At the end of that summary Rosier noted:

> One especiall thing is [the local Indians'] manner of killing the whale, which they call powdawe; and will describe his form; how he bloweth up the water; and that he is twelve fathoms long; and that they go in company of their king with a multitude of their boats and strike him with a bone made in the fashion of a harping iron fastened to a rope, which they make great and strong of the bark of trees, which they veer out after him: then all their boats come about him, and as he riseth above water, with their arrows they shoot him to death: when they have killed him and dragged him to shore, they call all their chief lords together, and sing a song of joy: and those chief lords whom they call sagamores, divide the spoil and give them to every man a share, which pieces so distributed, they hang up about their houses for provision; and when they boil them, they blow off the fat, and put to their pease, maize, and other pulse which they eat.[5]

Even if this account came directly to Rosier from a Native source, it should be treated with caution. Given the language barrier between him and the two Indians taken to Britain, the journal's rapid publication after returning to England, and the absence of any evidence that Weymouth or his crew had actually witnessed any Native whaling, the narrative may be a garbled interpretation of what the captives were trying to say. They may have been attempting to describe drive fishing for dolphins or small whales, which Maine Natives are now known to have undertaken, or perhaps they were explaining the practice of salvaging drifting carcasses. The account might even have been an attempt by the captives to describe Basque whaling they had observed or their efforts to copy it. These last two possibilities deserve some consideration, because according to Gabriel Archer, Maine residents had at least some contact with the Basques before the 1600s (see also chapter 8).

Archer was a journalist on an English expedition to the same area that was led by Bartholomew Gosnold in 1602 (a voyage on which Gosnold bestowed the name "Cape Cod" to the famous sandy hook dangling off the Massachusetts mainland). Upon reaching the Maine coast in mid-May, Archer offered the following description: "[There] came towards us a Biscay shallop with sail and Oares, having eight persons in it, whom we supposed to bee Christians in distress. But approaching us neere, we perceived them to be Savages. These coming within call hayled us, and wee answered. Then after signes of peace . . . they came boldly aboord us . . . One that seemed to be their Commander wore a Wastecoate of blacke worke, a paire of Breeches, cloth Stockings, Shooes, Hat and Band . . . made by some Christians . . . [and] spake divers Christian words."[6]

How these Gulf of Maine Natives came to possess a "Biscay shallop" is unknown, but it seems to confirm they had contact with either Basque whalers or cod fishermen before 1602. If the shallop came from a whaling vessel, it could indicate that Basque pelagic whalers based along the Strait of Belle Isle had reached the Gulf of Maine. As noted earlier, the Basques made pelagic whaling trips at least as far south as Cape Breton, Nova Scotia, in the late 1500s.[7] If they also reached the Gulf of Maine, Maine Natives may have watched Basques kill whales and then perhaps even experimented with whaling themselves. One way or another, this could somehow explain Rosier's 1605 description, without necessarily indicating an established whaling tradition among the indigenous people in what is now the state of Maine.

In their history of whaling on the eastern end of New York's Long Island, published in 1932, Everett Edwards and Jeannette Rattray also suggested that Native Americans engaged in whaling: "When the colonies were first settled [on Long Island], Indians pursued whales off the coast in canoes; attacking them with crude wooden harpoons attached to wooden floats. It was impossible, of course, for a canoe to fasten and tow behind a wounded whale as the regulation whale boats did later on. The Indians wore out their prey by incessant pricks, finally killing them by loss of blood."[8] Other historians have suggested Indians used dugout canoes, "crude" stone-tipped harpoons, lines of "Indian hemp" or deerskin thong, and wooden drogues to catch whales.[9] Those early historians may have simply assumed Natives in the New World had such capabilities due to the important role they played in advancing colonial era whaling on Long Island, Cape Cod, and Nantucket after the arrival of English colonists; however, the authors offer no direct evidence to support their assertions.

William Bradford, Plymouth Colony's second governor, recorded what may be the first eyewitness account of indigenous people making use of whales in his celebrated description of the Pilgrims' initial landing on Cape Cod in 1620. His journal describes the English refugees' first exploratory excursions (figure 10.2), including an encounter with 10 to 12 Natives cutting up a "grampus" (an early generic term for dolphins, porpoises, and pilot whales). Penned more than a decade after the *Mayflower*'s landfall, the journal offers no hint that the Cape Cod Natives actually killed the whale. Rather, they may have been salvaging stranded whales. According to Bradford, a *Mayflower* landing party "came allso to yᵉ place wher they saw the Indians yᵉ night before, & found they had been cutting up a great fish like a grampus, being some 2. inches thike of fate like a hogg, some peeces wher of they had left by yᵉ way; and yᵉ shallop found 2. more of these fishes dead on yᵉ sand, a thing usuall after storms in yᵉ place, by reason of yᵉ great flats of sand that lye of."[10] Bradford's observation seems to have occurred near the current town of Eastham on the bay side of the Cape's forearm, less than 10 miles south of today's Wellfleet, where marine mammal strandings are now common.

While references to possible precontact whaling by Natives are intriguing, they seem unconvincing.[11] Neither archeological evidence of whaling implements nor any firsthand accounts by early explorers or colonists exist to support the hunting by indigenous people of large whales in Maine, or anywhere else south of the Arctic. The absence of written accounts is particularly telling. Early explorers and colonists made a point of recording any interesting details about

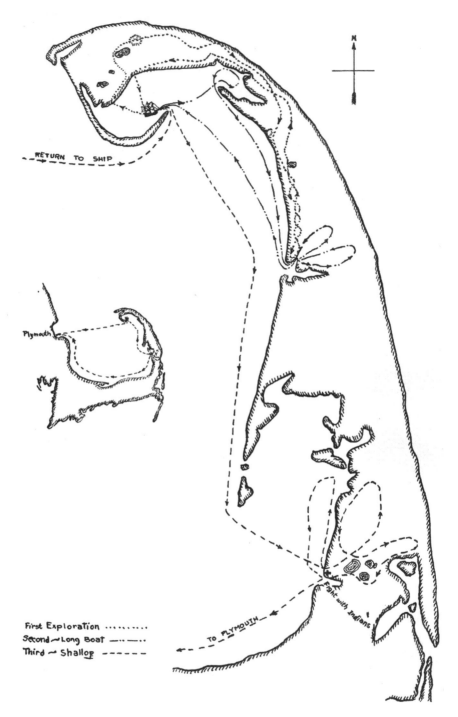

RETURN TO SHIP

N

Plymouth

First Exploration ···········
Second — Long Boat — ·— ··
Third — Shallop — — — — —

TO PLYMOUTH

Fight with Indians

Figure 10.2. Routes taken by the Pilgrims on their first three scouting expeditions after the *Mayflower* arrived at Cape Cod, in December 1620. (Image courtesy of the Brownell Collection, Providence Public Library Special Collections, Providence, RI. Image redrawn from SB Alexander and SM Magoun. 1928. *The Frigate Constitution and Other Historic Ships*. Southworth Press. Portland, ME, p. 42.)

Native customs, and whaling for large whales certainly would have caught their attention had they witnessed it. It therefore seems likely that Native "whaling" before the 1600s was limited to salvaging stranded animals and hunting small whales, such as porpoises and pilot whales, by driving them ashore or perhaps herding and shooting them with bows and arrows; these activities are more clearly documented by early colonists.

The First English Whaling Attempt

Although eyewitness accounts of Native whaling for large whales are absent from the journals of early explorers, they routinely recorded observations of "whales" and placed them high on lists of resources from which their sponsors or investors might profit. In the early 1600s, these references presumably were to right whales, which were still a primary target for whalers. For the colonists who followed, however, there were other priorities: establishing settlements, coping with hostile Natives, and surviving in an unfamiliar wilderness. In the few instances concerted whaling attempts were made, inept skills produced frustrating results.

The first to try was Captain John Smith. In 1614, he led two ships from the English colony at Jamestown on an expedition to explore New England, which was then still considered the northern part of their Virginia plantation. Just seven years after Jamestown's founding, Smith had yet to see the colony's territory to the north, but he was aware that Weymouth, Gosnold, Popham, and Gilbert had all reported whales. He therefore set out to investigate the region's economic potential, including whaling, for himself. Visiting Cape Cod and the central coast of Maine in April 1614, he apparently chased the wrong kinds of whales and his efforts were a complete failure. Smith recalled the voyage in his classic 1627 account of the early history of Virginia: "Our plot was there to take Whales, for which we had one Samuel Cramton and divers others expert in that facility, & also to make trialls of a Mine of gold and copper; if those failed, Fish and Furs were then our refuge to make our selves savers howsoever: we found this Whale-fishing a costly conclusion, we saw many and spent much time in chasing them, but could not kill any. They being a kinde of Jubartes [humpback whale], and not the Whale that yeelds Fins and Oile as we expected."[12]

The First Dutch Whaling Attempts

The next attempt at whaling was probably by the Dutch some 15 years later. The Dutch had claimed a stretch of coast on the strength of Henry Hudson's third attempt to find a sea route to China in 1609, this time for the Dutch East India Company. Hudson was following reports of a large river somewhere north of Virginia that led to an immense western sea he assumed to be the Pacific Ocean. Rumors of this vast sea—probably the Great Lakes—might have come from John Smith and his Native contacts at Jamestown. In any event, in 1609 Hudson sailed his small ship the *Half Moon* up two major East Coast rivers in search of the rumored passage west.

He first went a few miles into a wide bay at the mouth of a river he called the "South River," now the Delaware River, before deciding it didn't match the waterway for which he was looking. He then explored a second river he called the "North River" more methodically. Later renamed the Mauritius River after Maurits of Nassau, sovereign Prince of Orange, who later granted Dutch Charter to the area, this watercourse was eventually named the Hudson River once the English gained control over the territory at the end of the century. Hudson traveled more than 125 miles up this river to the site of present-day Albany before concluding it offered no passage to the Pacific. His journal duly noted the presence of "whales" near the mouths of both rivers.

With no settlements north of Jamestown, and Hudson's exploration on behalf of Dutch interests the basis for a claim to the area, the Dutch States General granted a seven-year charter in 1614 to the New Netherland Company to develop a broad swath of land stretching from what is now Delaware to southern New England. When the charter expired in 1621 with no colonies established, a new patent was awarded to the Dutch West Indies Company. Its first director, a wealthy Dutch merchant named Samuel Godyn, had made his fortune financing whaling trips to Spitsbergen and serving as administrator of the Noordsche Company. In 1620, a year before the new patent was approved, Godyn sponsored an exploratory voyage to the region led by Cornelis Jacobszoon Mey. When Mey returned and confirmed Hudson's observations of an abundance of whales off the "South River" and its large northern cape (now Cape May, New Jersey), Godyn reasoned that whaling could provide the financial footing for their planned colony of New Netherland.

Having no firsthand whaling experience himself, Godyn found an opportunity to set his plan in motion when he met a young navigator and ship captain named David Pieterszoon de Vries (figure 10.3). In only his late twenties, De Vries was experienced beyond his years. He had served as captain of a Dutch whaling ship sent to Jan Mayen in 1617, a military ship that saw action in the Mediterranean in 1618, and a fishing vessel off Newfoundland in 1620. The two men struck a deal, with De Vries joining the Dutch West Indies Company as a patroon (patron), and were thereby granted rights to colonize the southern part of their claim, as well as part of what is now Staten Island, across the Mauritius River from the new settlement of New Amsterdam (now New York City).

As part of the arrangements, De Vries agreed to send a ship and small yacht to the South River in 1630 to establish a colony that could guard the southern end of their claim and also conduct whaling. According to De Vries, Godyn had "represented that there were many whales that kept before the bay, and the oil, at sixty guilders a hogshead, . . . would realize a good profit;" De Vries therefore saw to it that his ships were equipped "to carry on the whale fishery in that region."[13] He didn't accompany the voyage himself and events of the trip are poorly recorded,

Figure 10.3. David Pieterz de Vries in a 1653 engraving by Cornelis Visscher. De Vries, a patroon with the Dutch East India Company, sponsored the unsuccessful colony of Zwaanendael (Valley of Swans) at the mouth of today's Delaware River, where he attempted to start a whaling business in 1631 and 1632. (Image courtesy of the Emmet Collection, Miriam and Ira D. Wallach Division of Art, Prints and Photographs, The New York Public Library, Astor, Lenox, and Tilden Foundations.)

but his crew successfully built a small stockade called Zwaanendael (Valley of Swans) inside the mouth of today's Delaware Bay, near the present town of Lewes, Delaware. When it was finished in the fall of 1631, they left behind a small force and returned to Holland with only "a sample of oil from a dead whale found on shore."[14] The leader of the expedition explained that the return was poor because they arrived too late in the year, after the whales had already left the bay.

Despite the disappointing oil yield, Godyn encouraged De Vries to try again, this time under his direct command. De Vries agreed and once more arranged for a ship and small yacht to return to Zwaanendael. This time he was intent on reaching the bay "in December in order to conduct whale fishing in winter, as the whales come in the winter and remain till March."[15] Shortly before leaving, word arrived from a returning Dutch vessel that their stockade at Zwaanendael had been demolished by local Natives. Undeterred, De Vries set sail late in May of 1632, taking a southern route through the Caribbean to trade in the West Indies.

When he arrived at Zwaanendael in early December, De Vries found an abundance of whales that he thought ripe for "royal work." However, he also found the charred ruins of the stockade, with the skulls of livestock and men strewn about the ground; the men were those who had stayed behind the previous year. His first order of business was therefore negotiating peace with the local Natives. With a display of level-headed tact and reserve few other colonial leaders were able to exercise, De Vries secured peace at the expense of most of the trinkets and goods he had brought for trading.

By January, with the larder running low, and unable to obtain supplies from local Natives, De Vries decided to take the yacht *Squirrel* upriver in hopes of trading with another group of Indians. His mission lasted a month and proved unsuccessful. The inland residents he met were enmeshed in war, with all sides destroying supplies of the others, leaving nothing to offer in trade. During his excursion, however, he noted several sightings of whales, including one for which he professed great surprise, because it occurred some 25 miles upstream in brackish water. When he returned to Zwaanendael, he found that his men had taken two whales in his absence, both yielded little oil.[16]

With their vegetable stores still low, De Vries decided to seek supplies from the English at Jamestown. On March 5, 1633, he set out in the *Squirrel* for the Chesapeake and arrived at Port Comfort (now Newport News) three days later in a late-winter snowstorm. After picking up a pilot to guide him up the James River to Jamestown, he dropped anchor on March 11. His small ship was the first non-English vessel to call at the colony. Despite disputes in Europe between the English and Dutch over territorial claims, De Vries was received cordially by the colony's governor, John Harvey. After exchanging news of the territory, De Vries's request for food was granted, and he retraced his path to Zwaanendael. When he arrived on March 29, he found that seven more whales had been taken, but there were "only thirty-two cartels of oil obtained." Clearly frustrated, he made the following conclusion: "The whale fishery is very expensive when such meager fish are caught. We could have done more if we had good harpooners, for they had struck seventeen fish and only secured seven which was astonishing. They had always struck the whale in the tail. I afterwards understood from some Basques who were old whale fishers that they struck the harpoons in the fore-part of the back."[17] It appears that despite his service as master of a whale ship in Jan Mayen, De Vries was not well versed in the art of catching whales, and neither was his crew. Given the low quantities of oil, perhaps they were hunting small cetaceans or only the calves of right whales.

Two weeks later on April 14, they dismantled their kettles, loaded their meager yield of oil, and headed north to New Amsterdam on the Mauritius River. De Vries arrived at New Amsterdam two days later and met with the Dutch colony's governor, Wouter Van Twiller, who asked about their whaling venture. Disappointed and feeling somewhat misled about the whaling prospects, De Vries replied that he had a sample of whale oil; however, he made the following statement to the governor: "They were foolish who undertook the whale-fishery here at such expense, when they could have readily ascertained with one, two, or three sloops in New Netherland, whether it was good fishing or not."[18]

After some tense confrontations with the governor and his officers, whom De Vries judged to be a raucous, drunken, and generally incompetent bunch, he left New Amsterdam on June 15, and arrived back in Holland on July 22, 1633. His voyage was again a losing venture. The greatest return came not from the meager quantity of whale oil, but from a load of salt acquired in the Caribbean at the start of the voyage.

De Vries sailed twice more to New Amsterdam in 1634 and 1638, but accounts of both trips make no mention of whaling. Curiously, the Dutch were unable to take advantage of New Amsterdam as a whaling base, even though they were well on their way to becoming world leaders in the pursuit through their Arctic ventures. From the time New Amsterdam was first colonized in 1625 until control was ceded to the English in 1664—as well as during their brief period of reoccupation between 1672 and 1674—there is little hint of any further whaling by the Dutch in New Netherland. Only a vague reference appeared in 1652, when directors of the Dutch West India Company in Europe wrote to New Netherland Governor Peter Stuyvesant, "In regard to the whale fishery we understand that it might be taken in hand during some part of the year. If this could be done with advantage, it would be a very desirable matter, and make the trade there flourish and animate many to try their good luck in that branch."[19] The only indication that this plea to take up whaling was not ignored entirely was a fleeting reference in April 1656 that the governing council of New Amsterdam "received the request of Hans Jough, soldure and tanner, asking for a ton of train-oil or some of the fat of the whale lately captured."[20] With more definitive references absent, it would appear any other whaling in the Dutch colony was limited at best.

The Early Pilgrims and Whales in Cape Cod Bay

Early English colonists in New England cast equally covetous eyes toward right whales. Some even suggested opportunities to hunt whales were a motivating factor for colonization.[21] When King James I issued his charter for the Plymouth Colony in 1620, he conveyed rights to "have and to hold, posess and enjoy, all . . . Lands, Territories, Islands, . . . Sea Waters, Fishings, with all Manner their Commodities, Royalties, Liberties, Preheminences and Profitts, that shall arise from thence."[22] Because whales were then considered fish, they were included under the term "fisheries." Whales were recognized more explicitly in 1629 when the heir to James I, King Charles I, issued another charter for lands immediately north of Plymouth Colony to the Massachusetts Bay Colony. It granted colonists "free Libertie of fishing in or within any the Rivers or Waters with the Boundes and Lymytts aforesaid, and the Seas thereunto adjoining; and all Fishes, Royal Fishes, Whales, Balan, Sturgions, and other Fishes of what Kinde or Nature soever."[23]

Whether or not right whales factored into the decision of English pilgrims to sail for America, the animals were one of the first natural wonders to greet the 101 sea-weary refugees when the *Mayflower* first reached land on November 11, 1620. That landmark event occurred not at the legendary rock in Plymouth Harbor, Massachusetts, as schoolchildren are often taught, but rather on the sandy shores of the broad half-moon cove on the tip of Cape Cod, where Provincetown now lies. An account of their first days on Cape Cod by an unidentified member of the voyage (possibly William Bradford or Edward Winslow[24]) tells us the following: "Everie day we saw Whales playing hard by us, of which in that place, if we had instruments and meanes to take them, we might have made a verie rich returne, which to our great grief we wanted. Our Master and his Mate, and others experienced in fishing, professed, wee might have made three or four thousand pounds worth of oil; they preferred it before Greenland Whale-fishing, and propose the next winter to fish for Whale here."[25]

Another reference to what must have been right whales upon arriving at the cape is from the journal of William Bradford, the Plymouth Colony's second governor. Bradford recounted what was certainly an inept attempt at whaling if not an idle-hour prank: "We saw daily great whales, of the best kind for oil and bone, come close aboard our ship and in fair weather swim and play about us. There was once one, when the sun shone warm, came and lay above water, as if she had been dead. For a good while together, within half a musket shot of the ship; at which two were prepared to shoot, to see whether she would stir or no. He that gave fire first, his musket flew in pieces, both stock and barrel; yet thanks be to God, neither he nor any man else was hurt with it, though many were there about. But when the whale saw her time, she gave a snuff, and away."[26]

The Pilgrims' stay on Cape Cod lasted two weeks while they repaired various leaks in their ship. The site of their first landing almost became the site of their first settlement, in part because of the multitude of whales frolicking around their ship.[27] To resolve differing opinions over whether to stay on the cape or set roots elsewhere, the colonists prepared a list of reasons for both. Prominent among their short list of reasons to stay was that "Cape Cod was like to be a place good for fishing; for we saw daily great whales of the best kind for oil and bone."[28] This could only mean right whales. In the end, however, a lack of fresh water and an urge to consider other options won out. After a series of exploratory excursions from their temporary base, they instead chose a mainland site on the opposite shore of Cape Cod Bay, where they founded New Plymouth, the first permanent English settlement north of Jamestown.

Although the new settlers had pledged to return to the cape to hunt whales the next winter, they had no chance to do so. By the following spring, nearly half their party had perished in a hard first winter. Although their numbers were replenished by the arrival of 35 new colonists in the summer of 1621, with 67 more in 1622, any thoughts of hunting whales went unfulfilled. They had no whaling equipment and were sorely pressed by more urgent needs for shelter, laying in stores of food, and dealing with their Native neighbors. The abundance of whales off the cape, however, would not be forgotten. Many residents of the new colony became dissatisfied with their location at Plymouth, and in 1640, some, perhaps including those who initially wanted to stay on the cape in 1620, sought a grant to settle on Cape Cod's outer arm. Their request was approved, and after reaching an agreement with local Natives to purchase a tract of land, they left Plymouth in 1643 to found the first English settlement on the outer arm of the cape, near what is now the town of Eastham.[29]

Even before then, however, colonists from both the Plymouth and Massachusetts Bay Colonies visited the cape regularly to fish and scavenge stranded whales. In 1629, the Reverend Francis Higgenson, one of the first residents of Salem in the Massachusetts Bay Colony, described the waters around Cape Cod: "[the] abondance of sea-fish is almost beyond beleeving, and sure I should scarce have beleeved it except I had seene it with mine owne Eyes. I saw great store of Whales, and Crampusse [grampuses, or porpoises and pilot whales]."[30]

Although both Plymouth and Massachusetts Bay colonists were tantalizingly close to prime right whale habitat, they were unable to take advantage. Credit for the first successful whaling on American shores would instead go to early settlers on what is now Long Island, in New York State. They would be followed in short order by colonists on Cape Cod and Nantucket. At each location, the legacy of shore whalers would leave a lasting mark on the history and character of those communities, and whaling still remembered through relics and exhibits at local museums.

NOTES

1. Allen, GM. 1916. *The Whalebone Whales of New England*. Memoirs of the Boston Society of Natural History Vol. 8, no. 2. Boston Society of Natural History. Boston, MA, US, pp. 109–175, 146.

2. Acosta, Joseph de (CR Markham ed.). (1604) 1880. *The Natural & Moral History of the Indies*. Hakluyt Society. London, UK, Vol. 1, p. 149.

3. Rosier, J. (1605) 1860. *Rosier's Narrative of Waymouth's Voyage to the Coast of Maine, in 1605: Complete. With Remarks by George Prince, Showing the River Explored to Have Been the George River*. Eastern Times Press. Bath, Maine, US.

4. Ibid., p. 38 (quotation).

5. Ibid., p. 39 (quotation).

6. Archer, G. (1626) 1905. The relation of Captaine Gosnols voyage to the north part of Virginia, begun the sixe and twentieth of March, Anno 42. Elizabethae Reginae 1602. And delivered by Gabriel Archer, a gentleman in the said voyage. *In* S Purchas, *Hakluytus Posthumus, or Purchas His Pilgrimes: Contayning a History of the World in Sea Voyages and Lande Travells by Englishmen and Others*. Reprint. James Maclehose & Sons. Glasgow, UK, Vol. 18, pp. 302–313, 304 (quotation).

7. Holmes, A. 1805. *American Annals; Or a Chronological History of America from Its Discovery in 1492 to 1845*. W. Hilliard. Cambridge, MA, US, Vol. 1, p. 133.

8. Edwards, EJ, and JE Rattray. 1932. *Whale Off! The Story of American Shore Whaling*. Frederick Stokes. New York, NY, pp. 195–196 (quotation).

9. Ashley, CW. 1926. *The Yankee Whaler*. Houghton Mifflin Co. The Riverside Press, Cambridge, Boston, and New York. Palmer, WR. 1959 [1981]. The whaling port of Sag Harbor. Ph.D. thesis, Faculty of Political Science, Columbia University. Publ. by University Microfilms

International, Ann Arbor, Michigan, USA. Both as cited in RR Reeves and E Mitchell. 1986. The Long Island, New York, right whale fishery: 1650–1924. *In* RL Brownell, PB Best, and JH Prescott (eds.), Right Whales: Past and Present Status. Special issue 10, Reports of the International Whaling Commission, pp. 201–213.

10. Bradford, W. (ca. 1650) 1898. *Bradford's "History of Plimoth Plantation" from the Original Manuscript*. Wright and Potter. Boston, MA, US, pp. 173–174 (quotation).

11. Little, EA. 1981. Indian contribution to shore-whaling at Nantucket. Nantucket Algonquin Studies 8:1–85, 59–60.

12. Smith, J. (1624) 1907. *The Generall Historie of Virginia, New England, and the Summer Isles With the Names of Adventurers, Planters, and Governours from Their First Beginning Ano 1584. to the Present 1624 . . . The Second Volume Containing the Sixth Booke of the Generall Historie of Virginia, New England, and the Summer Isles Together With the True Travels, Adventures, and Observations, and A Sea Grammar*. Reprint. Robert Maclehose & Sons. Glasgow UK, Vol. 2, p. 3 (quotation).

13. De Vries, DP. (1655) 1853. *Voyages from Holland to America, A.D. 1632 to 1644*. HC Murphy (trans.). Privately printed. New York, NY, p. 22 (quotation), https://archive.org/details/cu31924028729402.

14. Ibid., p. 23.

15. Ibid.

16. Ibid., pp. 39–40.

17. Ibid., p. 55.

18. Ibid., p. 57 (quotation).

19. Starbuck, A. 1878. *History of the American Whale Fishery from its Earliest Inception to the Year 1876*. Report of the US Commissioner of Fish and Fisheries, Part 1. US Government Printing Office. Washington, DC, p. 11 (quotation).

20. Ibid., p. 11 (quotation).

21. Ibid., p. 5.

22. Thorpe, FN (ed.). 1909. *The Federal and State Constitutions, Colonial Charters, and Other Organic Laws of the States, Territories, and Colonies Now and Heretofore Forming the United States of America*. US Government Printing Office. Washington, DC, Vol. 3, p. 1834 (quotation).

23. Ibid., p. 1850.

24. Young, A. 1844. *Chronicles of the Pilgrim Fathers of the Colony of Plymouth from 1602 to 1625*. 2nd ed. Charles C. Little & James Brown. Boston, MA, US, pp. vii–viii.

25. A relation or journal of a plantation setled at Plimoth in New England, and proceedings thereof; printed 1622 and here abbreviated. *In* Purchas, *Hakluytus Posthumus* (n. 6), Vol. 14, pp. 312–343, 312 (quotation).

26. Young, *Chronicles* (n. 24), p. 146.

27. Thatcher, J. 1832. *History of the Town of Plymouth from Its First Settlement in 1620, to the Year 1832*. Marsh, Capen & Lyon. Boston, MA, US, pp. 20–21.

28. Young, *Chronicles* (n. 24), p. 146 (quotation).

29. Freeman, F. 1862. *The History of Cape Cod: The Annals of the Thirteen Towns of Barnstable County*. G. E. Rand & Avery, Cornbill. Boston, MA, Vol. 2, pp. 347–350.

30. Higginson, F. 1630. *New England's Plantation or A Short and True Description of the Commodities and Discommodities of that Country*. Michael Sparke. London, UK, p. 8 (quotation).

11

LONG ISLAND WHALING

IT WAS SOME 50 to 60 years after the founding of Jamestown and 20 to 30 years after the *Mayflower* dropped anchor in Cape Cod Bay that the first signs of successful whaling appeared in the budding colonies along the eastern edge of North America. Surprisingly, it occurred not in the colonies surrounding Cape Cod Bay, where the Pilgrims first marveled at right whales cavorting around their ship and where a substantial portion of its remaining population now assembles every spring to feed. Rather, it was on the eastern tip of what is now Long Island in New York—where right whales today are seldom seen and almost never in large numbers. But in the 1600s, when right whales were more abundant and the climate was still chilled by the Little Ice Age, the waters off Long Island must have been far more appealing to right whales.

In the early 1600s, Long Island and Connecticut were hinterlands under nominal claim by the Dutch, entrenched at their colony of New Netherland on the lower Mauritius River (Hudson River) a hundred miles to the west, as well as by the English in the Plymouth and Massachusetts Bay Colonies, a hundred miles to the northeast. Both the Dutch and English had been content to leave the eastern end of Long Island and the mainland to the immediate north within the realm of hostile Natives. In 1633, however, shortly after a smallpox epidemic decimated the region's Native villages, the rival claimants began eyeing depopulated lands along the "Fresh River" (now the Connecticut River) as prospective settlement territory. In that competition, it was the English who struck first. To prevent Dutch incursions up river, John Winthrop Jr., first governor of the large tract of land that would become Connecticut, dispatched an English military engineer and adventurer named Lyon Gardiner to erect a fort at the mouth of the river in 1635.

After an unsuccessful settlement attempt in the fall of that year on the banks of the Connecticut River, a second attempt was made in June 1636 by a group of Puritans, who were finding other colonists in Cambridge, Massachusetts, increasingly intolerant of their religious beliefs. Led by a pair of prominent ministers, Thomas Hooker and Samuel Stone, they trekked through the green woods of the southern New England wilderness, "musical with birds and bright with flowers," to the Connecticut River. This second group of emigrants successfully established the towns of Hartford, Wethersfield, and Winsor, and shortly thereafter, New Haven. Within a year more than 800 colonists had descended on the area.[1]

Although the Pequots, the dominant indigenous people in what is now known as the Connecticut River valley, had been devastated by the smallpox outbreak, they were still a formidable force that considered the English particularly unwelcome interlopers. Therefore, when Gardiner set to work on his fort, one of his first orders of business was sorting out relationships among the Native people and forging alliances with other regional groups who welcomed any potential allies to help defend them from the Pequots' frequent raids. When confrontations erupted between the English settlers and the Pequots, the English organized what became known as the Pequot War of 1637—a war in which the English and their Native allies, principally the Narragansetts and Mohegans, all but exterminated the Pequots.

One of the groups enlisted for the campaign by Gardiner was the Montauketts, the most powerful people on Long Island. In the course of his dealings, Gardiner learned the Montaukett language and became fast friends with their "sachem," or chief, Wyandanch; this was a friendship no doubt nurtured by Gardiner's role in vanquishing the Pequots. His relationship with Wyandanch led to the purchase of a large island in 1639—an island that still bears Gardiner's name and lies between the two spits of land forming the forked tail on Long Island's eastern tip. His friendship with the Montauketts also may have emboldened a handful of English colonists to cross the narrow sound from Connecticut in 1640 to establish the "plantation" of Southold on Long Island's northeastern spit of land opposite mainland New England. That same year, another band of colonists from the Massachusetts Bay Colony of Lynn arrived on the southern fork fronting the Atlantic Ocean and established the "plantation" of Southampton.

There is no evidence that the seasonal presence of whales off Long Island was a factor drawing those first colonists to Long Island, yet one might wonder if it was, given the speed with which they began to exploit whales in one form or another. Their early interest in whaling, as well as tensions between them and the local Montauketts and Shinnecocks, is implied in a 1643 order adopted by leaders of the Southampton plantation. The leaders directed the town's blacksmith Robert Bond "not to make for the Indians any harping irons [harpoons] or fishing irons which are known to be dangerous weapons to offend the English."[2] There are no records confirming that the first settlers hunted whales, yet the reference to local manufacture of harping irons indicates a maritime livelihood and possibly some early thoughts of whaling just three years after their arrival. Despite tensions with their Native neighbors, the colonists and local Indians would soon proceed jointly to exploit the local whale resources.

Although it is unlikely that Long Island Natives hunted large whales before the arrival of the English, they clearly utilized and prized dead whales cast up by the sea. Early deeds for the purchase of coastal lands from local Natives often recognized and preserved their rights to use all or parts of any whales that happened to strand on the tract. In 1648, Connecticut Governor Theophilus Eaton purchased a large parcel that would eventually become the community of East Hampton, not far from Southampton, for "twenty coats, twenty four hoes, 24 knives, and other articles." The deed stipulated that the Montauketts would continue to "have the fynns and tayles of all such whales as shall be cast upp, and desire that they may be friendly dealt with in the other parts [of whales]."[3] The Montauketts reportedly roasted the fins and tails and offered them as gifts to their "gods" during religious ceremonies; they ate the meat with relish and found it delicious.[4] With English settlers most interested in whalebone and oil from the blubber, while the Native people coveted the tails, flukes, and meat, allocating parts of stranded whales, at least initially, may have been done with satisfaction to all concerned.

After the Montauketts and Shinnecocks learned of the trade value earned by the English from stranded animals, drift whales apparently become a more important negotiating point during the sale of coastal lands. In 1658, when Lyon Gardiner purchased lands along the beach at Southampton, Wyandanch ensured the contract specify his interests: "whales cast up shall belong to me and to other Indians within these bounds."[5] Another Southampton deed from 1662 made a point of reserving for the Native people "half of all profits and benefits of the beach on the south side of said Island in respect of fish, whale or whales that shall by Gods providence be cast up from time to time, and at all times."[6]

To prevent disputes between the colonists themselves over salvage rights, in 1644 the town of Southampton adopted an order that may be the first official act on American soil providing for the orderly exploitation of whales, albeit for stranded animals dying of natural causes. The order made the following proclamation:

> If by the Providence of God there shall bee henceforth within the bounds of this plantacon any whale or whales cast up, for the prevention of Disorder yt is Consented unto that there shall be foure Wards in this Towne eleaven persons in each ward. And by lott two of each Ward (when any such whale shall be cast up) shall be imployed for the cutting out of the sayd whale who for their paynes shall have a double share. And every Inhabitant with his child or servant that is above sixteen years shall have in the division of their part an equall proportion provided that such person when yt falls into his ward [be] a sufficient man to be imployed aboute yt.[7]

The order went on to list the members of each ward authorized to process whales. Some residents, however, apparently still attempted to keep the bounty of stranded whales for themselves. As a result, another order was passed the following year (1645) that imposed a 20 shilling fine on anyone carrying off parts of a whale without first notifying and receiving permission from a designated town magistrate or whale inspector. By 1653, the four "wards" had expanded to six "squadrons" (still with eleven men each); these were granted exclusive rights to cut up dead whales that stranded along their designated stretch of beach.[8] The proceeds were shared communally among town residents, with extra shares for those who processed the whales. Although some stranded whales may have been right whales, they undoubtedly included a range of both small and large cetaceans.

Almost immediately some ward members began eyeing live right whales spouting a short distance from shore in winter and spring. To catch them, small whaling companies were formed to maintain light cedar dories furnished with harpoons and whaling gear that could be launched through the surf. To encourage those efforts, local residents agreed that any struck whales that later stranded would belong to whomever could prove they had harpooned the animal, rather than becoming community property. This marked the start of efforts to actively hunt whales on American shores. Although town ordinances attest to this development, there are no details about its success. The ordinances were apparently modeled after a similar order approved in 1647 by the town of Hartford, Connecticut, that awarded a short-term whaling monopoly to a "Mr. Whiting" in hopes of encouraging him to develop the business. It provided, "Yf Mr. Whiting, wth any others shall make tryall and presecute a designe for the takeing of whale wthin these libertyes, and if uppon tryall within the termes of two yeares, they shall like to goe on, noe others shalbe suffered to interrupt the, for the tearme of seaven yeares."[9]

Following Hartford's example, in 1650 the town of Southampton awarded John Ogden exclusive rights to hunt whales for seven years, provided his "designe" for killing whales proved successful within the first two years. As part of those rights, Ogden was granted the privilege of landing and trying out the oil of whales at any location along the beach.[10] Evidently he was not successful, because the same privilege was granted to another town resident in 1654.[11]

There is nothing to suggest that these early Long Island settlers had any prior whaling experience. Rather, it seems that they and the local Natives hired to work their boats learned how to catch whales jointly through a process of trial and error. Indeed, a good argument could be made that most of the credit for the accomplishment rightfully belonged to the Natives' side of this partnership, because they provided most of the labor for early whaling ventures. Whereas the colonists provided boats, equipment, and probably advice based on a general understanding of European whaling methods from oral or written accounts, it was left largely to hired Natives to convert that conceptual knowledge into actual practice and skill.

Some 20 years of experimentation was needed before proficiency at killing whales improved sufficiently to rival reliance on drift whales. To encourage those efforts, New York's first English governor, Richard Nicolls, directed in 1664 that whereas a duty of one gallon was to be provided to the crown for every 16 gallons of oil produced from drift whales, any whales killed at sea would be exempt from such duties.[12] The period of the exemption is unclear, but after Governor Thomas

Dongan reauthorized it in the mid-1680s, the right to hunt whales at sea free of taxes was taken by Long Island residents as a permanent privilege. This would become a bitter bone of contention between Long Island whalers and a succession of New York governors who sought to impose duties on those whales as well.

Long Island Whaling Finds Its Footing

The early settlers considered Natives exceptionally skilled at capturing the "great fish" and quickly came to rely on them. One of the first indications of this reliance was a contract between Jacop Schellenger and a group of local Native American whalers. Schellenger and his stepson James Loper, the only two Dutchmen then living in East Hampton, were principals in one of Long Island's most prominent whaling companies. They had taken well to the whaling business and in 1668 contracted for the services of six local Natives to work on their whaling shallop. The crew was to "attend dilligently with all opportunitie for ye killing of whales or other fish, for ye sum of three shillings a day for every Indian: ye sayd Jacobus Skallenger and partners to furnish all necessarie craft and tackling convenient for ye designe."[13]

By 1675, Schellenger, Loper, and their other financial partners expanded the operation to two boats, and hired 12 Native whalers,[14] which suggests at least some level of success. Loper's reputation for catching whales had already spread throughout New England. The diary of John Hull, a Boston merchant and shipowner, who also served as mintmaster making pine-tree shillings for the colony, reported that in February 1671 "the men of Long Island this winter made a hundred or two tuns of oil of whales that they there killed."[15] In 1672, colonists on Nantucket offered Loper land and other inducements to move to their island and teach them the whaling trade. Perhaps fearing his loss to the community, the town of East Hampton gave Loper five acres of land in 1673 if he would stay and continue his trade as a cordwinder (shoemaker), which he practiced when not running his whaling business.[16] Loper apparently declined the Nantucket offer and continued his whaling business on Long Island.

Arrangements to hire local Natives were made by most Long Island whaling companies not only to catch whales (figure 11.1) but also to cut them up and cart their blubber to try houses for rendering. So great was demand for their services that Governor Francis Lovelace, New York's second English governor, approved an order adopted on May 2, 1672, by the town of Southampton that was designed to prevent bidding wars for the employment of experienced Native whalers. His order, which extended to "East Hampton and other Places if they shall find it practicable," required that "whosoever shall Hire an Indian to go a-Whaling, shall not give him for his Hire above one Trucking Cloath Coat, for each whale, hee and his Company shall Kill, or halfe the Blubber without the Whale Bone, under a Penalty therein exprest."[17]

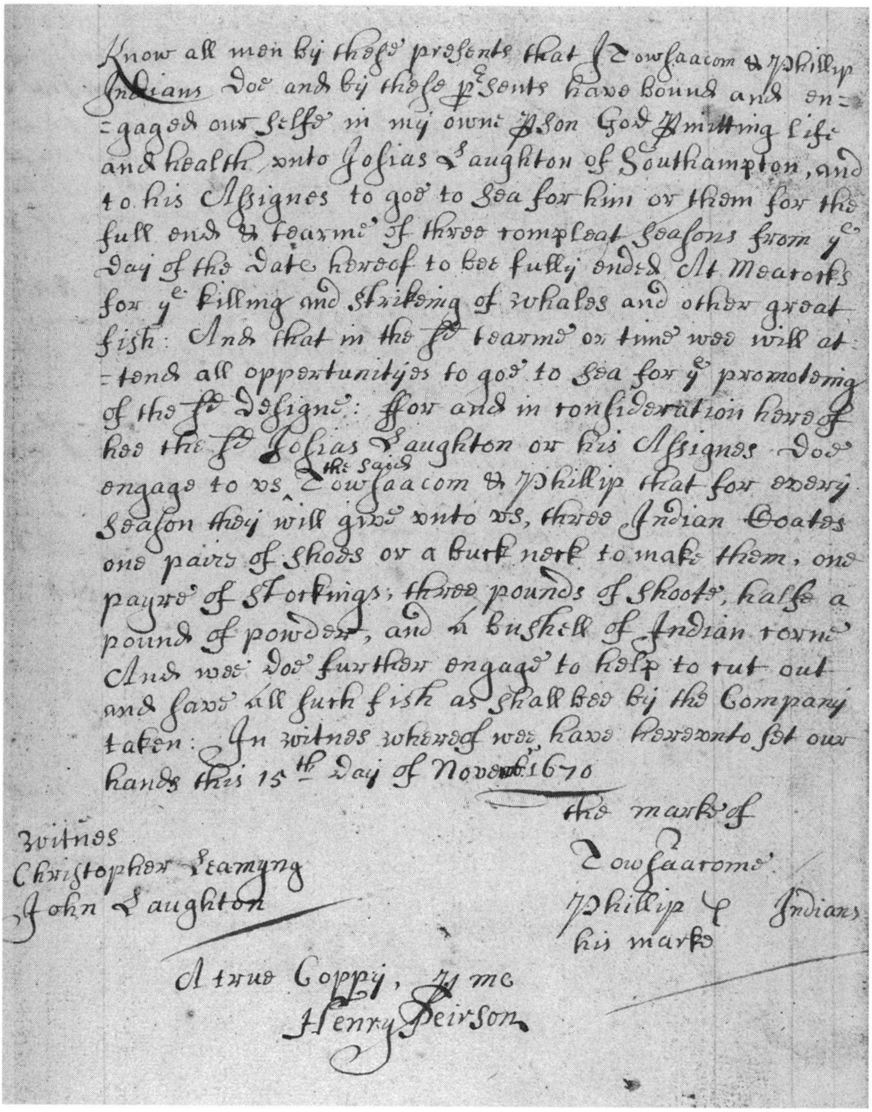

Figure 11.1. Whaling contract signed November 15, 1670, between Josais Laughton, owner of a whaling company in Southampton, and two "Indians" named Cowsaacom and Phillip, to "go to sea . . . for ye fishing and striking of whales and other great fish" for three seasons. (Image courtesy of the Office of Southampton Town Clerk, Historic Division Archives, Southampton, NY.)

Native involvement in Long Island whaling was so extensive that Charles Wolley, an Anglican chaplain who lived in New York from 1678 to 1680, thought it fitting to describe the activity in a treatise on the lives and habits of the area's Native Americans.[18] Wolley's appreciation of the fact that whales were not "fish," given that they nursed their young, suggests he was well read and a critical thinker. Interestingly, his description of Native whaling would have been no less accurate had it been written to describe Basque shore whaling two centuries earlier, or the last Southampton shore whaling two centuries later. After relating Native methods for "fishing and fowling," Wolley described their whaling activities:

It may not be improper to add some thing about [the Native] art of catching Whales, which is thus, two Boats with six Men in each make a Company, *vis* four Oars-men or Rowers; an Harpineer and a Steers-man; about *Christmas* is the season for Whaling, for then the Whales come from the North-east, Southerly, and continue till the later end of *March*, and then return they again; about the Fin is the surest part for the Harpineer to strike: As soon as he is wounded, he makes all foam, with his rapid violent Course, so that if they [the crew] be not very quick in clearing their main Warp to let him run upon the tow, which is the line fastened to the Harping-iron about 50 fathoms long, it is a hundred to one he over-sets the Boat: As to the nature of the whale, they copulate as Landbeasts, as evidence from the female Teats and Male's Yard, and that they spawn as other Fishes is a vulgar error. . . . A whale about 60 foot long having a thick and free Blubber may yield or make 40 or 50 Barrels of Oyl, every Barrel containing 31 or 32 Gallons at 20s. a Barrel if it hath a good large bone it may be half a Tun or a Thousand weight, which may give 25 /. Sterling old England Money.[19]

Each hunt offered a unique challenge. To the Natives working on most whaling boats, this unpredictability seemed exhilarating sport. In 1725, Paul Dudley, another chronicler and one of the first Americans to contribute to the scientific literature on whales off colonial shores, wrote, "The whale is sometimes killed with a single Stroke, and yet at other Times she will hold the Whale-men in Play near half a Day together with their Lances, and sometimes they will get away after they have been lanced and spouted blood with Irons in them, and Drugs [drogues] fastened to them which are thick Boards about fourteen Inches square."[20]

What may be the first evidence of multiple whales being caught appeared in an April 1669 letter cited by Alexander Starbuck in his authoritative history of American whaling. It suggested that Long Island whaling was not only becoming profitable, but that it had begun to range beyond the reach of shore-based dories by using small sailing vessels. The letter also indicated that whalers at least occasionally held fast to their quarry by using their dories, rather than drogues, to tire whales: "On ye Eastend of Long Island there were 12 or 13 whales taken before ye end of March, and what since wee heare not; here are dayly some seen in the very harbour, sometimes within Nutt Island. Out of the Pinnace the other week they struck two, but lost both, the iron broke in one, the other broke the warpe. The Governor hath encouraged some to follow that designe. Two shallops made for itt, but as yet we doe not heare of any they have gotten."[21]

Whales and Taxes

As whaling operations on Long Island were being perfected, regional Dutch rule came to an end in the bloodless conquest of New Amsterdam by the English in 1664. King James I had granted rights to the region first to Virginia and then to the Massachusetts colonies. Ignoring that precedent, King Charles I, son and successor to James I, issued a new charter for all former Dutch territories, including

Long Island (as well as Martha's Vineyard and Nantucket), to his brother James, the Duke of York. This turn of events sparked a long-running battle over taxes on whale oil produced on Long Island. Unlike Connecticut, with whom Long Island settlers had aligned themselves during earlier years of Dutch neglect, the new colony of New York had no representative form of government until 1683. Thus, Long Island whalers bristled when early New York governors attempted to impose duties on their whale oil and to require that it be brought to the port of New York for export. Accustomed to trading with New England ports where duties were lower and access to markets was more convenient, the whalers were outraged at this taxation without representation. Their anger foreshadowed deep-seated objections that would lead New England merchants to protest a similar "injustice" a century later by tossing a shipment of tea into Boston Harbor.

One of the first signs of opposition to taxes on whale products was a petition delivered in 1672 directly to the court of England that was responsible for colonial affairs. Bypassing the governor of New York and submitted on behalf of the towns of East Hampton, Southampton, and Southold, the petition asked that the towns either be allowed to continue under Connecticut rule, as had been the case when the Dutch occupied New Netherland, or to be made a free corporation. In addition to offering insight into the length of time it took to perfect their whaling business, one cannot help but detect a faint, yet distinct, similarity to another list of complaints delivered to English authorities a century later in the Declaration of Independence penned by Thomas Jefferson and others. In part, the petition read:

> Setting forth that they [the petitioners] have spent much time and paines, and the greater part of their Estates in settling the trade of whale fishing in the adjacent seas, *having endeavored it above these twenty years*, but could not bring it to any perfection till within these 2 or 3 yeares last past. And it now being a hopeful trade at New Yorke, in America, the Governor and the Dutch there do require ye Petitioners to come under their patent, and lay heavy taxes upon them beyond any of his Magesties subjects in New England, and will not permit the petitioners to have any deputys in Court, but being chiefe, do impose what Laws they please upon them, and insulting very much over the Petitioners threaten to cut down their timber which is but very little they have to Casks for oyle . . . and have till now [been subjects] under the Government and Patent of Mr. Winthrop, belonging to Conitycut Patent, which lyeth far more convenient for ye Petitioners assistance in the aforesaid Trade.[22]

When their plea was denied, most Long Islanders continued to ignore the whale tax, much to the frustration of a succession of New York officials.

When New York first convened an assembly of representatives from province communities in 1683, Long Island residents elected a prosperous East Hampton whaler named Samuel Mulford. For nearly 25 years he doggedly championed their cause, as well the development of a port on the eastern end of Long Island from which to ship whale oil. His blunt rhetoric attacking New York governors over tax matters and corruption twice resulted in his expulsion from the assembly. Each time, his fellow Long Islanders reelected him and sent him back where he would continue to be a thorn in the side of whatever governor happened to be in power at the time.[23]

Expanding Horizons

Once the final pieces of the whaling design fell into place in the late 1660s or early 1670s, the hunt for right whales on Long Island expanded rapidly—especially on the east end at Southampton and at East Hampton, where it became an integral part of community life. In some years, ministers and school teachers in these towns were paid in whale oil, and during the whaling season from December to April, school was suspended so children could help by watching for whales, cutting up blubber, or trying out oil.[24] According to one source, shore whaling was so widely practiced and so important that every able-bodied man in town was obliged to take a turn watching for whales from elevated positions along the beach.[25] For this purpose a series of huts was built along the shoreline with staffs planted at the highest points in the dunes. When a whale was sighted, a "weft" or flag was raised to summon whalers to launch their dories through the surf.

Over time some whaling companies expanded their operations by adding small sailing ships capable of voyages of a week or more. By day they patrolled nearshore waters beyond the range of shore whalers; at night they pulled into land. Long Island shore and coastal whaling probably peaked between 1670 and 1725. By 1687, at least 14 whaling companies of 12 men each kept seven try houses that were scattered among the communities stoked and busy processing blubber.[26] In April of that year, at the end of the winter season, whaling companies in Southampton alone reportedly had 2,148 barrels of oil "on hand," which presumably represented all or most of the year's catch; an additional 1,456 barrels was on hand in East Hampton.[27] If all of that oil came from right whales producing an average of 36 to 44 barrels per whale, their combined catch could have been between 82 to 100 whales during that season alone.[28] In 1690, at least 18 companies dotted a 50-mile stretch of coast along the east end of the Long Island.[29] Boat owners continued using experienced Native whalers, but as time passed and disease took its toll on the indigenous population, they were gradually replaced by white settlers, many coming from elsewhere in the colonies to take up the trade.

Whales were also caught elsewhere along Long Island as other communities sought to stake a claim in the growing business. The new community of Amagansett near East Hampton became a prominent whaling center in the 1680s, and by the early 1700s at least a few whales were being taken along the central and western parts of Long Island and Staten Island.[30] In 1668, whalers at Graves End, near present day Coney Island, "had pretty good luck" killing three large whales.[31] Perhaps the most successful new whaling company was at Smith Point near Fire Island, midway along the island's Atlantic coast. Its founder, Colonel William "Tangier" Smith, had close ties to New York governor Thomas Dongan, which helped him amass a vast 40,000-acre tract of land. His station was said to catch an average of 20 whales per year in the early 1700s.[32] On his death in 1705, his wife, Martha Tunstall Smith, took charge and became the only woman on Long Island known to have managed a whaling company. Like other company owners, she relied almost entirely on Native crews to carry out the whaling. Consistent with their historical roots, most Long Island whalers on the eastern tip shipped their oil to Connecticut and Boston markets, bypassing New York and its taxes.[33]

No time series of catch totals for Long Island whaling companies have been found from the late seventeenth or early eighteenth century. Perhaps their absence reflects efforts to conceal their catch and oil trade from New York officials in order to avoid taxes. Indeed, smuggling was a big business in the early days of English rule in the province of New York. Whale oil was undoubtedly part of the contraband. Thus, catch records may have been guarded jealously or underreported to avoid the attention of colony authorities. In 1688, the Duke of York asked one of his agents John Lewen to find out how many whales had been killed over the past six years, how much oil and bone had been produced, and what share was thus due. Lewen's answer advised that no records had been kept and the oil and bone were shared by members of the companies killing the whale.[34]

A later New York governor, Lord Cornbury, also cast a covetous eye toward elusive tax revenues from whale oil and whalebone. In the summer of 1708 he complained to authorities in England about the "illegal trade . . . carried on between New England, Connecticut, and the East End of Long Island." His complaint noted, "The quantity of Train Oyl made in Long Island is very uncertain, some years they have much more fish than others, for example, last year [1707] they made four thousand barrels of Oyl and this last season they have not made above Six hundred."[35] Using conversion factors noted above, those amounts could equate to between 91 and 111 right whales in the winter of 1706–07, but only 14 to 17 the following winter. Although production totals cited are little more than hearsay from an unknown source, the substantial difference between years could indicate that even in the best of times, Long Island whaling was a fickle business, subject to dramatic swings in success from year to year.[36] Alternatively, the difference might reflect the incomplete recordkeeping Cornbury had at his disposal. By 1711, however, at least some Long Island whale oil was being shipped through New York, while some went directly from Long Island to London.[37]

Reconstructing Long Island catch totals by shore whalers from the late 1600s is further complicated by catches from small whaling vessels usually operated by Native crews with white captains; these boats cruised coastal waters for a week or more at a time. Whether their catch is included in sparse catch records is unknown. Initially, they patrolled nearby Long Island coastal waters;[38] however, more distant voyages farther down the coast may have begun as early as the winter of 1666.[39] By the late 1680s, these vessels were often sailing south to the Delaware Bay and perhaps beyond, on trips lasting several weeks. William Penn's account of the founding of Pennsylvania included a report of 1683: "Mighty Whales roll upon the Coast, near the Mouth of the Bay of Delaware. Eleven caught and workt into Oyl one Season. We justly hope a considerable profit by a Whalery; they being so numerous and the Shore so suitable."[40] Successful catches off Delaware Bay enticed some Long Island whalers to move to Cape May between 1690 and 1710, just 60 years after De Vries's failed efforts in the same area.[41]

At some point in the late 1600s or early 1700s, a few vessels also began heading farther offshore to hunt sperm whales. Perhaps the first record documenting this intent is a 1668 petition by Timothens Vanderuen, captain of a New York brigantine, seeking a license to hunt "Sperma Coeti whales" on grounds off the Bahamas and Florida Cape.[42] Spermaceti, a waxy oil from the enormous heads of sperm whales, fetched a premium price and was distinctively labeled. Oil rendered from sperm whale blubber, however, was simply called "whale oil," thus raising doubts about whether reported quantities were from right whales. Nevertheless, right whales were still the primary target as the seventeenth century came to a close. Considering the number of whales struck and lost, the proliferation of small whaling companies all along Long Island, and the many companies using offshore and pelagic whaling methods locally and further south, it may be reasonable to assume that 100 or more right whales were killed annually during most of the boom years in Long Island shore and coastal whaling; this figure would encompass the period from 1690 to 1710 or even 1720.

The Decline of Shore and Coastal Whaling

By 1725 Long Island shore whaling and coastal whaling had taken a dramatic turn for the worse.[43] A 1722 newspaper reported that only four whales had been taken on Long Island that year and "a deficiency of whales is intimated . . . but little oil is expected from thence."[44] The few Long Island catch records reporting more than 10 whales after 1720 included 40 taken in 1721[45] and 11 caught in Southampton in 1726.[46] By the mid-1700s, the catch in local waters continued to decline. Most Long Island whalers had therefore turned to fitting out sloops and other small vessels for longer voyages in search of sperm whales, some as far away as Nantucket Shoals and the Bahamas.[47]

During the American Revolution, whaling came to a grinding halt. For eight years, Long Island was cut off from the colonies by a blockade as British forces occupied the port of New York. Any hope of continued whaling was further crippled as many able-bodied seamen snuck across Long Island Sound to join the Continental Army. Those that remained used their skills and whaling boats to harass the British in what was known locally as the "whale-boat war," during which local residents used their boats to mount raids on British encampments.[48] Indeed, the Continental Army prized whaleboats; they were the stealth weapons of their day. Used in military campaigns from the late 1600s through the revolutionary war,[49] their small size and light weight meant they could be carried over land. Their silent movements, high maneuverability, and shallow draft also made whaleboats ideal for covert operations carrying men and equipment across rivers, bays, and sounds for surprise attacks or secret escapes.

At the outset of the Revolutionary War in 1776, several men from Cape Cod were asked to deliver 60 whaleboats to Falmouth, Massachusetts, for use in the war.[50] The boat with the double prow depicted in Emanuel Leutze's iconic painting of Washington crossing the Delaware may have been a requisitioned whaleboat. Although the importance of these versatile craft for the war effort has received little attention, some research might reveal a compelling case for their instrumental role in crucial events. Without their availability for the miraculous escape of Washington's army from the British in August 1776, when they secretly crossed the East River under cover of fog; and again a few days later, when they made their way across the Hudson; and perhaps in the assault at Trenton in December of that year, the British might well have prevailed and fundamentally altered the course of American history.

When the war ended, Long Island shore whalers resumed whaling and killed at least a few right whales in the 1790s.[51] It was not until after the War of 1812, however, that whaling was again taken up in earnest. By then, shore whaling was a shadow of its former importance compared to the pursuit in distant waters. To be sure, right whales continued to be killed by shore whalers whenever opportunities arose; but in the 1800s, the species was hard to come by. For the most part, landings appear to have averaged no more than one to five per year,[52] although five right whales were taken on a single day at Southampton and East Hampton in mid-April of 1847.[53] Shore whaling required little investment, and with whalers then relying on fishing or other forms of employment for the principal part of their livelihood, they had the luxury of waiting for the infrequent appearance of a right whale (figure 11.2; and see plate 8). For them, right whales were an occasional bonus bestowed by nature's bounty to supplement their income—something always sought, never passed up, but by no means reliable. By the late 1800s, shore whaling was fast becoming an anachronism.

The decline in right whale abundance off their shores led Long Island whaling merchants to shift their attention to other species in other areas. Ironically, the advances in whaling technology in the 1700s may have been the saving grace for North Atlantic right whales. Throughout the 1600s, other large whale species were largely ignored. They were too fast and too elusive, and with an ample supply of right whales producing far more value per kill, there was little reason to chase anything else. By the late 1720s, however, the tide turned. Right whales were no longer plentiful and whaling equipment and methods had improved. To maximize their profit, whalers channeled financial returns into fitting out larger ships and hiring

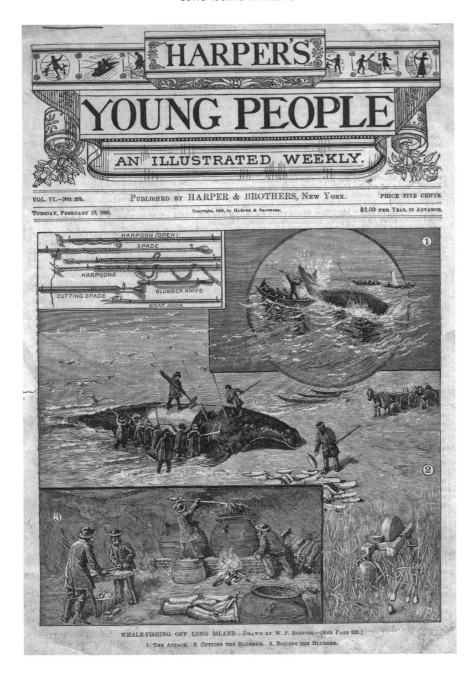

Figure 11.2. Cover of February 10, 1885, edition of *Harper's Young People*, featuring Long Island shore whaling for right whales, with prints by W. P. Bodfish. (Image courtesy of the New Bedford Whaling Museum, New Bedford, MA.)

experienced shore whalers for voyages to more distant areas in search of more abundant species, particularly sperm whales, bowhead whales, and right whales. As a result, most shifted attention away from the exhausted right whale population along the East Coast.

The transition to other species and other whaling grounds was well under way by the time of the American Revolution. The English statesman Edmund Burke acknowledged the dramatic change in a 1774 address to the British Parliament, which sounded a note of both begrudging admiration and disgust: There is "no sea, but what is vexed with their [American whale] fisheries; no climate that is not witness of their toils . . . A more hardy or enterprising race of mariners is nowhere to be met on the watery element."[54]

Long Island's Sag Harbor, southwest of Gardiners Island on Long Island's eastern tip, was an important part of the transition to hunting other whale species. Its local merchants pushed hard to improve the port for that purpose. In 1700 Abraham Schellenger, Jacop Schellenger's son, built a wharf for small whaling sloops, and in 1702 whaling merchant Samuel Mulford added another wharf and storehouse. About sixty years later a larger wharf with tryworks was constructed. By 1790, Sag Harbor hosted a greater tonnage of square-rigged sailing ships than the port of New York, with most engaged in trade for sugar, rum, and molasses in the Caribbean.[55] Although Long Island lagged a century behind Nantucket in sending out whaling ships to scour the world's oceans, Sag Harbor was home port for 6 whaling ships in 1820, 20 in 1832, and 29 in 1843. When Yankee whaling reached its zenith, with nearly 750 whaling ships flying the American flag in 1847,[56] 63 called Sag Harbor their home port; this made it America's fourth largest whaling port behind only Nantucket, New Bedford, and New London.[57]

When the Yankee whaling era came to a close in the late 1800s, many whalers returned to hamlets on eastern Long Island where some applied their rusting skills to shore whaling (figure 11.3). In doing so, they prolonged Long Island shore

Figure 11.3. Long Island whalers flensing a right whale caught circa 1900. (Image courtesy of the Nantucket Historical Association, Nantucket, MA, photo no. P18255.)

whaling into the early 1900s, even though the vocation had long since peaked by the last decades of the 1800s. By then, opportunities to catch right whales had dwindled to almost nothing, yet whaling attempts sputtered on, even though only five whales are known to have been landed on Long Island in the first two decades of the 1900s.

The last right whale taken on Long Island—and the last one landed by shore whalers anywhere along the eastern seaboard—was in 1918. By then shore whaling was less about generating income than paying homage to a bygone lifestyle. With less expensive substitutes then available for all whale products, economic returns from landings were meager at best. A few boats continued chasing whales unsuccessfully until 1924, but for practical purposes, the whale landed in 1918 was the end of American shore whaling. Fittingly, the last landing of a right whale occurred along the same stretch of Long Island beach where it began nearly 300 years earlier. For natural history museums, those last whales offered what some thought might be the last opportunity to secure a complete skeleton for museum collections before the species slipped silently into extinction.[58]

NOTES

1. Tyler, LG. 1904. *England in America, 1580–1652*. Harper and Brothers. New York, NY, p. 263 (quotation).

2. Edwards, EJ, and JE Rattray. 1932. *Whale Off! The Story of American Shore Whaling*. Frederick Stokes. New York, NY, Part 2, pp. 195–196.

3. Ibid., pp. 194–195 (quotations).

4. Gardiner, D. (1840) 2000. Chronicles of the town of Easthampton, County of Suffolk, New York. *In* T Twomey (ed.), *Exploring the Past: Writings from 1798 to 1896 Relating to the Town of East Hampton*. Newmarket Press. New York, NY, pp. 105–226.

5. Edwards and Rattray, *Whale Off!* (n. 2), p. 195 (quotation).

6. Howell, GR. 1887. *Early History of Southampton, L. I., New York with Genealogies Revised, Corrected, and Enlarged*. 2nd ed. Weed, Parsons. Albany, NY, US, p. 453 (quotation).

7. Ibid., p. 182 (quotation).

8. Ross, P. 1902. *A History of Long Island from Its Earliest Settlement to the Present Time*. Lewis Publishing. New York, NY, Vol. 1, p. 870.

9. Starbuck, A. 1878. *History of the American Whale Fishery from its Earliest Inception to the Year 1876*. Report of the US Commissioner of Fish and Fisheries, Part I. US Government Printing Office. Washington, DC, p. 9 (quotation).

10. Ross, *A History* (n. 8), p. 870.

11. Edwards and Rattray, *Whale Off!* (n. 2), pp. 197–198.

12. Starbuck, *History* (n. 9), pp. 26–27.

13. Ibid., p. 12 (quotation).

14. Ibid., p. 12.

15. Dow, FG. (1925) 1985. *Whale Ships and Whaling: A Pictorial History of Whaling during Three Centuries, with an Account of the Whale Fishery in Colonial New England*. Reprint. Dover. New York, NY, US, p. 16 (quotation).

16. Edwards and Rattray, *Whale Off!* (n. 2), pp. 182–183.

17. Starbuck, *History* (n. 9), p. 11 (quotation); and GB Goode (ed.). 1887. *History and Methods*

of the Fisheries. Section 5 of *The Fisheries and Fishing Industry of the United States,* Goode (ed.). 1884–1887. US Government Printing Office. Washington, DC, Vol. 2, p. 34.

18. Wolley, C. (1701) 1902. *A Two Years' Journal in New York and Part of its Territories in America . . . Reprinted from the Original Edition of 1701. With an Introduction and Notes by EG Bourne.* Borrows Brothers. Cleveland, OH, US.

19. Ibid., pp. 45–46 (quotation).

20. Dudley, P. 1725. An essay upon the natural history of whales, with particular account of the Ambergris found in the sperma ceti whale. Philosophical Transactions of the Royal Society of London 33:256–269, 263–264 (quotation).

21. Starbuck, *History* (n. 9), p. 11 (quotation).

22. Ibid., pp. 10–11 (quotation).

23. Edwards and Rattray, *Whale Off!* (n. 2), pp. 209–230.

24. Rattray, JE. (1953) 2001. East Hampton history including genealogies of early families. *In* T Twomey (ed.), *Discovering the Past: Writings of Jeannette Edwards Rattray, 1893–1974.* Newmarket Press. New York, NY, pp. 156–354.

25. Starbuck, *History* (n. 9), p. 12.

26. Edwards and Rattray, *Whale Off!* (n. 2), p. 201.

27. Howell, *Early History* (n. 6), pp. 180–181. Pelletreau, WS. 1903. *A History of Long Island from its Earliest Settlement to the Present Time.* Lewis Publishing. New York, NY, Vol. 2, pp. 495–496.

28. Reeves, RR, and E Mitchell. 1986. The Long Island, New York, right whale fishery: 1650–1924. *In* RL Brownell, PB Best, and JH Prescott (eds.), Right Whales: Past and Present Status, Special issue 10. Reports of the International Whaling Commission, pp. 201–214, 203.

29. Ross, *A History* (n. 8), p. 872.

30. Reeves, RR, JM Breiwick, and ED Mitchell. 1999. History of whaling and estimated kill of right whales, *Balaena glacialis,* in the Northeastern United States, 1620–1924. Marine Fisheries Review 61:1–36, 18.

31. Starbuck, *History* (n. 9), p. 15 (quotation).

32. Edwards and Rattray, *Whale Off!* (n. 2), p. 232.

33. Starbuck, *History* (n. 9), p. 13.

34. Ibid., p. 15.

35. Ibid., p. 26 (quotation). Edwards and Rattray, *Whale Off!* (n. 2), p. 217.

36. Reeves and Mitchell, The Long Island (n. 28), pp. 204–205.

37. Howell, *Early History* (n. 6), p. 181.

38. Edwards and Rattray, *Whale Off!* (n. 2), pp. 201–202.

39. Palmer, WR. 1858 (1981). The whaling port of Sag Harbor. Ph.D. Thesis, Faculty of Political Science, Columbia University, 1959. Publ. by University Microfilms International, Ann Arbor, Michigan, US; London, England. As cited in Reeves and Mitchell, The Long Island (n. 28).

40. Penn, W. (1685) 1912. A further account of the Province of Pennsylvania and its improvements for the satisfaction of those that are adventurers, and inclined to be so. *In* JF Franklin (ed.), *Original Narratives of Early American History.* Charles Scribner's & Sons. New York, NY, Vol. 12, pp. 259–268, 265 (quotation).

41. Rattray, East Hampton history (n. 24), p. 204.

42. Starbuck, *History* (n. 9), p. 15 (quotation).

43. Reeves and Mitchell, The Long Island (n. 28), p. 210.

44. Watson, JF. 1855. Annals of Philadelphia and Pennsylvania in olden times . . . by TH Mumford. In 2 Vols. Parry and McMillan, Successors to A. Hart (late Carey and Hart), as cited in Reeves and Mitchell, The Long Island (n. 28), p. 204 (quotation).

45. Anonymous. 1882. History of Suffolk County, New York, with illustrations, portraits, & sketches of prominent families and individuals. W. W. Munsell & Co. 36 Vesey Street, New York. As cited in Reeves and Mitchell, The Long Island (n. 28), p. 25.

46. Ross, *A History* (n. 8), p. 873.

47. Reeves and Mitchell, The Long Island (n. 28), p. 205.

48. Ross, *A History* (n. 8), p. 873.

49. Braginton-Smith, J, and D Oliver. 2008. *Cape Cod Shore Whaling: America's First Whalemen*. History Press. Charleston, SC, US, pp. 35–36.

50. Deyo, SL (ed.). 1890. *History of Barnstable County, Massachusetts*. Higginson Book Co. Salem, MA, US, p. 71.

51. Edwards and Rattray, *Whale Off!* (n. 2), p. 245.

52. Reeves and Mitchell, The Long Island (n. 28), pp. 202 and 205.

53. Allen, GM. 1916. *The Whalebone Whales of New England*. Memoirs of the Boston Society of Natural History Vol. 8, no. 2. Boston Society of Natural History. Boston, MA, US, pp. 109–175, 115–116, 136.

54. Freeman, F. 1862. *The History of Cape Cod: The Annals of the Thirteen Towns of Barnstable County*. G. E. Rand & Avery, Cornbill. Boston, MA, US, Vol. 2, p. 327 (quotation).

55. Edwards and Rattray, *Whale Off!* (n. 2), p. 245.

56. Bockstoce, J. 1986. *Whales, Ice, & Men: The History of Whaling in the Western Arctic*. University of Washington Press. Seattle, WA, US.

57. Edwards and Rattray, *Whale Off!* (n. 2), p. 179.

58. Ibid.

12

CAPE COD WHALING

WHILE LONG ISLAND settlers labored to perfect their whaling enterprise, colonists 100 miles (160 kilometers) to the northeast on Cape Cod Bay harbored similar ambitions; however, their skills lagged a decade behind. As already mentioned (chapter 10), when the Pilgrims dropped anchor in what is now Provincetown Harbor in 1620, the right whales cavorting about their ship instilled an immediate interest in whaling. It is therefore somewhat curious that colony sponsors anxious for returns on their investments and well aware of the right whales made no effort to provide either whaling equipment or experienced whalers. By 1620, and certainly by the 1630s and 1640s, the English commanded substantial whaling ventures in Spitsbergen. Yet there is no evidence that sponsors offered colonists anything but exhortations to take up the trade. Perhaps the lack of support reflected the Muscovy Company's self-interest in quashing potential competition for the whale products they were reaping in the Arctic. In any event, like the colonists on Long Island, those on Cape Cod were left to their own devices. As elsewhere, mastering whaling skills began by processing stranded whales.

John Winthrop, first governor of the Massachusetts Bay Colony, whose 1630 charter granted rights to lands immediately north of the Plymouth Colony, may have been the first to record efforts to salvage stranded whales on Cape Cod. In April 1635, he wrote: "Some of our people went to Cape Cod, and made oil of a whale, which was cast ashore. There were three or four cast up, as it seems there is almost every year."[1] Plymouth officials raised no objections when their new neighbors in the Massachusetts Bay Colony began visiting the peninsula to fish and salvage whales, even though the peninsula was part of the Plymouth Colony.

As residents of both colonies hacked out footholds for their new settlements, sponsors in England were growing anxious for returns on their investments. Letters from sponsors or their agents to the two new colonies routinely urged steps to exploit the resources granted in their charters, including those related to fishing. To early colonists, "fishing" referred to the catch of all forms of aquatic life, including whales. As an incentive to take up fishing, officials of the Massachusetts Bay

Colony adopted a measure in 1639 that provided "such ships & vessels & stock . . . properly employed and adventured in the taking, making, & transporting of fish . . . shall be exempt for 7 years from henseforth from all country charges."[2] There is no evidence the incentive led to hunting live whales at that early date, but it may have spurred the salvage of dead whales, because at least some whale oil was produced.

When the seven-year exemption expired, with "whaling" still limited to the salvage of dead animals, new orders were issued governing the disposition of whale oil. In 1654, the Massachusetts Bay Colony ordered that oil from drift whales in the town of Weymouth, a short distance south of Boston, be divided such that "the countrye shall have one third pte, the town of Weymouth another third pte, and the finder the other third pte."[3] The greatest opportunity for salvaging drift whales, however, was on Cape Cod. The occasional right whale may have been among the beach-cast whales, but both large and small cetacean species were undoubtedly processed and subject to the orders covering stranded whales.

It is unclear what part stranded whales played in encouraging early settlers on the mainland to move to Cape Cod, but they were likely a factor. The first towns of Barnstable, Yarmouth, and Sandwich were settled primarily by emigrants from the Massachusetts Bay Colony. All three communities were incorporated under the Plymouth Colony's charter in 1639 and 1640 and were located on the broad neck of land between Nantucket Sound and Cape Cod Bay, near the peninsula's connection to the mainland (figure 12.1).

To early colonists the name "Cape Cod" referred only to the outer arm fronting the open Atlantic, between the peninsula's elbow and northern hook. Although visited by fishermen and whale scavengers as early as the 1630s, the first permanent residents on this outer arm did not arrive until the 1640s, and its first town (Eastham) was not incorporated until 1651. The early homesteaders were quick to exploit stranded whales. Unlike communities on Long Island that agreed to share oil from stranded whales, Plymouth officials, at least initially, neither taxed nor regulated the harvested products and left all rights for their use to those who found them.[4] By the 1650s, however, disputes over stranded whales were becoming commonplace. At the same time, pressure was mounting to show financial returns to patrons in England. As a result, the Plymouth Colony's General Court and local communities like those in the neighboring Massachusetts Bay Colony adopted laws specifying duties and the disposition of stranded whales.

What may be the first official proclamation specific to whales on Cape Cod was a 1652 action by the Plymouth Court at the colony's seat of governance in New Plymouth. The court stipulated that any drift whale found on the beach, in the harbor, or within a mile of shore belonged to the town in which it was found and that for each one cut up, one barrel of oil was to be paid to the colony treasurer.[5] Whales towed ashore from beyond one mile, however, continued to belong free and clear to whoever brought them in. To collect its oil duties, each town was to designate an individual charged with that responsibility. Beginning in 1656, the towns also were to deliver their oil duties directly to colony agents in Boston.

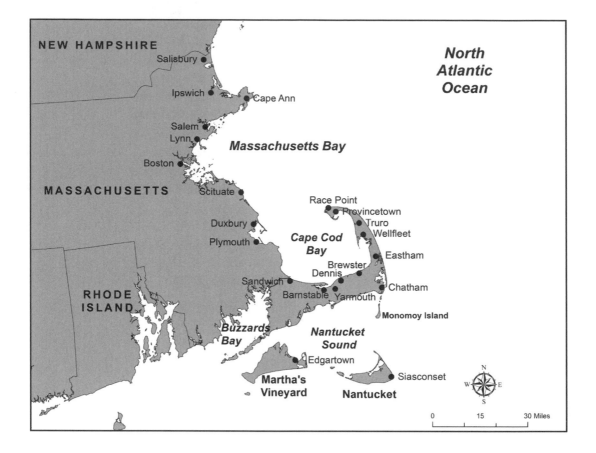

Figure 12.1. Cape Cod, Massachusetts, and surrounding areas. (Image prepared by Linda White.)

Because Plymouth Harbor was too shallow for the large ships needed to transport oil back to proprietors in England, Boston was the designated delivery point, even though it was part of the Massachusetts Colony and farther from Cape Cod.

Over the next 20 years, Plymouth officials and town councils passed a flurry of orders concerning drift whales. This spurt of official attention could reflect an increase in strandings caused by inefficient attempts to kill live whales. According to Fredrick Freeman, author of a voluminous tome published in 1860 on the history of Barnstable County, during the 1650s, "such was the activity of the whalemen that instances were frequent of whales escaping wounded from their pursuers and dying subsequently, being washed to the shore."[6] Whereas this might indicate early attempts to hunt right whales off Cape Cod, it also could reflect hunting for small whales or "grampuses."

In either case, all towns sprouting up along the cape, including Eastham, were salvaging stranded whales by the early 1650s.[7] Unfortunately, most early town records have been destroyed or lost. Among those that survive is an order from a town meeting in Sandwich on January 24, 1652: "the pay of all the whales shall be divided to every householder and to every young man that has his own disposing equally."[8] Those who reported a stranded whale received a double share, but anyone found processing a whale without having reported it was fined. The town of Eastham, which claimed lands along most of the outer cape, opted for a different approach. For unstated reasons, its residents were divided in half (perhaps representing different geographic areas within its boundaries), with all oil from each stranded whale awarded to one or the other group on an alternating basis.[9] On November 7, 1652, the town of Sandwich went a step further and appointed six men to "take care of all fish that Indians shall cut up within the limits of the town, as to provide casks for it, and dispose of it for the town's use."[10] Yarmouth and Eastham soon followed suit, appointing individuals to cut up whales and disperse the oil on behalf of their towns.[11]

The one-barrel tax on each stranded whale was clearly unpopular. This would have been particularly onerous if applied to small whales, such as porpoises or pilot whales that may have produced only two to three barrels. As a result, duties often went unpaid or were delivered in undersized, leaking, or partially empty barrels. Annoyed at the impudence of those practices, the Plymouth Court ordered in 1656 that "all oyle bee delivered att Boston vis a full barrel of merchantable oyle."[12] To track down violators of that standard, the court further stipulated, "all such Caske as are made by any Cooper within this Government shall have the two first letters of his name set upon such Caske he makes by a burnt mark [and they] are to make all their Caskes according to London Gages upon the like penaltie."[13]

Towns nevertheless continued to shirk the whale tax. In 1660 the Plymouth Colony treasurer reported that its towns owed 8 barrels of oil, and in 1661, that they owed more than 15 barrels.[14] It is unclear if the 1661 total included arrears carried over from previous years, but it implies that perhaps 15 whales were cut up in those years. Growing ever more incensed at the evasion of oil duties, colony officials attempted to force the issue with a new order in June 1661. They proclaimed that henceforth all drift whales no longer belonged to towns and that two hogsheads per whale (an amount equal to about four barrels) were to be delivered to the colony, while the rest could be kept by the towns as payment for cutting up the whale.[15] The officials further proclaimed that one-half of all whales towed in from waters farther than a mile offshore also belonged to the colony. If payment of one barrel per whale was contentious before 1661, one can imagine the outrage over these new measures. The colony's treasurer at the time, Constant Southworth, attempted to calm the seething unrest with a letter to the towns of Barnstable, Sandwich, Eastham, and Yarmouth in October of that year. Opening with the traditional soothing salutation "Loving friends," he continued, "if you will duely and trewly pay to the country for every whale that shall come, one hogshead of oyle att Boston, where I shall appoint, and that for current and merchantable, without any charge or trouble to the countrey—I say for peace and quietness sake you shall have it for theis present season, leaveing you and the Election Court to settle it soe it may bee to satisfaction on both sides."[16]

Because a hogshead was nearly twice the size of a barrel, this still represented a doubling of the earlier one-barrel tax. Yarmouth officials agreed to the two-barrel tax, but it was alone in capitulating; no other towns agreed, and even Yarmouth refused to accept the proposition that any parts of the whales belonged to the colony. Colony officials therefore retreated from their former position with a new order in October 1661: "That the towns where any [whale] shall come on shore may rent them for three years att the rate of two hogshead for a fish yearly to bee payed att Boston full and merchantable."[17] Hoping to allay the storm of discontent, the order encouraged town councils to provide a part of the oil from each whale (presumably part of their tax) to support the local ministry. The reason was explained: "There hath been much controversye occationed for want of a full and cleare settlement of matter relateing unto such whales as by Gods providence doe fall unto any pte of this Jurisdiction . . . [the towns] should agree to sette apart some pte of every such fish or oyle for the Incurragement of an able Godly Minnester amongst them."[18]

When opposition to the increased tax persisted, the Plymouth General Court retrenched and further reduced the amount. In June of 1662, it issued yet another order reducing the quantity to be paid to one full hogshead of oil for every whale cast ashore, and if the "ffish" was torn or wasted so that one-fourth of it was gone, only half a hogshead was due.[19] In still further deference to irate townspeople, a note was added in the margin of the order, "This Court doth order for the prevention of any discontent or controversy for the future and for final Issue and settlement . . . the Countreys due of every such whale was altered from a hogshead to a barrel."[20] With the tax restored to its pre-1661 level, matters settled down and at least some towns agreed to honor the colony's order by providing a portion of each whale to the local ministry. The Town of Eastham voted to do so in 1662,[21] but others were slower to act. Yarmouth waited until 1692 to adopt a similar order.[22]

The Beginning and the Peak of Hunting of Live Whales

What appears to be the first clear evidence of hunting live whales in the Massachusetts Bay and Plymouth Colonies comes from the diary of Reverend Simon Bradstreet of New London, in what would become the state of Connecticut. His source of information is not identified, but Bradstreet mentioned the death of Jonathan Webbe, who drowned in Boston Harbor in October 1668 while "catching a whale below the Castle." The castle was an earthen-walled fort built in 1634 to defend Boston Harbor. "In coiling up ye line unadvisedly he did it about his middle thinking the whale to be dead, but suddenly shee gave a Spring and drew him out of the boat, he being in ye midst of the line, but could not be recovered while he had any life."[23] The account indicates that at least some experimentation in whale hunting had begun by 1668. By the 1670s it was surely underway.[24] In 1675, the Boston merchant and shipowner John Hull sent his ketch *Sea-Flower* to London to sell a cargo of whale oil and whalebone.[25]

By the 1670s, Cape Cod whalers were still learning about their craft and suffering high struck-and-loss rates. In some cases, the whales struck and then lost were found dead, drifting at sea, and were towed ashore by others; in other cases, they were claimed by coastal land owners when washed ashore. This led to disputes that officials were often unable to resolve. In 1672, for example, the court at Plymouth made an order: "In reference unto a whale brought on shore to Yarmouth from sea, the Court leaves it to the Treasurer to make abatement of what is due to the countrey therof, by law, as hee shall see cause, when hee treated with those that brought it on shore."[26] Drift whales not killed by whalers, however, remained an important source of income for town coffers. In 1680, Yarmouth assigned responsibility for securing and processing stranded whales to either four or five men in each of three shoreline sectors; the men received a payment for each whale processed.[27]

Perhaps the first sign that Cape Cod whalers were becoming proficient is from records kept by Massachusetts Bay Colony Treasurer Samuel Sewall, who commented on the regular shipments of whale oil being sent to England in the early 1680s.[28] Although some of that oil was surely from stranded whales, including whales other than right whales, and some may have been oil delivered to Boston by Long Island whalers, any doubt that Cape Cod whalers were successfully hunting large whales is eliminated by the later part of that decade. By then the Crown of England had briefly assumed direct control over colonial affairs in New England and installed Edmond Andros as the colonies' governor. On January 24, 1687, Massachusetts Bay Colony Secretary Edward Randolph sent a letter to authorities in England: "Since the governor's arrival, New Plymouth colony have great profitt by whale killing. I believe they will have nigh 200 tons for to send to England and will be one of our best returns, now that beaver and peltry fayle us."[29] Using conversion factors to translate tons of oil first into barrels, and then to whales, 200 tons would equate to between 40 and 50 right whales if all of the oil was from that species.[30] In 1690, Boston merchants sent at least two shipments of 144 and 152 barrels of whale oil to London, and another 748 barrels of oil "of New England fishing" to Amsterdam.[31] Together those shipments could have represented 24 to 29 whales.

Further progress at hunting live whales is apparent in a decree from March of 1688 by the Massachusetts Bay Colony that made the whaler's creed ("craft claims the whale"[32]) the law of the land. This new law affirmed that a whale belonged to whoever killed it (the "craft"). Its intent was squarely aimed at resolving disputes over whales that died after being struck and lost and were then later claimed as drift whales by other mariners and shore owners. Its enumerated sections proclaimed the following:

> Furst: if aney persons shall find a Dead Whael on the stream And have the opportunity to toss her on shore; then ye owners to allow them 20 shillings; 2ly : if thay cast her out & secure ye blubber and bone then ye owners to pay for it 30s (that is if ye wheal ware lickly to be loast;). 3ly, if it proves to be floatesom not killed by man then ye Admirall to Doe Thaire as he shall please; 4ly, that no persons shall presume to cut up any wheal till she

be vewed by toe persons not concarned; that so ye Right owners may not be rongged of such whael or whaels; 51y, that no whael shall be needlessly or fouellishly lansed behind ye vitall to avoid stroy; 61y, that each companys harping Iron and lance be Distinckly marked on ye heads & sockets with a poblick mark: to ye prevention of strife; 71y, that if a whale or whalls be found & no Iron in them: then they that lay ye nearest claime to them by their strokes & ye natural markes to have them; 81y, if 2 or 3 companyes lay equal claimes, then thay equally to share.[33]

In November 1690, a year after the unpopular regime of Governor Andros ended and a representative government was restored to the Massachusetts Bay and Plymouth Colonies, the general court for Plymouth adopted a similar law; this included the following provision: "1. This Court doth order, that all whales killed or wounded by any man & left at sea, sd whale killers that killed or wounded sd whale shall presently repaire to some prudent person whome the Court shall appoint, and there give in the wounds of sd whale, the time & place when & where killed or wounded; and sd person so appointed shall presently commit it to record, and his record shall be allowed good testimony in law."[34]

Accompanying its decree, the Plymouth Court appointed whale "inspectors" to ensure the "prevention of suits by whalers."[35] The new posts were established at most towns ringing the shores of Cape Cod Bay, including Plymouth, Sandwich, Barnstable, Yarmouth, Eastham, Scituate, and Duxbury. Appointees were responsible for recording circumstances and wounds associated with any whales reported struck and lost by whalers and for inspecting all stranded whales to determine rightful owners. For assigning ownership, the inspectors relied on their records of when and where whales were reported struck, the wounds reportedly inflicted, and evidence of embedded harpoons. Those officials were also charged with noting any taxes due to the colony. Apparently a great number of whales were killed at this time, with many still being struck and lost. The position of whale inspector was maintained by at least some towns until the 1720s. One surviving report signed by Joshua Soule read as follows:

Decembr ye 3 1724 I was desired by Obedia Lamson to obsarve the marks on a whale cut up by him in Duxborough beach ware as folleth. A short bone with 2 iarnholes one on her rite side about six feete abaft her spouts w'h iarn I gug [judged] mortal the other about 4 feete back of that & sumthing on the left side of her back bone not mortal and I gudg pricked with a lance one on the rite fin near the joint. I suppose this fish to be killed 4 or 5 days be fore this date.[36]

In October 1691 the English Crown combined charters for the Plymouth and Massachusetts Bay Colonies. Shortly thereafter at least some towns reasserted ownership rights to whales cast up within their jurisdictions. In November 1693 the town of Sandwich did then order and appoint their trusty Friends Samuel Prince and Thomas Smith both of said Town, Gentlemen, their agents to defend the said Towns right and interest in the whale fish or part of a fish which come on shore at the Town Neck Beach in said Town of Sandwich in the month of October last past that was seized for their Majesty by William Basset Sheriff of this County of Barnstable . . . and all other whales or part of such fish that shall come or be cast on shore within the limits of our Town of Sandwich.[37]

Reliable data on the number of whales killed during this period are scarce, but the glory days for Cape Cod shore whaling apparently began in the 1690s, or perhaps a few years earlier. By then the success of shore whaling on the cape was well known throughout New England. In 1690, Nantucket colonists successfully recruited a Cape Cod whaler named Ichabod Paddock of Yarmouth to teach them the whaling trade. Unfortunately, if colony officials kept records on the number of whales killed by Cape Cod whalers, they have been lost. What survives offers only a vague outline of possible catch levels. Cotton Mather provided a 1697 description of Plymouth Colony whaling:

They have since passed on to the catching of whales, whose oil is become a staple commodity of the country; . . . And within a few days of my writing this paragraph, 1697, a cow and calf were caught at Yarmouth. The cow was 55 ft. long, the bone was 9 or 10 in. wide; a cart upon wheels might have gone into the mouth of it. The calf was 20 ft. long, for unto such calves the sea monsters draw forth their breasts. But so does the good God here give these people to suck the abundance of the seas.[38]

While hunting large whales assumed a prominent place in the lives of Cape Cod residents, drift whales were still an important source of community revenue. In 1702, Sandwich passed a town ordinance directing that proceeds from stranded whales be set aside for their local minister, Reverend Roland Cotton.[39] The order stated that he was to receive "all such drift whales as shall, during the time of his ministry in Sandwich, be driven or cast ashore within the limits of the town, being such as shall not be killed with hands."[40] That same year Reverend John Cotton in Yarmouth realized 40 pounds from a similar measure adopted by that town in 1692.[41]

Among the few references offering any insight into the number of whales killed is in a letter by Wait-Still Winthrop (grandson of John Winthrop, the first governor of the Massachusetts Bay Colony) to his brother on January 24, 1699. This reported that 29 whales were killed in a single day and that, on a previous trip to Plymouth, Winthrop had learned of a group of whalers who killed 6 whales within a few days.[42] With the whaling season extending into April and 29 whales caught in a single day in January, the total number of whales killed that season, including those struck and lost, easily could have approached 50 and perhaps even 100 whales.

A comparable or even greater catch seems plausible in the early 1700s, considering the number of men employed during the whaling season. In 1715, the town of Barnstable divided a stretch of beach on Cape Cod Bay called Sandy Neck into 60 lots, with "four spots or pesses [places] for the setting up of four try houses of about half an acre to each for ye laying blubber barrels, wood, & other nessesceries for ye trying of oil as need may require."[43] Other lots were used as staging centers by owners of whaleboats or to house whalers. According to Reverend John Mellen, whale houses built to put up whalers along this stretch of beach "employed nearly two hundred men for three months."[44] A few miles east in Yarmouth, at a place called "Black Earth" near the current town of Dennis, 36 whaleboats were operating in 1713 from barracks housing another 200 whalers,[45] many of whom were Native Americans.[46] Still other whaling crews were based in Plymouth, Sandwich, and towns farther out on the cape, including Eastham,[47] Wellfleet (whose name may be a corruption of "whale fleet"),[48] and Truro.

As early as 1670, whalers also operated from a lawless outpost on the unsettled tip of Cape Cod between Wood End and Race Point.[49] Dubbed "Hell Town" and located just a few miles from the *Mayflower*'s first landfall near what would become Provincetown in 1727, the area harbored a motley crew of seasonal invaders who set up camp to fish, salvage drift whales, and possibly hunt whales. William Clapp, a homesteader who set down roots in 1701 some 15 miles down the cape in what would become the town of Truro, provided some insight into their activities. On July 13, 1705, he wrote to Joseph Dudley, governor of the Massachusetts Bay Colony, about illegal whaling carried out in the area. Requesting authority to enforce colony whaling laws, his letter claimed that 130 men had taken up residence at the tip of Cape Cod during the 1705 whaling season:

> I have very often every year sien that her magiesty has been very much
> wronged of har dues by these country peple and other whall men as coms
> here a whalen every year which tacks up drift whals which was never killed
> by any man which fish I understand belongest to har magiesty and had i
> had power i could have seased severl every year and lickwies very often here
> is oportunyty to seas vesels and goods which are upon a smoglen [smug-
> gling?] acompt i believe had i had a comishon so to do I could have seased
> a catch this last weak which had most of thar men out landish men i judge
> portges she lay hear a week and a sloop i believe thar bisnes for them.[50]

The "out landish" men he judged to be "portges" may have been Portuguese or Basque fishermen or whalers, or perhaps the French privateers reportedly prowling the cape's waters at this time.[51]

By the turn of the century, Cape Cod whalers began using sloops and schooners that could carry a pair of shallops to kill whales beyond the range of shore whalers. In 1713, Barnstable had a fleet of seven vessels totaling 170 tons.[52] During the whaling season, many of those vessels likely chased right whales inside Cape Cod Bay and along the cape's outer coast. This could have significantly increased their catch, possibly keeping annual catch totals near or even above 100 whales per year for at least some years; according to one source, 5,000 barrels of "fish oil" per year were being exported from New England around the time of 1718.[53] If three-quarters of that oil was from right whales, at conversion rates of 36 to 44 barrels per whale, the catch would equate to between 65 and 108 of that species. Additional amounts of oil may have been sold or traded locally.

The large number of whale carcasses, or "lean," left by whalers after stripping blubber and whalebone became an object of concern to the cape's residents in the early 1700s. The standard disposal practice apparently was to do nothing; carcasses were simply left to rot and wash away with the tide. According to one contemporary comment, whale carcasses were left "for Indians and Swyne to eat."[54] Some residents scavenged meat or other parts to fertilize the cape's poor soils, and once the flesh had rotted away, the bones were salvaged for various purposes. An archeological excavation at the site of a tavern built in the 1680s or 1690s on Grand Island in Wellfleet unearthed a large whale vertebra that had been used as a chopping block.[55] Frederick Freeman's exhaustive history of Barnstable County published in 1862 noted that people alive when he wrote his tome had memories of places on the cape strewn with the bones of large whales, and that fifty years earlier whale ribs had commonly been used for fence posts.[56]

The accumulating carcasses were a nuisance to some, but an entrepreneurial opportunity to others. In 1704, two men from Eastham and one from Nantucket proposed to process the remains into saltpeter and petitioned colonial authorities in Boston for exclusive rights to "take away the leftover whale after blubber striped, and not let any others do it."[57] The petition was granted for a four-year term. Whether the petitioners followed through on their plan is unknown. If they did, they must have made hardly a dent in the surplus of carcasses. In 1706, whalers in Eastham became concerned that rotting carcasses were making it difficult, or at least unpleasant, to process new kills. As a result, they petitioned the town in support of a permit requested by Thomas Haughton to take away the remains: "all or most of us are concerned in fitting out Boats to Catch and take Whales when ye season of ye year Serves; and whereas when we have taken any whale or whales, our Custom is to Cutt them up and to take away ye fatt and ye Bone of such Whales as are brought in and afterwards to let ye Rest of ye Boddy of ye Lean of whales Lye on shoar in lowe water to be washt away by ye sea, being of noe vallue nor worth any Thing to us."[58]

The Culmination of Cape Cod Shore Whaling

Cape Cod's best shore-whaling days ended in the 1720s. In early parts of the decade, whaling was still an economic mainstay that merited special dispensations. At the outbreak of hostilities between colonists and some of the region's Native Nations in 1724 and 1725, colony officials passed a law providing that any Cape

Cod Indians agreeing to fight with the settlers would be dismissed from service for the whaling season so they could easily return to the cape to crew local boats.[59] By then, however, whales in Cape Cod Bay were becoming scarce and catch levels were declining precipitously. As a result, forward-thinking whalers had begun investing their profits in larger vessels to pursue whales on more distant grounds. Alluding to this transition, an article in the March 20, 1727, *Boston News-Letter* reported, "we hear from the Towns on the Cape that the Whale Fishery among them has failed much this Winter, as it has done for several Winters past, but having found out the way of going to Sea Upon that Business and having had much Success in it, they are now fitting out several Vessels to sail with all Expedition upon that dangerous Design this Spring."[60]

By 1737, the same paper reported that a dozen whaling vessels were operating out of Provincetown, some of which may have been hunting right whales east of the cape. The May 27, 1737, edition reported the following in regard to a whaling sloop out of Chatham that had returned from a voyage on May 11: "About forty leagues to the eastward of Georges Banks, they struck and wounded two Whales, which lay upon the Water seemingly in a dying Posture: but one of them suddenly rush'd with great Violence over the midst of one of their Boats, and sank both the Boat and Men in the sea; one Man was thereby kill'd outright, and two others wounded: Tis a wonder they were not all destroy'd, for the whale continued striking and raging in a most furious manner."[61] Although the two whales may have been sperm whales, it seems more likely they were right whales given that the Great South Channel and waters 20 to 40 miles east of Cape Cod along the northern flank of Georges Bank are now a late spring feeding area for large numbers of right whales.

At least some Cape Cod vessels followed the lead of Dutch whalers to Arctic waters in the Davis Strait between Greenland and Canada to hunt bowheads and possibly right whales. A 1737 edition of the *Boston News-Letter* mentioned that a whaling ship from Eastham had been sent to those grounds.[62] According to William Douglass, an early historian of the American colonies, bowhead whales were hunted in Davis Strait north of 70° latitude (above the Arctic Circle), while right whales were taken at more southerly latitudes. A summary of colonial whaling activities in his history of the colonies published in 1760 noted the following in regard to the whales caught north of 70° latitude:

> [They] may yield 150 puncheons being 400 to 500 barrels of oil and bone of eighteen feet and upwards; they are heavy loggy fish, and do not fight, as New England whalers express it; they are easily struck and fastened, but are not above one third of them recovered; by sinking and bewildering themselves under the ice, two thirds of them are lost irrecoverably; the whalebone whales killed upon the coast of New England, Terra de Labrador, and the entrance of Davis's straits, are smaller; do yield not exceeding 120 to 130 barrels of oil, and of nine feet bone 140 lb. wt. they are wilder more agile and do fight.[63]

John Osborn memorialized voyages to those northern grounds in verse extolling the adventure and romance of the whaler's life. Osborne, born in 1713 to the minster in Eastham, eventually entered the medical profession. Whether he ever shipped out on a whaling vessel is unknown, but his song, probably written sometime in the 1730s, was said to be a favorite of Cape Cod whalers. Notwithstanding its starry-eyed tone, the verse offers interesting insights that are consistent with Douglass's account and suggest it is more than musings by an uninformed romantic:

When spring returns with western gales, And gentle breezes sweep,
The ruffling seas, we spread our sails To plow the wat'ry deep;
For killing northern whales prepar'd, Our nimble boats on board
With craft and rum (our chief regard,) And good provisions stor'd;
Cape Cod, our dearest native land, We leave astern, and loose
Its sinking cliffs and lessening sands, Whilst Zephyr gently blows,
Bold, hearty men, with blooming age, Our sandy shore produce;
With monstrous fish they dare engage, And dangerous calling choose.
Now towards the early dawning East, We speed our course away
With eager minds and joyful hearts, To meet the rising day;

. . . . .

When eastward, clear of Newfoundland, We stem the frozen pole,
We see the icy islands stand, The northern billows roll.
As to the North we make our way, Surprising scenes we find,
We lengthen out the tedious day, And leave the night behind.
We see the northern regions where Eternal winter reigns.
One day and night fill up the year, And endless cold maintains;
We view the monsters of the deep, Great Whales in numerous swarms,
And creatures there that play and leap, Of strange, unusual, forms.
When in our station we are placed, And Whales around us play,
We launch our boats into the main, And swiftly chase our prey.
In haste we ply our nimble oars, For our assault designed;
The sea beneath us foams and roars. And leaves our wake behind;
A mighty whale we rush upon, And in our irons throw;
She sinks her monstrous body down, Among the depths below;
But when she rises out again, We soon renew the fight,
Thrust our sharp lances in amain, And all her rage excite.
Enraged, she makes a mighty bound, Thick foams and whiten'd sea,
The waves in circles rise around, And wid'ning roll away.
From numerous wounds, with crimson flood, She stains the froathy seas,
And gasps, and blows her latest blood, While quiv'ring life decays.
With joyful hearts we see her die, And on the surface lay,
While all with eager haste apply, to save our deathful prey.[64]

While pelagic whalers found increasing success, Cape Cod shore whalers faced increasing struggles. In 1738, the *Boston News-Letter* reported that Provincetown

207

whalers had caught only two small whales as of January 5, and that as of February, Yarmouth whalers had landed just one.[65] The following year, Provincetown whalers took just six small whales and a large whale, while Sandwich whalers caught only two small whales by February 15, well into the local whaling season.[66] In 1739, when it was clear the local whale fishery was dying, some Provincetown whalers gave up and moved to Casco Bay, Maine.[67] Those that stayed began targeting humpback whales[68] and small whales called "blackfish" or pilot whales.

By the mid-1700s, the mainstay of Cape Cod shore whalers had shifted to these blackfish and large whales other than right whales. In 1746, only three or four large whales were caught on all of Cape Cod,[69] whereas reports of blackfish landings became common. Producing only up to two barrels of oil each, what blackfish lacked in size and quantity of blubber, they made up for in numbers. They sometimes occurred in pods of hundreds of whales, and on rare occasions in pods of more than 1,000. Shore whalers herded these pods ashore by encircling them with boats and creating a great commotion pounding sticks on the side of their boat, shouting at the top of their voices, and blowing bugles (figure 12.2). The largest pod known to have been taken by cape whalers was driven ashore on November 17, 1874; it numbered 1,405 whales and produced 870 barrels of oil.[70]

In the first decade of the 1800s, right whale catch rates dwindled to two or three per year.[71] After that, landings became even smaller and more erratic. Between

Figure 12.2. This 1875 print by E. Morse was captioned "A Remarkable Catch of Blackfish in Cape Cod Bay." The blackfish referred to are pilot whales. (Image courtesy of the New Bedford Whaling Museum, New Bedford, MA.)

1840 and 1858 at least half a dozen right whales were caught,[72] including one notable "right whale" caught by a 50-ton sloop off the tip of Cape Cod in May 1843. Believed to be the largest whale ever taken off the cape, it was thought to have had the potential to yield 200 barrels of oil and 2,000 pounds of whalebone. But because it was flensed at sea and cargo space was limited, whalers were able to collect only some of its blubber and whalebone before casting the remainder adrift. Nonetheless, it produced 125 barrels of oil and 300 pounds of bone (baleen) over 14 feet long.[73] If the estimated oil quantities and reported length of the whalebone are accurate, this whale almost certainly was a stray bowhead from the Arctic. Such an occurrence of this species so far south is extremely rare. Indeed, no sightings had ever been confirmed south of Newfoundland until the summer of 2012, and again in the spring of 2013, when a lone bowhead—the same individual in both years—was photographed in Cape Cod Bay.

A fleeting resurgence in right whale landings occurred in Cape Cod Bay in the spring of 1888, when the pelagic steamer *A. B. Nickerson* managed to kill five right whales with a bomb lance (figure 12.3). Two were large whales killed off Provincetown on May 20 yielding 170 barrels of oil, and a third was shot but not recovered until it was found adrift several days later. A few weeks later in early June, the steamer's crew killed a mother and her calf. Although the calf sank, the large cow, estimated to have been 60 feet long, produced 100 barrels of oil and 1,500 pounds of bone.[74] At least some whales were also taken by shore whalers during the 1880s (figure 12.4).

Figure 12.3. This engraving of a whale being killed with a bomb lance from a steamer appeared in an 1883 issue of *The Picture Magazine*. (Image courtesy of the Picture Collection, The New York Public Library, Astor, Lenox, and Tilden Foundations.)

Figure 12.4. Cape Cod whalers flensing a right whale near Provincetown, circa 1880. (Image courtesy of the Nantucket Historical Association, MA, photo no. P3193.)

The last right whale killed on Cape Cod was a 35-foot-long calf. After becoming entangled in a fish trap in Provincetown harbor on January 15, 1909, it was dispatched with a bomb lance. So unusual was the catch of a right whale by then that an enterprising undertaker injected it with formalin and had it shipped to Boston to be put on display.[75]

NOTES

1. Hosmer, HK (ed.). 1908. *Winthrop's Journal, "History of New England," 1630–1640.* Charles Scribner's Sons. New York, NY, p. 148 (quotation).

2. *Records of the Governor and Company of Massachusetts Bay.* Vol. 1, 8, p. 257, as cited in J Braginton-Smith and D Oliver. 2008. *Cape Cod Shore Whaling: America's First Whalemen.* History Press. Charleston, SC, US, p. 43 (quotation).

3. *Records of the Governor and Company of Massachusetts Bay.* Vol. 4, 2, 192. As cited in Braginton-Smith and Oliver, *Cape Cod Shore* (n. 2), p. 43 (quotation).

4. King, RH. 1994. *Cape Cod and Plymouth Colony in the Seventeenth Century.* University of America Press. Boston, MA, US, p. 205.

5. Ibid., p. 206.

6. Freeman, F. 1862. *The History of Cape Cod: The Annals of the Thirteen Towns of Barnstable County.* G. E. Rand & Avery, Cornbill. Boston, MA, US, Vol. 2, p. 50 (quotation).

7. King, *Cape Cod* (n. 4), p. 206.

8. Sandwich Town Meeting Records. Vol 1, 1A, 2. As cited in Braginton-Smith and Oliver, *Cape Cod Shore* (n. 2), p. 45 (quotation). The reference to "every young man who has his own" may be in regard to the sons of property owners in line to inherit.

9. King, *Cape Cod* (n. 4), p. 206.

10. Freeman, *The History of Cape Cod* (n. 6), p. 50 (quotation).

11. Allen, GM. 1916. *The Whalebone Whales of New England.* Memoirs of the Boston Society of Natural History Vol. 8, no. 2. Boston Society of Natural History. Boston, MA, US, pp. 109–175, 115–116, 149.

12. 11/Plymouth Colony Records/66. As cited in Braginton-Smith and Oliver, *Cape Cod Shore* (n. 2), p. 44 (quotation).

13. 11/Plymouth Colony Records/60. As cited in Braginton-Smith and Oliver, *Cape Cod Shore* (n. 2), p. 44 (quotation).

14. Plymouth Colony Records 8:99 and 8:104. As cited in King, *Cape Cod* (n. 4), p. 206.

15. King, *Cape Cod* (n. 4), p. 207.

16. Starbuck, A. 1878. *History of the American Whale Fishery from Its Earliest Inception to the Year 1876*. Report of the US Commissioner of Fish and Fisheries, Part I. US Government Printing Office. Washington, DC, p. 7 (quotation).

17. 4/Plymouth Colony Records/9 Court Orders. As cited in Braginton-Smith and Oliver, *Cape Cod Shore* (n. 2), p. 50 (quotation).

18. 11/Plymouth Colony Records/132-5. As cited in Braginton-Smith and Oliver, *Cape Cod Shore* (n. 2), pp. 49–50 (quotation).

19. Allen, *The Whalebone Whales* (n. 11), p. 151.

20. 11/Plymouth Colony Records/207. As cited in Braginton-Smith and Oliver, *Cape Cod Shore* (n. 2), p. 53.

21. Thoreau, HD. (1864) 1908. *Cape Cod*. Reprint. Thomas Y. Crowell. New York, NY, p. 51.

22. Braginton-Smith and Oliver, *Cape Cod Shore* (n. 2), p. 50.

23. New England Historical and Genealogical Register, 1885, vol. 9, p. 44. As cited in Allen, *The Whalebone Whales* (n. 11), p. 154 (quotation).

24. Reeves, RR, JM Breiwick, and E Mitchell. 1992. Pre-exploration abundance of right whales off the eastern United States. *In* J Hain (ed.), *The Right Whale in the Western North Atlantic: A Science and Management Workshop*, 14–15 April 1992, Silver Spring, MD. Reference Document 92-05. Northeast Fisheries Science Center, National Marine Fisheries Service. Woods Hole, MA, US, p. 8.

25. Dow GF. (1925) 1985. *Whale Ships and Whaling: A Pictorial History of Whaling during Three Centuries, with an Account of the Whale Fishery in Colonial New England*. Reprint. Dover. New York, NY, US, p. 16.

26. Allen, *The Whalebone Whales* (n. 11), p. 151 (quotation).

27. Ibid., pp. 151–152.

28. Ibid., p. 152.

29. Edward Randolph to M. Povey, Boston, January 24, 1687. *In* T Hutchinson (ed.). (1687) 2010. *A Collection of Original Papers Relative to the History of the Colony of Massachusetts-Bay*. Reprint. Applewood Books. Carlisle, MA, US, pp. 556–558, 558 (quotation).

30. Lindquist, O. 1992. Comments concerning old whaling statistics: the British whale oil measures gallons and tun, the German and Dutch Quardeelen and the ratio between them. Reports of the International Whaling Commission 42:475–477.

31. Dow, *Whale Ships* (n. 25), p. 17.

32. Starbuck, *History* (n. 16), p. 8.

33. Massachusetts Archives, Vol. 125, Leaf 80, as cited in Dow, *Whale Ships* (n. 25), p. 10 (quotation).

34. Records of the Colony of New Plymouth. 1856, vol. 6, p 252. As cited in Allen, *The Whalebone Whales* (n. 11), pp. 152–153 (quotation).

35. Braginton-Smith and Oliver, *Cape Cod Shore* (n. 2), p. 58 (quotation).

36. Dow *Whale Ships* (n. 25), p. 11 (quotation).

37. Sandwich Town Records, Vol. 2, as cited in Braginton-Smith and Oliver, *Cape Cod Shore* (n. 2), p. 60 (quotation).

38. Cotton Mather 1697. As cited in Freeman, *The History of Cape Cod* (n. 6), p. 631 (quotation).

39. Thoreau, *Cape Cod* (n. 21), p. 51.

40. Goode, GB (ed.). 1887. *A Geographical Review of the Fisheries Industries and Fishing Communities for the Year 1880*. Section 2 of *The Fisheries and Fishing Industries of the United States*, Goode

(ed.). 1884–1887. US Government Printing Office. Washington, DC, p. 731 (quotation).

41. New England Historical and Genealogical Register, 1885, vol. 9, p. 44. As cited in Allen, *The Whalebone Whales* (n. 11), p. 154.

42. Wait-Still Winthrop to Fitz-John Winthrop, Boston, 27 January 1699/1700, Collec. Mass. Hist. Soc., Ser. 6,5(1862): 54–55. As cited in King, *Cape Cod* (n. 4), p. 205.

43. McLauthlin, JR (ed). 1935. *Records of the Proprietors of the Common Lands in the Town of Barnstable, Massachusetts, 1703–1795.* Barnstable Planning Board. Revised edition with index by Andrea Leonard, Bowie, MD. Heritage Books, p 130. As cited in Braginton-Smith and Oliver, *Cape Cod Shore* (n. 2), p. 31 (quotation).

44. Trayser, D. 1971. *Barnstable, Three Centuries of the Cape Cod Town.* Parnassus Imprints. Yarmouth, MA, pp. 326–327. As cited in Braginton-Smith and Oliver, *Cape Cod Shore* (n. 2), p. 31 (quotation).

45. Thomas Prince Howes, Letter to the *Yarmouth Register*, February 3, 1860, 1. As cited in Braginton-Smith and Oliver, *Cape Cod Shore* (n. 2), p. 31. A badly deteriorated occipital bone from the skull of a right whale perhaps killed by these whalers in the barracks near Dennis, or perhaps even by earlier Basque whalers, was unearthed at Ellis Landing in 2011, a few miles east of Dennis; see E. Williams, Whale Skull Removed from Brewster Beach, *Cape Cod Times*, September 22, 2011. Later analyses revealed its age at 350 to 500 years old, dating it between 1500 and 1650.

46. Swift, CF. (1884) 1995. *History of Old Yarmouth: Comprising the Present Towns of Yarmouth and Denis.* C. F. Swift and Son. Yarmouth Port, MA. Preservation photocopy and binding by Acme Bookbinding. Charlestown, MA, US, p. 114.

47. Braginton-Smith and Oliver, *Cape Cod Shore* (n. 2), p. 66.

48. Brigham, SP. 1920. *Cape Cod and the Old Colony.* G. P. Putnam's Sons. New York, NY, p. 105.

49. Kingor, *Pirates of Race Point*, 77. As cited in Braginton-Smith and Oliver, *Cape Cod Shore* (n. 2), p. 66.

50. Freeman, F. 1858. *The History of Cape Cod, The Annals of Barnstable County, including the District of Mashpee*, Vol I, Geo. C. Rand and Avery, Boston, MA, p. 342 (quotation).

51. Smith, WC. (1923) 2009. *A History of Chatham, Massachusetts.* Reprint. Bibliolife. Charleston, SC, US, pp. 174–178.

52. Waters, JJ, Jr. 1968. *The Otis Family.* University of North Carolina Press. Chapel Hill, NC, pp 51–53. As cited in Braginton-Smith and Oliver, *Cape Cod Shore* (n. 2), p. 61.

53. Douglass, W. 1760. *A Summary, Historical, and Political, of the First Planting, Progressive Improvements, and Present State of the British Settlements in North America.* J. S. Dodsley. London, Vol. 1, p. 298.

54. Braginton-Smith and Oliver, *Cape Cod Shore* (n. 2), pp. 64–65 (quotation).

55. Deetz, J, and P Deetz. 2000. *The Times of Their Lives–Life, Love, and Death in the Plymouth Colony.* Freeman and Co. New York, NY, p 253. As cited in Braginton-Smith and Oliver, *Cape Cod Shore* (n. 2), p. 32.

56. Freeman, *The History of Cape Cod* (n. 6), p. 623.

57. Province Laws Vol. 3, chapter 119, Acts of 1706–1707, 207. As cited in Braginton-Smith and Oliver, *Cape Cod Shore Whaling* (n. 2), p. 64 (quotation).

58. Starbuck, *History* (n. 16), p. 30 (quotation).

59. Ibid., p. 31.

60. Ibid.

61. Ibid., p. 32 (quotation).

62. Clark, AH. 1887. History and present condition of the fishery. In GB Goode (ed.). 1887. *History and Methods of the Fisheries.* Section 5 of *The Fisheries and Fishing Industry* (n. 40) Vol. 2, pp. 3–218, 27.

63. Douglass, *A Summary* (n. 53), p. 56 (quotation). The high struck-and-loss rate may be due

to early American whaling methods used in this area. According to Douglass, while the Dutch handled long warps of line that kept whales held fast when they dove under the ice, the New Englanders used drogues.

64. Freeman, *The History of Cape Cod* (n. 6), p 89 (quoted verse).

65. Starbuck, *History* (n. 16), pp. 32–33.

66. Starbuck, *History* (n. 16), p. 33.

67. Reeves, RR, JM Breiwick, and ED Mitchell. 1999. History of whaling and estimated kill of right whales, *Balaena glacialis*, in the northeastern United States, 1620 to 1924. Marine Fisheries Review 61(3):1–36, 8.

68. Clark, History and present condition (n. 62), p. 27.

69. Douglass, *A Summary* (n. 53), p. 59.

70. Braginton-Smith and Oliver, *Cape Cod Shore* (n. 2), p. 82.

71. Ibid., pp. 76–79.

72. Ibid. Allen, The *Whalebone Whales* (n. 11), p. 136.

73. Allen, *The Whalebone Whales* (n. 11), p. 135.

74. Nantucket Journal, June 7, 1888. As cited in GM Allen, *The Whalebone Whales* (n. 11), p. 139.

75. Allen, *The Whalebone Whales* (n. 11), p. 140.

13

NANTUCKET,
MARTHA'S VINEYARD,
AND CAPE MAY

PERHAPS NO PLACE on earth is more closely associated with whaling than Nantucket. Ask a random stranger to name a whaling port, and odds are it will be Nantucket. At first blush, the island's stature as the world's preeminent whaling center seems totally out of proportion to its diminutive size. It is only 14 miles long and 5 miles wide. On a nautical chart, Nantucket's crescent shape, lying 30 miles south of Cape Cod, creates the illusion of a comet streaking east into the open Atlantic (figure 13.1). Although the island owes its existence to rubble dropped 20,000 years ago as the Laurentide Ice Sheet began to retreat, its global renown is based on whaling. For nearly 100 years between the mid-1700s and 1840s, Nantucket was the whaling capital of the world. In 1774, when Yankee whalers sent out some 360 sailing vessels to comb the far corners of the globe for sperm whales and right whales, nearly half called Nantucket home.[1]

That this speck of land is still recognized as the best-known American whaling port more than 100 years after its last ship set sail to catch whales is a tribute to the industrious, frugal character of its merchants. Most were descendants of Quaker farmers who moved to the island to escape a rising tide of religious persecution elsewhere in the New England colonies. When the island's crops quickly faltered due to poor soils, its settlers had little choice but to turn to the sea. In doing so, it was North Atlantic right whales that led them to their destiny. This species made up a miniscule part of their catch over Nantucket's 200-year whaling history, but it was the most important part. Right whales were the islanders' first target, and it was profits from their oil and baleen that underwrote expansion into a global whaling power. Even

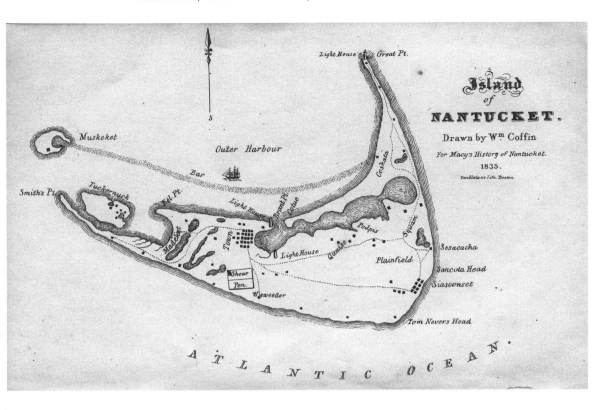

more important, the hunt for right whales taught them their craft and forged their marketing networks. Without right whales off its shores in the late 1600s and early 1700s, Nantucket may never have become the whaling center it did.

Figure 13.1. Map of Nantucket from Obed Macy's 1835 *History of Nantucket* (Boston, MA). (Image courtesy of the Nantucket Historical Association, Nantucket, MA.)

The Start of Whaling on Nantucket

Although Nantucket was part of Plymouth Colony's original charter and barely 50 miles from the colony's capital, nearly 40 years passed between the *Mayflower*'s landing and the arrival of the island's first English settlers. Its colonial history began in 1640 when Thomas Mayhew of Watertown, a settlement in the Massachusetts Bay Colony, acquired a deed to both Nantucket and Martha's Vineyard from colony officials. To secure development rights, however, English law required that colonists execute a deed of purchase from local Native residents, who were recognized as the rightful original owners. By 1659, Mayhew had only an oral commitment from a local sachem that he would sell part of the island. Perhaps finding it too difficult to complete the purchase agreements with the Native residents as required under English Law to secure full and unencumbered ownership of the island, Mayhew decided to sell most of his rights. He found a group of willing buyers in Thomas Macy and several Quakers, including Edward Starbuck, who then lived in the growing northern Massachusetts Bay settlement of Salisbury. Except for two small parcels of land, Mayhew sold his rights for the sum of 30 pounds sterling and two beaver hats.[2]

The new landholders had become increasing dissatisfied with oppressive ordinances pushed by Salisbury's new residents, who had engineered a pronouncement that declared it a crime to follow certain religious practices, including those of the Society of Friends. Nantucket seemed a convenient yet isolated location to escape the growing persecution. Therefore, immediately after purchasing the deed, Edward Starbuck and either Tristram Coffin or Thomas Macy arranged for a small, open boat to take them around the horn of Cape Cod to Nantucket to inspect their new claim. When they finally pulled up on Nantucket's northern shore, they found it covered with a patchwork forest, abundant in fish and fowl, and populated by a Native community of 1,500 to 3,000 Wompanoags.[3] Greeting the Englishmen cordially and with intense curiosity, the Native residents agreed to let either Macy or Coffin build a shelter for their family. Meanwhile, Starbuck returned to Salisbury to report on conditions to his fellow land grant holders and encourage them to move to the island. His report was well received and the following spring Starbuck, with his family and 8 to 10 other Quaker families, moved to the island and formed the nucleus of Nantucket's first English settlement; this was at Siasconset on the island's eastern shore.

Their fledgling community grew rapidly in its first years as other Quaker families moved to the island. By 1664, the settlers managed to finally purchase rights to the entire island from various village sachems under arrangements that allowed the local Wompanoags to continue hunting and farming wherever they pleased. That same year, the English wrested control over New Netherland and its largest colony, New Amsterdam (now New York City) from the Dutch. King Charles I immediately exercised his power over the territory by awarding control of all former Dutch lands, as well as Nantucket and the nearby island of Martha's Vineyard, to his nephew, the Duke of York. Under that authority, in 1672, New York Governor Thomas Lovelace certified the legitimacy of the Nantucket colonists' land purchases, which gave them free and clear title to all of Nantucket Island.

During their first few decades on Nantucket, English settlers cleared most of the island's forest for farms and grazing land. In doing so, the island's poor soils were quickly depleted and the agricultural base on which they hoped to make their livelihood was undermined. Surrounded by water, it is hardly surprising that they turned seaward. According to an oft-repeated local legend, in the late 1600s, an unnamed resident standing atop a coastal crest called Folly House Hill, a mile southwest of Siasconset, gazed across a panoramic ocean vista when a burst of inspiration flashed through his mind. With whales spouting in the distance, he raised an arm in a sweeping gesture and vowed, "there is the green pasture, where our children's grandchildren will go for bread."[4] The years that followed certainly fulfilled that prophecy.

None of Nantucket's early settlers was a whaler, yet they were quick to take advantage of dead whales cast up by the sea. Like most whaling traditions, salvaging stranded whales was their introduction to the trade. In the late 1600s, the prospect of finding stranded whales might have been greater due to the whales struck and lost by shore whalers who were still honing their skills a short distance away on Long Island and Cape Cod.[5] There is no evidence of Nantucket colonists killing live whales until 1668 when, according to another local legend, they took their first large whale as much by providence as design.[6] An early history of Nantucket published in 1835 by Obed Macy (figure 13.2), a direct descendant of Thomas Macy, included the following description of the event:

> A whale, of the kind called "scrag," came into the harbor and continued there three days. This excited the curiosity of the people, and led them to devise measures to prevent his return out of the harbor. They accordingly invented, and caused it to be wrought for them a harpoon with which they attacked and killed the whale. This first success encouraged them to undertake whaling as a permanent business; whales at that time being numerous in the vicinity of the shores. In furtherance of their design, they made a contract with James Lopar [see chapter 11], to settle on the island and engage in the business.[7]

This reference to a "scrag" whale has attracted special interest from today's whale experts. Some believe the name was used to identify the North Atlantic gray whale, a species that went extinct a few decades later. If so, this account could represent one of the species' last records. Even if it was a gray whale, there is no doubt that the species of primary interest to the Nantucket colonists was the one that would serve as their mainstay for developing whaling skills—the North Atlantic right whale.

As already mentioned (chapter 11), Nantucket residents tried unsuccessfully to entice Long Island whaler James Loper to move to their island and teach them how to whale. Terms of their offer, approved at a town meeting on April 4, 1672, promised Loper one-third ownership in any whaling company he established and a monopoly on all Nantucket whaling. To sweeten the deal he was offered a free plot of land and free grazing rights for his livestock. In part, their offer spelled out the following terms:

> [If] James Lopar doth Ingage to carry on a designe of Whale Citching on the Island of Nantuckket, that is the said James Lopar Ingage to be a third in all respeekes, and som of the Town Ingage Also to Carrey on the other two thirds with him in like manner, . . . the Town do also Ingage that no other Company shal be allowed hereafter; Also whosoever [of Loper's Company] Kil any whale . . . they ar to pay the Town for every such Whale five Shillings—and for Incorragement of the said James Lopar the Town doth grant him Ten Acres of Land in som convenant place, that he may Chuse in, (Wood land excepted) and also Liberty for the Commonoge.[8]

At the same meeting, town leaders tendered an offer to a mainland cooper named John Savage to move to the island and build barrels for their anticipated whaling business.[9] Savage accepted the offer and moved to Nantucket in 1672,[10] but soon

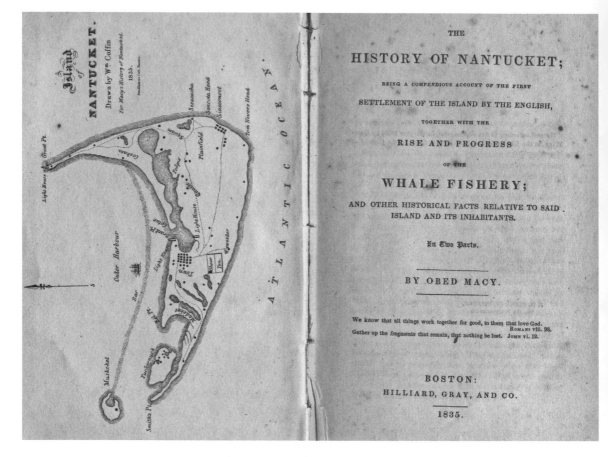

THE

HISTORY OF NANTUCKET;

BEING A COMPENDIOUS ACCOUNT OF THE FIRST

SETTLEMENT OF THE ISLAND BY THE ENGLISH,

TOGETHER WITH THE

RISE AND PROGRESS

OF THE

WHALE FISHERY;

AND OTHER HISTORICAL FACTS RELATIVE TO SAID
ISLAND AND ITS INHABITANTS.

In Two Parts.

BY OBED MACY.

We know that all things work together for good, to them that love God.
ROMANS viii. 28.
Gather up the fragments that remain, that nothing be lost. JOHN vi. 12.

BOSTON:
HILLIARD, GRAY, AND CO.
1835.

Figure 13.2. Title page and overleaf map of Obed Macy's 1835 book *History of Nantucket* (Boston, MA). (Courtesy of the Nantucket Historical Association, Nantucket, MA.)

left for Monomoit (now Chatham) on Cape Cod because he found no demand for his trade after Loper rejected the offer.

If there were attempts to hunt whales over the following decade, there is no record of it. The next reference to whaling on Nantucket appeared nearly 20 years later when town leaders again tried to entice a whaler to the island. As Macy explained in regard to Nantucket residents, "the people of Cape Cod had made greater proficiency in the art of whale catching than themselves," and they again sought an experienced whaler to teach them the trade.[11] This time they asked Cape Cod whaler Ichabod Paddock of Yarmouth "to instruct them in the best manner of killing whales and extracting their oil."[12] Paddock accepted their offer and moved to Nantucket in 1690. Again, no records survive of his early efforts, but apparently he was successful and founded a whaling business that launched dories through the surf to kill right whales.

Within two years of Paddock's arrival, Nantucket's south shore was divided into four whaling sectors, each with a company of six men supplied with a dory and associated gear for the purpose of chasing any whales that came within sight. At the center of each sector, a "hut" was constructed to shelter whalers and their equipment. A tall spar with cleats was also erected at a high point in the dunes to scan for passing whales and signal adjacent crews for help if one was sighted.[13] As on both Long Island and Cape Cod, local Natives (the Wampanoags in this case) were hired to carry out most of the whaling and were said to find the challenge exhilarating.[14] Virtually every whaling boat carried at least some Wampanoags, and some boats were composed entirely of these crews.[15] The practice meant that initial development of the island's whaling business was as much an accomplishment of the Native community as it was of the English colonists.[16] In good weather, the crews ventured almost out of sight of land. When a whale was caught and towed ashore, portable kettles were erected at the landing site to render oil.

Always willing to experiment with ways to improve effectiveness and profit, Nantucket entrepreneurs soon followed the lead of Long Island and Cape Cod whalers building sloops to chase whales farther afield. The first was built in 1694, and within a few years several others were added. By 1715 they had six whaling sloops, each about 40 tons in size.[17] Although used for multiple purposes, each of these vessels might have caught at least one or two whales per year in waters surrounding Nantucket, returning each time with raw blubber packed in barrels. Instead of continuing to rely on portable kettles set up on the beach to render this oil, separate try houses were built by enterprising vessel owners.

At some point around 1710, Christopher Hussey, captain of a Nantucket whaling sloop, became the first islander credited with catching a sperm whale. As Obed Macy described the event in his 1835 history of Nantucket, Hussey "was cruising near the shore for Right whales, and was blown off some distance from land by a strong northerly wind." Macy continued: "He fell in with a school of that species of whales [sperm whales], and killed one and brought it home. At what date this adventure took place is not fully ascertained, but it is supposed to be not far from 1712. This event gave new life to the business for they immediately began with vessels of about thirty tons to whale out in the "Deep," as it was then called, to distinguish it from shore whaling. They fitted out for cruises of about six weeks, carried a few hogsheads, enough probably to contain the blubbers of one whale."[18]

Although not fully appreciated at the time, Hussey's feat was a defining moment in the history of Nantucket whaling. It not only convinced Nantucket whalers they could successfully hunt sperm whales, it lit a path to continue whaling as right whales became harder to find. By the mid-1700s, Nantucket shore whaling for right whales was largely replaced by hunting sperm whales in the "deep." Although shore whaling persisted as an important effort from about 1690 to 1760, the highest catches were between approximately 1695 and 1730.[19] The shift to other species, however, afforded right whales at least a modicum of relief.

Information on how many right whales were landed is frustratingly poor, with few surviving records to indicate its scale.[20] In 1715, the 6 sloops working from Nantucket apparently produced 600 barrels of oil and 11,000 pounds of whalebone, which gave them the princely sum of 1,100 pounds sterling.[21] At an equivalent of 36 to 44 barrels of oil per right whale, this catch may have included 14 to 17 whales. Those returns were reinvested in the building of larger vessels capable of longer voyages to more distant grounds. By 1730, some 25 vessels produced oil and bone valued at 3,200 pounds sterling.[22] Although part of those returns likely came from sperm whales caught "on ye deep," in winter and spring at least some vessels continued targeting right whales near Nantucket and to the south along the coast, particularly in the Carolinas.

One of the few surviving references that links catch results to whaling grounds is a ragged sheepskin-covered ledger used to track credits owed to 17 Native laborers for various kinds of work, including whaling.[23] Started in 1662 by Nantucket resident Mary Starbuck and continued to 1768 by her son Nathaniel and another unidentified person, 23 of its 177 pages are missing. Elizabeth Little of the Nantucket Historical Association carefully transcribed 198 entries listing credits to Wompanoag whalers for their part in successful hunts between 1721 and 1758. Although most entries list only shares of oil and/or whalebone owed to laborers and represent a small proportion of the Native work force, 70 entries identify 8 whaling grounds where hunts occurred. Sixteen were listed as "along shore" or at "Wewedah" (a beach on the Nantucket's south shore), beginning in 1728; 11 were at "bobel" or" Bowbel" (a location thought to be a short distance southeast of Nantucket), beginning in 1728; 19 were on "ye deep," starting in 1729; 9 were at "Canso" (presumably off Nova Scotia), beginning in 1731; 10 were off "Newfoundland," starting in 1738; and one each was off "Greenland" and "Cariliner" (Carolina) in 1738 and 1753, respectively.

These entries cover the years that Nantucket whaling sloops were extending their range and seem to be consistent with a shift in effort to the north as right whales became scarce close to Nantucket between the mid-1720s and early 1730s. The only entries listing both whaling locations and shares of "bone," which would indicate a catch of right whales rather than sperm whales, were taken "along shore," at "Wewedah," and for the lone whale taken in the Carolinas. Thus, none of the whales taken on "ye deep," at "bobel" or in the northern grounds could be definitively identified as a right whale or other baleen whale.

Other available catch records that seem to be right whales are generally attributed to Nantucket shore whalers. In 1726, Nantucket shore whalers were credited with landing 86 right whales, including a record 11 whales in a single day.[24] That year 28 different boat owners caught from 1 to 6 whales each. This may have been the high-water mark for Nantucket shore whalers; no larger one-year total is known either before or after that date. An even larger catch was possible if whales taken from island sloops working further offshore or down the coast were not included in the total.

Based on the meager records available, it appears possible that 50 to 80 right whales per year (not including whales struck and lost) were landed by Nantucket shore whalers during their peak years between 1695 and 1730. After that, catch totals tailed off rapidly. If sloops working farther offshore killed an additional 10 to 20 whales per year between 1710 and 1730, Nantucket shore and pelagic whalers may have caught 60 to 100 right whales annually, with perhaps another third or more struck and lost.

For most of their shore-whaling years, Nantucket merchants sold their oil and whalebone through Boston markets, even though they were part of the New York Charter and subject to its duties. Nantucket whalers, however, refused to pay those duties to the great irritation of New York officials. The dispute over their allegiance was resolved in 1693 when the English Crown, then occupied by William and Mary, approved a request by residents of both Nantucket and Martha's Vineyard to return jurisdiction of their islands to the Massachusetts Bay Colony. By 1745, when it became apparent that Boston merchants were shipping the oil they purchased to London where prices were higher, Nantucket businessmen began shipping it directly to European buyers in their own boats, thereby cutting out the middlemen and increasing their profits.[25]

As shore whaling on Nantucket withered for lack of right whales, pelagic whaling in offshore and distant coastal waters flourished. In 1730, Nantucket merchants sent out at least 25 whaling sloops ranging from 30 to 50 tons; by 1748 they had 60 vessels ranging from 50 to 70 tons; and in 1756 the number reached 80 ships, including a few schooners, averaging 75 tons.[26] With limited resources and manpower on Nantucket, new vessels had to be ordered from mainland shipbuilders and their whaling crews had to be recruited from Long Island, Cape Cod, and elsewhere. With each passing year, the Nantucket vessels set sights on ever more distant whaling grounds in ever larger ships.

An indication of the geographic area of their operations was contained in a petition submitted to the Massachusetts Court in 1757 by whalers from both Nantucket and Martha's Vineyard. The petition sought exemption from an embargo imposed on ships by officials responsible for colonial affairs in England due to military hostilities with France over territory in Canada. The petition argued that the islands' whalers should be exempt because their winter voyages were to the south on grounds "westward of the Virginia Capes and southward of that until the month of June."[27] This would have placed them in the northern part of today's right whale calving grounds off the Carolinas and within an area known to have been a winter whaling ground for at least some right whales.

Although shore whaling continued on Nantucket well into the 1800s, it was sporadic at best after the late 1700s. One of the last successful hunts was in mid-April 1886, when a small group of right whales was spotted off the island's west end and some 25 boats were dispatched to chase them. Most boats, including the schooner *Glide*, failed to catch any whales.[28] Nevertheless, at least three right whales were landed to yield a total of 125 barrels of oil and 1,500 pounds of whalebone. A fourth was struck and immediately began towing the whalers out to sea at a lively clip. When some 30 miles from shore in dense fog, the whale had to be cut free. The whalers made it back to shore only with great difficulty; the whale was found floating at sea a few days later and towed to New Bedford for processing.[29] If this was not the last time Nantucket shore whalers landed a right whale, it was the last year in which their annual catch total was known to have exceeded one whale.

Martha's Vineyard

Right whales also attracted the attention of colonists on Martha's Vineyard, just a dozen miles west of Nantucket. Its first settlers arrived on the heels of Thomas Mayhew, who moved to the island in 1641 to develop what remained of his Plymouth Colony grant after he had sold most of his rights to Nantucket. Like their neighbors, interest in whaling by Vineyard colonists evolved from the salvage of drift whales. On April 13, 1653, town fathers in Edgartown, the Vineyard's largest settlement, assigned responsibility for cutting up whales to various town residents, depending on where the carcasses were cast up.[30]

Drift whales again factored into the purchase of lands from local Natives. When Thomas Mayhew purchased a tract from a local sachem in 1658, the deed stipulated in part that he was "to have four spans round in the middle of every whale that comes upone the shore and of this quarter part and no more."[31] The length of four spans, presumably about two feet, suggests that most of the "whales" prompting this provision were small pilot whales or porpoises. Proceeds from drift whales were considered common property divided among Edgartown's residents according to the amount of land they owned. Their shares were routinely listed in the town's records describing the major assets of its residents. For example, assets listed for Richard Sarson in December of 1663 included "the Six and Twentyth part of fish and whale that Belongs to this town."[32]

Shore whalers on Martha's Vineyard learned the craft from their Nantucket neighbors at about the same time Ichabod Paddock arrived on Nantucket in 1690. The first evidence of hunting live whales on the Vineyard comes from a dispute in 1692 over a whale cast ashore at Edgartown. The animal was claimed by the town as a drift whale, but it was also claimed by John Steel, a harpooner, who asserted he had killed it.[33] More definitive evidence that colonists were hunting right whales appeared a decade later in early 1703 when Edgartown records listed a catch of three right whales by John Butler and Thomas Lothrup; these were said to have been "all killed about the middle of February last passed, all great whales between six and seven and eight foot bone."[34] Although they may have been taken during a voyage to southern grounds rather than in local waters, the reference to whalebone of that length clearly indicates they were right whales. Whaling off Martha's Vineyard never developed to the extent that it did on Nantucket, but at least a few shore whalers were active in the early 1700s. Perhaps up to 10 or more whales per year were killed, including whales struck and lost.

At some point early in the 1700s, a small number of Vineyard whaling vessels joined the growing fleet of New England ships hunting right whales in distant waters down the coast. In 1720, Samuel Butler, who may have been the same Samuel Butler listed as a "mariner" in Martha's Vineyard records (see the following chapter),[35] is known to have sailed to North Carolina to seek a license to catch whales along the Outer Banks.[36] In 1724, another Vineyard whaler, Pain Mayhew Jr., filed a lawsuit charging a Barnstable whaler with defrauding him of proceeds in

NANTUCKET, MARTHA'S VINEYARD, AND CAPE MAY

a joint whaling trip to "Barnstable Bay" (the southern part of Cape Cod Bay) on "a great voyage,"[37] which suggests multiple whales were killed. By the 1760s and 1770s, Vineyard-based whaling vessels, like those elsewhere, had redirected their efforts to other species, principally sperm whales "on the deep," bowheads in the Davis Strait and Baffin Bay, and right whales in the Gulf of St. Lawrence.[38] At least some boats also traveled south to the Virginia Capes in search of right whales.

New Jersey

A more substantial whaling effort developed south of New England along the shores of southern New Jersey. Indeed, whaling opportunities appear to have drawn some of the region's first settlers to the area in the late 1650s, just 30 years after the Dutchman David Pietersz De Vries unsuccessfully tried his hand at whaling at the mouth of the Delaware Bay. According to several historical references, 14 "skilled pilots" were leading whalers to Cape May by 1658.[39] With what success this was undertaken is unknown, but at that early date the whalers must have been from Long Island. The next reference to whaling in the area was not until the 1680s, when Long Island whalers were making profitable runs down the coast in small sloops to Cape May. Long Island whalers reportedly found right whales in abundance off the mouth of Delaware Bay in late winter and early spring, and were clearly visiting the area by 1683; at that time, a letter by William Penn advised that 11 whales were taken around the bay's capes.[40]

Other evidence implicating Long Island as the source of early whalers on Cape May comes from testimony in a 1685 dispute over a stranded whale. During the case, Edward Pynde testified in the New Jersey Province's court at Burlington that he had been asked by one of the litigants to view the disputed whale before it was cut up: "comeing to ye s'd ffish sayth it was a whale ffish and yt hee saw an Iron (with warp thereat) in ye s'd fish, which Iron & Warp ye s'd depon't knowing them to belong to s'd Caleb Carmen & Company."[41]

Caleb Carmen was a whaler living on Long Island at the time, and identification of his iron in an animal stranded on Cape May indicates he may have been whaling off New Jersey. At least two other court cases heard by the Burlington court in 1688 involved claims to whales filed by New England whalers that winter.[42]

Successful whaling and inexpensive land prices encouraged approximately 20 Long Island whalers to move to Cape May in 1690 and 1691, where they founded Cape May Town (near what is now Cape May, NJ), the cape's first settlement. The seaside location of their settlement, however, was chosen poorly. By the mid-1700s most of the town's buildings had been lost or abandoned due to an encroaching sea. Notwithstanding those misfortunes, shore whaling provided a good income to local residents despite a seasonal invasion of other vessels from the north. Local concerns about revenues lost to northern interlopers prompted the assembly for the province of New Jersey to pass "An Act Relating to Fishing" in October 1693. It stipulated, in part:

Whereas the Whalery in the *Delaware* Bay has been in so great a Measure invaded by Stranger and Foreigners, that the greatest Part of Oyl and Bone, recovered and got by that imploy hath been Exported out the Province, to the great detriment thereof; to obviate such mischief, BE IT ENACTED . . . that all Persons not residing within the Precincts of this Province, or the Province of *Pennsylvania*, who shall kill or bring on shore any Whale, or Whales within *Delaware* Bay, or elsewhere within the Boundaries of this Government, shall pay one full and entire Tenth of all the Oyl and Bone made out of the said Whale, or Whales unto the present Governor of this Province for the Time being.[43]

In 1696, an agent was appointed to collect the fee,[44] but any records of revenues collected from this tax or numbers of whales killed seem to have been lost.

Among the first whalers emigrating to Cape May from Long Island were Christopher Leaming and his 18-year-old son Thomas. The elder Leaming moved to the Cape in 1691. He was a whaler during the winter and a cooper during the rest of the year. His son later wrote that the occupation of cooper at the time was a good one "by reason of the great number of whales caught in those days, [which] made the demand and pay for casks certain."[45] One of the few surviving documents providing insight into the number of whales caught by a local Cape May whaler is the diary of the younger Leaming, who had followed his father to Cape May in 1694. The winter after he took up permanent residence in Cape May, Leaming described his experience of whaling: "We got eight whales and five of them we drove to the Hoar Kils and we went there to cut them up and staid a month and the first day of May we Came home to Cape May and my father was very Sick and on the third day 1695 he departed this Life at the house of Shamgar Hand."[46] His whaling descriptions continued the following year when he "got an old Cow and Calf," and the winter after that (1697–1698) when he landed "one small whale." Thomas returned to New England in the summer months and for at least one summer joined the crew of a pelagic whaling vessel.

With several whaling crews operating on Cape May at the end of the 1600s, the catch recorded by Thomas Leaming presumably was only a part of the town's total landings. In 1698, an account attributed to Gabriel Thomas described the natural assets of "West Jersey," and noted, "the commodities of Cape May County are oyl and whalebone, of which they make prodigious quantities every year; having mightily advanced that great fishery taking great numbers of whales."[47] Any significant catch levels, however, must have been short lived. According to one historian, by 1700, many Cape May whalers had grown discouraged by shrinking catches and either returned to Long Island or New England, or took up other lines of work.[48]

Other contemporary accounts suggest that at least some whaling spread up the coast of southern New Jersey with moderate success in the early 1700s. In 1704, the governor for the provinces of New Jersey and New York issued a license for whaling to Joseph and James Lawrence; this directed "to fish for, kill, and cut up

... what whales and Other Royal fish you Can or may find on the Coust of this Province of New Jersey betwixt Sandy hook and barnegate Inlett."[49] The whalers at that time must have had some success, because the March 17–24, 1718, edition of the *Boston News-Letter* reported that "the whale men catched six whales at Cape May and twelve at Egg Harbor," which is just north of present-day Atlantic City.[50] Given the number of experienced whalers that moved to Cape May in the early 1690s and what is available that describes catch levels, those living along the Jersey Shore might have killed up to 20 or 30 whales per year (including the ones struck and lost) in the last decade of the 1600s and 15 to 25 in good years during the first two decades of the 1700s.[51]

After that, catch levels plunged sharply in New Jersey as they did elsewhere. The *Philadelphia Gazette* of March 13–19, 1730, reported the stranding of a 50-foot female whale near Cape May on March 5 that was apparently struck and lost by an unknown whaler. On March 19, 1739, the same paper reported two whales taken on Cape May that were expected to yield 40 barrels each, with sightings of several other whales offering hope for further catches.[52] Two whales also were killed at Cape May in early April of 1742. Similar levels of spotty success continued through the mid-1700s. The diary of south New Jersey whaler Lewis Cresse recorded what was likely a calf killed in 1764 near Townsend Inlet that produced "Twenty three Barrals of oyl her bone four feet eight inches long."[53] The following year he suggested in his diary that at least four whales were taken.[54] Cresse noted the following after seeing some whales at Seven Mile Beach: "the owners of that beach made so much disturbance in our [whaling] company that we left the chance and went [north about ten miles] to Peck Beach . . . and next morning after we got there we kild a whale and after kild another and struck another and she runaway and we follow'd her along Ludlams Beach and that company came off and joined us and we kild her. The Ludlams Beach Company kild a whale the same season."[55]

In 1775, Aaron Leaming, Christopher Leaming's youngest son, was awarded a 30-day lease for landing and trying out whales on Seven Mile Beach near Egg Harbor.[56] Shore whaling sputtered along at even lower levels through much of the 1800s until 1882, when probably the last right whale taken off New Jersey was landed by an experienced crew at Egg Harbor.[57] The 48-foot whale was put on display for several days before processing to allow time for scientists, who by then were keenly interested in any right whale specimens, to take its measurements.

NOTES

1. Starbuck, A. 1878. *History of the American Whale Fishery from Its Earliest Inception to the Year 1876*. Report of the US Commissioner of Fish and Fisheries, Part I. US Government Printing Office. Washington, DC, p. 57.

2. Macy, O. 1835. *The History of Nantucket; Being A Compendious Account of the First Settlement of the Island by the English, Together with the Rise and Progress of the Whale Fishery; and Other Historical Facts Relative to Said Island and Its Inhabitants*. Hilliard, Gray. Boston, MA, US, pp. 4–5.

3. Folger, IH. 1875. *Handbook of Nantucket Containing a Brief Historical Sketch of the Island with Notes of Interest to Summer Visitors*. Island Review Office. Nantucket, MA, US.

4. Starbuck, *History of the American* (n. 1), p. 17 (quotation).

5. Allen, GM. 1916. *The Whalebone Whales of New England*. Memoirs of the Boston Society of Natural History, Vol. 8, no. 2. Boston Society of Natural History. Boston, MA, US, pp. 109–175, 115–116, 163.

6. Folger, *Handbook* (n. 3), p. 27.

7. Macy, *The History of Nantucket* (n. 2), p. 28 (quotation).

8. Ibid., pp. 28–29 (quotation).

9. Starbuck, *History of the American* (n. 1), pp. 16–17.

10. Edwards, EJ, and JE Rattray. 1932. *Whale Off! The Story of American Shore Whaling*. Frederick Stokes. New York, NY, Part 2, p. 178.

11. Macy, *The History of Nantucket* (n. 2), pp. 29–30 (quotation).

12. Ibid., pp. 29–30 (quotation).

13. Allen, *The Whalebone Whales* (n. 5), p. 165. Starbuck, *History of the American* (n. 1), p. 19.

14. Folger, *Handbook* (n. 3), p. 28.

15. Macy, *The History of Nantucket* (n. 2), p. 30.

16. Little, EA. 1988. Indian contribution to along-shore whaling at Nantucket. Nantucket Algonquian Studies 8:1–85, 60.

17. Macy, *The History of Nantucket* (n. 2), p. 37.

18. Ibid., pp. 36-37 (quotation).

19. Ibid., p. 36.

20. Reeves, RR, JM Breiwick, and ED Mitchell. 1999. History of whaling and estimated kill of right whales, *Balaena glacialis*, in the northeastern United States, 1620 to 1924. Marine Fisheries Review 61(3):1–36, 13–15.

21. Macy, *The History of Nantucket* (n. 2), p. 233.

22. Starbuck, *History of the American* (n. 1), pp. 21–22. Spears, JR. (1908) 1922. *The Story of the New England Whalers*. MacMillan. New York, NY, p. 55.

23. Little, EA. 1988. Nantucket whaling in the early 18th century. *In* José Mailhot (ed.), *Papers of the Nineteenth Algonquian Conference*. Carleton University. Ottawa, Ontario, Canada, pp. 111–131.

24. Macy, *The History of Nantucket* (n. 2), p. 31. Starbuck, *History of the American* (n. 1), p. 22. Scheville, WE, WA Watkins, and KE Moore. 1986. Status of *Eubalaena glacialis* off Cape Cod. *In* RL Brownell, PB Best, and JH Prescott (eds.), Right Whales: Past and Present Status, Special issue 10, Reports of the International Whaling Commission, pp. 79–82.

25. Macy, *The History of Nantucket* (n. 2), p. 51.

26. Ibid., pp. 232–233.

27. Starbuck, *History of the American* (n. 1), p. 38 (quotation).

28. Scheville, WE, KE Moore, and WA Watkins. 1981. *Right Whale*, Eubalaena glacialis, *Sightings in Cape Cod Waters*. Report prepared for the Office of Naval Research. Woods Hole Oceanographic Institution Technical Report WHOI-81-50. Woods Hole, MA, US.

29. Allen, *The Whalebone Whales* (n. 5), p. 138.

30. Banks, CE. 1911a. *The Whale Fisheries*. Vol. 1 of *The History of Martha's Vineyard, Dukes County, Massachusetts in Three Volumes*. George H. Dean. Boston, MA, US, pp. 430–451, 431.

31. Ibid., p. 432 (quotation).

32. Banks, CE. 1911. *Annals of Edgartown*. Vol. 2 of *The History of Martha's Vineyard, Dukes County Massachusetts*. George H. Dean. Boston, MA, US, pp. 1–209, 106 (quotation).

33. Starbuck, *History of the American* (n. 1), p. 18 (quotation).

34. Banks, *The Whale Fisheries* (n. 30), pp. 433–434 (quotation).

35. Ibid., pp. 435–438.

36. Minutes of the North Carolina Governor's Council, December 3, 1720. In WL Saunders

(ed.), *The Colonial Records of North Carolina*. P.M. Hale. Raleigh, NC, US, Vol. 2, (1713–1728), p. 397.

37. Banks, *The Whale Fishery* (n. 30), p. 435 (quotation).

38. Ibid., pp. 435–438.

39. Cook, GH. 1856. *Geology of the County of Cape May, State of New Jersey*. New Jersey Geological Survey. Printed at the Office of the True American. Trenton, NJ, US, p. 199. Stevens, LT. 1897. *The History of Cape May County, New Jersey, from the Aboriginal Times to the Present Day*. Lewis T. Stevens. Cape May, NJ, US, p. 30 (quotation).

40. Cope, ED. 1865. Note on a species of whale occurring on the coast of the United States. Proceedings of the Academy of Natural Sciences of Philadelphia 144:168–169. As cited in JB Holder. 1883. The Atlantic right whales: a contribution embracing an examination . . . to which is added a concise résumé of historical mention related to the present and allied species. Bulletin of the American Museum of Natural History 1(4):99–137, 100.

41. Stevens, *The History of Cape May* (n. 39), pp. 30–32 (quotation).

42. Ibid., pp. 32–35.

43. Allen, JA. 1908. The North Atlantic right whale and its near allies. Bulletin of the American Museum of Natural History 24:277–329, 318 (quotation).

44. Stevens, *The History of Cape May* (n. 39), p. 51.

45. Ibid., p. 98 (quotation).

46. Ibid., pp. 45–46 (quotation).

47. Ibid., p. 57 (quotation).

48. McMahon, W. 1973. *South Jersey Towns: History and Legends*. Rutgers University Press. New Brunswick, NJ, US, p. 8.

49. Stevens, *The History of Cape May* (n. 39), pp. 62–63 (quotation).

50. Ibid., p. 63.

51. Reeves et al., History of whaling (n. 20).

52. Stevens, *The History of Cape May* (n. 39), pp. 63–64.

53. Schaad, J. 2015. Bizarre history of Cape May: early settlers sought whales at sea and God on land. Cape May Gazette, January 15.

54. Reeves, RR, and E Mitchell. 1987. *Shore Whaling for Right Whales in the Northeastern United States*. Southeast Fisheries Center, National Marine Fisheries Service. Miami, FL, US, p. 26.

55. Schaad, Bizarre history (n. 53).

56. Stevens, *The History of Cape May* (n. 39), p. 175.

57. Holder, The Atlantic right whales (n. 40), pp. 106–108.

14

WHALING FROM
THE CAROLINAS
TO FLORIDA

THE FIRST HARPOON to kill a large whale south of the Chesapeake Bay was probably thrown in the late 1600s off what is now known as the Outer Banks of North Carolina. With its sandy capes elbowing their way 50 miles out to sea, this was a good place to find migrating whales, but a perilous place to catch them. A witch's brew of powerful storms, tricky currents, and treacherous shoals conspired to make this region a deadly spot for ships. Indeed, waters off the Outer Banks' easternmost point, Cape Hatteras, have come to be called the graveyard of the Atlantic because of the countless ships driven to ruin on its constantly shifting sandbars. Although the first English settlers to take up residence on its slender barrier islands may well have been attracted by opportunities to salvage wreckage from foundering ships, they surely found the frequent strandings of whales and porpoises an additional incentive to move to its shores. But completely lacking in whaling skills, the banks' early settlers could only watch as New England sloops sailed down the coast to North Carolina shores to kill whales and then carried the bounty back north. Carolina Colony officials persistently urged local residents to take up whaling, but it would take them decades to learn the necessary skills. In the interim, they were consigned to processing stranded carcasses.

The first permanent residents on this windswept strand of the Carolina coast arrived in the 1650s from the Virginia Colony. They followed a failed attempt by Sir Walter Raleigh to found a colony on Roanoke Island in 1587—a calamity that spawned innumerable legends about the fate of the region's enigmatic "Lost Colony"—and another futile charter granted in 1629 by King Charles I to Robert Heath.[1] Heath's patent covered a broad swath stretching from the northern end of the Outer Banks to southern Georgia. (It was Robert Heath who named the territory "Carolana," after King Charles I using a derivation of the Latin word for Charles, *Carolus*.) But Heath was unable to mount a colonization attempt before being stripped of his patent as a consequence of political intrigue that forced him to flee England for France in 1645.

Under more favorable political conditions, the movement of Virginia colonists to the Outer Banks might have provided grounds for annexing the northern region of Carolina under the Virginia Colony, but that was not to be. In 1663, King Charles II issued yet another charter for the "Carolanas" to eight proprietors and their heirs as a reward for their help in restoring the monarchy that was deposed when his father, Charles I, was beheaded in 1649.[2]

Boundaries for the new charter were identical to those under the 1629 Charter and included what by then was standard language conveying rights to "fishing of all sorts of fish, whales, sturgeons and all other royal fishes of the sea."[3] At the colony's northern end, abutting the Virginia Colony, lay the territory of Albemarle; this was assigned to one of the proprietors, Edward Earle. When it was learned that Virginia colonists had been moving into their territory, the proprietors feared that their northern neighbor had designs on annexing the area and appealed to Charles II to clarify the boundary of their "plantation." He did so in 1665, and amended the Carolina charter to fix its northern boundary at what is now the Virginia–North Carolina border.

Whaling was a matter of great interest to colony proprietors even before their first settlement was established in 1670 near what would become Charleston, South Carolina. In 1666, Peter Carteret, the newly appointed governor and a distant cousin of one of the proprietors of the Albemarle territory, issued a whaling license to Humphrey Hughes, a Boston merchant with two partners from Southampton, Long Island. The license authorized rights to "enjoy the privilege to make use of all whales that shall be caste up" or that could be killed or destroyed "between the inlet of Roanoke and the inlet of Caretuck."[4] In 1667, Hughes and his partners commissioned the ship *Speedwell* to "go themselves or by competent agent to Roanoak to those parts upon the designe of killing or getting whales or great fish for the procuring of oyle."[5] As was common practice at that time, such a "design" involved sailing down the coast and setting up camp on a beach. These camps typically included a temporary shelter, a crow's nest planted on a high dune to spot whales, and portable caldrons to try out any whales they found or caught. Unfortunately, no records survive describing the outcome of the expedition.[6]

The first sign that Albemarle residents were profiting from whales, albeit presumably from salvaging drift whales, comes a year later when colony records reported that some 80 barrels of "oyle" were produced at Colleton Island, near what is now Kitty Hawk on the Outer Banks' northern arm.[7] Whether from animals dying of natural causes or victims of deliberate hunting, the oil was welcomed as an encouraging sign for local whaling prospects by Peter Colleton, son of another proprietor. Colleton wrote Carteret on October 22, 1668: "I am glad you have found out a Whale ffishing I doe desire you to Continue it ffor our Shipps yt [that] went to Greeneland tooke but one Whale last season soe that Oyle & Whalebone is at psent a great Comodity here [in England] & I conceive London is a better markett than Barbados, soe yt I desire you to send all you take here. And consigne it to mee, all with I shall Lay out in supplyes for yor plantacon."[8]

The following year the proprietors approved a document entitled the "Fundamental Constitutions of Carolina."[9] The manifesto set out guiding principles for governing the colony and divided the Carolina plantation into four precincts. The northern two precincts—Albemarle and Carteret—included the 200-mile-long arc of barrier islands that now constitute North Carolina's Outer Banks. Among the document's long list of provisions was Article 114. This stipulated that all "whale-fishing, and one-half of all ambergris, by whomsoever found, shall wholly belong to the Lord Proprietors."[10]

Over the next decade, the exploitation of whales was modest at best. An inventory of plantation goods for 1672 by Peter Carteret reported, "Att severall tymes I have Shipped one hundred nynty five Barrels of Whale Oyle for London and Consigned it to Sir Peter Colliton for your Lordship and Company, wch I Conceive may have Cleered about 25. s p barrel."[11] This was an amount then equivalent to 243 pounds sterling.[12] Disappointed by those returns, colony proprietors concluded that their claim to "all whale fishing" was discouraging colonists from taking up the trade more aggressively. Accordingly, in July 1681, proprietors of the Albemarle Precinct wrote to their colonial governor John Jenkins urging further steps to promote local whaling. As encouragement, they ordered that all duties on whales be waived for seven years: "We being informed that there are a great many Whales upon the Coast of Carolina, which fish being by our Fundamentall Constitutions reserved for us: we have notwithstanding (for the incouragement of Carolina) thought fitt to give all persons whatsoever that are inhabitants of our Province free lease for the space of seaven yeares to commence from Michaelmas next [late September] to take what whales they can and convert them for their owne use and this our concession you are to make publick."[13]

Despite their interest in promoting whaling and connections with Arctic whaling, the proprietors offered neither equipment nor an experienced whaler to teach the colonists how to catch whales. Not surprisingly, the colonists were at a loss as to how to proceed other than to continue salvaging stranded animals. The few surviving colony records from the 1680s refer only to shipments of small amounts of oil, generally less than 30 barrels, to Barbados, Jamaica, Massachusetts, and Rhode Island.[14] Dissatisfied with those returns, yet aware that New England whalers were beginning to make a good profit on trips to colony waters in winter and spring, the proprietors wrote to their latest governor, Philip Ludwell, on May 13, 1691, again urging development of local whaling.[15] This time they offered their colonists all rights to any whale oil or whalebone produced for a 20-year period (through 1711), provided one-tenth of the yield from each whale was paid to their "Receiver General" (as the proprietors' share). The new instructions directed the following:

Whereas we are informed that there are many Whales that frequent the Coast of Carolina but that noe persons endeavours to take ye same by reason ye taking of all Royall ffish & the fish are granted to us by our Charter and from ye Crowne & reserved to ourselves by our Constitutions and wee being willing to Incourage the Inhabitants of our Province & for ye Increase in that trade Doe hearby grant free leave & authority to all persons whatsoever that are Inhabitants of Carolina that Shall first have given bond to Paule Grimball Esqr our Receiver Generall for the time being That they shall & will give a true acc^tt: to him are Impowered by us [to have] ... what whale oile, or whalebone, or whaleffin, or Spermacceti the s^d whale or whale ffish soe taken Shall have produced & shall pay or have delivered one Tenth part thereof to ye said Paule Grimball ... untill Michaelmass [September] in the years of our Lord 1711.[16]

The new provisions may have had some success, as indicated by a marked increase in disputes over whales in colony courts during the 1690s.[17] In addition, whalebone or "whale oyle" became an accepted form of colony currency.[18] For example, on winning a lawsuit in 1697, Thomas Harvey asked that the debt owed to him be paid in "whale Oyle."[19]

Despite a growing attention to whales at the end of seventeenth century, John Lawson, the colony's surveyor-general, reported that the people living on the Carolina coast were still limited to salvaging stranded whales. Lawson's observation was related in a journal that included a description of his reconnaissance trip in 1700 from "Charles-town" to northern parts of the province to take stock of its inhabitants, natural resources, and opportunities for trade. His account, presented to the proprietors a year later, advised the following, "As for the Whale-fishing, it is no otherwise regarded than by a few People who live on the Sand-Banks; and those only work on dead Fish cast on shoar, none being struck on our Coast, as they are to the Northward; altho' we have Plenty of Whales there."[20] Another part of his account described four sorts of whales "of prodigious Bigness" from which "bone and Oil was made," as well as smaller "bottle noses, crampois, and porpoise."[21] Other than the "Sperma-cæti" or sperm whale, said to be highly prized but rarely found, his descriptions were too vague to identify the species exploited. Nevertheless, the journal provides useful information. "Whales are very numerous on the Coast of North Carolina, from which they make Oil, Bone, &c., to the great Advantage of those inhabiting the sand banks, along the ocean, where these whales come ashore, none being struck or killed with a harpoon in this place as they are to the northward and elsewhere; all those fish being found dead on the shore, most commonly by those that inhabit the banks, and sea side, where they dwell for that intent, and for the benefit of wrecks which sometimes fall upon that shoar."[22]

Perhaps heartened at hearing of the abundance of whales, in March 1709 the proprietors wrote to their governor, then Edward Tynte, asking for a report on "the state of the Whale fishery and what further encouragement is proper & fitting for us to give to increase the same."[23] The governor's response seems to have been lost, but subsequent letters from proprietors to succeeding governors Edward Hyde in 1712 and Charles Eden in 1713 continued to press for the advancement of whaling. Those letters were followed by another in 1715, apparently prompted by new information that northern fishermen were continuing to catch whales along their coast without paying the one-tenth fee owed to the colony. On March 26, 1715, the proprietors wrote to Eden directing that New England whalers be allowed to whale along their coast. "We think it proper to give all due Encouragement to such persons as are willing to come and settle among you, and we do therefore hereby require you to give a Power or Liberty to any New England Men or other to catch Whale, Sturgeon or any other Royal Fish upon your coast during the Term of three years, they paying only two Deer Skins yearly to the Lords as an acknowledgement to them for the same."[24]

The objective of their largesse apparently was hope that experienced Yankee whalers would somehow share their whaling know-how with local residents. This could explain why colony officials requested only a token fee in deer skins from northern whalers. The incentive was short-lived but apparently effective. By 1720, local residents as well as Yankee whalers were actively hunting whales along the north coast of the Carolinas, particularly around Cape Lookout at the southern end of the Outer Banks near Beaufort.

Catch records at this time are again sparse, but documents confirm that New England whalers continued visiting the Outer Banks. In 1719, at least four New England boats worked the Carolina coast and returned to Boston with 500 barrels of oil and considerable whalebone after just three weeks of whaling.[25] At least some New Englanders adhered to colony licensing requirements (figure 14.1). Among the more interesting evidence in this regard is a pair of affidavits describing an incident in which Virginia officials misrepresented their authority over whaling in the Carolinas.[26] In the winter of 1720, a New England whaling vessel under the command of Samuel Butler—possibly the same Samuel Butler mentioned in the previous chapter as a "mariner" in Martha's Vineyard—was en route to the Carolinas on a whaling expedition when a mid-November storm forced it to shelter in the port of Hampton at the mouth of the Chesapeake Bay. There Butler found two other sloops riding out the storm; these were the *George* from Rhode Island under the command of Mathew Wolfe and the *Hopewell*, with Thomas Predeaux as master. Both vessels may also have been bound for the Carolinas on whaling trips given Wolfe's claim that Butler was "Bound likewise to North Carolina in order to procure a License to Whale."[27]

While in Hampton, Wolfe and Predeaux overheard a naval officer acting as the port's custom official advise Butler that "the Governor of North Carolina had no power to grant any Licenses for whaleing but the Governor of Virginia had the sole authority of Whaleing as far [south] as Cape Fear and that all vessels ought to have their orders from him for Whaling."[28] Butler was warned that if he was found without Virginia orders, any oil or whalebone he might have would be confiscated.

According to Wolfe, this "so Terrified the sd Butler that he was at a loss for some time [as to] what to doe in the business." Wolfe, who was familiar with North Carolina whaling requirements, advised Butler to seek his license in North Carolina, which he presumably did.

When Wolfe and Predeaux arrived in North Carolina, they reported the incident to North Carolina officials, who were so angered by the deceit that they asked both to sign statements detailing what they had overheard. The two did so willingly. The officials then noted "the ill consequences of such threats to the Trade and Welfare of the [North Carolina] Government" and ordered that the treachery recounted in the affidavits be brought to the attention of the colony's proprietors.[29] Whether this was done is unknown.

Despite the increase in Carolina whaling, the proprietors remained frustrated with the low production of whale oil, and in June 1723 wrote to the governing council and assembly in more urgent tones. They were apparently counting on whale taxes to help turn a profit on their investments:

We earnestly recommend to you the care of the Fishery, the carrying on of which must be not only a great Improvement of Trade but will also cause a great Number of Inhabitants to come into the Province and settle among you. We have upon this Account granted a Lease of the Whale Fishery to our Govr and two others who have applyed to us for the same reserving only the tenths of the Products to Our Use. We hope therefore that you will jointly with us give all due Encouragement to those two Branches of Trade and we doubt not but with your Assistance we shall in little time be enabled to put North Carolina in as flourishing a Condition as any of his Majesty's Colonies in America.[30]

Figure 14.1. A whaling license issued to Samuel Chadwick in 1726 by the governor of North Carolina authorizing "three boats to fish for Whale or Other Royall fish on ye Seay Coast of this Government . . . you paying to ye Honorable Governor one tenth parts of ye oyle and bone Made." (Image courtesy of the State Archives of North Carolina, Raleigh, NC.)

233

Although the proprietors were still discouraged by the small scale of whaling—or at least by the poor returns in whale duties—northern whalers continued to find that the opportunities made the annual trips to the Carolinas worthwhile. Some were sufficiently encouraged to resettle in North Carolina (or at least to purchase tracts of land to meet residency requirements). In the process, local residents finally acquired skills necessary to hunt whales on their own. This apparently occurred after 1715 and made the 1720s perhaps the most productive decade ever for local whaling in the Carolinas.

One of the first and most successful local whalers was John Records, who had begun whaling by at least 1714 and was working the "Sea Coast of port Beaufort" near Cape Lookout by the 1720s. In one year during the early 1720s, he and his partners paid a one-tenth tax of 60 barrels of oil and 800 pounds of whalebone.[31] Assuming a rate of 36 to 44 barrels of oil per right whale, the implied 600-barrel yield would have amounted to a total catch of 14 to 17 right whales. In a two-year period later that decade, "Capt. John Records and Company" reportedly caught whales yielding 670 barrels,[32] or 8 to 10 right whales each year. Overall, between 1714 and 1722, Records paid whale duties amounting to some £2,000.[33] All of these taxes came to light not from tax records, but from allegations filed before the Court of the Vice-Admiralty in 1730 by Governor Richard Everard; these were filed during a bitterly contested transfer of authority over the Carolinas from the proprietors back to the Crown. Everard's complaint charged that those entrusted with collecting whale taxes had improperly diverted payments for their own use or for friends. The credibility of alleged tax amounts in such a dispute may be questionable, but it suggests they were sufficiently large to tempt county officials to commit fraud and that official receipts were incomplete or unreliable.

Figure 14.2 (*facing page*). Detail from Edward Moseley's 1733 map "A New and Correct Map of the Province of North Carolina," showing, from northeast to southwest, Cape Hatteras, Cape Lookout, and Cape Fear, and the coastal geography of barrier islands at the time. (Image courtesy of the J. Y. Joyner Library, East Carolina University, Greenville, NC.)

While Records and others used sloops to carry their whaling paraphernalia to nearby beaches each season, shore whaling by permanent residents beginning to settle on barrier islands at this time is also possible. One of the first to do so may have been John Shackleford, who had moved from the mainland to the barrier island between Beaufort and Cape Lookout by at least 1712 and served as a member of the colony's militia for close to twenty-one years. In 1712, Colonial records report that Shackleford had been assigned an official duty—he was to keep a sharp eye out for "every ship drawing eight feet of water anchoring at the . . . Shackleford Banks to charge three shillings six pence per foot." At least some of those ships were surely New England whalers.[34]

Whether Shackleford owned the land on which he lived in 1712 is uncertain; the first known deed to property on the southern banks of Carteret Precinct was issued to a John Powell the following year. In 1723, however, Shackleford and Enoch Ward purchased 7,000 acres from Powell and split ownership of the Outer Banks' southernmost barrier island, whose expanse was then shaped like an unbroken V; Cape Lookout anchored the island's midpoint with a pair of sandy arms, one stretching northeast toward Cape Hatteras and the other west to Beaufort (figure 14.2). Shackleford took possession of Cape Lookout and the western arm,

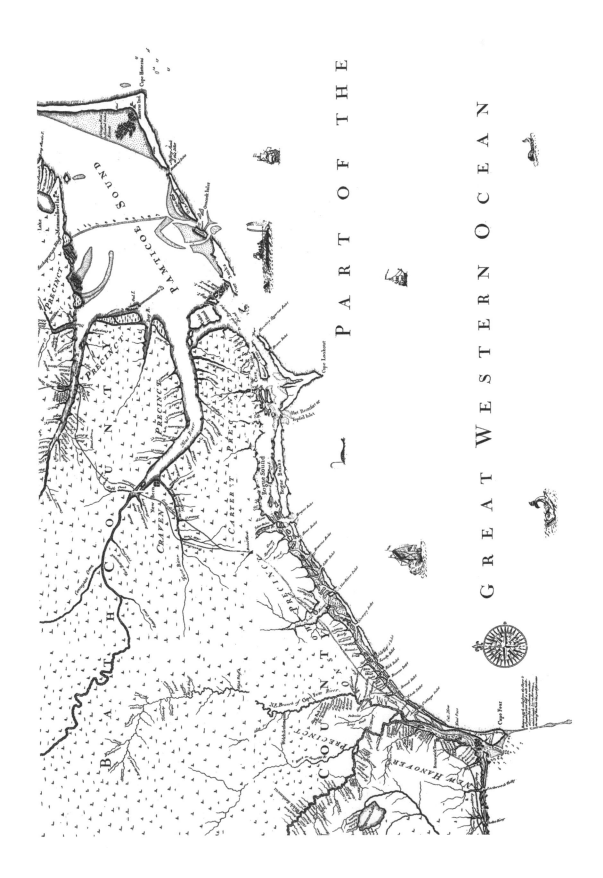

which is now an island called Shackleford Banks separated from Cape Lookout by Barden Inlet. For the next two centuries, Shackleford Banks and Cape Lookout would be the center of North Carolina whaling. Shackleford himself was surely involved in whaling, but how is unclear. He may have been an active whaler, or perhaps he simply charged fees to other whalers wishing to camp on his land. In any case, when he died in 1734, he bequeathed to his daughter's husband, Joseph Moss, the right "to whaile off the Banks he paying yearly to My Beloved wife Ann Shackleford during her life and no longer the rent of two barrels of oyle for his share of one half of one . . . of my boats.[35]

When Carolina whale taxes reverted to a one-tenth fee in 1718, they were paid by some whalers, but by no means all. In 1727, colony officials filed a complaint in the Carolina court at Edenton that charged Josiah Doty with failing to pay his whale tax. Doty was one of the most successful whalers on the Carolina coast. Although born in the Plymouth Colony, he owned land in Carolina, had a whaling license from that government, and caught several whales off Cape Lookout. The complaint began by putting Doty's violation in a regional context. It noted that whaling licenses had been issued to "Sundry Inhabitants of New England and elsewhere as well as the Inhabitants of this Government for severall years past [and they] . . . made considerable quantities of Oyle and Bone and have constantly payd the tenth part thereof annually." The complaint then got down to specifics:

> In the year of our Lord 1725 on or about the month of December Josiah Doty, Master of a Sloop or Ship with a great company of men and Severall Boats with Tackle & stores under his care . . . did enter his sayed Vessel and stores in Port of Beaufort within this Government by virtue of the aforesaid Toleration or lycence . . . to go on the sayd employ of Whaling on the Sea Coasts of this Government for the winter Season [and] during the sayd season did kill, take, and save a great number of Whales . . . and thereof did save and make the quantity of three hundred barrels of Oyle and one thousand weight of whalebone yet nevertheless the said Josiah Doty did not pay and render . . . to the said Palantine and the rest of ye true & absolute Lords Proprietors nor their Receiver the sayd Tenth part of the said Oyle and Bone.[36]

Assuming an average yield of 36 to 44 barrels of oil per right whale, the 300 barrels of oil Doty's crew was said to have made could have involved a catch of 7 or 8 right whales.

Another example of illicit whaling is cited in John Brickell's remarkably lucid description of the flora and fauna of North Carolina published in 1737. Brickell noted that two northern vessels working off Ocracoke Island, south of Cape Hatteras, killed whales producing 340 barrels of oil. This would have amounted to perhaps 8 to 10 whales. Interestingly, it also suggests that the island's local residents were still limited to salvaging drift whales—possibly including animals struck and lost by Yankee whalers:

These Monsters are very numerous on the Coasts of North Carolina, and the Bone and Oil would be a great Advantage to the Inhabitants that live on the Sand-Banks along the Ocean, if they were as dexterous and industrious in Fishing for them as they are Northwards; but as I observed before, the People of these parts are not very much given to the Industry, but wait upon Providence to throw those dead Monsters on Shoar, which frequently happens to their great advantage and Profit. For which Reason abundance of Inhabitants dwell upon the Banks near the Sea for that Intent and the benefit of Wrecks of Vessels which are sometimes driven in upon these Coasts. Not many Years ago there were two Boats that came from the Northward to Ocacock Island, to fish, and carried away that Season Three Hundred and Fourty Barrels of Oil, besides the Bone, but these Fishermen going away without paying the Tenths to the Government.[37]

Problems collecting the one-tenth share of whales—whether from whalers evading taxes or dishonest tax collectors—apparently contributed to a decision by colony proprietors to terminate their interests in the colony. In a petition to the King's Privy Council in 1728, they proposed surrendering all rights and privileges to their Carolina plantation in return for a payment of £9,500 from the Crown to cover outstanding debts owed to them by province residents.[38] Among the claimed arrears appended to their petition was four years of back payments on whales killed under colony licenses: "The tenths reserved upon the whale fishery of North Carolina granted four years ago which according to the Accounts received must have been considerable valued for the four years at . . . £800."[39]

In July 1731, just a few years after submitting their petition, colony officials adopted a bill to standardize the value of goods and services produced against the English pound; whale oil was set at a value of £2.5s per barrel.[40] If all the whales for which back taxes were claimed were right whales and all of the lost tax reflected oil, £800 in taxes at £2.5s per barrel over a four-year period would have represented a catch of 20 to 25 right whales per year. Recognizing that some of the claimed tax would have been paid in tenths of whalebone and that some oil may have come from species other than right whales, a catch of 10 to 12 right whales per year might be a more likely estimate. Of course, the accuracy of claims in a settlement request are open to some question, but if it was a fair representation of arrears and at least some duties were successfully collected, perhaps 20 to 30 right whales per year might have been killed annually (including whales struck and lost) during peak whaling years in the 1720s. Such a level seems consistent with other information on the number of whalers and catch levels from the thin record available for the decade. It also is consistent with a peak catch of 16 to 26 whales per year (not including whales struck but lost) postulated by Marcus B. Simpson Jr. and Sally W. Simpson in their excellent historical review of North Carolina whaling.[41]

Shortly after the proprietors filed their petition to cede rights to the colony, Thomas Lowndes, provost marshal for the Carolina Province, prepared a memorandum dated February 1729 noting "the Tenths reserved upon the Whale Fishery" as one of the advantages to the Crown if it purchased proprietor rights.[42] In 1730 the Crown approved the purchase, but that December King George II directed that all duties on whales taken in the province be abolished. His instructions to the province's first royal governor, George Burrington, directed the following: "Whereas for some years past Governors of some of our Plantations have seized and appropriated to their own use the produce of whales of several kinds taken upon those coasts upon the pretense that whales are royal fishes which tends to discourage this branch of fishery.... It is therefore Our Will & Pleasure that you do not pretend to any such claim nor give any discouragemt to our subjects upon the coasts of the Province of North Carolina."[43]

With all whale duties waived, official references to the activity petered out abruptly. Some whaling continued off the Carolinas, but by then it included a greater mix of sperm whales and perhaps other species, because right whales were becoming scarce. A journal kept by whaling Captain Ashley Bowen of Marblehead, Massachusetts, records a 1753 voyage by his ship the *Susannah* to Cape Lookout, where he found three other sloops also intent on whaling during the coming winter season. After an unsuccessful start with no landings, the ships parted company and spent varying amounts of time at Cape Fear to the south. Although any success by the latter vessels is unknown, Bowen grumbled in his journal that he "did not get a drop of oil."[44] To make matters worse, whalers at this time had to contend with the growing scourge of piracy and privateers. A whaling schooner sailing out of Boston, for example, was captured off the Virginia Capes by a Spanish privateer in 1747.[45]

Despite at-sea interdictions and declining catches, a few Yankee and local whalers must have met with some success. A letter dated 1755 from Governor Arthur Dobbs of North Carolina to the Board of Trade in England described a reconnaissance trip to Cape Lookout to assess its prospects as a port. Dobbs noted in passing that the area was "where the whale fishers of the northward have a considerable fishery from Christmas to April, when the whales return to the northward."[46] Jacop Lobb also noted the presence of "Whalers huts" on the east end of Shackleford Banks on a map he drew in 1764 during another scouting mission to assess Cape Lookout's potential as a port.[47] A year later, a visiting Frenchman again reported seeing "whale fishers tents" near Cape Lookout,[48] and in 1768 the port of Beaufort exported 1,126 gallons of whale oil and 150 pounds of whalebone,[49] amounts that could have come from a single right whale.

Whaling off the Carolinas all but stopped during the Revolutionary War when the English blockade of ports threatened colonial vessels with confiscation of their goods and impressment of their crews into service for the Royal Navy. After the war, whaling resumed, but with limited success. Between 1785 and 1789, a score of merchant vessels were recorded leaving North Carolina ports with whale oil listed among their cargo.[50] Neither the amounts of oil nor the species they came from are clear, but around this time bottlenose dolphins, as well as sperm whales, were becoming increasingly common targets in the Carolinas as right whale numbers continued to plunge. Before long, coastal whaling again came to a halt when war broke out between the British and Americans in 1812. After that conflict ended, coastal whaling resumed with little success. Yankee whalers still cruised the "Hatteras" and "Charleston" grounds, but most apparently were either passing through on their way to more distant areas in hopes of picking up a whale or two, or were searching offshore for sperm whales. Among the few with any success was the whaling vessel *Winslow* out of New Bedford, Massachusetts, which took what was said to have been a 90-barrel whale off Cape Fear in 1838.[51]

Information on shore whaling leading up to the Civil War is sparse, but North Carolina whalers interviewed after the war indicated that a few crews were operating near Beaufort on a nearly continuous basis throughout the early 1800s.[52] In 1869, Elliot Coues, a physician and naturalist stationed at Fort Macon on the east end of Bogue Banks (figure 14.3), informed Spencer F. Baird, assistant secretary of the Smithsonian Institution, that he had just returned from a trip to Shackleford Banks where local whalers had landed a 45-foot right whale. Aware that the Smithsonian was interested in right whale specimens, he advised that the skeleton was available for collection, although it would require considerable effort to clean; he

Figure 14.3. The southern end of North Carolina's Outer Banks. (Map prepared by Linda White.)

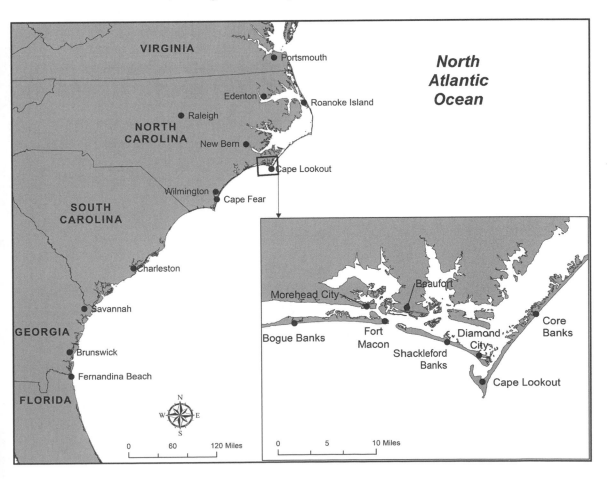

also commented that several other well-beached jaws and sundry bones were lying on the beach.[53] In a separate account on the region's natural history published in a scientific journal two years later, Coues reported that area fishermen caught whales they called "right whales" and occasionally also took two other kinds, including humpback and "scrag" whales; they usually landed two or three whales of varying combinations of species each spring.[54]

In the mid- to late 1870s, most but not all whaling in the Carolinas was based on Shackleford Banks, where some 500 people had taken up residence in a small cluster of homes called Diamond City. The settlement was named after the diamond-shaped spit making up Cape Lookout. Virtually all vestiges of Diamond City are now gone, save a cemetery and a few rare photographs of its whalers (figure 14.4). During its heyday, however, residents lived off the sea, extracting oil from whales, dolphins, and mullet as their principal source of income. For most of the 1870s, 2 to 3 whaling camps operated on the banks, each with 3 dories and a crew of 18. In 1879, the whaling force grew to four camps that caught five right whales and netted a total of $4,000. In 1880, an even larger contingent of six crews tried their hand, but managed to land only one whale. This bigger group caught what was thought to have averaged about 4 whales per year over the next several years, with each animal producing about 1,800 gallons of oil and 550 pounds of whalebone.[55] Multiple crews were still operating in 1894, when five crews were said to be operating on Shackleford Banks and its neighbor to the west, Bogue Banks.[56]

The whaling methods used on Shackleford Banks were similar to those used by shore whalers elsewhere (figure 14.5). Although the catch for much of the 1800s may have averaged two or more whales per year, a review of surviving records by Edward D. Mitchell and Randal R. Reeves confirms only a score of known or suspected right whale landings by North Carolina shore whalers over the entire nineteenth century. All of these were landed south of Cape Hatteras, principally off

Figure 14.4. Devine Guthrie, a whaler, preacher, and boatbuilder of Diamond City, North Carolina, stands beside one of the last whaling boats constructed on Shackleford Banks, circa 1890. (Image courtesy of the Core Sound Waterfowl Museum, Harker's Island, NC.)

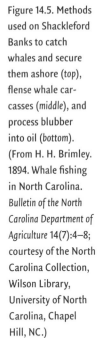

WHALE ON BEACH AT BEAUFORT.

Figure 14.5. Methods used on Shackleford Banks to catch whales and secure them ashore (*top*), flense whale carcasses (*middle*), and process blubber into oil (*bottom*). (From H. H. Brimley. 1894. Whale fishing in North Carolina. *Bulletin of the North Carolina Department of Agriculture* 14(7):4–8; courtesy of the North Carolina Collection, Wilson Library, University of North Carolina, Chapel Hill, NC.)

CUTTING BLUBBER.

"TRYING OUT" OIL.

Cape Lookout and Shackleford Banks.[57] Except for a brief surge in the 1870s, and again in 1898, they could confirm no years with more than a single landing during the entire 1800s. The few years with multiple landings were 1874 (two whales), 1879 (five), 1894 (two), and 1898 (three).

Yankee whalers continued visiting the Carolinas through the 1880s in search of both sperm whales and right whales. How many tried their luck is unknown, but repeated visits suggest at least some success. One of those vessels, the *Daniel Webster*, searched the coast in the winter of 1874–75 and was said to have introduced the whaling gun to local whalers.[58] Although others attribute its introduction to local innovation or simply a purchase from northern suppliers,[59] this shoulder gun became a standard piece of whaling equipment for North Carolina shore whalers after 1875. In the winter of 1878–79, another vessel, the *Seychelle*, worked its way along the coast scouting for whales, apparently without success. The following August while still looking for whales, it was caught in a hurricane and foundered a half mile from Cape Lookout, resulting in the ship's total loss.[60]

As part of their review of North Carolina whaling, Reeves and Mitchell uncovered records kept by four shore whalers that listed lifetime tallies of their successful hunts. All four lived and worked in Diamond City on Shackelford Banks between 1830 and 1890, with their careers overlapping one another's. Together their lists totaled 114 successful hunts; however, some entries undoubtedly involved the same whales. Indeed, two of the four whalers lived in adjacent houses for parts of their careers. Based on those records, Reeves and Mitchell conservatively estimated that the 4 whalers were involved in landing at least 62 right whales, amounting to about 1 per year.[61] Of course, some whales undoubtedly died after being struck and lost. In 1894, when two right whales were known to have been killed on Shackleford Banks, the whalers reportedly chased six.[62]

By the late 1800s, as artisanal shore whaling was becoming a disappearing way of life, whalers on Shackleford Banks adopted the practice of giving landed whales separate names. Like scientists today who give whales individual names based on the shapes of their markings and scars to help reidentify them, the names assigned by whaling communities effectively etched successful hunts into the memories of local residents and became part of the community lore.[63]

One of those whales was a 53-foot-long behemoth that produced 40 barrels of oil and 700 pounds of whalebone. Landed on May 4, 1874, it was named "Mayflower." Still remembered as one the most epic battles between a whale and whalers on the North Carolina coast, 2 crews of 3 boats each fought the whale for 6 hours, between the time it was first spotted in a group of 10 whales a mile off the beach and the whale's death (figure 14.6).[64] The whale remained submerged for up to a half hour between surface bouts with whalers and managed to capsize one boat with its thrashing tail despite sustaining multiple harpoon wounds and six shots from a shoulder whale gun. When it finally expired, the whalers had been drawn 10 miles offshore and had to labor long into the night to tow their prize ashore, guided by fires lit in the dunes. Mayflower's skeleton was salvaged by the

Figure 14.6. Diamond City whaler Josephus Willis Sr., circa 1880, repairing a fishing net. Willis was captain of one of the two crews of boats that participated in landing the right whale named "Mayflower." (Image courtesy of the State Archives of North Carolina, Raleigh, NC.)

North Carolina Museum of Natural Sciences in Raleigh, where it can still be seen today (figure 14.7). The same day that Mayflower was killed, other crews on the banks landed two other whales—one dubbed "Lady Hayes" produced 35 barrels of oil and 600 pounds of whalebone, while the other, eventually called by default "Haint Yet Been Named," yielded 25 barrels and 450 pounds of whalebone.

Another whale landed in mid-March 1894 was killed by a contingent of five or more boats filled with inexperienced boys, and was therefore named "Little Children." "Mullet Pond," a 46-foot whale, was killed on February 18, 1898, and butchered at a cove known locally as Mullet Pond; it produced 27 barrels of oil and 650 pounds of whalebone. Like Mayflower, its skeleton was collected by scientists and is now on display in the natural history museum at the University of Iowa. A small 39-foot whale named "Big Sunday" was landed on Sunday, May 17, 1908, towed to Beaufort, and put on exhibition with only some of its baleen and none of its oil saved. Another, called "Cold Sunday," was said to have been killed on a Sunday "so cold that flying ducks froze solid in flight."[65] The last right whale killed in North Carolina was a mammoth 57-footer taken at Cape Lookout on March 16, 1916. The following year, after losing much of their whaling gear in a fire, the last remaining whaling crew disbanded, ending North Carolina's long history of shore whaling.

Figure 14.7. The skeleton of "Mayflower," a right whale landed in an epic 1876 whaling battle off Shackleford Banks, on display at the North Carolina Museum of Natural Sciences in 1894 (*top*) and in 2007 (*bottom*). (Courtesy of Margaret Cotrufo, North Carolina Museum of Natural Sciences, Raleigh, NC.)

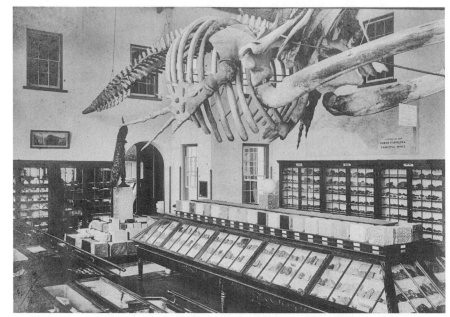

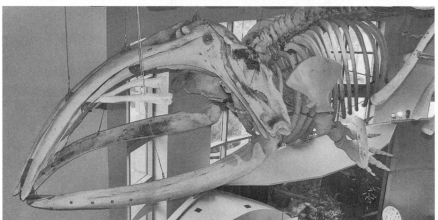

South Carolina, Georgia, and Florida

Moving further down the coast, records of whaling drop off sharply. Nearshore waters of Georgia and Florida are now the center of the North Atlantic right whale population's calving grounds, but they seem to have been spared direct exploitation during the colonial era. This part of their range was under Spanish dominion until 1763, when Spain ceded what are now Florida and parts of Georgia to the English in exchange for control over Havana, Cuba. Before then, periodic skirmishes between the English colonists in the Carolinas and the Spanish in Florida, resulting from their disputed boundary, apparently dissuaded Yankee whalers from moving south of present-day South Carolina. One of the few hints those occasional hostilities did not deter everyone is a license issued to a New York whaler by English authorities in 1688 to fish "about the Bahames Islands, And Cap florida, for sperma Coeti whales and Racks" (sperm whales and whalebone from right whales).[66] Whether the license was exercised and with what success is unknown.

British rule in the region lasted just 20 years before it was returned to Spain in 1783 as part of the Treaty of Paris at the end of the American Revolution. The second period of Spanish rule lasted until 1819, when the burden of maintaining control in the face of Native unrest and incursions by American settlers became too much. John Quincy Adams, then secretary of state, convinced Spain to cede its rights to the territory to the United States in return for assuming $5 million in claims by US citizens against Spain. By then, however, most whalers had already redirected their focus to species other than North Atlantic right whales.

The first firm evidence of whaling south of North Carolina does not appear until January 7, 1880, when a 40-foot right whale killed in Charleston Harbor was collected for display by G. E. Manigault, curator of a local museum. A letter by Manigault describing the "hunt" depicts a chaotic scene one could hardly consider "whaling."[67] It involved no experienced whalers. The whale was harpooned in the morning and by the afternoon, "a much larger attacking force started in pursuit, consisting of four steam tugs, between fifty and sixty row boats, and a few sailing craft."[68] After repeated blows from the bow of a tug, the whale finally succumbed. It was put on display at the Charleston docks for several days before being turned over to the museum (figure 14.8). At the time of its collection, fewer than half a dozen complete right whale skeletons had been preserved on either side of the Atlantic.

Around the same time, at least a few experienced whalers were working off South Carolina. Manigault's letter tells us that just a few weeks after the Charleston Harbor whale was killed, "a whale sixty feet in length was cast ashore on the beach at Sullivan's Island [at the mouth of Charleston Harbor], which had already been stripped of its blubber and baleen at sea."[69] He added that one or two schooners fitted out for whaling at Port Royal, South Carolina, and Brunswick, Georgia, had made several captures. Brunswick, Georgia, and Fernandina on Amelia Island, Florida, also served as resupply bases for Yankee whalers hunting sperm whales off the Bahamas. While stopping at those ports they found and at least occasionally took right whales. When the whaling ship *Golden City* returned to New England after an 1875–1876 voyage to the Bahama grounds, it returned with 440 barrels of sperm oil and 40 barrels of oil from a right whale caught off Fernandina on February 26, 1876.[70] Catch records from the late nineteenth century are rare, but at least 25 right whales were killed off South Carolina, Georgia, and northeast Florida between 1868 and 1897.[71] If most of those whales were adult females and calves, it would have had a devastating effect on the remaining population.

The last right whale deliberately killed off the southeastern United States— and the only record from that area in the 1900s—was a calf killed off Florida on March 24, 1935.[72] Like the whale killed in Charleston Harbor in 1880, it would be a stretch to call it a victim of "whaling." The calf and its mother were discovered at 9 a.m. by Frank Merritt, captain of the charter boat *Caliban* out of Pompano Beach,

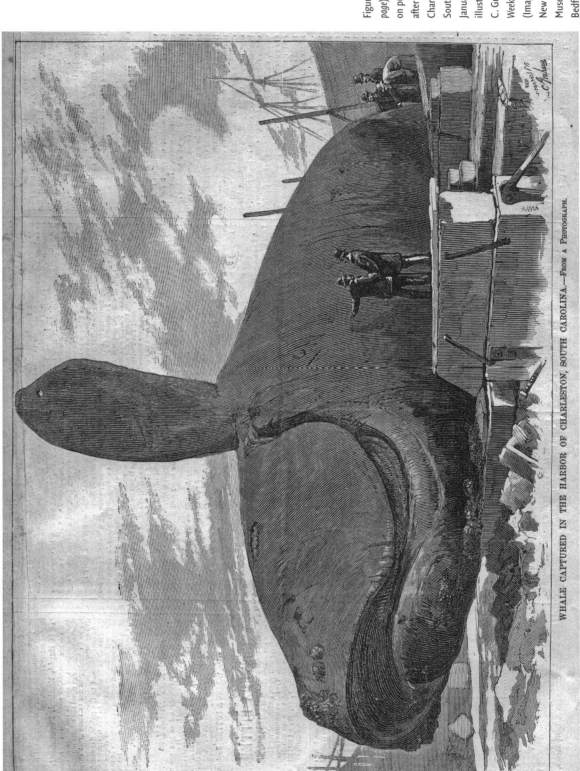

WHALE CAPTURED IN THE HARBOR OF CHARLESTON, SOUTH CAROLINA.—From a Photograph.

Figure 14.8 (facing page). A right whale on public display after being killed in Charleston Harbor, South Carolina, on January 7, 1880, as illustrated by engraver C. Graham in Harper's Weekly, January 1880. (Image courtesy of the New Bedford Whaling Museum, New Bedford, MA.)

while on a trip to harpoon tuna. He was less than a mile off the beach near Hillsboro Lighthouse, north of Fort Lauderdale, when he found the pair.[73] Interested only in bragging rights, Merritt decided to attack. Recognizing the adult female was far too large to handle, he tried to drive her off and isolate the calf by pumping more than 100 rounds from a high-powered rifle into the mother. When that failed to chase her off, he set his sights on the calf anyway.

As the attack wore on, crowds gathered on the beach and recreational boats surrounded the *Caliban* to watch the action. After a six-hour assault during which the calf was stabbed repeatedly with a harpoon, and two boxes of high-powered shells were fired into the besieged animal, it finally died. Only then did the mother abandon her calf. It took the *Caliban* and two other boats three hours to tow the 25-foot carcass to the dock, where it was put on display (figure 14.9). After a few days, the skull was salvaged for mounting and sale, and the rest of the carcass was towed offshore and dumped. A newspaper reporter who happened to be aboard the *Caliban* at the time filed an article with photographs detailing the episode in the *New York Herald Tribune*.[74]

Five decades later, when building a photo-identification catalogue for North Atlantic right whales in the 1980s, Amy Knowlton at the New England Aquarium in Boston came across photos of the encounter. To her amazement, the callosity pattern of the mother photographed during the 1935 incident matched a whale listed as catalogue #1445. Having survived her gunshot wounds, she was photographed nearly 25 years later in Massachusetts Bay in the spring of 1959 by Bill Schevill and Bill Watkins, two marine mammal scientists at the Woods Hole Oceanographic Institute, and again in Cape Cod Bay in April 1980. Although never again seen with a calf, she was photographed several more times until 1995, when she was found drifting listlessly near Georges Bank off Cape Cod with a

Figure 14.9. A right whale calf, mislabeled as a humpback whale, on display in Pompano Beach, Florida, after being killed as a trophy by charter boat captain Frank Merritt on March 24, 1935. (Image courtesy of the Fort Lauderdale Historical Society, FL.)

massive propeller gash on the side of her head. The wound exposed part of the skull and she was found dead the next day. Assuming she was at least 8 to 10 years old when photographed with her calf in 1935, she would have been at least 70 years old when she died, making her the oldest known right whale.

NOTES

1. Sir Robert Heath's patent, 1629. *In* WL Saunders (ed.), 1886. *The Colonial Records of North Carolina*. 10 vols. P. M. Hale. Raleigh, NC, US, Vol. 1 (1662–1712), pp. 5–13.

2. First Charter to the Lords Proprietors of Carolina, 1663. *In* Saunders, *The Colonial Records* (n. 1), Vol. 1, pp. 20–23.

3. Saunders, *The Colonial Records* (n. 1), Vol. 1, p. 22 (quotation).

4. Palmer, WR. 1959. The whaling port of Sag Harbor. Ph.D. thesis. Columbia University Microfilms Int., Ann Arbor, MI. 327p, as cited in RR Reeves and E Mitchell. 1988. *History of Whaling in and near North Carolina*. NOAA Technical Report NMFS 65. National Oceanic and Atmospheric Administration. Washington, DC.

5. Reeves and Mitchell, *History of Whaling* (n. 4), p. 3 (quotation).

6. Simpson, MB, Jr., and SW Simpson. 1990. *Whaling on the North Carolina Coast*. Division of Archives and History, North Carolina Department of Cultural Resources. Raleigh, NC, US, p. 6.

7. Ibid.

8. Peter Colleton letters to Peter Carteret, October 22, 1668, docs. 9 and 10 (quotation), Ye countie of Albemarle in Carolina. *In* J-M Poff (ed.), The Colonial Records Project. Office of Archives and History, North Carolina Department of Cultural Resources, Raleigh, NC, US, http://www.ncpublications.com/colonial/bookshelf/countie/countie1.htm#5. Last updated July 23, 2001.

9. The fundamental constitutions of Carolina, drawn up by John Lock, March 1, 1669. *In* Saunders, *The Colonial Records* (n. 1.), Vol. 1, pp. 187–206.

10. Ibid., p. 205 (quotation).

11. Parker, Mattie Erma Edwards (ed). 1968. *North Carolina Higher Court Records,* Volume II, 1670–1696 of *Colonial Records of North Carolina*, Second Series, edited by MEE Parker and Others. Division of Archives and History. Department of Cultural Resources, projected multi-volume series, (1963–), as cited in Simpson Jr. and Simpson, *Whaling* (n. 6), pp. 6–7 (quotation).

12. Carteret, P. 1673. Financial accounts of Peter Carteret 1666-1673 in Carolina, doc. 24 (quotation). *In* Poff, The Colonial Records Project (n. 8), http://www.ncpublications.com/colonial/bookshelf/countie/countie1.htm#24.

13. Instructions to the Governor of Albemarle County concerning whales. *In* Saunders, *The Colonial Records* (n. 1), Vol. 1, p. 338 (quotation).

14. Cain, RJ (editor). 1981. North Carolina higher-court minutes 1724–1730. The Colonial Records of North Carolina (Second ser.) Vol. VI. Dep. Cult. Res., Div. Archives and Hist. Raleigh. 791p, as cited in Reeves and Mitchell. 1988. *History of Whaling* (n. 4), p. 3.

15. Instructions to the Governor (n. 13), p. 338 (quotation).

16. Salley, AS, Jr. (ed.). 1916. *Commissions and Instructions from the Lords Proprietors of Carolina to Public Officials of South Carolina, 1685–1715.* Printed for The Historical Commission of South Carolina by the State Co. Columbia, SC, US, p. 26 (quotation), http://catalog.hathitrust.org/Record/001263862.

17. Court Records, September 28, 1694. *In* Saunders, *The Colonial Records* (n. 1), Vol. 1, pp. 418–420, p. 419.

18. Simpson Jr. and Simpson, *Whaling* (n. 6), pp. 7–9.

19. Parker, *North Carolina Higher Court* (n. 11), pp. 122, 204–205, 207.

20. Lawson, J. (1714) 1860. *The History of Carolina, Containing the Exact Description of the*

Natural History of that Country: Together with the Present State Thereof. And a Journal of a Thousand Miles, Traveled through Several Nations of Indians. Giving a particular Account of their Customs, and Manners, &c., &c. Strother & Marcom. Raleigh, NC, US, p. 146 (quotation).

21. Ibid., pp. 251–254.

22. Ibid., p. 251.

23. Instructions to Colonel Edward Tynte Governor of Carolina, March 24, 1709. *In* Saunders, *The Colonial Records* (n. 1), Vol. 1, p. 706 (quotation).

24. Letter to Governor Eden, March 26, 1715. *In* Saunders, *The Colonial Records* (n. 1), Vol. 2 (1713–1728), pp. 175–176, p. 176 (quotation).

25. Cain, RJ (ed). 1981. North Carolina Higher Court Records, 1724–1730. The Colonial Records of North Carolina (Second series), Vol. VI. Department of Cultural Resources. Division of Archives and History. Department of Cultural Resources, NC, xxii, as cited in Simpson Jr. and Simpson, *Whaling* (n. 6), p. 11.

26. Statements of Mathew Wolfe and Thomas Predeaux, 3 December 1720. *In* Saunders, *The Colonial Records* (n. 1), Vol 2, pp. 396–398.

27. Ibid., Vol. 2, pp. 397 (quotation).

28. Ibid.

29. Ibid., Vol. 2, p. 398 (quotation).

30. Ibid., Vol. 2, p. 490 (quotation).

31. Vice Admiralty Papers, Vol. I, fo. 28, State Archives, Raleigh, NC. As cited in Reeves and Mitchell, *History of Whaling* (n. 4), p. 19.

32. Vice Admiralty Papers. Vol. I. fo. 22, State Archives, Raleigh NC. As cited in Reeves and Mitchell, *History of Whaling* (n. 4), p. 19.

33. Petition of Richard Everard to Edmund Porter, July 23 1730, Vice-Admiralty Papers CCR 142. As cited in Simpson Jr. and Simpson, *Whaling* (n. 6), p. 10.

34. Stick, D. 1958. *The Outer Banks of North Carolina.* University of North Carolina Press. Chapel Hill, NC, US, p. 311 (quotation).

35. Fleet, B. (1937–1949) 1988. *Virginia Colonial Abstracts.* Reprint. Vol. 3. Genealogical Publishing. Baltimore, MD, US; and Simpson, T. 1995. Shackelford Banks history notes. Mailboat 2(2): no pagination (quotation).

36. Hathaway, JRB (ed.). (1901) 2002. *The North Carolina Historical and Genealogical Register.* Reprint. Genealogical Publishing. Baltimore, MD, US, Vol. 1, pp. 187–188 (quotations).

37. Brickell, J. 1737. *Natural History of North-Carolina with an Account of the Trade, Manners, and Customs of the Christian and Indian Inhabitants.* James Carson. Dublin, Ireland, pp. 219–220 (quotation).

38. Saunders, *The Colonial Records* (n. 1), Vol. 2, pp. 771–772.

39. Ibid., Vol. 2, p. 772 (quotation).

40. Ibid., *The Colonial Records* (n. 1), Vol. 3 (1728–1733), p. 168.

41. Simpson Jr. and Simpson, *Whaling* (n. 6), p. 16.

42. Saunders, *The Colonial Records* (n. 1), Vol. 3, pp. 10–12.

43. Ibid., Vol. 3, p. 99 (quotation).

44. Reeves and Mitchell, *History of Whaling* (n. 4), p. 7 (quotation).

45. Starbuck, A. 1878. *History of the American Whale Fishery from its Earliest Inception to the Year 1876.* Report of the US Commissioner of Fish and Fisheries, Part 1. US Government Printing Office. Washington, DC, p. 171 (quotation).

46. Arthur Dobbs to the Board of Trade of Great Britain, May 19, 1755. *In* WL Saunders, *The Colonial Records* (n. 1), Vol 5, (1752–1759), pp. 344–347 (quotation).

47. Lobb, J. 1764. A plan of the harbor of Cape Lookout surveyed and sounded by His Majesty's sloop Viper, Captain Lobb in Sepr. 1764. Gm72003573. Geography and Map Division, Library of Congress. Washington, DC.

48. "Journal of a French Traveller in the Carolinas, 1765," American Historical Review 26 (July 1921), 733, as cited in Simpson Jr. and Simpson, *Whaling* (n. 6), p. 20.

49. Kell, JB and TA Williams (eds). 1975. North Carolina's Coastal Carteret County during the American Revolution, 1765–1785. Era Press. Greenville, NC. p. 105. As cited in Simpson Jr. and Simpson, *Whaling* (n. 6), p. 107.

50. Simpson Jr. and Simpson, *Whaling* (n. 6), p. 24.

51. Dennis Woods. "Abstracts of Whaling Voyages [1835-1875]" an unpublished manuscript, 5 volumes, n.d., New Bedford Free Public library, New Bedford, Massachusetts, as cited in Simpson Jr. and Simpson, *Whaling* (n. 6), p. 27.

52. Stick, *The Outer Banks* (n. 34), p. 185.

53. Elliot Coues to Spencer F. Baird, April 15, 1869, Incoming Correspondence, Assistant Secretary, Smithsonian Institution Archives, Washington, DC. As cited in Simpson Jr. and Simpson, *Whaling* (n. 6), p. 31.

54. Coues, E. 1871. Notes on the natural history of Fort Macon, N.C., and vicinity (no. 1). Proceedings of the Academy of Natural Sciences of Philadelphia 23(May):12–48, 13.

55. Clark, AH. 1887. History and present condition of the fishery. In *History and Methods of the Fisheries*, Section 5 of GB Goode (ed.), *The Fisheries and Fishing Industries of the United States*. US Government Printing Office. Washington, DC, Vol. 2, pp. 1–218, 49.

56. Brimley, HH. 1894. Whale fishing in North Carolina. Bulletin of the North Carolina Department of Agriculture 14(7):4–8, 4.

57. Reeves and Mitchell, *History of Whaling* (n. 4), pp. 12–13.

58. Clark, History and present condition (n. 55), p. 49.

59. Simpson Jr. and Simpson, *Whaling* (n. 6), p. 34.

60. Reeves and Mitchell, *History of Whaling* (n. 4), p. 7.

61. Ibid., p. 19.

62. Simpson Jr. and Simpson, *Whaling* (n. 6), p. 47.

63. Reeves and Mitchell, *History of Whaling* (n. 4), p. 17. Simpson Jr. and Simpson, *Whaling* (n. 6), pp. 37–41.

64. *A North Carolina Naturalist: H. H. Brimley: Selections from His Writings*. 1949. EP Odum (ed.). University of North Carolina Press. Chapel Hill, NC, US, pp. 109–116.

65. Stick, *The Outer Banks* (n. 34), p. 190 (quotation).

66. Starbuck, *History* (n. 45), p. 15 (quotation).

67. Clark, History and present condition (n. 55), pp. 49–52.

68. Manigault, GE. [1880]. Letter describing a whale taken in Charleston Harbor, S.C. in 1880. *In* JB Holder. 1883. The Atlantic right whales: a contribution embracing an examination . . . to which is added a concise resume of historical mention related to the present and allied species. Bulletin of the American Museum of Natural History 1(4):99–137, 102–104.

69. Ibid., p. 104 (quotation).

70. Reeves, RR, and E Mitchell. 1986. American pelagic whaling for right whales in the North Atlantic. *In* RL Brownell, PB Best, and JH Prescott (eds.), Right Whales: Past and Present Status, Special issue 10, Reports of the International Whaling Commission, pp. 221–254.

71. Clark, History and present condition (n. 55), p. 49. Reeves and Mitchell, American pelagic whaling (n. 70), pp. 227–229, 235–236.

72. Moore, JC. 1953. Distribution of marine mammals to Florida waters. American Midland Naturalist 49(1):117–158.

73. Gore, J. 1935. Six ton whale is captured off Ft. Lauderdale Coast. Fort Lauderdale Daily News. March 25.

74. Kraus, SD, and RM Rolland. 2007. *The Urban Whale: North Atlantic Right Whales at the Crossroads*. Harvard University Press. Cambridge, MA, US, pp. 1–3.

15

ESTIMATING PRE-EXPLOITATION POPULATION SIZE

HISTORICAL WHALING RECORDS are important for many reasons. Among other things, they offer valuable clues to former habitat, how habitat preferences might have changed over time, and where unrecognized habitat might still exist today. The pattern of right whale landings by colonial whalers is somewhat at odds with their distribution and movements today. Whereas winter-spring whaling in Cape Cod Bay at its height between the 1690s and 1720s is fully consistent with the January to mid-May right whale assemblages now seen in the Bay, the success of shore whalers in those same months off Nantucket, Long Island, Cape May, and North Carolina is not. Indeed, the sparse sighting records in those areas today seem entirely incongruous with the notion that their abundance in those locations and seasons could have once supported a viable local whaling business. Conversely, few or no historical catch records exist in other places where seasonal aggregations are now the norm, such as the Great South Channel east of Cape Cod, the Bay of Fundy at the head of the Gulf of Maine, or Browns Bank off the southern tip of Nova Scotia.

Several explanations for these disparities seem possible, none of which is entirely satisfying. Perhaps whaling off Long Island, Nantucket, Cape May, and North Carolina caused right whales to avoid those areas and establish new habitat-use patterns that persist to this day. Or perhaps colder conditions in the Little Ice Age during the colonial period were more conducive to overwintering whales. Alternatively, the winter-spring abundance in those areas during colonial times could reflect the broader distribution of a larger right whale population that might once again emerge. Yet another possible explanation is that more whales occur in those areas today than is currently recognized, due to limited efforts to count whales in those seasons.

As for the scarcity of past whaling records for current aggregation sites, such as the Great South Channel and Bay of Fundy, perhaps whalers never discovered those areas, or records for those grounds were lost or never written down. Different climatic and oceanographic conditions also could have made those areas less attractive to whales centuries ago. Further research into historical whaling records and past climate patterns may rule out some of these explanations or perhaps identify new ones. At this point, the distribution discrepancy between the seventeenth and eighteenth centuries and the present can only be bookmarked as an unresolved question that needs further investigation. It should not be forgotten, particularly when one considers the possible effects of climate change.

Low Ebb

Historical catch records are important for another reason. Two questions asked by almost anyone interested in an endangered species are "how many animals are left?" and "how many were there before their populations became endangered?" Whether you are a curious layperson, an involved scientist, or a responsible resource manager, the difference between those numbers is an obvious yardstick against which to measure their current status. Estimating current abundance is invariably the easier question to answer. Although rarely as simple as it sounds, persistence and cooperation between researchers usually yields a useful number. For North Atlantic right whales, the best current estimate of abundance comes from photos of individuals known to have been alive in a given year. In 2015, that number was about 500 whales for the entire North Atlantic right whale population, a number that had been rising slowly since 2000.[1]

How many survived when hunting ended in the early 1900s is less certain. Between the mid-1800s and early 1900s, whaling effectively eliminated what remained of the eastern North Atlantic right whale population. Basque whalers may have been responsible for most of its decline, but they were not the cause of its ultimate demise. The Basques stopped sending out ships to hunt right whales in the North Atlantic before the 1800s. Although Basque shore whalers continued to chase right whales sporadically off their homeland coasts through the late 1800s, fewer than a dozen landings can be confirmed for the entire nineteenth century. Far more culpable were pelagic whalers from other countries who opportunistically scoured right whale habitats off Europe and North Africa, as well as commercial shore whalers strategically located at coastal stations in northern Europe.

What must have been a crippling blow leading to the assumed extirpation of the eastern population was delivered by pelagic whalers, including a few dozen Yankee vessels and an unknown number of northern European boats, all working off North Africa. In the 1850s, they discovered a right whale calving ground in

Cintra Bay off Western Sahara, on the bulge of northwestern Africa. These whalers were attracted to the region mainly by sperm, humpback, and pilot whales, but logbooks of the New England whalers reveal a catch of at least 90 right whales between 1855 and 1858.[2] How many more might have been taken by European vessels is unknown; a search for their logbooks, if they still exist, has yet to be made. To the north, pelagic whalers took at least 25 right whales off Cape Farwell on the southern tip of Greenland between the 1860s and 1898,[3] and 23 more off Iceland and in the Denmark Strait between Iceland and Greenland between 1889 and 1908 (figure 15.1).[4] Those animals, however, may have included a mix of right whales from both the eastern and western populations.

The final blow to eastern North Atlantic right whales was probably delivered by commercial shore whalers based at two dozen or so whaling stations built between the 1880s and early 1900s. Their principal targets were species other than right whales, but their locations in Norway, Iceland, the Faroe and Shetland Islands, the Hebrides, Ireland, the Azores, and Madeira cast an expansive web across most of the population's core feeding grounds as well as strategic way points along its migratory path. Waters up to 100 miles around those whaling stations were tended vigilantly by steam-powered catcher boats equipped with bomb lances. When a whale was killed, it was lashed to the vessel's side, towed back to the station, and winched up a slipway for butchering. Blubber was then stripped and rendered into oil, while the meat, viscera, and bones were ground into fertilizer, which left no part unused.

Records on the numbers of right whales landed at whaling stations were far better than those of artisanal shore whalers, but they still left much to be desired. Thousands of processed whales were not identified to species. Those that were listed included at least 24 right whales in the last decade of the 1800s at stations

Figure 15.1. Photograph of a 46-foot-long "Nordkaper" (right whale) possibly taken by Captain L. Berg in Dyre Fjord, Iceland, during a research expedition circa 1891. With jaws agape, this view reveals the deep cleft at the front of the mouth through which prey is taken when feeding, the massive tongue, and the upper jaw with dislodged baleen plates overlapping the high sideboard cheek (in life these plates would be inside the cheek). (Image from J. A. Guldberg. 1893. Zur Kenntniss des Nordkapers (*Eubalaena biscayensis* Eschr.). Zoologische Jahrbücher. Abteilung für Systematik, Geographie und Biologie der Tiere 7:1–22, panel 1, fig. 3; courtesy of the National Library of Norway.)

on Iceland and the Faroe Islands; those sites and more in Hebrides, Shetlands, and Ireland listed an additional 131 to 137 killed between 1900 and 1926.[5] Most were killed in the Hebrides off northwest Scotland, where at least 91 right whales were landed between 1906 and 1914. After 1926, only four more records can be found—all from stations farther south. Three were killed off Madeira in 1967, including a cow-calf pair. The last one killed in the eastern North Atlantic was in the Azores in either 1969 or 1970, although there is some doubt if it was in fact a right whale.[6] With no strandings and virtually no sightings off Europe that have not been identified as wayward stragglers from the western population, eradication of the eastern population seems to have been complete.

In the western North Atlantic, pelagic and artisanal shore whalers continued catching right whales in low numbers until the 1920s, as described in previous chapters. Unlike in the eastern North Atlantic, at least a few survived. Their saving grace may have been the lack of commercial shore whaling stations located strategically throughout their feeding range. More than a dozen of these stations were built in Labrador and Newfoundland between 1898 and 1904,[7] but they were all north of the population's principal foraging range. The only right whale listed among their catch—and the last right whale deliberately killed in the western North Atlantic—was taken in May 1951 at a station in Trinity Bay, Newfoundland, not far from the former Basque whaling center in Red Bay.[8] The only commercial whaling station south of Newfoundland—the Blandford station near Halifax, Nova Scotia—did not become operational until the 1950s, well after a ban on hunting right whales had been adopted. The Blandford station logged numerous right whale sightings, including groups of up to 10 individuals,[9] but with adherence to the ban on killing the species, none was taken. Their restraint may well have saved the species from extinction.

Estimates of the number of survivors when the last significant catches were made in the early 1920s are uncertain, but a range of possibilities has been calculated from recent abundance estimates and assumed population growth rates. By the early 1990s, two estimates of population size were available; these included (1) 380 right whales, generated by an intensive aerial survey program in the early 1980s,[10] and (2) 300–350 whales, based on photo-identification records that had started to be collected in the early 1980s.[11] Estimates of population growth from those sources ranged from 2.5 percent to 3.8 percent per year.[12] Assuming those rates remained constant throughout the 1900s and that the population numbered 300 whales in 1990, the number alive in 1935 when a ban on hunting right whales went into effect would have ranged between 50 and 100 whales.[13] If right whale abundance bottomed out a few years earlier in the late 1920s, when the last significant catch was recorded, an even lower number is possible.

Whether a highly migratory species that had been reduced to a few tens of animals could mount a successful recovery begins to stretch the limits of credulity. Although the biblical story of Noah's Ark would have us believe just a single male and a single female of each species is all that is needed to repopulate the world, scientists know otherwise. When populations fall to extremely small numbers, a series of factors referred to as "Allee effects" (also called "small population effects") begin to erode viability and push discrete populations into an "extinction vortex" that accelerates their final downward spiral toward extinction.[14]

Factors contributing to Allee effects include (1) an inability of animals in breeding condition to find one another; (2) inbreeding, which reduces reproductive fitness and introduces genetic deformities; (3) extreme weather events or disease outbreaks; (4) random fluctuations in numbers of female offspring; and (5) the collapse of social groups that facilitate reproductive success or protection from predators. Even seemingly innocuous events, such as occasional ship collisions or entanglements in fishing gear that kill a disproportionate number of mature females, can tip the balance of small populations toward extinction. The point at which populations become "too small" to survive varies by species and their life-history traits. For large whales that roam entire ocean basins and produce just a single calf every three to five years, Allee effects can begin exerting their influence when numbers fall to the low hundreds, and become critical when they drop to tens of animals. Just how small a right whale population might get and still recover may still be learned from the tragic situation befalling North Pacific right whales.

North Pacific right whales, the only species of large whale that may be more endangered than North Atlantic right whales, were also devastated by commercial whalers in the late 1800s and early 1900s. Their decline occurred so rapidly that their calving grounds were never discovered and remain a mystery. Similar to North Atlantic right whales, they are thought to occur in two relatively discrete populations on either side of the Pacific. By one estimate, the western North Pacific population off Russia now numbers 900 animals,[15] but few experts consider that realistic and believe a number in the low hundreds is far more likely.

The eastern North Pacific population off Alaska is in far more desperate shape. It was apparently making good progress toward recovery; then, between 1963 and 1967 at least 529 right whales were killed in the southeastern Bering Sea and Gulf of Alaska by illegal Soviet whaling.[16] This plunged the eastern population to the brink of extinction. When revelations of the illicit practice came to light in 2000 after the fall of the Soviet Union,[17] it immediately became clear why sightings from Alaska to Mexico plummeted to barely half a dozen between the 1960s and the mid-1990s.

Then, in the summer of 1996, scientists with the National Marine Fisheries Service discovered four right whales feeding together in the southeastern Bering Sea.[18] Over the next 15 years, agency scientists scoured the area each summer searching for more animals. The largest number they saw was 24 whales in 2001. As of 2009, photo-identification data and genetic samples gathered over a 12-year span had identified only 31 individuals—8 females, 20 males, and 3 of unknown sex. The findings suggest that the eastern North Pacific right whale population is the smallest of any large whale population in the world.[19] With no reliable way of counting animals annually, it will take decades to determine if the population is increasing or decreasing. Once known, however, scientists will have another reference point for judging whether North Atlantic right whales might have been reduced to a few tens of animals early in the nineteenth century, yet later still mounted initial steps toward recovery.

Estimating Pre-exploitation Abundance

Estimating the size of whale populations hundreds of years ago before they were first hunted is a daunting challenge, particularly for a species like the North Atlantic right whale whose history of exploitation by humans spans more than 1,000 years. Nevertheless, the question is important for reasons beyond simple curiosity. For biologists, an answer offers insight into the role right whales once played in the balance of ocean life when the North Atlantic was still pristine. For resource managers charged with restoring decimated populations of endangered species and maintaining productive ocean ecosystems, a reliable answer is the gold standard for measuring the animal's current status.

To attack the question of pre-exploitation population size, the scientist's weapon of choice is a mathematical model based on reliable estimates of current population size, annual mortality and reproductive rates, and historical catch records. From those data, possible trends in population abundance over the course of the animal's whaling history can be estimated; then, the number of whales likely alive when whaling began can be back calculated. For North Atlantic right whales and their millennium-long persecution, the one thing that is clear is that this proposition is murky at best. While estimates of natural mortality, potential reproduction rates, and current abundance are possible, resurrecting their catch history is confounded by a mind-boggling array of data gaps, assumptions, uncertainties, and complete unknowns.

Only a dedicated few have waded into the archival abyss, hoping to resurrect enough records to piece together something resembling a coherent North Atlantic right whale catch history. To do so, they must rely on hard-to-find documents ranging from old newspaper accounts, whalers' diaries, log books from whaling voyages, and tax records for imported or exported whale products; to even harder-to-find references that might include wills, genealogical records, transcripts of court suits contesting whaling disputes, and contracts to build, equip, or insure whaling vessels. To further complicate their task, records are often in multiple modern and archaic languages, handwritten with varying degrees of legibility, and scattered around the world in archives that may lack indexes to search for relevant information. Moreover, many records have been irretrievably lost and those that remain may be from second- or thirdhand sources passed down by persons of uncertain veracity. And of course some whaling activity simply may never have been recorded at all.

Even seemingly reliable sources cannot be taken for granted. For much of the twentieth century, whale experts assumed that North Atlantic right whales were once distributed as an unbroken band arching across the entire North Atlantic from Europe to North America. This accepted view was based on a series of whaling charts prepared in the 1850s under the direction of US scientist Matthew Fontaine Maury, known as the "Father of Oceanography" (see plate 7). Based on whalers' logbooks, the charts reportedly depicted where and when right whales were killed in the early nineteenth century. These were long assumed to be accurate, even though no scientist had ever seen a right whale in the center of the North Atlantic. Of course, scientists in the late nineteenth and mid-twentieth centuries had virtually no chance of seeing a right whale anywhere. They therefore had little reason to doubt the work of such a respected scientist.

In the 1980s and 1990s, however, scientists became suspicious of the transatlantic distribution on Maury's charts—a swath that came to be known as "Maury's smear" (see plate 8). His charts were at odds with new sightings and satellite tracking studies that revealed right whales as coastal denizens who rarely strayed beyond the continental shelf. As doubts about his charts grew, whaling historians decided to reexamine the original logs Maury and his associates used to prepare the charts. They found that catch locations on Maury's charts were erroneous and apparently based on faulty data extraction and transcription procedures.[20] As a result, the current consensus holds that North Atlantic right whales are, and always have been, a species principally occupying continental shelves. Whether Maury's chart was sloppy work by an otherwise careful scientist, or influenced by the prevailing view of his day that whales had entire oceans in which to hide, it illustrates the importance of keeping an open mind even with seemingly reliable records.

Notwithstanding such cautionary tales, all hope for understanding a species' history rests on interpreting accounts from past reports and integrating them with current information on the species' "life-history." Mathematical models are powerful tools for this purpose, yet they too have their limitations and elements of alchemy. Modelers use complex formulas that leave people unversed in higher mathematics to rely on peer reviewers to ensure the calculations are appropriate for the task at hand. Modelers also must make judgments about the values they plug into formulas. Those judgments are not always clear and are often debatable. In a self-deprecating acknowledgement of their limitations, modelers routinely invoke the quip by statistician George Box; this cautions that "all models are wrong, but some are useful."[21]

With so many data gaps and uncertainties in the long catch history for North Atlantic right whales, modeling to back calculate pre-exploitation population size has been considered futile. Until a more complete and reliable catch history can be patched together—if, indeed it ever will be, given the poor state of surviving records—the only alternative is relying on "expert opinion." Expert opinion is little more than an educated guess by knowledgeable individuals—in this case, people familiar with historical references, the species' biology and ecology, and whaling in general. An important responsibility for scientists is exercising their best judgment to make educated guesses on important questions that can't be answered in preferred ways. However, because they are trained to avoid conclusions based on anything less than well-substantiated data and reproducible methods, few have ventured to put an estimate of right whale pre-exploitation population size in writing.

The first to do so was Canadian naturalist and environmental writer Farley Mowat. Based largely on catch records for Arctic whaling from the seventeenth to nineteenth centuries, he ventured a guess in the early 1970s that some 200,000 "right whales" once roamed the world's oceans.[22] However, he made no attempt to break down the estimate by ocean basin or distinguish between right whales and bowheads. Shortly after his guess, another was put forward by Canadian Edward Mitchell, a respected expert on whales and whaling who rarely traveled to professional meetings without a trunk full of books and scientific papers in tow for ready reference. Noting that early Arctic whalers rarely distinguished between right whales and bowheads, which rendered Mowat's estimate an uncertain amalgamation of both, and recognizing that reproduction and natural mortality over the long whaling period had to be considered, Mitchell speculated that the global pre-exploitation abundance of right whales and bowheads combined was perhaps 300,000 whales.[23] Unsatisfied with the scientific rigor of either estimate, a 1983 review of endangered whales by expert Howard Braham with the US National Marine Fisheries Service concluded that no useful estimates existed for the abundance of any right whale population prior to whaling.[24]

In the early 1990s, investigations into Basque whaling in the Strait of Belle Isle, Canada, prompted two new estimates of right whale abundance before whaling began. As described earlier (chapter 8), archival studies suggested that 25,000 to 40,000 whales were caught by Basque whalers in Canada over the 80-year span between 1530 and 1610,[25] half of which (12,500 to 20,000) were initially thought to be right whales.[26] From those findings, scientists and resource managers preparing a right whale recovery plan adopted in 1991 by the National Marine Fisheries Service used their expert opinion to estimate that 10,000 right whales must have been alive in the western North Atlantic at the onset of Basque whaling in the 1500s; this number accounted for the Basque landings plus the later catch of several thousand more by colonial whalers.[27] That same year Canadian whale expert David Gaskin at the University of Guelph arrived at a similar estimate—12,000 to 15,000 whales[28]—a number again based largely on expert opinion.

In 2004, however, genetic analyses revealed that only one of the 21 whale bones found at Red Bay was from a right whale, while all the rest were from bowheads.[29] With that information, the pre-exploitation population size estimate of 10,000 right whales was thrown into doubt, and the Fisheries Service removed it from the North Atlantic right whale recovery plan when an updated plan was adopted in 2005 with no new estimate.[30]

Which Whales Were the Basques Hunting in Terranova?

As described in chapter 8, in 2008 scientists double-checked the 2004 finding that only one of the Red Bay bones was from a right whale. They did this by collecting and analyzing additional whale bones found lying around old Basque stations.[31] Using the same genetic techniques, they examined 218 bones, including all of the 21 bones initially analyzed in the 1980s. Again, scientists found only one right whale bone—the same flipper bone identified in the 2004 study. They concluded that the Basques caught negligible numbers of right whales, which must have had an insignificant effect on the western population, and the pre-exploitation population size must therefore have been much smaller than the earlier estimates of 10,000 to 15,000 whales. To account for those killed by later colonial whalers, the investigators also concluded that the population probably numbered at least a few thousand.

Analyses were then taken a step further in 2010 when some of the same scientists compared genetic diversity from DNA in the lone sixteenth-century right whale bone from Red Bay with DNA from the current population.[32] Prior research had found the genetic diversity of living right whales to be very low, presumably the result of a "population bottleneck" created when sixteenth-century Basque whalers greatly reduced the size of the population, thereby eliminating many alleles that were once part of the species' genome. Scientists conducting the new genetic study reasoned that if Basque whalers had indeed significantly reduced right whale abundance, the lone sixteenth-century bone would likely have had alleles not found in the current population.

They therefore compared a small sample of alleles from living whales with those from the lone sixteenth-century bone. For this they examined alleles at specific sites on the chromosomes of nuclear DNA (genetic material found only in a cell's nucleus and derived from both the male and female parents). They found no "lost" alleles in the older bone and determined its level of genetic variability in the 1500s was comparable to that in the existing right whale population. Therefore, they reasoned that their findings were contrary to past suggestions that the Basques had significantly reduced right whale numbers.

New methods for analyzing genetic code offer exciting prospects for improving insights into the history of species such as North Atlantic right whales. The new findings offer strong evidence that Basque whalers almost exclusively caught bowhead whales in the Strait of Belle Isle. However, investigators studying DNA from the lone right whale bone recovered at Red Bay may have made more of their findings than is justified. Curiously, their conclusions regarding genetic similarities between the sixteenth-century bone and the current population ignored their own finding that mitochondrial DNA (genetic material from a cellular body called the mitochondria inherited only from the mother) from the older bone included a haplotype, or "matriline" (all descendants whose lineage can be traced back to a single female) unknown in the current population. The loss of a haplotype from a species' genetic code would require the random elimination of an enormous number of right whales in order to eradicate all the descendants carrying it. While its loss could have been caused by later colonial whaling, the haplotype's presence in a sample of one whale from the sixteenth century suggests that other matrilines also may have existed when the Basques were in the Strait of Belle Isle and that substantial numbers of right whales could have been killed during the 1500s.

With just seven haplotypes identified in the current North Atlantic right whale population,[33] the loss of a single haplotype and an entire matriline is itself a significant reduction of genetic diversity. Because more than two-thirds of the current population has been genetically sampled, the possibility that the haplotype from the sixteenth-century bone simply has yet to be found in the current population is exceedingly small. For purposes of comparison, studies of southern right whales off Argentina—a population now numbering several thousand whales—found 37 unique haplotypes in a sample of just 146 living right whales.[34] Those results indicate that the Argentinean population retained a far greater proportion of its genetic diversity after whaling, probably because it was not reduced to the extent of North Atlantic right whales.

To pin down whether and to what extent genetic diversity of North Atlantic right whales might have been reduced by whaling, further analysis of DNA from the bones of more right whales killed during different whaling eras on both sides of the Atlantic is needed; this work should ideally examine alleles from larger sample of chromosome sites. Such analyses could also help resolve questions about the extent to which western North Atlantic right whales migrated to European waters (as some still do today[35]), whether whalers in Europe might have begun to reduce that western population even before the Basques reached the Strait of Belle Isle, and the extent to which right whale numbers were reduced in different whaling eras. Opportunities for such analyses certainly exist. Right whale bones left by seventeenth-century whalers along Europe are housed today in collections at several museums and new discoveries periodically come to light. For example, a headless right whale skeleton thought to have been killed by whalers in the seventeenth century was discovered in 2010 during a wharf construction project on the banks of the Thames in Greenwich, England.[36] Still others might be collected from sites of known whaling stations.

In any case, genetic evidence from studies of the lone sixteenth-century bone from Red Bay seems too tenuous to assert that Basque whalers did not reduce the genetic diversity of North Atlantic right whales. By the same token, although historical evidence is also thin, it seems premature to dismiss the possibility that the Basques reduced the size of the western North Atlantic right whale population during their Terranova whaling era by the pelagic whaling noted in chapter 7. If they did catch right whales in more open ocean areas south of the Strait of Belle Isle, and half of the oil they shipped back to Europe between 1520 and 1610 was from those animals, they may have killed as many as 18,000 to 29,000; if only a third came from right whales, they still could have killed some 11,600 to 18,000 (including whales struck and lost) over the 80-year period.

Reexamining Catch Levels and Pre-exploitation Abundance

Regardless of how many right whales might have been taken by the Basques before 1610, a substantial number must have survived to account for the thousands of right whales caught by English colonists in the late 1600s and early 1700s. To estimate how many survived, it is necessary to know how many were caught during the colonial era when the last substantial catches were made and how many were taken in subsequent years before the size of the right whale population reached its lowest point in the early 1900s.

Archival searches by dedicated whaling historians have revealed the broad outlines of whaling effort from colonial days up to the early 1900s. The fruit of their labors, much of which is reviewed in earlier chapters, provides a basis for estimating overall colonial catch totals. Randy Reeves, Tim Smith, and Elizabeth Josephson undertook the daunting task of compiling all well-documented catch records of right whales from 1620 to 1951 along the entire eastern seaboard of North America. Their results produced an estimated kill of at least 5,500 right whales (including those struck and lost) over the entire 330-year period, with 80 to 90 percent of that—some 4,400 to 4,950 whales—killed over just a 50-year period from the late 1680s to 1730.[37] In an earlier analysis, they estimated 3,839 whales were killed in just 38 years from 1696 to 1734 between New England and Delaware Bay.[38] To generate these estimates, they used only well-documented landings and assumed an additional 20 percent died after being struck and lost. Recognizing the conservative nature of their approach, they were careful to couch their estimate as a minimum number of whales killed.

How many more whales were killed by whalers whose records were lost or never recorded in the first place, and whose crude methods required decades to perfect, will never be known with certainty. Further work to scour obscure documents could yield additional records, but this is a slow process that seems to have reached a point of diminishing returns. Thus, for the near future, any further attempts to estimate colonial catch totals during peak years will necessarily rest on "expert opinion" that considers a region-by-region review of both landing records and information on whaling effort.

Chapters 10 through 14 provide the basis for a subjective assessment of colonial whaling that takes into account information on both catch records and whaling effort. By reviewing this disjointed record, a qualitative assessment of total kills by decade and by area can be stitched together to provide an approximate scale of right whale landings. Figure 15.2 and table 2 in the appendix provide results of such an exercise by the author. They suggest that some 7,500 right whales were killed during the peak of colonial whaling between 1680 and 1730. This is roughly a third greater than the minimum number estimated by Reeves and his colleagues.[39] Over the 200 years following 1730, the total number killed fell to about 1,000, suggesting that the number surviving after the peak of colonial whaling probably did not exceed a few hundred. To catch 7,500 whales over a 50-year period and leave only

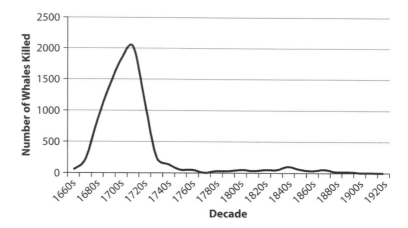

Figure 15.2. Estimated number of North Atlantic right whales killed by decade (including those struck and lost) along the shores of North America, 1660–1930. See also appendix, table A.2.

a few hundred whales by the mid-1700s, the western North Atlantic right whale population might have numbered between 4,000 to 5,000 whales in the 1680s, when colonial whalers first became proficient.

Stepping back further in time to the end of the Basque Terranova whaling era in 1610, we can speculate about the size of the right whale population by accounting for population growth during the intervening 70-year period. A generally accepted estimate for the maximum growth rate of large whale populations is 4 percent per year.[40] Assuming that rate, approximately 200 whales in 1610 could have produced a population of between 4,000 and 5,000 right whales by 1680. If the growth was closer to 3 percent per year to account for a low level of Basque whaling in the western North Atlantic after 1610 and the effects of experimental colonial whaling before 1680, a right whale population of about 500 in 1610 could have reached 4,000 to 5,000 by 1680.

Finally, if we take one more step back to the years immediately preceding Basque Terranova whaling and assume they killed some 18,000 to 29,000 right whales (including 20 percent struck and lost) between 1530 and 1610, leaving at least a few hundred alive in 1610, abundance levels approaching earlier pre-exploitation estimates of 10,000 to 12,000 whales in the western North Atlantic may still be plausible. This would be even more likely if significant numbers of western right whales crossing the Atlantic to feed off Europe in the summer were killed by medieval whalers even before the Basques reached Terranova. If a comparable number of right whales once lived in the eastern North Atlantic, the total number throughout the North Atlantic once might have approached 15,000 to 20,000 whales. Interestingly, a recent study comparing historical whaling records with available habitat for North Atlantic right whales also concluded that a pre-exploitation population size of between 10,000 and 20,000 right whales might have been possible given what is now known about the species' range and potential foraging habitat.[41]

Firm estimates of the pre-exploitation size of right whale populations on either side of the North Atlantic may never be known with much certainty. Of course, research always offers hope for new insights and better estimates. However, despite the large range of uncertainty in former numbers, there is no doubt that North Atlantic right whales are now one of the world's most endangered large whales with a long road ahead if they are to recover to former population levels.

NOTES

1. Pettis, HM, and PK Hamilton. 2015. North Atlantic Right Whale 2015 Annual Report Card. Report to the North Atlantic Right Whale Consortium, http://www.narwc.org/pdf/2015%20 Report%20Card.pdf. Accessed May 2016.

2. Reeves, RR, and E Mitchell. 1986. American pelagic whaling for right whales in the North Atlantic. *In* RL Brownell, PB Best, and JH Prescott (eds.), Right Whales: Past and Present Status, Special issue 10, Reports of the International Whaling Commission, pp. 221–254, 229–230, 236.

3. Schevill, WE, and KE Moore. 1983. Townsend's unmapped North Atlantic right whales (*Eubalaena glacialis*). Brevoria 476:1–8. Reeves and Mitchell, American pelagic whaling (n. 2), pp. 225–227, 235.

4. Collett, R. 1909. A few notes on the whale *Balaena glacialis* and its capture in recent years in the North Atlantic by Norwegian whalers. Proceedings of the Zoological Society of London 1909:91–98. Reeves and Mitchell, American pelagic whaling (n. 2), pp. 230–231.

5. Brown, SG. 1986. Twentieth-century records of right whales (*Eubalaena glacialis*) in the northeast Atlantic Ocean. *In* Brownell et al., Right Whales (n. 2), pp. 121–127.

6. Ibid., p. 123.

7. Sanger, CW, and AB Dickenson. 1991. Expansion of regulated modern shore-station whaling in Newfoundland and Labrador, 1902–3. The Northern Mariner 1(2):1–22.

8. Sergeant, DE. 1966. *Populations of Large Whale Species in the Western North Atlantic with Special Reference to the Fin Whale*. Circular no. 9. Arctic Biological Station, Fisheries Research Board of Canada. Ste. Anne de Bellevue, Quebec, Canada.

9. Mead, JG. 1986. Twentieth-century records of right whales (*Eubalaena glacialis*) in the northwestern Atlantic Ocean. *In* Brownell et al., Right Whales (n. 2), pp. 109–119.

10. Cetacean and Turtle Assessment Program (CETAP). 1982. *A Characterization of Marine Mammals and Turtles in the Mid- and North-Atlantic Areas of the U.S. Outer Continental Shelf*. Final Reports for 1980. Contract no. AA551-CT8-48. Bureau of Land Management, US Department of the Interior. Washington, DC.

11. National Marine Fisheries Service. 1991. *Final Recovery Plan for the Northern Right Whale, Eubalaena glacialis*. Office of Protected Resources, National Oceanic and Atmospheric Administration, US Department of Commerce, Silver Spring, MD, US, p. 1.

12. Knowlton, AR, SD Kraus, and RD Kenney. 1994. Reproduction in North Atlantic right whales (*Eubalaena glacialis*). Canadian Journal of Zoology 72(7):1297–1305. Kenney, RD, HE Winn, and MC Macaulay. 1995. Cetaceans in the Great South Channel, 1979–1989. Continental Shelf Research 15(4/5):385–414.

13 Kenney et al., Cetaceans (n. 12), p. 408. Reeves, RR. 2001. Overview of catch history, historic abundance and distribution of right whales in the western North Atlantic and Cintra Bay, West Africa. *In* PB Best, JL Bannister, RL Brownell Jr., and GP Donovan (eds.), Right Whales: Worldwide Status, Special issue 2, Journal of Cetacean Research and Management, pp. 187–192, 188.

14 The term "Allee" comes from the name of the scientist Warder Clyde Allee, who proposed that a species' potential for population growth declines as their numbers approach zero.

15. Miyashita, T, and H Kato. 1998. Recent Data on the Status of Right Whales in the NW Pacific Ocean. Unpublished document submitted to the Workshop on the Comprehensive

Assessment of Right Whales. Scientific Committee Document SC/M98/RW11. International Whaling Commission. Cambridge, UK.

16. Ivashchenko, YV, PJ Clapham, and RL Brownell Jr. 2011. Soviet illegal whaling: the devil and the details. Marine Fisheries Review 73:1–19. Ivashchenko, YV, and PJ Clapham. 2012. Soviet catches of bowhead (*Balaena mysticetus*) and right whales (*Eubalaena japonica*) in the North Pacific and Okhotsk Sea. Endangered Species Research 18:201–217.

17. Doroshenko, NV. 2000. *Soviet Whaling Data (1949–1979)*. Center for Russian Environmental Policy. Moscow, pp. 96–103.

18. Goddard, PD, and DJ Rugh. 1998. A group of right whales seen in the Bering Sea in July 1996. Marine Mammal Science 21(1):169–172.

19. Wade, PR, A Kennedy, R LeDuc, J Barlow, J Carretta, K Shelden, W Perryman, et al. 2011. The world's smallest whale population? Biology Letters 7:83–85, doi:10.1098/rsbl.2010.0477.

20. Reeves, RR, E Josephson, TD Smith. 2004. Putative historical occurrence of North Atlantic right whales in mid-latitude offshore waters: 'Maury's smear' is likely apocryphal. Marine Ecology Progress Series 282:295–305.

21. Box, GEP. 1979. Robustness in the strategy of scientific model building. *In* RL Launer and GN Wilkinson (eds.), *Robustness in Statistics: Proceedings of a Workshop*. Academic Press. New York, NY, pp 201–236, p. 202 (quotation).

22. Mowat, FO. 1972. *A Whale for the Killing*. McClelland and Steward. Toronto, Canada.

23. Mitchell, ED. 1973. The status of the world's whales. Nature Canada 2(4):9–25.

24. Braham, HW. 1983. The status of endangered whales: an overview. Marine Fisheries Review 46(4):1–6.

25. Aguilar, A. 1986. The black right whale, *Eubalaena glacialis* in the Cantabrian Sea. *In* RL Brownell et al., Right Whales (n. 2), pp. 454–459.

26. Cumbaa, SL. 1986. Archaeological evidence of the 16th century Basque right whale fishery in Labrador. *In* Brownell et al. Right Whales (n. 2), pp. 187–190.

27. National Marine Fisheries Service. *Final Recovery Plan* (n. 11), p. 9.

28. Gaskin, DE. 1991. An update on the status of right whales, *Eubalaena glacialis*, in Canada. Canadian Field Naturalist 105:198–205.

29. Rastogi, T, MW Brown, BA McLeod, TR Frasier, R Grenier, SL Cumbaa, J Nadarajah, and BN White. 2004. Genetic analyses of 16th-century whale bones prompts a revision of the impact of Basque whaling on right and bowhead whales in the western North Atlantic. Canadian Journal of Zoology 82:1647–1654.

30. National Marine Fisheries Service. 2005. *Recovery Plan for the North Atlantic Right Whale* (Eubalaena glacialis). Office of Protected Species, National Oceanic and Atmospheric Administration, US Department of Commerce. Silver Spring, MD, US.

31. McLeod, BA, MW Brown, MJ Moore, W Stevens, SH Barkham, M Barkham, and BN White. 2008. Bowhead whales and not right whales were the primary target of 16th to 17th century Basque whalers in the western North Atlantic. Arctic 61(1):61–75.

32. McLeod, BA, MW Brown, TR Frasier, and BN White. 2010. DNA profile of a sixteenth century western North Atlantic right whale (*Eubalaena glacialis*). Conservation Genetics 11:339–345.

33. McLeod, BA, and BN White. 2010. Tracking mtDNA heteroplasmy through multiple generations in the North Atlantic right whale (*Eubalaena glacialis*). Journal of Heredity 101(2):235–239.

34. Patenaude, NJ, VA Portway, CM Schaeff, JL Bannister, PB Best, RS Payne, VJ Rowntree, et al. 2007. Mitochondrial DNA diversity and population structure among southern right whales (*Eubalaena australis*). Journal of Heredity 98(2):147–157.

35. Knowlton, A, J Sigurjonsson, JN Ciano, and SD Kraus. 1992. Long distance movements of North Atlantic right whales (*Eubalaena glacialis*). Marine Mammal Science 8:397–405; and Jacobsen, K-O, M Marx, and N Øien. 2004. Two-way trans-Atlantic migration of a North Atlantic right whale (*Eubalaena glacialis*). Marine Mammal Science 20(1):161–166.

36. Firth, N. 2010. Titan of the Thames: Resurfacing After 200 Years, the Whale that Met a Bloody End. Mail Online, September 10, London, UK, http://www.dailymil.co.uk/sciencetech/article-1310489/Skeleton-19th-century-whale-London-met-bloody-end.html.

37. Reeves, RR, TD Smith, and EA Josephson. 2007. Near-annihilation of a species: right whaling in the North Atlantic. *In* SD Kraus and RM Rolland (eds.), *The Urban Whale: North Atlantic Right Whales at the Crossroads.* Harvard University Press. Cambridge, MA, US, pp. 39–74, 63–65.

38. Reeves, RR, JM Briewick, and ED Mitchell. 1999. History of whaling and estimated kills of right whales, *Balaena glacialis*, in the northeastern United States, 1620–1924. Marine Fisheries Review 63(1):1–36.

39. Reeves et al., Near-annihilation of a species (n. 36).

40. Barlow, J, SL Swartz, TC Eagle, and PR Wade. 1995. *U.S. Marine Mammal Stock Assessments: Guidelines for Preparation, Background, and a Summary of the 1995 Assessments.* NOAA Technical Memorandum NMFS-OPR-6. Office of Protected Species, National Marine Fisheries Service, National Oceanic and Atmospheric Administration, Department of Commerce. Silver Spring, MD, US.

41. Pennino, M, T Smith, R Reeves, C Meynard, D Kaplan, and A Rodrigues. 2015. Modeling the pre-whaling distribution and abundance of North Atlantic right whales. Paper presented at 21st Biennial Conference of the Society for Marine Mammalogy, December 14–18, 2015. San Francisco, CA, US.

16

A SECOND CHANCE

BY THE 1920s, as the last embers of commercial interest in North Atlantic right whales were finally dying, the species' plight could hardly have been worse. Centuries of whaling as relentless as it was remorseless had whittled their ranks to no more than 100,[1] and possibly not even half of that. Most, and perhaps all, survivors were part of the western population calving along the shores of North America. If any still existed off Europe, they probably numbered fewer than 10. Except for fleeting encounters with others of their kind on feeding and calving grounds, remaining individuals spent most of their existence roaming the ocean alone. The situation was only slightly better for right whales in other oceans. In the Southern Hemisphere where Europeans first hunted them in the seventeenth century, and in the North Pacific Ocean where a blitzkrieg by Yankee whalers overwhelmed the region in the late 1800s, their numbers had been slashed to the low- to mid-hundreds.

That any North Atlantic right whales survived to see the twentieth century is something of a miracle. Ironically, their saving grace may have been the thoroughness of past whaling. By the time of the American Revolution, their rarity had already forced the growing armadas of whaling ships to redirect their attention to more abundant species, thereby sparing the few remaining North Atlantic right whales from a more focused pursuit. Yet even then they remained vulnerable to shore whalers who could afford to wait for the odd straggler to appear at their doorstep. Like a battered prizefighter wobbling on exhausted legs, right whales reeled against the ropes of extinction after a long, one-sided battle.

As fate would have it, however, a proverbial bell was about to ring in the form of a new conservation ethic. Around the world, species once abundant were vanishing under the assault of unrestrained hunting. Gone or nearly gone from air and land were such iconic species as the dodo, passenger pigeon, Carolina parakeet, Tasmanian tiger, American buffalo, and eastern elk. Absent or disappearing from the seas were great auks, Steller's sea cows, West Indian manatees, sea otters along the rim of the North Pacific, sea minks along the northeastern Atlantic, the three species of monk seals (Caribbean, Mediterranean, and Hawaiian), northern elephant seals, and North Pacific gray whales. As extinctions or impending extinctions mounted, a handful of government officials began to exercise a new sense of responsibility to intercede in the shortsighted carnage of species whose abundance had once seemed limitless but was rapidly evaporating. Initially, their concerns were advanced on the grounds of restoring lost economic potential. If there were any sense of moral obligation or concern for environmental integrity, it was rarely raised as justification for action.

Right whales were an obvious focal point for attention by new conservation proponents. By the early 1900s, commercial whalers were well aware they had reaped all the profit they could from the species and had already set their sights elsewhere. With new technological advances—steam-powered catcher boats, harpoon guns and exploding harpoons, and factory ships—other species, such as blue, fin, humpback, and sei whales, had taken center stage in a new whaling frontier. This new frontier was the Antarctic. Its enormous potential compared to the exhausted status of right whales opened a door for ending the hunt for all right whales, with little fear of objection from powerful whaling interests concerned only about economic profits.

It certainly occurred to whalers long before 1900 that whales were not an inexhaustible resource (figure 16.1). They had witnessed them disappear from a succession of whaling grounds. Seas once alive with misty spouts erupting like so many miniature white volcanoes had become empty expanses. The explosive deep whoosh of exhaling whales had gone silent. Yet even then, it seemed inconceivable to most whalers that an animal as ubiquitous as the whale with entire oceans in which to hide could ever be exterminated.

Figure 16.1. A cartoon from the April 20, 1861, edition of *Vanity Fair*, published shortly after the first successful oil well was drilled in Titusville, Pennsylvania, in 1859. (Image courtesy of the New Bedford Whaling Museum, New Bedford, MA.)

Herman Melville read much of what had been written by naturalists and others about the great whales while researching his masterpiece, *Moby-Dick*, published in 1851. Combining the fruits of his research with his own experience aboard a whaling vessel, he wove descriptions of right whales and sperm whales into his novel that were as accurate and thoughtful as any scientific accounts in his day. In one of his many chapter-length asides, Melville pondered the fate of whales and compared them to "humped herds of buffalo, which, not forty years ago, overspread by tens of thousands the prairies of Illinois and Missouri." Observing that "at the present day [1851] not one horn or hoof of them remains in all that region," he asked if the same fate could befall the ocean's whales. The question was surely one that many a whaler turned philosopher must have mused in their idle moments. Capturing these thoughts in an oft-quoted passage, Melville asked the following question: "Whether owing to the almost omniscient look-outs at the mast heads of the whaleships, now penetrating even through Behring's straits, and into the remotest secret drawers and lockers of the world, and the thousand harpoons and lances darted along all continental coasts; the moot point is, whether Leviathan can long endure so wide a chase, and so remorseless a havoc; whether he must not at last be exterminated from the waters, and the last whale, like the last man, smoke his last pipe, and then himself evaporate in the final puff."[2]

In Melville's time this fate seemed impossible. After considering the proposition that "the hunted whale cannot now escape speedy extinction," Melville ultimately summoned the opposing view that still prevailed in his day—one that lacked appreciation or understanding of the limits of whale biology. In his final analysis, he confidently reassured any fatalists that whales "are only being driven from promontory to cape." Melville continued:

> And if one coast is no longer enlivened with their jets, then be sure, some
> other and remoter strand has been recently startled by the unfamiliar
> spectacle. [Whales will always] . . . have two firm fortresses, which in all
> human probability, will forever remain impregnable; . . . hunted from
> the savannas and glades of the middle seas, the whale-bone whales can
> at last resort to their Polar citadels, and diving under the ultimate glassy
> barriers and walls and there, come up among icy fields and floes and
> in a charmed circle of everlasting December, bid defiance to all pursuit
> from man.[3]

By the late 1800s and early 1900s, this false perception finally began to crumble (figure 16.1). Advances in transportation carrying people to the most remote corners of the Earth at ever faster speeds were beginning to shrink the world in the mind's eye, much as the first pictures of our planet from space would do a century later. At the same time, naturalists were gaining new insights into the biological limits of wildlife—even those in the sea. Applying this new knowledge to old questions, the logic that had assured Melville and others of the ultimate persistence of whales became less assuring. The fallacy of earlier reasoning was becoming more and more apparent as more and more species disappeared under the dual barrage of relentless hunting and habitat loss. As it did, government officials began to act.

The First Efforts to Protect Right Whales

In 1893, growing concern that manatees were on the verge of extinction prompted the state of Florida to establish a fine of up to $500 and three months in jail for killing one.[4] For similar reasons, in 1911 the Convention for the Preservation and Protection of Fur Seals banned oceanic hunting of northern fur seals and all hunting of sea otters throughout the North Pacific.[5] The first major proponent for restraints on exploiting ocean species was a quasi-official group called the International Council for the Exploration of the Sea, or ICES (pronounced "Ice-eez"). Established in 1902 on little more than a handshake between scientists from a half dozen countries in northern Europe, ICES sought to encourage and coordinate oceanographic research.[6] When strong management action was warranted, however, the organization did not shy away from urging international arrangements to promote harvesting constraints.

If it was Melville who etched an enduring fascination for whales into the public conscience, it was ICES and a forgotten visionary from Argentina named José Leon Suárez whose conviction seeded the first concerted efforts toward their conservation. At the request of a committee chartered by the League of Nations in 1924, Suárez was commissioned to explore options for international agreements to rein in the excessive exploitation of vanishing marine species.[7] Armed with scientific studies by ICES detailing the decline of great whales, Suárez became the leading voice for designing an international framework to regulate whaling. He argued that because whale migrations respected no national boundaries, whale conservation would likewise require a "counterpart in a legal solidarity in the sphere of international law."[8]

Based on his report, the committee recommended that an international conference draft an overarching convention to strengthen the ineffective patchwork of national laws that allowed a virtually unrestrained slaughter of whale populations. The result was the Convention for the Regulation of Whaling. After its final text was drafted and signed by delegates from 26 nations in 1931, it was submitted to their respective governments for formal approval. By 1935, the necessary threshold of eight nations officially ratified its terms, allowing the convention to "enter into force." This formally obligated those eight signatory nations, and any other "parties" subsequently filing their formal acceptance, to abide by its provisions.[9]

Article 4 of the convention called on its parties to ban all hunting of right whales by their nation's whalers. It also banned the killing of calves, immature whales, and females with calves of all whale species. Other sections sought to reduce wasteful taking by requiring that whaling ships be equipped to process all parts of whales and that whalers be paid fixed salaries in lieu of wages based on the number of whales killed; the latter practice had led to the killing of more whales than could be processed. The convention failed, however, to address the most fundamental need—limits on the numbers of whales killed. With many parties recognizing the deficiency, almost as soon as the 1935 convention entered into force, negotiations began on a new agreement—the International Agreement for the Regulation of the Whaling of 1937. That agreement extended the ban on hunting right whales to include gray whales, expanded its scope to include shore whalers (the 1931 convention applied only to pelagic whalers), required parties to appoint whale inspectors on all whaling ships, and limited the sizes of blue and humpback whales that could be taken.[10]

In hindsight, the most important achievement of the first two conventions was the precedent they set for bringing all whaling nations and all forms of commercial whaling under the umbrella of a single set of international standards—even though they were woefully inadequate. The 1937 convention still lacked authority to limit catch levels, and in the late 1930s, ICES, whose whaling committee continued its role as conscientious watchdog, voiced alarm at the plummeting number of blue whales. Faced with continuing opposition to any set limits on the numbers of whales that could be caught, ICES argued that nothing less than a limit on the annual production of whale oil would prevent the economic and eventual biological extinction of blue whales.[11] With many members agreeing with that assessment, yet another round of negotiations began in 1939, but these were soon suspended by World War II.

When talks resumed in 1944, a new convention was drafted—the International Convention for the Regulation of Whaling of 1946. Signed by 15 nations at a conference in Washington, DC (figure 16.2), it entered into force in November 1948, once the requisite number of nations filed their formal acceptances.[12] This is the convention that remains in effect today as the international framework for managing the commercial harvest of whales. Article III of the convention established the International Whaling Commission, a body best known by its acronym "IWC" and composed of representatives from all Convention parties. The IWC met for the first time in the spring of 1949. Chief among its duties is setting limits on annual whaling. Its initial currency for doing so was the "blue whale unit" (an amount of oil equal to that produced by a single adult blue whale regardless of which species produced it), but within a few years, it was agreed that quotas should be set in terms of numbers of whales taken from individual "stocks" (a term synonymous with populations) of each great whale species.

After several decades of quotas still far too large to stem the excessive slaughter, the IWC adopted a controversial new measure in 1982 that went into full effect in 1986—a moratorium on all commercial whaling. The moratorium suspended all commercial whaling for an indefinite period to allow time for depleted whale populations to recover. During the hiatus, the status of all whale stocks was to be assessed and a "new regulatory procedure" was to be adopted to ensure safe catch

Figure 16.2. The signing ceremony for the International Convention on the Regulation of Whaling in Washington, DC, on December 2, 1946. (Photo courtesy of the International Whaling Commission [www .iwc.int], Impington, Cambridge, UK.)

quotas. Once those steps were completed, the moratorium could be lifted upon approval of three-fourths of the IWC's membership. As stocks began to recover, assessments of their status were completed and a new formula was approved to calculate conservative catch limits that would ensure sustainable harvests.

Since the mid-1990s, however, acrimonious debate over lifting the moratorium has split the IWC into two bitterly divided factions with opposing views. Its annual meetings have pitted pro-whaling nations against anti-whaling nations in an ongoing, rancorous feud. In one camp, the parties are convinced that some whale stocks have recovered sufficiently to renew hunting with catch quotas calculated using the new management formula. In the other, a coalition of nations is no less committed to maintaining the moratorium on grounds that either (1) stocks need more time to recover; (2) the IWC's harvest-monitoring provisions are insufficient to prevent underreporting of catch levels; or (3) commercial whaling should be ended altogether because it is neither ethically justified nor necessary.

The ban on killing right whales, however, has remained above the fray. Reduced to such low numbers that they are still far below historical levels, protection of all right whales has been carried forward without opposition—at least without *formal* opposition. The caveat is necessary in light of recent revelations brought to light of illegal whaling. Before 1972, when the IWC adopted new international observer provisions, compliance with catch limits relied entirely on an honor system that left member states to monitor and enforce their own whalers. Even after 1972, when international observers were appointed from a pool of people nominated by member nations, IWC enforcement measures proved to be unreliable.

As described earlier (chapter 14), Soviet whalers illegally killed hundreds of right whales in the southeast Bering Sea and Gulf of Alaska between 1963 and 1967.[13] That violation was but a small part of a massive government-condoned secret whaling campaign carried out by Soviet whalers in total disregard of IWC regulatory measures. Between 1948 and 1978, Soviet representatives to the IWC officially reported a catch of 185,000 whales, while failing to report the catch of some 200,000 others.[14] Based on records of actual catch totals leaked after the collapse of the Soviet Union in the early 1990s, their unreported catch included at least 3,349 southern right whales taken off Argentina, South Africa, and Crozet Island in the southwestern Indian Ocean between 1950 and 1972.[15] The revelations of illegal Soviet whaling and continued resistance to an independent monitoring scheme have further undermined confidence in the IWC's ability to enforce its whaling quotas and the wisdom of lifting the whaling moratorium.

Rediscovery of the North Atlantic Right Whale

In the decades following the 1935 ban on killing right whales, it seemed everything possible had been done to protect them by virtue of that one measure. No one knew whether, much less where, any North Atlantic right whales might still be found. Even if it were known, scientists and managers assumed with understandable justification that there would be little more they could do than record the sighting and wonder if they would be the last to see one alive.

Yet to protect a species, one must first come to know it. Any hope of protection requires a large measure of blind trust in that maxim. Precisely what information might inspire or support new conservation measures is impossible to predict. In the early 1930s, despite centuries of exploitation, little was known about North Atlantic right whales other than where they had once been caught and that they had effectively vanished. Obvious questions had no answers. How many, if any, survived? How many were there in the past? When and where did they migrate, feed, and reproduce? What factors other than hunting might affect their survival and recovery?

For the first half of the 1900s, any hope of answering such questions was moot. Robert Collett, a Norwegian zoologist and director of the Zoological Museum in Oslo, did glean some information on right whale prey and calving by examining some of the last whales killed by commercial shore whalers in the Hebrides off northwest Scotland during the first decade of the twentieth century.[16] However, from the mid-1920s to 1950, there was only one reliable right whale "sighting" anywhere in the North Atlantic. That was the mother-calf pair encountered off Florida in March 1935 that ended in the calf's death at the hand of a charter boat captain (chapter 14).[17] During World War II, with whaling ships requisitioned for the war effort and hostile submarines stalking the oceans, any thoughts of whaling or studying whales had to be shelved.

It was not until 1950, when the species was "rediscovered," that the possibility of research again became a viable option. In early March of that year, five right whales, including two mother-calf pairs, were spotted off northeastern Florida by Harry Kritzler, a Duke University scientist participating on a research cruise organized by the world's first oceanarium, Marineland of Florida in St. Augustine.[18] His discovery and a series of late-winter sightings off Florida from then until the mid-1970s were squirreled away in the files of David and Melba Caldwell, a husband and wife team of marine biologists working at Marineland. An unpretentious pair not opposed to testing the odd fresh-stranded animal on their barbeque, the Caldwells were true pioneers in their field when Florida was still relatively undeveloped. Their own right whale sightings were almost exclusively between January and March. Although never including more than a few whales, their observations often included mother-calf pairs. However, because the sightings were never published, they remained largely unknown to other scientists at the time.[19]

A few years later, Bill Schevill, a scientist at the Woods Hole Oceanographic Institution, discovered right whales in Cape Cod Bay. For Schevill, his first sighting in the spring of 1955 was as astonishing as finding an ivory-billed woodpecker in one's backyard. While conducting an aerial survey of the bay's cetaceans, he spotted a pair of adult right whales barely two miles off the town of Barnstable,[20] where colonial shore whalers had launched their dories to hunt right whales nearly 250 years earlier. Subsequently finding at least a few right whales feeding in the bay each year in late winter and spring, Schevill and a coworker, Bill Watkins, became the first scientists to carry out true observational studies of free-swimming North Atlantic right whales. Between the 1950s and 1970s, they saw as many as 70 right whales in a single day[21] and recorded a series of scientific firsts for the species. They were the first to describe the species' feeding and diving behavior,[22] the first to identify an individual North Atlantic right whale by its callosity patterns (a technique developed in the early 1960s for North Pacific right whales by a Soviet scientist, S. K. Klumov[23]), the first to document the calving interval of a mature female, and the first to record right whale vocalizations.[24]

Another early clue to the species' survival came from a pair of Canadian scientists, David J. Neave and B. S. Wright. They were studying the seasonality and movements of harbor porpoises in the Bay of Fundy at the northern end of the Gulf of Maine and had asked watchkeepers aboard three different ferries crisscrossing the bay to record sightings of harbor porpoises and any other cetaceans on their trips between December 1965 and November 1966. The results were published two years later. In what was little more than a side note on sightings of cetaceans other than harbor porpoises, they reported a single observation 30 miles from shore of 15 right whales on an August run by the *Bluenose* ferry from Bar Harbor, Maine, to Yarmouth, Nova Scotia.[25]

The sighting was a new record for the species in those waters and was remarkable for its large number. It also sparked some harsh criticism from Bill Schevill. Noting how difficult it was to identify cetacean species at sea, he questioned whether it was reasonable to expect hard-working sailors, competent as they may be at their profession, to become instant experts in cetacean identification—particularly for a species as rare as the right whale. Accordingly, Schevill panned the study for its uncritical reliance on untrained observers.[26] Neave and Wright, however, stood their ground. They explained in their rebuttal that the watch officers had been provided with the best available field guide to cetaceans in the region and that they had accompanied each of the observers on a trip to familiarize them with the key.[27] Neave and Wright agreed, however, that the sighting, as with all sightings of rare whales, should be considered unconfirmed until a specimen was actually taken.

Except for a few scientists, the scattered sightings between 1950 and 1970 were largely ignored or overlooked. A dedicated research and management program for North Atlantic right whales was still years away. And as it would turn out, its development would require the proverbial—and in this case literal—act of Congress.

NOTES

1. Kenney, RD, HE Winn, MC Macauley. 1995. Cetaceans in the Great South Channel, 1979–1989. Continental Shelf Research 15:385–414, 408.

2. Melville, H. (1851) 1943. *Moby-Dick; or, the Whale*. Limited Editions Club. Heritage Press. New York, NY, p. 491 (quotations).

3. Ibid., pp. 492–493 (quotation).

4. Florida Laws 145, 1993, ch. 42.

5. Dorsey, K. 1998. *Dawn of Conservation Diplomacy: U.S.-Canadian Wildlife Protection Treaties in the Progressive Era*. Weyerhaeuser Environmental Book. University of Washington Press. Seattle, WA, US.

6. Birnie, P. 1985. The regulation of whaling during the League of Nations period 1919–1946. In Vol. 1 of Birnie, *International Regulation of Whaling: From Conservation of Whaling to Conservation of Whales and Regulation of Whale Watching*. 3 vols. Oceana Publications. New York, NY, Vol. 1, pp. 105–142.

7. The League of Nations was established in 1919 at the behest of President Woodrow Wilson after World War I to work for the peaceful resolution of international disputes. Initially composed of 28 nations, it was the forerunner to the United Nations. Its Committee of Experts for the Progressive Codification of International Law asked Suárez to evaluate "whether it is possible to establish by way of international agreement, rules regarding exploitation of the products of the sea." Birnie, *International Regulation* (n. 6), p. 110 (quotation).

8. Suarez, JL. 1926. Report on the exploitation of products of the sea. American Journal of International Law 20:231–241. As cited in Birnie, *International Regulation* (n. 6), Vol. 1, p. 111.

9. Birnie, *International Regulation* (n. 6), Vol. 1, pp. 105–109.

10. Ibid., pp. 125–127.

11. Went, AEJ. 1972. *Seventy Years a Growing: A History of the International Council for the Exploration of the Sea, 1902–1972*. No. 165 in the Reports and Proceedings of the International Council for the Exploration of the Sea. International Council for the Exploration of the Sea. Copenhagen, Denmark, p. 92.

12. Birnie, *International Regulations* (n. 6), Vol. 1, pp. 168–204.

13. Ivashchenko, YV, and PJ Clapham. 2012. Soviet catches of bowhead (*Balaena mysticetus*) and right whales (*Eubalaena japonica*) in the North Pacific and Okhotsk Sea. Endangered Species Research 18:201–217.

14. Ivashchenko, YV, PJ Clapham, and RL Brownell Jr. 2011. Soviet illegal whaling: the devil and the details. Marine Fisheries Review 73:1–19.

15. Tromosov, DD, YA MiKhalev, PB Best, VA Zemsky, K Sekiguchi, and RL Brownell Jr. 1988. Soviet catches of southern right whales, *Eubalaena glacialis*, 1951–1971; biological data and conservation implications. Biological Conservation 86(2):185–197.

16. Collett, R. 1909. A few notes on the whale *Balaena glacialis* and its capture in recent years in the North Atlantic by Norwegian whalers. Proceedings of the Zoological Society of London 1909:91–98.

17. Mead, JG. 1986. Twentieth-century records of right whales (*Eubalaena glacialis*) in the northwestern Atlantic Ocean. Brown, SG. 1986. Twentieth-century records of right whales (*Eubalaena glacialis*) in the northeast Atlantic Ocean. Both in RL Brownell, PB Best, and JH Prescott (eds.), Right Whales: Past and Present Status, Special issue 10, Reports of the International Whaling Commission, pp. 109–119, 112–114; and 121–127.

18. Moore, JC. 1953. Distribution of marine mammals in Florida waters. American Midland Naturalist 49(1):117–158.

19. Winn, HE. 1984. *Development of a Right Whale Sighting Network in the Southeastern U.S.* Report No. MMC-82/05 for the US Marine Mammal Commission. PB84-240548. National Technical Information Service. Springfield, VA, US.

20. Schevill, WE. 1959. The return of the right whale to New England waters. Journal of Mammalogy, 40(4):639 (reference to a WHOI film), as cited in WE Schevill, KE Moore, and WA Watkins. 1981. *Right Whale*, Eubalaena glacialis, *Sightings in Cape Cod Waters*. Woods Hole Oceanographic Institution Technical Report WHOI-81-60. Woods Hole, MA, US.

21. Schevill et al., *Right Whale* (n. 20).

22. Watkins, WA, and WE Schevill. 1976. Right whale feeding and baleen rattle. Journal of Mammalogy 57(1):56–66.

23. Klumov, SK. 1962. Gladkie (yaponskie) kity Tichogo Okeana (Japanese right whales in the Pacific Ocean.) Trudy Instituta Okeanologii. Akad. Nauk, Soviet Socialist Republic of Russia 58:202–297, as cited in Schevill et al., *Right Whale* (n. 20), pp. 2–3.

24. Schevill, WE, and WA Watkins. 1962. Whale and porpoise voices, a phonograph record and report. Contribution No. 1320. Woods Hole Oceanographic Institution. Woods Hole, MA, US.

25. Neave, DJ, and BS Wright. 1968. Seasonal migrations of the harbor porpoise (*Phocoena phocoena*) and other cetaceans in the Bay of Fundy. Journal of Mammalogy 49(2):259–264.

26. Schevill, WE. 1968. Sight records of *Phocoena phocoena* and of cetaceans in general. Journal of Mammalogy 49(4):794–796.

27. Neave, DJ, and BS Wright. 1969. Observations of *Phocoena phocoena* in the Bay of Fundy. Journal of Mammalogy 50(3):653–954.

17

A DEDICATED
RECOVERY PROGRAM

THE ROOTS OF environmental consciousness in the United States are often traced back to the first decade of the 1900s when President Theodore Roosevelt began setting aside vast tracts of federal land as national parks and monuments to preserve the nation's dwindling wildlife and natural areas. It was not until the late 1960s and early 1970s, however, that the movement truly came of age. Stirred by a series of widely acclaimed classics, such as *Sand County Almanac* by Aldo Leopold, *The Sea Around Us* and *Silent Spring* by Rachel Carson, and *The Silent World* by Jacques Cousteau, a groundswell of public concern arose between the end of World War II and the 1970s over the insidious effects of pollution and the ever expanding reach of human activity in both marine and terrestrial environments.

The US Congress responded with a wave of environmental legislation between the late 1960s and mid-1970s. Those laws still form the core of the nation's wildlife protection programs. The crest of that wave may well have occurred in October 1972. Within a span of barely two weeks, Congress passed four major statutes subsequently signed into law by President Richard M. Nixon: the Federal Water Pollution Control Act (now the Clean Water Act); the Coastal Zone Management Act; the Marine Protection, Research and Sanctuaries Act (the law banning ocean dumping and establishing the National Marine Sanctuaries Program); and the Marine Mammal Protection Act. Following in short order were several others, including the Endangered Species Act of 1973, and the Fisheries Management and Conservation Act of 1976 (now the Magnuson-Stevens Act).

Most important for right whales were the Marine Mammal Protection Act (MMPA) and Endangered Species Act (ESA). The former was the first law in any country to recognize and incorporate the concept of ecosystem protection. In an eloquent opening paragraph, the act made it US policy to restore marine mammal populations to their "optimum sustainable population level" so they would not

276

"diminish beyond the point at which they cease to be a significant functioning element in the ecosystem of which they are a part." The centerpiece for achieving its lofty goal was "a moratorium on the taking and importation of marine mammals and marine mammal products."[1]

The sweeping intent of the moratorium was underscored by the act's definition of "take." It covered not only deliberate hunting, killing, and capturing, but also harassing marine mammals intentionally or incidentally. Exceptions were carved out for activities believed to be justified and that posed only minor, manageable risks. Those exceptions included hunting by Alaska Natives for traditional subsistence and handicraft purposes, commercial fishing that took marine mammals unintentionally, scientific research, and public display. For other activities posing potential threats, the act authorized waivers to the moratorium once regulations were developed by administering agencies to ensure that their impacts were negligible.

Primary responsibility for protecting most, but not all, marine mammals was assigned to the US Department of Commerce. Under its charge were whales, seals, and dolphins. However, lead responsibility for walruses, sea otters, polar bears, and manatees was given to the US Department of the Interior. This jurisdictional split now seems odd, but in 1970 steps had just been taken to consolidate all federal ocean programs under a single super agency to bring a more coherent approach to managing the nation's ocean resources.[2] That agency was called the National Oceanic and Atmospheric Administration, or NOAA. Where to place this new agency in the federal bureaucracy sparked a hot political squabble between Secretary of Interior Walter Hickel and Secretary of Commerce Maurice Stans. Each sought control over the massive new agency and argued that it best fit within their respective departments. After Congress rejected a proposal for a new Department of the Environment that would have included NOAA, Stans prevailed, and the new agency was placed in the Department of Commerce where it now makes up 60 percent of the department's budget.

When the Marine Mammal Protection Act passed two years later, in 1972, the turf war was still a sore subject. Stans contended administrative responsibility should go to NOAA based on the goal of keeping ocean-related matters in one agency, while Hickel argued that keeping wildlife management authority together under the Interior Department's Fish and Wildlife Service was more sensible. Given its recent decision to create NOAA, Congress decided to place responsibility for most marine mammals in the Department of Commerce, but as a concession to Hickel and in recognition of the Department of Interior's previous work on the four species noted above, they were assigned to that entity. Stans subsequently delegated responsibility for the marine mammals under his jurisdiction to the National Marine Fisheries Service, which was NOAA's only branch with experience in managing living resources.

The Marine Mammal Protection Act also included something of an insurance policy to ensure its directives were met. For that purpose it established the Marine

Mammal Commission. Set up as an independent federal agency in the Executive Office of the president, it consists of three commissioners appointed by the president and a nine-member Committee of Scientific Advisors on Marine Mammals selected by the commissioners. The commission is charged with providing independent, science-based advice and recommendations on any federal agency actions that could affect marine mammals; the purpose of that advice is to ensure that agencies comply with the act's policies and provisions. All federal agencies are directed to either follow the commission's recommendations or respond to it within 120 days explaining why its recommendations were not followed.

The second law underpinning right whale protection is the Endangered Species Act. Passed in 1973, it replaced an earlier version called the Endangered Species Conservation Act of 1969. Like its predecessor, the 1973 act called for the development of a list of "endangered" and "threatened" species to be covered by its provisions. Its initial list was carried over from the earlier act and included right whales. As with the Marine Mammal Protection Act, responsibility for endangered or threatened marine mammals was split between agencies in the Departments of Commerce (whales, dolphins, and seals) and the Interior (walruses, manatees, sea otters, and polar bears). The act also directs that all federal agencies use their authority to further the conservation of listed species.

Like the Marine Mammal Protection Act, the Endangered Species Act bans the taking of species covered by its provisions. However, it also includes other measures for restoring listed species. To ensure recovery work is well thought out, it directs the two lead agencies to prepare recovery plans for each listed species unless, for some reason, doing so would not improve recovery prospects. Recovery plans must identify the specific research and management tasks necessary to promote the species' recovery. The act also authorizes the formation of recovery teams composed of species experts to help prepare those plans.

Another provision of the Endangered Species Act—one receiving little attention when it was passed but that has since become a central pillar of its protection—was a brief two-sentence paragraph designated as section 7. It requires all federal agencies to consult with the agency responsible for a listed species if any action they plan to take or authorize might jeopardize it or adversely affect habitat critical to its survival. In government parlance, these interagency deliberations have come to be called "section 7 consultations." Their purpose is to ensure that decisions on any proposed federal action carefully consider effects on listed species. If severe effects seem possible, the agency responsible for the listed species (e.g., the National Marine Fisheries Service in the case of right whales) must prepare a "biological opinion" recommending steps that the "action" agency could take to avoid either jeopardizing the species or adversely affecting its critical habitats.

Together, the two acts provide federal agencies with the direction, regulatory power, and funding authority needed to tailor recovery programs for individual species. Work on right whale recovery under auspices of the acts began in the late 1970s, but its path ahead was anything but straightforward (table 17.1). The development of right whale protection offers an intriguing glimpse into the ebb and flow of politics, policies, people, and procedures influencing a species' recovery. For anyone contemplating a career in wildlife conservation or who is simply interested in what can be involved in addressing particularly thorny issues affecting a species' survival, the course of events shaping right whale recovery offers valuable and instructive insight.

Understanding the Challenge

In the mid-1970s almost nothing was known about North Atlantic right whales other than that they were on the brink of extinction and were occasionally found along the East Coast of the United States. When research in the 1980s revealed that perhaps as few as 300 whales still survived, they were placed in the unenviable company of the world's rarest mammals, whether on land or in the sea. Indeed, for several reasons their risk of extinction was even greater than it was for other species of comparable abundance.

In the 1980s, notable advances were made to improve recovery prospects for endangered species. Among them were new methods for (1) relocating animals to reestablish or supplement populations lost or disappearing from the wild; (2) captive breeding to provide stock for restoring wild populations; (3) artificial insemination to increase reproductive success and minimize debilitating inbreeding effects; (4) rescuing and rehabilitating sick and injured animals; (5) nurturing young in captivity through early life stages when survival rates in the natural habitat are low, and then releasing them back to the wild ("head start" programs); and (6) restoring or manipulating habitat to compensate for its loss. For highly migratory waterfowl, such as cranes, resourceful biologists even trained fledgling chicks to follow ultralight aircraft and thereby resurrect migratory routes covering thousands of miles; knowledge of those routes had been extinguished when wild populations had been eliminated, but now the chicks could later teach their offspring.

Unfortunately, none of those options was practical for a species as enormous as a right whale. Moreover, most species mature faster and produce more offspring than large whales, which makes faster recovery rates possible. Whooping cranes, for example, mature in four to five years and produce two chicks annually; black-footed ferrets mature in one to two years and have litters of three to four kits; and Florida panthers can breed at the ages of two to three and raise an average of two kittens per year. Right whales, however, don't reach maturity until the age of eight to ten years, and females bear just a single calf every three to five years. Thus, under the best of circumstances, a right whale population of just a few hundred individuals can face a 100-year recovery period. And because interventions available for other species are impractical for right whales, managers are left with only one option—reducing human-caused deaths and injuries. Recognizing the very

Table 17.1. Events leading to the formation of a dedicated North Atlantic right whale recovery program. (MMC = Marine Mammal Commission; IWC = International Whaling Commission; NMFS = National Marine Fisheries Service; Right Whale Recovery Team = RWRT.)

Date	Event
Oct. 1972	Congress passes the Marine Mammal Protection Act
Dec. 1972	Congress passes the Endangered Species Act
Oct. 1974	President Nixon appoints first commissioners to MMC
1976	MMC funds a Gulf of Maine whale sighting network
1978–1982	Bureau of Land Management funds the Cetacean and Turtle Assessment Program (CeTAP)
Oct. 1979	Congress appropriates $100,000 to MMC for East Coast cetacean research program
June 1983	IWC workshop held at the New England Aquarium to assess the status of right whales
1984	MMC funds Southeast US right whale sighting network
April 1985	Scientists testify before Congress urging support for 5-year right whale research initiative
Oct. 1985	Congress adds $500,000 to NMFS budget for the first year of 5-year research initiative
Dec. 1985	MMC completes an interim right whale recovery plan
Apr. 1986	Georgia Conservancy holds Southeast US Right Whale Workshop
Sept. 1986	NMFS releases funds for first year of 5-year research initiative to cooperating scientists
July 1987	RWRT formed by NMFS to draft a recovery plan
1987–1990	Congress adds $250,000 per year to NMFS budget for right whale recovery
1987–present	US Navy, Coast Guard, and Army Corps of Engineers fund Early Warning System for right whale calving grounds
Dec. 1988	NMFS convenes first meeting of the RWRT
1989	Minerals Management Service funds Satellite telemetry research
Nov. 1989	MMC provides NMFS draft right whale recovery plan
Jan. 1990	Right Whale Recovery Team submits draft recovery plan to NMFS
Feb. 1990	RWRT submits petition to NMFS to designate three critical habitat areas
Dec. 1991	NMFS adopts a final right whale recovery plan
1992	NMFS begins requesting funds to support right whale recovery
May 1993	NMFS proposes three right whale critical habitat areas
Aug. 1993	NMFS disbands RWRT and appoints the Southeast Implementation Team for recovery plan
Jul. 1994	NMFS designates three right whale critical habitat areas
Aug. 1994	NMFS appoints the Northeast Implementation Team for recovery plan

small size of the population, this means reducing rare events to becoming even more infrequent.

When legislative authority was passed, North Atlantic right whale recovery also faced a bureaucratic headwind. Leadership at the National Marine Fisheries Service initially viewed conservation responsibilities under the Marine Mammal Protection Act as something less than a welcome challenge. At times, agency officials deliberately obstructed conservation actions if they appeared to conflict with what they considered their primary mission—promoting and managing commercial fisheries.

For Fisheries Service staff and administrators trained in optimizing fish harvests, the ban on taking marine mammals seemed a self-fulfilling mandate. They saw little point in investing time or money on species that were to be left alone. And with no marine mammals subject to harvesting along the East Coast, as some still were in Alaska, they had little interest in devoting attention to them on the eastern seaboard. This left it largely to three other entities to press for right whale research and conservation from the mid-1970s through the 1980s. Those were the Marine Mammal Commission, the Bureau of Land Management (the agency then responsible for leasing offshore oil and gas resources), and a small but dedicated community of nongovernmental whale researchers. With no guide for developing a whale conservation program, they pushed ahead, trusting that research findings would light the way. In the process, they gradually induced Fisheries Service officials to take a more responsive leadership role.

Where to Begin?

The first to press for attention to right whales was the Marine Mammal Commission, then under the direction of John R. Twiss Jr., the agency's first executive director. Twiss was a rare breed. Over his 25 years of service in that position, he probably had more influence on translating the spirit and provisions of the Marine Mammal Protection Act into meaningful conservation programs than any other individual. Respected by all, he was a dedicated conservationist, a meticulous administrator, and a charismatic leader with an inborn ability to make others want to do what he thought was needed. Moving to the commission from the Office of Antarctic Programs at the National Science Foundation, he was drawn to the post by the act's cutting-edge challenge of merging ecosystem principles with wildlife management objectives.

Sharing the commission's leadership for most of Twiss's tenure was Robert J. Hofman, the agency's equally dedicated Scientific Program director. A former high school science teacher with a doctoral degree earned studying Weddell seals in the Antarctic, he had a gift for reducing complex scientific questions into a logical progression of bite-sized research pieces. Together, Twiss and Hofman guided efforts by the commission and its Committee of Scientific Advisors and helped fill a federal leadership void left by Fisheries Service officials with little enthusiasm and a jaundiced view of their assigned marine mammal management responsibilities. To do so, the commission supported small, well-placed investments designed to seed larger, long-term research initiatives. Particularly important was support for planning meetings to reach agreement on research and management priorities. From the outset, the commission recognized that the first task for right whale recovery was determining when and where the species occurred. Only then would it be possible to begin answering fundamental questions, such as those involving population abundance and trends, what threats might be affecting the species' recovery, and how those threats might be mitigated.

The commission's work on right whales began in a roundabout fashion in 1976 when it funded a large-whale sighting network in the Gulf of Maine. Contracted to Steven Katona, an energetic young marine biologist at the College of the Atlantic in Bar Harbor, Maine, the project sought to identify important whale habitats about which almost nothing was then known. By distributing sighting forms to fishermen, yacht club members, other seafarers, and scientists throughout New England, Katona hoped to learn what marine mammals were found where and when. The results included a startling 40 definite or probable right whale sightings. Most were from Cape Cod Bay and an open ocean area east of the cape called the Great South Channel.[3] Those findings brought the species occurrence off New England to the attention of agency officials and scientists beyond the few researchers who had been quietly documenting their presence since the 1950s.

Anxious to extend those studies, Congress provided funds to the Fisheries Service in 1978 specifically for East Coast cetacean studies. However, to the consternation of congressional appropriators, Fisheries Service officials redirected the money to work on whales in Alaska. Upset at the unauthorized reallocation but still committed to work along the East Coast, Congress provided the Marine Mammal Commission a special one-time appropriation of $100,000 in fiscal year 1980 for cetacean studies on the eastern seaboard.[4] Immediately upon receiving the funds in October 1979, the commission organized a planning meeting with Fisheries Service staff and independent scientists.[5]

At the time of the meeting, a new oil refinery was being planned near the town of Eastport, in northeastern Maine. To evaluate the project's potential effect on marine mammals, participants at the planning meeting agreed that a priority use of the funds should go to field surveys near the proposed refinery. They also agreed on both the need to determine whether natural marks on whales could be used on a broad scale to identify and track individuals over time, and the importance of further studies on right whales in Cape Cod Bay. Recognizing that the Fisheries Service had no staff with experience in conducting marine mammal surveys and that it would need to generate the relevant expertise and interest in carrying out such surveys, the commission transferred most of the funds to the agency with detailed specifications on how they were to be used. At the same time, it retained some funds for several smaller projects.

The money transferred to the Fisheries Service was used to support aerial surveys of the Bay of Fundy near the United States–Canada border; these were contracted to the New England Aquarium. The result was the discovery of right whale feeding grounds along the international border near the site of the proposed refinery. Between July and October, the aquarium sighted 25 to 53 right whales, including several mother-calf pairs.[6] To follow up on this unexpected finding, aquarium staff began a program of annual late summer–early fall vessel surveys in the Bay of Fundy. Although often having to scramble to patch together funding from government agencies, private foundations, and other sources, the aquarium has continued those surveys to the present day. In the process, it revealed a late summer feeding area used in most years by a large percentage of the species' mother-calf pairs.

Another part of the commission's appropriation went to Charles "Stormy" Mayo at the Provincetown Center for Coast Studies to compile right whale sighting records in Cape Cod Bay.[7] His findings confirmed that the bay was also a regularly used feeding area, with some whales arriving as early as January, but with the greatest numbers in March and April. Like the New England Aquarium, the Center subsequently mounted annual studies that have continued to the present, contributing vital data for life history analyses. In the process, Mayo and his colleagues brought to light ecological relationships between right whales and their prey—the copepod *Calanus finmarchicus*; this knowledge now enables Center scientists to predict peak arrivals and departures of right whales based solely on assessments of their prey composition and density.[8] Finally, the commission contracted Roger Payne to describe techniques he had been refining since 1971 to identify individual southern right whales off Peninsula Valdez, Argentina, by callosity patterns.[9] Payne's report provided a guide for scientists still learning how to use new photo-identification techniques to track individual North Atlantic right whales.

After 1980, the commission continued funding right whale studies. In 1984, it contracted Howard Winn to organize a right whale sighting network for the southeastern US coast. A crusty New Englander with a heart of gold at the University of Rhode Island's School of Oceanography, Winn was one of the nation's leading authorities on whales. His project brought to light nearly 200 right whale sightings from Virginia to Florida.[10] Many were culled from unpublished records filed away by David and Melba Caldwell at Marineland of Florida and the Scientific Event Alert Network, a compilation of marine mammal strandings and unusual sightings kept by Jim Mead at the Smithsonian Institution. With more than three-fourths of the recorded sightings south of North Carolina between January and March and many including cow-calf pairs (53 of 148 reports), Winn's report provided compelling evidence that waters off the southeastern United States served as a winter calving ground for right whales.

The report prompted the Georgia Conservancy, a nongovernmental conservation group, to host a workshop in 1986 on right whale research and management needs in the southeastern United States. Organized by Hans Neuhauser, the conservancy's director, and partially funded by the Marine Mammal Commission, workshop attendees urged support for aerial surveys off Florida and Georgia, the examination of every stranded right whale, and designation of the region's calving grounds as a right whale critical habitat under the Endangered Species Act.[11] The commission also used its modest research budget to meet funding shortfalls for right whale field studies in the Bay of Fundy and the Great South Channel,[12] to refine right whale photo-identification methods,[13] and to convene research and management planning meetings.[14]

Concurrent with the commission's efforts was a groundbreaking project that began in 1978 with funding from the Department of the Interior's Bureau of Land Management. The bureau was then planning to lease offshore areas along the East Coast for oil and gas development. Because so little was known about the region's marine wildlife, it decided to support a region-wide survey to collect data on the distribution and abundance of marine species, particularly those listed as endangered or threatened, in order to evaluate possible effects of its leasing program. In 1980, when the department created the Minerals Management Service to oversee offshore leasing, administration of the project was transferred to the new agency.

Best known by its acronym CeTAP, short for Cetacean and Turtle Assessment Program, the project deployed aerial observers from the Canadian border to Cape Hatteras, North Carolina. Organized under a contract to Howard Winn at the University of Rhode Island, aerial surveys were flown from November 1978 through January 1982 that covered a quarter million miles of track lines (specified survey routes) from shore to the outer edge of the continental shelf in all seasons of the year. The results exceeded even the most optimistic expectations of its designers.[15] Project scientists discovered a diversity and abundance of marine mammals that was totally unexpected; the study marked a watershed moment in understanding the region's marine fauna. In the process, the surveys afforded some of the first professional experience for a host of budding marine mammal scientists who would become many of today's leading authorities on North Atlantic right whales, including James Hain, Robert D. Kenney, Scott Kraus, Stormy Mayo, Carol Price-Fairfield, and Randy Reeves.

Over the three-year study, they logged 380 sightings involving 988 right whales, although some proportion of these was presumed to be resightings of the same animals. After crunching the numbers with statistical models, project scientists calculated an abundance ranging between 70 and 1,000 right whales with a best estimate of 380 whales. The best estimate proved remarkably close to a later estimate generated from photo-identification records in the late 1980s that placed right whale abundance at between 200 and 400. CeTAP surveys also revealed important spring and summer feeding aggregations in the Great South Channel (an area 20 to 50 miles southeast of Cape Cod), the Bay of Fundy, and Browns Bank off the southern tip of Nova Scotia.

Deciphering the massive amount of survey data in the age before modern computers was a Herculean task, and project scientists were not always correct. Based on sightings of right whale calves off Cape Cod in spring at the same time calves were seen off Florida and Georgia, they hypothesized that there were two calving areas—one in the north and one in the south. It was a reasonable interpretation given the lack of information on whale migrations at the time, but later studies revealed that virtually all right whales give birth off the southeastern US coast. Because trips from Florida to New England can take right whales as little as 11 days,[16] the temporal overlap noted by CeTAP scientists simply reflected the early arrival of some mother-calf pairs in Cape Cod Bay. Their erroneous conclusion is yet another reminder of the need for caution in accepting hypotheses based on limited data, and the importance of continuing research to resolve uncertainties.

Along with the Marine Mammal Commission and Bureau of Land Management, the third group contributing significantly to the formation of dedicated right whale recovery work was a coalition of nongovernmental scientists—many first brought together by CeTAP and commission planning meetings—that was eventually called the North Atlantic Right Whale Consortium. The consortium still plays a central role in right whale research. Established on little more than a handshake between researchers sharing similar concerns about right whales, they sought to promote complementary studies throughout the species' range. At the consortium's core was an agreement to share right whale photographs and sighting data in a centralized database available to all its members.

For scientists whose credentials rely on publishing analyses of the data they collect, sharing data at the risk of others publishing it before they do is no trivial matter. As a result, scientists typically treat their data as proprietary information and are very cautious about sharing it with others before they have a chance to analyze the data and publish the results. By the early 1980s, however, the need to share photographs of recognizable right whales was readily apparent. Several groups had begun collecting and copyrighting right whale photos from different parts of the population's range; yet without knowing whether the whales in one area were the same as those seen elsewhere, individual research teams were the proverbial blind men trying to describe an elephant by touching its different parts.

It was Howard Winn and his colleague John Prescott, then director of research at the New England Aquarium, who engineered a path forward to overcome this obstacle. Their delicate task began when Prescott and his new research associate, Scott Kraus, submitted a paper on right whales in the Bay of Fundy to a meeting of the International Whaling Commission's Scientific Committee in June 1982.[17] The IWC was anxious to learn whether overexploited whale stocks were recovering. With hunting right whales banned since 1935, new information on their status offered an opportunity to evaluate the question. Recognizing that several research groups had been studying *Eubalaena* in other oceans, Prescott, Kraus, and their colleagues urged the scientific committee to endorse a workshop on the status of right whales worldwide. Prescott sweetened the recommendation with an offer to host the meeting at the New England Aquarium. The scientific committee agreed, and the IWC subsequently approved partial funding for the workshop.

With additional funds from the Marine Mammal Commission, the Fisheries Service, and the World Wildlife Fund, the workshop was held a year later, in June 1983. It brought together scientists studying right whales in all the world's oceans, including researchers who had been collecting photographs in the North Atlantic. The workshop was particularly notable with regard to North Atlantic right whales for two reasons. The first was an admission by the Fisheries Service that it had been allocating less than $50,000 a year for right whale studies from an annual budget of just $500,000 for all marine mammals nationwide. The agency's representative also noted that there were no plans to increase funding. The second was a recommendation urging that "wherever possible, national groups and individuals involved in [right whale studies] should cooperate in comparing photos from different areas to evaluate stock identity and movements."[18]

Appalled at the low level of funding for right whale research from the Fisheries Service and armed with the recommendation for cooperative studies developed under the aegis of the IWC, Winn and Prescott spearheaded a cooperative nongovernmental research initiative and lobbied Congress for its support. The initiative eventually led to formation of the North Atlantic Right Whale Research Consortium. Sharing concern for the plight of right whales, members of the House Appropriations Subcommittee responsible for the National Marine Fisheries Service budget (the Subcommittee on Commerce, State, Justice, the Judiciary, and Related Agencies) invited Winn and Prescott to testify on right whale research needs at a hearing on April 16, 1985.

Together, Winn and Prescott urged support for a $500,000 per year, five-year program to collect data on right whales from Florida to Canada. Less than a month later, scientists from the University of Rhode Island, the New England Aquarium, the Center for Coastal Studies, the Woods Hole Oceanographic Institution, and Marineland of Florida submitted a joint research proposal to the Fisheries Service for the first year of the five-year "integrated program for research on the northern right whale off the eastern United States." It proposed field studies to collect data on population abundance, distribution, demographics, migration patterns, feeding behavior, and threats.

Having requested no money for right whale research but aware of congressional interest in providing funds for that purpose, the Fisheries Service convened a planning meeting in May of 1985, a month after the hearing. The meeting brought together scientists from the commission and other federal agencies to review information on right whales, as well as the research proposal submitted by nongovernmental scientists. With congressional funding still uncertain, meeting participants considered what might be done with amounts ranging from $250,000 to $500,000. Congress subsequently earmarked $500,000 for the initiative in the Fisheries Service's fiscal year 1986 appropriation. Because the federal fiscal year runs from October 1 through September 31, the beginning of fiscal year 1986 was on October 1, 1985.

At the start of the fiscal year, the Fisheries Service signed a funding agreement with the research group. The full appropriation, however, was not provided to the group. After a mandatory across-the-board reduction imposed on all federal spending as part of a general federal deficit reduction measure, the Fisheries Service diverted nearly $100,000 to unrelated agency projects. As a result, only $381,000 was eventually provided to carry out the right whale initiative, and that was not made available until the end of the fiscal year in September of 1986, thereby precluding the start of field work.

With planned research yet to begin at the end of the fiscal year, Congress appropriated an additional $250,000 for fiscal year 1987. That allowed for a fully funded first year of field work, albeit a year later than planned. In 1987, scientists undertook photo-identification and monitoring studies in four summer feeding areas in the Gulf of Maine: Cape Cod Bay, the Great South Channel, the Bay of Fundy, and Browns Bank, off the southern tip of Nova Scotia. With additional funds provided by the US Navy, US Coast Guard, and US Army Corps of Engineers, the Fisheries Service also contracted with the New England Aquarium to begin what would become an annual winter survey program off the southeastern US coast to document calf production and help direct vessels away from whales. For the two previous years, the aquarium had patched together a survey of the calving grounds off Georgia and northeast Florida on a shoestring budget with a heavy reliance on donated private planes and pilot services generously offered by off-duty Delta Airline pilots. The new funding, however, improved regional data collection efforts enormously.

Although the right whale research program was planned as a five-year initiative, the Fisheries Service requested no funds to continue work, leaving it to other agencies and groups to find funding. At the urging of the researchers and other groups, Congress continued adding $250,000 per year to the agency's budget in 1988 and 1989 specifically for right whales. This allowed nongovernmental scientists to continue their studies, but only on a limited basis. Some of the shortfall was made up with supplemental funding from private foundations and other agencies.

As photo-identification techniques were perfected, the New England Aquarium volunteered to assume responsibility for curating a photo-identification catalogue and matching right whale photos contributed by cooperating scientists in all areas. Robert D. Kenney at the University of Rhode Island also agreed to serve as custodian for data from all aerial and shipboard surveys reporting right whales. Using new computer technology, he archived all new survey results with CeTAP survey data to create a database that could be used to estimate seasonal densities and habitat use trends for right whales, as well as other large whale sightings from all along the eastern seaboard.

As the added value of cooperative research efforts became apparent, right whale researchers at the New England Aquarium, Provincetown Center for Coastal Studies, University of Rhode Island, and the Woods Hole Oceanographic Institution moved to formalize their cooperative compact. For the first 15 years they had worked together with no formal structure, but in 1998 they adopted bylaws to manage access and use of their collective data; they also dedicated their efforts to eliminating human causes of right whale mortality, protecting the species' habitat, and monitoring its reproductive success and population trends. The agreement established the name North Atlantic Right Whale Consortium as their official title.[19]

Consortium members also expanded their efforts to necropsy all dead right whales, disentangle whales caught in fishing gear, and hold annual meetings. Its annual meetings—open to all interested parties, including students, agencies, nongovernmental groups, and the public—proved to be a particularly effective way to update everyone on recent research findings, future research plans, and new management developments. The meetings also proved an invaluable catalyst for arranging collaborative studies. As interest in right whale recovery increased, so did the size of its meetings. Whereas thirty or forty people attended meetings in the late 1980s, more than 175 participants met in 2014.

In just a few years the cooperative studies revealed the broad outlines of right whale abundance, movements, habitat-use patterns, and reproduction rates. And they painted a grim picture. By the end of the 1980s, photos of unidentified individuals tailed off sharply, indicating that most of the whales were known, and the population likely numbered only 200 to 350.[20] Matches of whales photographed in different parts of their range showed that virtually all pregnant females traveled south in early winter, gave birth along the southeastern coast, and returned north with newborn calves to feed off New England and southeastern Canada in early spring. Surveys of calving grounds coupled with resightings of mother-calf pairs on feeding grounds also revealed a birth rate averaging just 11 to 12 calves per year, a perilously slender thread on which to hang a species' future.

To sharpen understanding of right whale movements, the Minerals Management Service contracted with Bruce Mate at Oregon State University. The agency wanted to apply his pioneering satellite telemetry techniques to track the movements of right whales in the northern Gulf of Maine during the summers of 1989 and 1991.[21] The results revealed a startling range of movement. Of the half dozen animals tagged, three traveled more than 1,200 miles, including a mother-calf pair that swam some 2,300 miles from the northern Gulf of Maine to the coast of New Jersey and back in just 42 days. Another whale traveled 1,900 miles in 42 days to the edge of the continental shelf south of Nova Scotia, where it skirted the edge of a warm-water ring for several days. As habitat-use patterns emerged, it became apparent that right whales spent most of their time over the continental shelf, where interactions with human activities were most likely.

Research results also revealed serious threats from both vessel traffic and fishing gear. Wounds from ship collisions and entanglement were implicated as the cause of death for a third of all discovered right whale carcasses. Considering the small size of the population and the unknown number of dead animals that were not observed, the findings were alarming. Photographs of live whales also revealed scars from ship propellers and fishing gear. Other threats of concern included oil spills from planned offshore oil and gas development (plans that were later withdrawn); and pollution from sewage outfalls, land runoff, and off-shore dredge spoil disposal.

By the end of the 1980s, efforts by the Marine Mammal Commission, Minerals Management Service, and Right Whale Research Consortium had formed the nucleus of a solid research program that was beginning to answer vital questions about the population's status, trends, and threats. With the species' plight now clear, the Fisheries Service began to embrace its conservation responsibilities and request funds for right whale work in its annual budget. As in the past, most funds were passed to the consortium but were still insufficient, requiring its members to find supplemental support from private foundations, corporate sponsors, state agencies, and other federal sources.

Several initiatives were taken to fill shortfalls in the Fisheries Service budget requests. In 1994, the Massachusetts Environmental Trust, a nonprofit organization established by the state legislature to raise funds for coastal and marine environmental protection, began offering automobile license plates featuring a right whale tail for an added charge. Under the energetic guidance of Executive Director Robbin Peach, a portion of the proceeds from license plate sales was funneled to right whale research and education projects. More than $1 million was provided by the Trust for right whale studies by the early 2000s.[22] The state of Florida through its Marine Research Institute, now part of the Florida Fish and Wildlife Commission, also stepped up to fund research in the right whale calving grounds.

With the US Congress then fighting a large budget deficit and unwilling to supplement right whale funding, it encouraged nongovernmental support. Acting on a recommendation by the Marine Mammal Commission, Senators Judd Gregg and Ted Stevens cosponsored a bill in 1998 directing the National Fish and Wildlife Foundation to establish a National Large Whale Conservation Fund to raise private contributions for work on the country's most endangered whales, particularly right whales.[23] The foundation subsequently raised several hundred thousand dollars for research and conservation programs for right whales and other endangered whales before the fund was terminated in 2005 due to difficulty in securing large private donations.

A Plan for Recovery

One of the cornerstones of the Endangered Species Act is its recognition of the need for thoughtful planning to map out a species recovery strategy. For this purpose, section 4(f) of the act requires the lead agencies "to develop and implement [recovery] plans for the conservation of and survival [of listed species] . . . unless such a plan will not promote the conservation of the species."[24] Although factors threatening North Atlantic right whales were still poorly understood in the mid-1980s, there was no doubt that their small number and low calving rate made them vulnerable to almost any adverse effects. The Marine Mammal Commission, therefore, wrote to the Fisheries Service in December 1984 recommending that it develop a right whale recovery plan to identify and rank recovery tasks and project long-range funding needs.[25] The Fisheries Service, still skeptical of the need for anything more than the ban on hunting, replied that once recovery plans were completed for species it felt more likely to benefit, including Hawaiian monk seals and sea turtles, it would turn its attention to planning needs for endangered whales, including right whales.[26]

Convinced that a plan was urgently needed and should not be deferred, the Marine Mammal Commission sponsored two workshops in February and June of 1985 to identify recovery priorities. Seeking a broad perspective, it invited officials from the Fisheries Service and other federal and state agencies, nongovernmental scientists, and conservationists. After reviewing the species' status and exchanging views on recovery strategies, the commission prepared a report listing nearly 40 tasks under three major objectives: (1) characterize and monitor important habitats; (2) evaluate and mitigate threats from human activities and pollution; and (3) improve information on current and historic right whale distribution and abundance.[27] In part, it recommended steps to improve methods for disentangling whales, prepare outreach materials for fishermen and mariners, conduct surveys that would alert mariners as to where right whales were likely to be encountered, and mark high-use right whale habitats on nautical charts.

In December 1985, the commission sent copies to the Fisheries Service recommending that this be used as an interim recovery plan until the agency was able to convene a recovery team and prepare a formal plan.[28] Recognizing that activities of several agencies could affect right whales and that each was obligated under the Endangered Species Act to protect the species, the commission recommended that the Fisheries Service determine which tasks it should carry out and which were more appropriate for other agencies to support. In particular, it urged immediate support for surveying high-use right whale habitats, compiling a photo-identification catalogue, reviewing historical catch records, examining all carcasses to determine cause of death, and evaluating threats from fishing, marine debris, and contaminants.

By the end of 1986, the Fisheries Service had yet to reply to the commission and had taken no steps to develop a recovery plan. However, a new administrator, William E. Evans, had been appointed to lead the Fisheries Service. A scientist with two decades of experience studying dolphins and a former chair of the Marine Mammal Commission, Evans was the agency's first administrator with a marine mammal background; he brought a new perspective to the agency. After the commission wrote to the new administrator in December 1986 reiterating its recommendations for forming a recovery team and preparing a plan,[29] the agency agreed. In July 1987, it invited a group of scientists, agency officials, and conservationists to participate on a right whale recovery team and prepare a recommended plan.

Progress, however, bogged down almost immediately. To meet deficit reduction constraints imposed by Congress, the Fisheries Service targeted work on marine mammals for particularly severe budget cuts. As a result, recovery team meetings were deferred until agency staff completed a draft plan for team review. A year elapsed before that was done and at its first meeting in December 1988, the team found the draft incomplete and in need of substantial revisions. A new plan outline was therefore developed and team members were assigned sections to draft or redraft. With computer technology and e-mail not yet in common use, the process was painfully slow. Matters were further complicated by funding constraints that allowed just one more team meeting. That was held in March of 1989 to review preliminary draft sections. After the meeting, members labored to redraft sections, exchanging text in the mail.

Nearly two years after its first meeting, the team was still struggling at its task. The slow pace of progress and uneven quality of the emerging draft led the Marine Mammal Commission to draft and forward a suggested plan to the Fisheries Service in November 1990. This spurred the team on to complete its work, and two months later, in January 1991, it submitted a draft. After some modifications, the Fisheries Service circulated this for public review in early March.[30] However, given limited time for frank, face-to-face discussions, the team's draft was penny wise and pound foolish. It was not well organized and identified actions that were poorly explained and not well justified.[31] After the public comment period closed in the spring of 1991, the Fisheries Service spent the remainder of the year making changes in response to comments and merging parts of the commission's draft into the team's draft.

In December 1991, nearly five years after starting on the plan, the Fisheries Service adopted a final right whale recovery plan.[32] Despite all the difficulties, it identified core recovery actions that remain at the heart of current right whale conservation work. By then, necropsies of dead whales and photographs of injured whales clearly identified commercial fishing gear and collisions with ships as the two most immediate threats. Even though some carcasses were seen only briefly floating offshore and others were too decomposed to identify a cause of death, more than a third of all dead right whales found between the 1970s and early 1991 were attributed to ship collisions (7 of 28 carcasses) or entanglement in fishing gear (3 of 28).[33] Major sections in the plan therefore focused on reducing those threats. Other parts called for designating "critical habitat" under the Endangered Species Act, continuing the ban on commercial hunting, and developing guidelines to minimize disturbance by whale-watching boats and other vessels.

Building on the Right Whale Consortium's work, the plan also outlined a full slate of research tasks. Chief among them was monitoring right whales in high-use habitats, determining if there were other important habitats not yet identified, maintaining the photo-identification catalogue, elucidating feeding behavior, clarifying physical and biological features attracting whales to preferred habitats, and examining photos of scars to better elucidate human interactions.

Other research called for evaluating possible effects of contaminants, examining all right whale carcasses, obtaining reports of dead whales (particularly those seen floating offshore), scouring historical records for clues about other habitats and past abundance, and refining satellite tracking technology to document travel routes and movements.

The plan projected a five-year work program costing nearly $2.5 million in the first year and declining to $1.1 million by the fifth. Most tasks fell to either the Fisheries Service or the consortium, but it was envisioned that other agencies and groups would share or assume responsibility for some tasks. However, the Fisheries Service's subsequent funding requests were woefully short of projected needs. For the plan's first year in 1992, the agency requested $230,000,[34] less than 10 percent of projected costs. Although funding increased over the five-year planning period, by 1996 it was still only $600,000.[35] The largest commitments by other agencies were from the Army Corps of Engineers, coast guard, and navy, each of which continued contributing $100,000 per year for annual aerial surveys in the calving grounds, and the Minerals Management Service, which provided over $100,000 for a satellite tracking study.

Once the plan was adopted, the Fisheries Service disbanded the team. In its place, it appointed two "Implementation Teams" composed of agency officials, scientists, and representatives of conservation and industry groups more familiar with regional issues. Each team was asked to provide advice and assistance in carrying out recovery plan actions in their respective regions. The first team was formed in August 1993 to address issues in the southeast calving grounds; the second was convened a year later to focus on feeding grounds in New England.

As research filled information gaps, task priorities and budget projections in the 1991 recovery plan became outdated. The Fisheries Service therefore adopted a revised plan in 2005.[36] Cost estimates for 150 tasks listed in this ranged from $7.7 million to $9.9 million per year during its 2006–2011 planning period. The new plan noted that there had been no signs of recovery since the first plan was completed, and with threats unchanged, its major focus continued to be reducing entanglements and ship strikes. The only new issue was the possible effect of loud noise from sources such as sonars, ship engines, seismic air guns, pile driving, and underwater explosions. The jury is still out on the extent to which such sounds affect right whales; however, studies of other cetaceans have shown that noise can mask whale communication,[37] cause animals to avoid areas,[38] and in some extreme cases even cause physical injury and death.[39]

Critical Habitat

The Endangered Species Act authorizes the designation of "critical habitat" in areas under US jurisdiction. Once designated, federal agencies must ensure that their activities avoid impacts to features that make identified habitats vital to a species.

To start the designation process recommended in its draft recovery plan, the right whale recovery team submitted a petition to the Fisheries Service in January 1990 proposing the designation of three areas: the calving and nursing grounds off Florida and Georgia, and two feeding grounds off Massachusetts—one in Cape Cod Bay and the other in the Great South Channel. The team was unanimous in its support for the idea, but like the recovery plan, the team's petition was poorly justified and all team members did not have a chance to review it. The petition, submitted on behalf of the team by its chair, was barely more than two pages, most of which listed a long set of boundary coordinates.[40]

Notably missing was an explanation of how the proposed boundaries were selected and what activities might require special management. After copies of the petition were sent to all team members, some pointed out those deficiencies and questioned the basis for some boundaries. Four months later, after further consideration, the team's chair submitted a revised petition with new boundaries for the southeast calving grounds and a somewhat longer, but still brief, seven-page review of the scientific literature.[41] The Fisheries Service responded by announcing receipt of the petition and inviting public comments in July 1990.[42]

Because of the limited options for protecting right whales, critical habitat is particularly important. Unfortunately, although the petition offered respected expert opinion, it still did not explain the basis for the recommended boundaries. After raising this concern with the Fisheries Service, the Marine Mammal Commission contracted for a study in September 1990 to map and evaluate right whale sightings data in and around each proposed area using the sightings database maintained by Robert D. Kenney at the University of Rhode Island.[43] The results were forwarded to the Fisheries Service in May 1991. Considering all sightings from 1950 through 1989 in and around the proposed boundaries, the study found that roughly 90 percent were inside the team's revised habitat boundaries.[44] Grouping sightings by month, the commission's report also identified the seasons of peak right whale occurrence. However, because most sightings lacked information on the amount of time spent searching for whales, it was impossible to tell whether sparse detection outside proposed boundaries reflected a true scarcity of whales or simply a lack of search effort. Nevertheless, the report reflected the best available data; when the commission forwarded the results to the Fisheries Service, it recommended designating all three areas but also that search efforts be evaluated to determine if adjacent areas should be designated as well.[45]

After receiving the report, the Fisheries Service waited two years to announce its decision. When it did, it agreed with the petitioned action, and in May of 1993 published a proposed rule to designate all three areas.[46] At that time there was no set policy on what regulations to designate critical habitat might include. As discussed in later chapters, many reviewers urged restrictions to reduce ship collision and entanglement risks. However, when a decision was announced to designate the three areas in July 1994 (figure 17.1),[47] the Fisheries Service included no restrictions on fishing or vessel traffic. Instead, it advised that specific restrictions

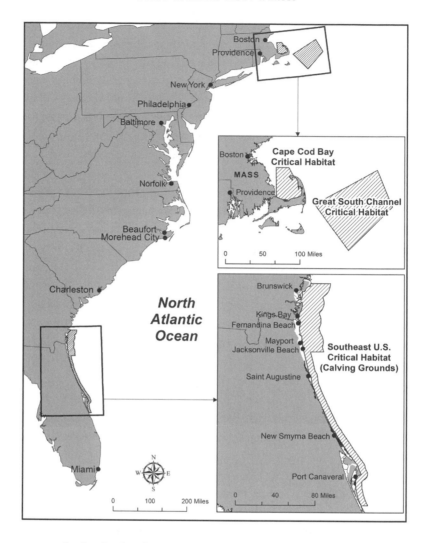

Figure 17.1. Three critical habitat areas were designated in 1994 under authority of the Endangered Species Act to protect North Atlantic right whales. Included were feeding areas in Cape Cod Bay and the Great South Channel off Massachusetts and the calving and nursery area off Florida and Georgia. (Map courtesy of Linda White.)

on activities had to be developed through separate rule-making actions. To do so, the agency said that it would ask the newly established regional implementation teams for advice.

Since then, this approach has been followed for all critical habitat designations. That is, rules for designating areas have been limited to identifying area boundaries with an accompanying description of special habitat features and management concerns for each area. Although designations add a layer of protection for species, public perceptions of their significance and scope are invariably inflated, which makes them lightning rods for controversy. Whereas supporters of designation proposals routinely overstate the protection benefits, critics typically counter with inaccurate claims that public access within the boundaries will be barred or all fishing in the areas will be shut down. The result is often a polarized debate over false perceptions.

What critical habitat designation actually does is expand the scope of impacts that federal agencies must consider when they fund or authorize activities in those areas. Thus, in addition to considering direct effects on a species—something already required under section 7 of the Endangered Species Act in any area—designations also stipulate that agencies avoid impacts on habitat features essential to species within those areas. For example, if a critical habitat is vital for feeding purposes, federal agencies must also consider effects of their actions on the species' prey. To advise federal agencies what features are essential, those features must now be identified in the designation process along with any activities that might require special management. However, activities by the public or state agencies that do not require federal action or approval, such as public access, swimming, diving, or navigation, are unaffected by designations except to the extent that they help make nonfederal agencies, groups, and the public aware of the area's special importance to a rare species.

The First 20 Years

By the mid-1990s, nearly 20 years after work began in earnest to determine North Atlantic right whale distribution, abundance, and ecology, much had been accomplished, yet much remained to be done. A solid research program was in place that produced vital information on the population's status and conservation threats. A recovery plan had been adopted that identified useful measures needed to reduce ship collisions and entanglements, and three critical habitat areas had been designated. Although Fisheries Service leadership was slow to embrace its responsibilities for right whale recovery and other groups had to provide an initial leadership role, the agency had begun to use its authority and request funding for right whale conservation, albeit at woefully inadequate levels. It would be another decade before funding would reach appropriate levels, and even then it would depend largely on congressional initiatives at the behest of entities other than the Fisheries Service (figure 17.2). The Fisheries Service also had begun hiring staff to work specifically on right whale recovery. These new hires brought a refreshing commitment to their task. However, marine mammal and endangered species conservation was still considered a secondary mission within the agency, and high-level support for right whale recovery work and funding was spotty.

New information also revealed troubling waters ahead for right whale recovery. When carcass salvage and necropsy protocols were improved early in the 1990s under the recovery plan, they revealed vessel and fishing gear–related deaths were occurring at higher rates than previously thought (figure 17.3). Between 1990 and 1995 more than half of all known right whale deaths were attributed to either ship strikes (5 of 10) or entanglement (1 of 10), and the total number of deaths due to both was increasing. In 1996, a record high six deaths were discovered, including at least five in the first three months of the year.

In the late 1990s, the situation took yet another alarming turn when calf counts plummeted to near zero. Whereas researchers had counted an average of 11 calves

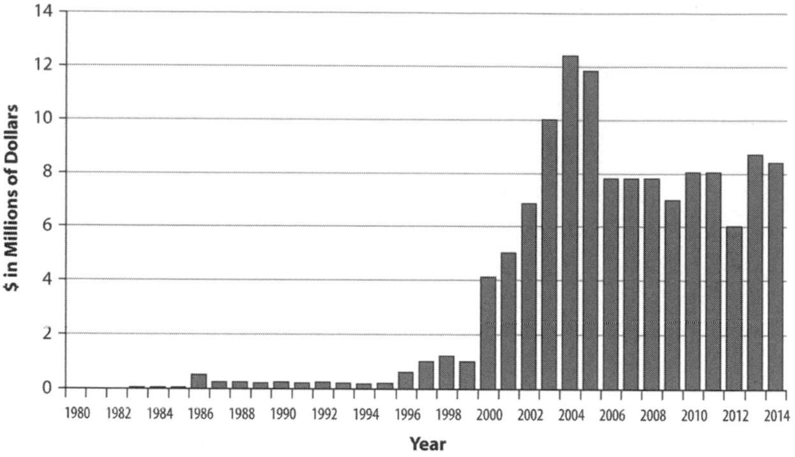

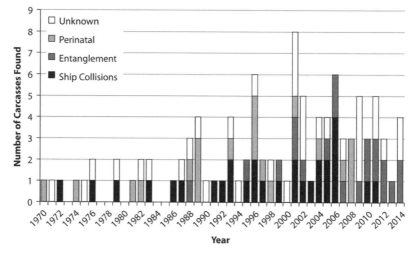

Figure 17.2. Funding levels appropriated by the US Congress to the National Marine Fisheries Service specifically for the recovery of North Atlantic right whales, 1980 to 2012.

Figure 17.3. Number of North Atlantic right whale carcasses and the causes of death, 1970–2014.

per year between the early 1980s and 1997, the numbers plunged to a series of record lows: 6 in 1998, 4 in 1999, and just a single calf in 2000. Accompanying the decline were signs of poor health. Many adults were unusually thin and had patches of white lesions from an unknown cause scattered over much of their body.

By then, the species' desperate plight was recognized as the nation's most urgent marine mammal conservation issue. It raised an outpouring of calls for reducing known causes of human-related right whale deaths—ship strikes and entanglements. Although commercial fishermen and vessel operators were sympathetic to the whale's troubles, they were unwilling to alter customary ways of doing business. As a result, virtually all management proposals were met with fierce opposition from industry representatives. As will be seen, the Fisheries Service's response to threats from the shipping and fisheries industries would differ markedly.

NOTES

1. Marine Mammal Protection Act of 1972. 16 USC §1361 (1972).

2. Stratton Commission. 1969. *Our Nation and the Sea: A Plan for National Action.* Commission on Marine Science, Engineering, and Resources. US Government Printing Office. Washington, DC.

3. Katona, SK. 1983. *The Gulf of Maine Whale Sighting Network: 1976.* Final report prepared for the Marine Mammal Commission. Contract no. MM6AC018. Report no. PB83-151290. National Technical Information Service. Springfield, VA, US.

4. Marine Mammal Commission. 1981. *Marine Mammal Commission Annual Report: Calendar Year 1980.* Report no. PB81-247 884. National Technical Information Service. Springfield VA, US, pp. 14–16.

5. Prescott, JH, SD Kraus, and JR Gilbert. 1980. *East Coast / Gulf Coast Cetacean and Pinniped Research Workshop.* Final report prepared for the Marine Mammal Commission. Contract no. MM1533558-2. Report no. PB80-160104. National Technical Information Service. Springfield, VA, US.

6. Kraus, SD, and JH Prescott. 1981. *Distribution, Abundance and Notes on the Large Cetaceans of the Bay of Fundy, Summer and Fall 1980.* Final Report prepared for the National Marine Fisheries Service under contract NA-80-FA-00048. Woods Hole, MA, US.

7. Mayo, CA. 1982. *Observations of Cetaceans: Cape Cod Bay and Southern Stellwagen Bank, Massachusetts: 1975–1979.* Final report prepared for the Marine Mammal Commission. Contract no. MM18009225-5. Report no. PB82-136263. National Technical Information Service. Springfield, VA, US.

8. Mayo, CA, and MK Marx. 1990. Surface foraging behavior of the North Atlantic right whale, *Eubalaena glacialis,* and associated zooplankton characteristics. Canadian Journal of Zoology 68(10):2214–2220, doi:10.1139/z90-308.

9. Payne, R, O Brazier, E Dorsey, J Perkins, V Rowntree, and A Titus. 1983. *External Features of Southern Right Whales* (Eubalaena australis), *and Their Use in Identifying Individuals.* Final report prepared for the Marine Mammal Commission. Contract no. MM6AC017. Report no. PB81-161093. National Technical Information Service. Springfield, VA, US.

10. Winn, H. 1984. *Development of a Right Whale Sighting Network in the Southeastern U.S.* Final report prepared for the Marine Mammal Commission. Contract no. MM2324805-6. Report no. PB84-240548. National Technical Information Service. Springfield, VA, US.

11. The Georgia Conservancy. 1986. *Report of the Southeastern U.S. Right Whale Workshop, 19–20 February 1986, Jekyll Island, Georgia.* Georgia Conservancy. Athens, Georgia.

12. Kraus, SD, JH Prescott, AR Knowlton, and GS Stone. 1986. Migration and calving of right whales (*Eubalaena glacialis*) in the western North Atlantic. *In* RL Brownell, PB Best, and JH Prescott (eds.), Right Whales: Past and Present Status. Special issue 10, Reports of the International Whaling Commission, pp. 148–157. Winn, HE, EA Scott, and RD Kenney. 1985. *Aerials Surveys for Right Whales in the Great South Channel, Spring 1984.* Final report prepared for the Marine Mammal Commission. Contract no. MM2324805-6. Report no. PB85-207 926. National Technical Information Service. Springfield, VA, US.

13. Kraus, SD, KE Moore, CA Price, MJ Crone, WA Watkins, HE Winn, and JH Prescott. 1986. The use of photographs to identify individual North Atlantic right whales (*Eubalaena glacialis*). *In* Brownell et al., Right Whales (n. 12), pp. 130–144.

14. Prescott et al., *East Coast / Gulf Coast* (n. 5). Scott, GP. 1985. *Report of the Working Group on NEFC/SEFC Marine Mammal Research: Results of the Meeting held 8–9 January, 1985.* Southeast Fisheries Center. Miami, FL, US. Kraus, SD. 1986. *A Review of the Status of Right Whales* (Eubalaena glacialis) *in the Western North Atlantic with a Summary of Research and Management Needs.* Final report prepared for Marine Mammal Commission. Contract no. MM2910905-0. Report no.

PB86-154-143. National Technical Information Service. Springfield, VA, US.

15. Cetacean and Turtle Assessment Program (CeTAP). 1982. *A Characterization of Marine Mammals and Sea Turtles in the Mid- and North Atlantic Areas of the Outer Continental Shelf.* Final report of the Cetacean and Turtle Assessment Program. Prepared for the Bureau of Land Management, US Department of the Interior. Contract no. AA551-CT8-48. Report no. PB83-215855. National Technical Information Service. Springfield, VA, US.

16. Brown, MW, and MK Marx. 2000. *Surveillance, Monitoring, and Management of North Atlantic Right Whales* (Eubalaena glacialis), *in Cape Cod, Massachusetts: January to Mid-May, 2000, Final Report.* Contract no. SCFWE3000-8365027. Massachusetts Division of Fisheries. Boston, MA, US, p. 16.

17. Kraus, SD, JH Prescott, and PV Turnbull. 1982. Preliminary notes on the occurrence of the North Atlantic right whale, *Eubalaena glacialis,* in the Bay of Fundy. *In* International Whaling Commission. *Thirty-second Report of the International Whaling Commission Covering the Thirty-second Financial Year 1980–1981.* Cambridge, UK, pp. 407–411.

18. Reeves, RR. 1983. *Report of Workshop on the Status of Right Whales, Boston, Mass, USA, 17–13 June 1983.* Report provided to the International Whaling Commission. Cambridge, UK.

19. Kenney, RD. 2001. The North Atlantic Right Whale Consortium databases. Maritimes 43(2):3–5.

20. Marine Mammal Commission. 1992. *Marine Mammal Commission Annual Report: Calendar Year 1990.* Report no. PB91-164236 National Technical Information Service. Springfield, VA, US, pp. 55–59.

21. Mate, BM, SL Nieukirk, and SD Kraus. 1997. Satellite-monitored movements of the North Atlantic right whale. Journal of Wildlife Management 62:1393–1405.

22. Burk, E, and D McKiernan. 2006. *Massachusetts Division of Marine Fisheries Right Whale Conservation Program: 2005 Projects and Accomplishments.* Report submitted to the National Marine Fisheries Service, National Fish and Wildlife Program, and Massachusetts Environmental Trust. Massachusetts Division of Marine Fisheries. Boston, MA, US.

23. Marine Mammal Commission. 2001. *Annual Report of the Marine Mammal Commission: Calendar Year 2000.* Bethesda, MD, US, pp. 24–25. The bill, called P.L. 105-277, and based on Senate Bill 2172, was introduced by Senators Judd Gregg and Ted Stevens.

24. Endangered Species Act of 1973. 16 USC § 1533 (1973).

25. From JR Twiss Jr., executive director of the Marine Mammal Commission, to WG Gordon, assistant administrator for Fisheries, National Marine Fisheries Service, December 11, 1984, on results of a commission review of the National Marine Mammal Laboratory research program at the Northwest and Alaska Fisheries Science Center.

26. From WG Gordon, assistant administrator for Fisheries, National Marine Fisheries Service, to JR Twiss Jr., executive director of the Marine Mammal Commission, March 13, 1985, on results of a commission review of the National Marine Mammal Laboratory research program at the Northwest and Alaska Fisheries Science Center.

27. Kraus, SD. 1985. *A Review of the Status of Right Whales* (Eubalaena glacialis) *in the Western North Atlantic with a Summary of Research and Management Needs.* Final report to the Marine Mammal Commission. Contract no. MM2910905-0. Report no. PB86-154143. National Technical Information. Springfield, VA, US.

28. From JR Twiss Jr., executive director of the Marine Mammal Commission, to WG Gordon, assistant administrator for Fisheries, National Marine Fisheries Service, December 31, 1985, on the development of a right whale recovery plan.

29. From JR Twiss Jr., executive director of the Marine Mammal Commission, to WE Evans, assistant administrator for Fisheries, National Marine Fisheries Service, December 23, 1986, on the development of a right whale recovery plan.

30. National Marine Fisheries Service. 1990. *Draft Recovery Plan for the Northern Right Whale*

(Eubalaena glacialis). Prepared by the Right Whale Recovery Team for the National Marine Fisheries Service. Silver Spring, MD, US.

31. Marine Mammal Commission. 1992. *Annual Report to Congress Calendar Year 1991*. Report no. PB92-139930. National Technical Information Service. Springfield, VA, US.

32. National Marine Fisheries Service. 1991. *Final Recovery Plan for the Northern Right Whale* (Eubalaena glacialis). Prepared by the Right Whale Recovery Team for the National Marine Fisheries Service, Silver Spring, MD, US.

33. Ibid., p. 11. A subsequent review concluded that two cases could not be confirmed as right whale entanglements. In one, it was unclear if entangling line caused the animal's death, and in the other, the whale could not be confirmed as a right whale. Accordingly, those deaths are not listed as entanglements (see figure 17.2).

34. Marine Mammal Commission. 1993. *Annual Report to Congress Calendar Year 1992*. Report No. PB93-154995. National Technical Information Service. Springfield, VA, US.

35. Marine Mammal Commission. 1997. *Annual Report to Congress Calendar Year 1996*. Report no. PB97-142889. National Technical Information Service. Springfield, VA, US.

36. National Marine Fisheries Service. 2005. *Recovery Plan for the North Atlantic Right Whale* (Eubalaena glacialis). Revision. Office of Protected Resources. National Marine Fisheries Service. Silver Spring, MD, US.

37. Clark, CW, WT Ellison, BL Southall, L Hatch, SM Van Parijs, A Frankle, and D Ponirakis. 2007. Acoustic masking in marine ecosystems: intuitions, analysis, and implication. Marine Ecological Progress Series 395:201–225.

38. Goldbogen, JA, BL Southall, SL DeRuiter, J Calambokidis, AS Friedlaender, EL Hazen, EA Falcone, et al. 2013. Blue whales respond to simulated mid-frequency military sonar. Proceedings of the Royal Society B, Biological Sciences 280(1765), doi:10.1098/rspb.2013.0657.

39. Fernández, A., JF Edwards, F Rodriguez, A Espinosa de los Monteros, Herráez, P Castro, JR Jabber, et al. 2005. "Gas and fat embolic syndrome" involving a mass stranding of beaked whales (Family Ziphiidae) exposed to anthropogenic sonar signals. Veterinary Pathology 42(4):446–457.

40. From H Neuhauser, chair of the Right Whale Recovery Team, to W Fox, administrator for Fisheries, National Marine Fisheries Service, February 5, 1990, concerning a petition to designate critical habitat for northern right whales.

41. From H Neuhauser, chair of the Right Whale Recovery Team, to W Fox, administrator for Fisheries, National Marine Fisheries Service, May 18, 1990, concerning a petition to designate critical habitat for northern right whales.

42. National Marine Fisheries Service. Endangered and threatened species and designation of critical habitat: petition to designate critical habitat for the northern right whale. 55 Fed. Reg. 8670. (July 12, 1990.)

43. From JR Twiss Jr., executive director of the Marine Mammal Commission, to N Foster, chief of the Office of Protected Resources, National Marine Fisheries Service, September 16, 1990, on the designation of critical habitat for northern right whales.

44. From JR Twiss Jr., executive director of the Marine Mammal Commission, to N Foster, chief, Office of Protected Resources, National Marine Fisheries Service, May 31, 1991, on the designation of critical habitat.

45. Kraus, S, and R Kenney. 1991. *Information on Right Whales* (Eubalaena glacialis) *in Three Proposed Critical Habitats in U.S. Waters of the Western North Atlantic*. Report prepared for the Marine Mammal Commission. Contract nos. T75133740 and 75133753. Report no. PB91-194431. National Technical Information Service. Springfield, VA, US.

46. National Marine Fisheries Service. 1993. Designated critical habitat: northern right whale. 58 Fed. Reg. 29186. (19 May 1993.)

47. National Marine Fisheries Service. 1994. Dedicated critical habitat; northern right whale. 59 Fed. Reg. 28793. (3 June 1994.)

18

NOBODY WANTS
TO HIT A WHALE

IT WAS A LANGUID November day in 1820, in an empty spot in the middle of the tropical South Pacific. Neither George Pollard nor his crew of 21 men could have possibly known they were about to become part of maritime lore. The drama that was to unfold would be one of the most remarkable tales any mariner would experience and live to tell about.[1] Their 87-foot schooner *Essex*, built in 1799, was a year out of Nantucket in search of sperm whales, and its crew had already experienced their share of misfortune. Just a few days into the voyage three of their seven whaleboats and part of their rigging were lost in a storm. After a perilous trip rounding the southern tip of South America through the Strait of Magellan and into the Pacific, Captain Pollard found his intended destination along the coast of Chile barren of whales. Hearing rumors of a new hunting ground in open waters to the west, he decided to give it a try. The rumors had substance. Some 2,000 miles west of northern Chile, the *Essex* finally fell in with a "shoal" of sperm whales and the crew felt their fortunes were about to turn. And indeed they were—but not for the better.

Lowering their three remaining whaleboats, the crew of each set their sights on a different whale and finally got down to the business that had brought them halfway around the world. Before long, one boat was dashed by a flailing fluke and limped back to the *Essex*, leaking badly. Another, under Pollard's command, had better luck latching onto a large whale. The skeleton crew aboard the *Essex* immediately turned to follow Pollard's boat as it was carried off into the empty horizon on a Nantucket sleigh ride. As it did, a "very large spermaceti" was spotted 100 yards off the *Essex*'s bow lying motionless and alone; it seemed to be eyeing the *Essex* as if sizing up an alien intruder. After sinking briefly into the electric blue expanse, it resurfaced seconds later heading straight for the ship, picking up speed. With the *Essex* now moving at a bare 3 knots, Owen Chase, the first mate in charge in Pollard's absence, shouted orders to pull hard up in an attempt to veer away from the charging whale. No sooner had the ship begun its turn, then the whale slammed its massive brow hard into the bow. The *Essex* shook to its core as a crack opened in the bow's 20-year-old, worm-weakened planks.

Stunned by the impact, the whale drifted listlessly beside the *Essex* as an equally stunned crew rushed to the sides to inspect for damage. Staring down over the rail they beheld a whale they judged to be an impossible 85 feet long—about as long as the ship itself—but well above the 60-foot length now cited as the maximum size for sperm whales. Notwithstanding doubts about its estimated size, it must have been an enormous specimen. Fearing the worst, Chase ordered that the ship's boat, their only remaining seaworthy shallop, be made ready just in case. As that was being done and other crew hastened to the pumps, the whale stirred from its stupor and retreated some 500 yards ahead of the vessel, where it turned to again face the object of its ire. Chase's narrative of events, written a year later, described the following:

> I was aroused (from my duties) by the cry of a man at the hatchway "Here he is – he is making for us again." I turned around and saw him about one hundred rods directly ahead of us, coming down apparently with twice his ordinary speed, and to me at that moment, it appeared with tenfold fury and vengeance in his aspect. And his course towards us was marked by white foam a rod in width, which he made with the continual violent thrashing of his tail; his head was about half out of the water, and in that way he came upon us and again struck the ship.[2]

Another unsuccessful attempt was made to veer away, but the rogue whale again smashed the bow, this time with a force that stove in the weakened timbers completely. Seemingly satisfied at having made whatever point it had intended, the whale turned and slowly swam off, never to be seen again. Meanwhile, water gushed through the breach far too rapidly for the pumps to keep up. The crew had no choice but to hastily salvage what food, water, and navigational aids they could and abandon the *Essex* in the ship's boat. Joining up with the remaining two whaling boats, the dumbfounded crew could only watch as their ship slowly slipped beneath the easy swell.

So began a desperate fight for survival by 22 men crammed into three open boats adrift in the middle of the South Pacific. Had they headed for the Marquesas some 300 to 400 miles southwest—the nearest land on their charts—they all may have survived. However, due to fearful rumors of savage cannibals on those islands, Pollard made a fateful decision; they would try for the far more distant shores of South America. But with steady trade winds beating them back from a direct route, Pollard chose to head a thousand miles south, parallel to the continent's coast, to catch the prevailing easterly winds at higher latitudes in hopes they would propel them east on a 3,000-mile leg to the continent. After a three-month ordeal, one boat and three men, including Pollard and Chase, would survive to reach the continent. Ironically, they were reduced to the very same cannibalism whose frightful prospect at the hands of "savages" kept them from making for the Marquesas at the outset. The fact that any lived to tell the tale is no small tribute to the hardy breed and expert seamanship of the Nantucket whaler.

After Chase penned an immensely popular narrative in 1821 that recounted the dramatic events surrounding the *Essex* demise, he retired (figure 18.1). Several of his sons, however, carried on the family tradition by shipping out in various capacities aboard Nantucket whaling ships bound for the Pacific. In 1841, a twenty-one-year-old Herman Melville, intent on satisfying his own desire for a firsthand taste of whaling, signed on as a seaman aboard the whaling ship *Acushnet* bound for the Pacific from Fairhaven, Massachusetts. As fate would have it, somewhere in the vast Pacific, his path crossed that of one of Chase's sons. Details of the encounter are murky. Some suggest Melville was trying to track down the elder Chase for details of the *Essex*'s terrifying fate. Recollecting the meeting in his declining years, Melville was convinced he met with Owen Chase himself. Other sources, however, confirm that Chase had given up life at sea long before Melville reached the Pacific. In any case, Melville struck up a friendship with one of Chase's sons, who loaned him a copy of his father's 1821 account and told him all he knew of the incident.[3] And so the seeds of *Moby-Dick* were sown, forever binding together the factual and fictional tales that cemented the last voyage of the *Essex* in the annals of maritime history.

Ship Strikes: A New Threat

Examples of whales deliberately attacking sailing ships are extremely rare. More common, but still rare considering the millions of miles logged by sailing ships before the 1900s, are accounts of accidental collisions with whales. Among them is the report of a whale struck in 1578 by the 120-ton English galleon *Solomon of Weymouth*, one of 15 ships under the command of Martin Frobisher on his third expedition to Baffin Island in northeastern Canada. Frobisher hoped to mine what he thought was a rich deposit of gold ore found on an earlier trip. It would turn out to be a worthless vein of pyrite, or fool's gold. An account of the expedition by George Best, captain of one of the vessels in Frobisher's fleet, included the following when his fleet was several weeks into its outbound voyage from England, somewhere west of Greenland: "On Monday, the last of June we met with many great whales, [acting] as [if] they had been porpoises. This same day the *Salamander*, being under courses and bonnet, happened to strike a great whale with her full stem, with such a blow the ship stood still, and stirred neither forward or backward. The whale thereat made a great and ugly noise, and cast up his body and tail, and so went underwater; and within two days after there was found a great whale dead, swimming above the water, which we supposed was that which the *Salamander* struck."[4] The description of the collision seems convincing, but

Figure 18.1. Owen Chase in retirement on Nantucket, circa 1870. Chase was first mate on the whaling ship *Essex*, which was sunk by a charging sperm whale in 1821. The incident became the inspiration for Herman Melville's novel *Moby-Dick*. (Image courtesy of the Nantucket Historical Association, MA, photo no. GPN4448.)

the fate of the whale is more suspect. Given that the ship was underway in the open ocean when it struck the whale, it seems implausible that the same animal would have been encountered dead or injured after sailing two more days. Such collisions, however, were rarely recorded in the era of tall ships.

On May 22, 1819, the SS *Savannah*, a modified sailing packet equipped with a steam engine and collapsible side-mounted paddle wheel, left Savannah, Georgia, bound for Liverpool, England, on the first transatlantic crossing by a "steamship." The *Savannah* ran its engine just 80 hours on its 23-day crossing, but it marked the beginning of the end for the age of sail. Sailing ships continued to be an economical means of transporting cargo into the 1920s, but by the late 1800s, the speed and reliability of steamships had already started to push the dwindling fleets of wind-powered vessels into an ever-shrinking group of specialty markets. During most of the 1800s, as steamships gradually replaced sailing vessels, shipping companies that were engaged in transatlantic passenger service vied with each other to build the fastest, largest, and most luxurious motorized vessels.[5]

Ships making the fastest transatlantic passage earned bragging rights to an unofficial prize called the "Blue Riband."[6] Steamships with side paddles similar to the *Savannah* were the first mechanically powered ships to claim the Blue Riband; their average crossing speeds increased from 8 knots (9 mph) in 1838 to 14.5 knots (16.7 mph) in 1862. In the 1870s, side paddles gave way to single-screw propulsion systems that quickly advanced in the 1890s to the modern double-screw design.

The White Star Line's *Adriatic* was the first single-screw ship to capture the prize, with a crossing that averaged 14.7 knots in 1872. The first double-screw steamer to win was the *City of Paris*, raising the bar on four separate occasions in the early 1890s, when its fastest crossing averaged 20.7 knots in 1892. In the first decade of the 1900s the record fell to a pair of coal-fired Cunard Line ships—the *Lusitania* and the *Mauretania*. The *Mauretania's* 1909 standard of 26 knots held for two decades until the oil-fired turbine-powered liner *Bremen* exceeded it in 1929 with a crossing at 27.8 knots. Thereafter, speeds increased slowly until 1952 when the *United States*, the last ship to claim the prize before the introduction of jet-powered watercraft, made the crossing at an average of 34.5 knots.

As the fastest steam ships reached speeds of 14 to 16 knots in the 1880s, the first records of motorized ships hitting whales began to appear.[7] According to Lloyd's *Register of Ships*, some 10,000 motorized vessels plied the world's oceans in the 1890s, but the maximum speed of the vast majority was several knots slower than the fastest liners. Interestingly, the fastest ships also account for most, if not all, of the early whale collision records. The earliest record found to date was in 1885 when the pilot boat *Alexander M. Lawrence*, traveling at 13 knots, hit an unidentified whale 20 miles east of Nantucket.

A search for early collision records found only a dozen more between then and 1950. Of those ships whose maximum speeds could be determined, all were passenger vessels that were among the fastest ships of their day.[8] Up to the early 1950s, however, most ships were still traveling at 12 to 14 knots or less, and collisions with whales, or "ship strikes," were considered remarkable oddities (see plate 9). Indeed, from the perspective of individual mariners, they were just that—and still are, considering the millions of miles that vessels travel the oceans today. Yet, as scientists increased their efforts to examine dead and stranded whales, they began finding evidence of ship strikes on a regular basis. At first, few thought to look at how frequently they occurred, but it soon became apparent it was far more often than previously thought. Concern for right whales brought this new revelation to light.

Outreach

Between 1970 and 1990, nearly a third of all dead right whales discovered along the eastern seaboard of North America (7 of 24) had suffered massive propeller slashes, implicating ship strikes as their cause of death (figure 18.2). In the early 1990s, when scientists began flensing right whale carcasses to the bone to glean all they could about their deaths, animals with no external signs of injury were found with fractured skulls and vertebrae that could only have been broken by the powerful, blunt force of a ship strike. As a result, the proportion of carcasses attributed to ship strikes rose to nearly 50 percent. And because some carcasses were seen only briefly floating offshore and others were too decomposed to assess their cause of death, even that proportion was believed to be a minimum.

Figure 18.2. A North Atlantic right whale calf struck by the coast guard cutter *Point Francis* on January 5, 1993, off Jacksonville, Florida. (Photo by Robert K. Bonde, US Geological Survey.)

For most people, the notion that a species as rare as the right whale could be struck accidentally in the open ocean so frequently seemed impossible. Yet, necropsy results argued otherwise. Thus, when the National Marine Fisheries Service adopted a right whale recovery plan in December 1991, ship strikes were flagged as a primary obstacle to their recovery. To reduce vessel-related deaths, the plan proposed several actions: alerting mariners to the problem, restricting vessel speeds in important right whale habitats, moving shipping lanes away from areas in which right whales occurred most often, and conducting research on both acoustic alarms to warn whales away from ships and sonar to detect them from ships.

After the plan was adopted, concerned parties fell into two camps. One group, including environmental organizations, scientists, and some government agencies (particularly the Marine Mammal Commission), urged use of the full range of measures identified in the recovery plan as quickly as possible. They believed that constraints on human activities threatening right whales were urgently needed and long overdue as indicated by the population's extremely small size. The second group included shipping interests and some other government agencies, particularly the US Navy and the US Coast Guard. They were all skeptical that random unintentional ship collisions could threaten the species or that the identified measures would be effective. Almost all vessel operators steadfastly maintained that their activities were not a threat to whales and argued that any regulatory measures were unwarranted, would cause dire economic consequences, and would threaten both navigational safety and national security. The leadership of the Fisheries Service sided with the second group and deferred all measures, other than mariner outreach, hoping that once aware of the problem, vessel operators would be able to avoid hitting whales.

Between 1991 and 1993, four of seven dead right whales discovered along the US East Coast were attributed to ship strikes, and calls for more forceful action intensified. In May 1993, when the Fisheries Service proposed designating right whale "critical habitat" under the Endangered Species Act and asked for public and agency comments, many groups responding to the request urged that the designation include regulations to reduce collision risks. Noting the use of speed zones to reduce deaths of Florida manatees resulting from collisions with recreational boats, the Marine Mammal Commission recommended that critical habitat regulations seasonally restrict the speed of large ships in right whale calving grounds.[9] It also recommended that ships crossing calving grounds be required to post whale lookouts and travel perpendicular to shore when entering or leaving port to minimize travel times and distances when crossing these nearshore waters preferred by mothers and calves. Others urged 100- to 300-yard vessel approach limits around right whales. Still other suggestions included the designation of mandatory vessel traffic lanes and the creation of an "early warning system" to alert ships to whale locations when crossing calving grounds.

As described in the previous chapter, however, when the Fisheries Service designated right whale critical habitat in June of 1994, it deferred all specific protection measures. Instead, the agency advised that a separate rule-making process would be necessary to develop such measures and that a new advisory group established in August 1993—the Southeastern Right Whale Recovery Plan Implementation Team—would be asked to provide recommendations for reducing ship-strike risks in the calving area. The new team consisted of regional shipping interests; the navy, which had three major facilities adjacent to the calving grounds; the Army Corp of Engineers, whose dredges maintained port access channels; the coast guard; regional port authorities; state wildlife officials from Florida and Georgia; and a conservation representative from the University of Georgia. Shortly after designating the right whale's critical habitats, the Fisheries Service appointed a similar implementation team for feeding areas off New England.

Although some groups continued to press for regulatory measures, the two teams supported only nonregulatory measures. Noting that nobody wanted to hit a whale, they argued that vessel operators would avoid whales if they knew about the risk and where whales occurred. Everyone acknowledged education and outreach were important and that vessel operators did not want to hit whales; however, some groups believed the teams' advice was motivated more by a desire to avoid measures unpopular with their constituencies rather than the wish to implement a proactive whale protection strategy. Nevertheless, the Fisheries Service accepted the teams' advice with no critical assessment of its own.

Thus, for the remainder of the 1990s, the Fisheries Service, agencies on the two implementation teams, and other groups devoted much effort to prepare and distribute education materials, many of which were recommended at a workshop on ship collisions with right whales held by the New England Aquarium in 1997. Sharing the outreach task, the involved entities blanketed East Coast ports with placards, posters, brochures, videos, articles for mariner magazines, notes for navigation charts, curricula for professional mariners, and additions to *Coast Pilots* (a regional publication describing local navigation conditions and hazards). The materials described the threat ships posed to whales, explained how to distinguish right whales from other whales and where right whales were most likely to occur, and offered suggestions for avoiding them. Suggested avoidance tactics included posting lookouts to keep a sharp eye out for whales, assuming whales would not get out of the way of ships, and steering clear of any observed whales. Although the two implementation teams balked at a suggestion that slower speeds should also be an identified option to avoid collisions, at the urging of other groups, the Fisheries Service included "reduced" speed as a recommendation in its materials, but without offering advice as to what speed might be prudent.

The cornerstone of their efforts, however, was an early warning system, or EWS, recommended by right whale scientists and identified as an action in the recovery plan. Its objective was to warn mariners transiting the calving grounds where right whales were located on a near real-time basis.[10] The EWS was essentially an expansion of the regional aerial surveys started by the New England Aquarium in the mid-1980s to count calves and photograph whales for the photo-identification catalogue. Also contracted to the aquarium, EWS surveys allowed scientists to expand their data collection, but with the added task of passing on whale sightings to vessel operators as soon as they were made.

Begun in 1994 with funding provided principally by the navy, Army Corps of Engineers, and coast guard, EWS surveys systematically swept nearshore waters where right whales were seen most often. Weather permitting, planes flew every day of the calving season from the first of December through the end of March. They followed 22 track lines set three miles apart, running perpendicular to shore and out to sea 15 nautical miles (nmi), or 17 statute miles, along a 70-mile stretch of coast from St. Augustine, Florida, to St. Simons Island, Georgia. By 1996, state wildlife agencies in Florida and Georgia added supplemental survey flights covering areas north and south of the core EWS survey area.

At first, whale sightings were relayed by pager to officials in the navy, coast guard, corps of engineers, state wildlife agencies, and port pilots, who then passed on locations to their respective vessels. In 1997, when members of the public also began reporting whales, thereby increasing the number of sightings, the US Navy's Fleet Area Surveillance and Control Facility in Jacksonville, Florida (FASFACJAX in abbreviated navy lingo) agreed to serve as a clearinghouse to expedite the distribution of sighting reports. After filtering out duplicate sightings, a navy dispatcher passed on whale sighting coordinates to its vessels, as well as to the coast guard and the NOAA Weather Service. At the height of the calving season between January and February, right whale advisories peppered the calving ground airwaves with broadcast notices to mariners, voice radio transmissions, alerts over NOAA weather broadcasts, and NAVTEX—a satellite-linked telex system for communicating with large ships. Over a five-year period, communications were honed to a point where whale sightings by aerial observers could be passed to transiting ships in as little as 10 minutes.

As an outreach tool, the EWS was a smashing success. Frequent articles described the system in regional print media, trade journals, and TV broadcasts. Both mariners and the general public were amazed at the lengths to which agencies and scientists were going to notify vessels where right whales were located. The favorable publicity and success at relaying sighting locations so quickly prompted the Fisheries Service and its Northeast Right Whale Implementation Team to start a similar program for southern New England in March 1997. However, as the sole management approach, the EWS and accompanying outreach efforts had many flaws. When researchers evaluated how well aerial observers could detect right whales, they found that, even under the best sighting conditions, only a third to a half of all those present were actually seen. Factors such as observer error, sun glare, sea state, and animals that were submerged at the time planes passed overhead made it impossible to see all or even most whales.[11]

More important, the approach assumed that vessel operators could avoid right whales once they were armed with the information on animal locations. Most large ships, however, have limited maneuverability. Because of their momentum, tankers and other large ships, once underway, can travel a mile or more between the time orders are given to make a turn and the point at which turns actually begin. As a result, operators of large vessels need to know exactly where a whale is located two miles or more ahead of the ship in order to begin an evasive maneuver. Even then whales could simply move into the vessel's new path by the time a turn is executed. Near collisions seen by EWS observers highlighted the dilemma. When they happened to see a ship closing in on a whale, they attempted to contact the vessel directly over voice radio. On the one hand, pilots asked to alter their course almost always responded positively, indicating they immediately recognized what was being asked and why. On the other hand, EWS observers found that vessel crews were often unable to spot whales reliably, even when they had been alerted to where the animals were. Thus, for all its effectiveness as an awareness tool, the EWS had critical limitations as a mitigation measure and was also expensive to carry out.

The Prince of Whales

As outreach measures were implemented, a lone environmental warrior to some and a menace to others mounted an aggressive one-person legal campaign against virtually all agencies and groups involved in right whale protection. Calling himself "citizen prosecutor" and "Prince of Whales," Richard Max Strahan spoke for a group called GreenWorld that, as best as anyone knew, included one member—Richard Max Strahan. Operating on a shoestring budget collected on the streets of Boston, Strahan combined an unrestrained passion for protecting endangered species with a take-no-prisoners attitude. Holding an unwavering conviction that no involved agencies or groups were pursuing protection efforts commensurate with the species' dire plight, he let them know his thoughts about the urgency for stronger right whale protection at every opportunity in no uncertain terms.

At public meetings, he unleashed salty invectives against anyone expressing less than open disdain for impotent government efforts to prevent ship traffic and commercial fishing from killing and injuring right whales. His unflinching use of the legal system to target all involved agencies and groups soon had factions on all sides calling him "Mad Max"—a nickname he seemed to relish as a badge of honor. For some scientists and agency officials his verbal attacks raised fears for their physical safety, compelling them to seek restraining orders barring him from buildings where they worked or met. With no travel funds, Strahan rarely ventured out of Massachusetts. But when meetings were held in Boston, as they often were, meeting organizers often felt it necessary to arrange for security guards to escort Max out of the building should he appear and disrupt proceedings.

Notwithstanding his abrasive manner, Strahan's lawsuits, threats of lawsuits, petitions, and Freedom of Information Act requests captured the attention of officials in the Fisheries Service and other agencies to an extent that others' pleas for action did not. The results of his lawsuits included several actions that otherwise likely would not have been taken. Most of his lawsuits were aimed at commercial fishing, but some targeted vessel operations. His suits usually sought idealistic goals, such as the complete shutdown of fishing or shipping operations along the entire eastern seaboard. One of his first lawsuits was filed in June of 1994 against the commandant of the coast guard after two of its patrol vessels struck and killed right whale calves.

The first collision involved the coast guard cutter *Chase* while it was conducting sea trials off Delaware Bay on July 6, 1991; the second was on January 5, 1993, off Jacksonville, Florida, when the cutter *Point Francis* struck and killed one of only eight calves born that winter. Coast guard accounts of the accidents reported that both vessels were traveling at 15 knots when they hit the whales. Strahan's suit, filed in the District Court for the District of Massachusetts (*Strahan v. Linnon*), charged that the collisions constituted illegal takes under the Endangered Species Act. He therefore asked the court to bar the coast guard from operating its own vessels in known right whale habitats and to also prevent it from issuing safety inspection certificates that other vessels needed to operate lawfully in US waters.

In May 1995, the court ruled that the coast guard had indeed violated the Endangered Species Act by neither obtaining an incidental take permit authorizing unintentional takes of endangered whales nor consulting with the Fisheries Service on ways to avoid hitting endangered whales. The court therefore ordered the coast guard to consult with the Fisheries Service under provisions of section 7 of the Endangered Species Act and to seek an incidental take permit. The court, however, did not agree that the coast guard's vessel safety inspection program violated the act. On that charge, it ruled the coast guard was statutorily obligated to issue safety certificates if vessels met safety standards and that it lacked discretion to withhold certificates for whale protection purposes.

In response to the ruling, the coast guard promptly initiated the consultations with the Fisheries Service, which in turn provided its advice in a formal "Biological Opinion" on September 15, 1995.[12] The opinion concluded that coast guard vessels operating along the Atlantic Coast would be unlikely to hit right whales if certain precautions were followed. Specifically, it recommended that the coast guard train and post dedicated whale lookouts on vessels operating in high-use right whale habitats, routinely monitor EWS alerts for right whale sighting locations, operate at the slowest safe speed when close to whales or when visibility was limited, and stay at least 100 yards away from any whale seen, regardless of species. Agencies are not legally bound to adopt recommendations made in biological opinions, but they rarely refuse to do so. Because of the deference given by courts to such advice, agencies would have difficulty in defending an alternative course of action in the event of a lawsuit. The coast guard accepted the recommendations.

Less than a month after the biological opinion was completed, the coast guard cutter *Reliance* struck what was thought to be a humpback whale while on a fishery enforcement mission in the Great South Channel right whale critical habitat. Despite searching for the whale immediately after the collision, it could not be located and its fate was unknown. Nevertheless, the incident demonstrated that the coast guard was unable to avoid whales despite following the Fisheries Service recommendations. The incident triggered another round of section 7 consultations, and a new biological opinion was then issued in July 1996.[13] In addition to measures in its earlier opinion, the new opinion recommended that the coast guard also use "slow safe speed" when operating in right whale critical habitats in nonemergency situations and whenever its vessels were within 5 nautical miles of any EWS right whale sighting less than 12 hours old. It also recommended that coast guard vessels stay at least 500 yards away from any whale unless it was confirmed to be a species other than a right whale, in which case they were to stay at least 100 yards away.

There is nothing inherently unique about coast guard vessels or the way they operate that makes them more likely to hit whales. Rather, their crews are simply more aware of reporting requirements than those on commercial and recreational vessels and more responsible in recording collisions and associated details. In doing so, as will be seen later, their forthcoming reports contributed vital data that helped identify ways to reduce whale collision risks.

A few months after Strahan filed his lawsuit against the coast guard, he petitioned the Fisheries Service for a rule prohibiting vessels from approaching right whales closer than 500 yards. Modeled after a 500-yard approach limit adopted by the State of Massachusetts in 1985—an action also prompted by a lawsuit brought by Strahan—the petition sought to apply the measure to US waters all along the Atlantic Coast. The Fisheries Service had previously banned vessels from approaching humpback whales closer than 100 yards in some areas, but with no approach limit having been established for right whales, the agency decided the even wider buffer sought by Strahan was warranted given the species' highly endangered status. In December 1994, it therefore requested public comments on the petition and in August 1996, the agency proposed a rule requiring vessels to maintain a distance of at least 500 yards from any right whale.[14]

The proposal elicited objections from an unexpected source. Right whale researchers in New England argued that a 500-yard approach limit would hamper conservation efforts. They pointed out that whale-watching vessels would no longer be able to get close enough to detect entangled right whales or get good photographs, which had been a valuable source of pictures for the right whale identification catalogue. The Fisheries Service, however, concluded that the added protection outweighed those considerations, and in February 1997 it adopted a final rule prohibiting vessel operators from knowingly approaching right whales closer than 500 yards or positioning their vessel in the path of a swimming right whale.[15] If operators found themselves closer than 500 yards, they were to steer clear of the 500-yard perimeter at a slow and safe speed.

In practice, the rule was largely unenforceable. Detecting a right whale at 500 yards is challenging for even trained observers. Moreover, proving intent to knowingly approach one closer than that distance is rarely possible, except for whale-watching boats. Nevertheless, coming six years after adopting the recovery plan, this was the first compulsory regulation adopted by the Fisheries Service to reduce the risk of ships striking right whales. Up to that point, the agency had relied entirely on voluntary actions, assuming vigilant mariners would be able to avoid hitting whales.

Evaluating Technological Options as Pressure Mounts

As the Fisheries Service deliberated over its 500-yard vessel approach rule, right whale recovery prospects were dealt another serious blow. In early 1996, five or possibly six dead right whales were found in the calving grounds over a ten-week span. Three were calves, one was an adult male hit by a ship, and another that was photographed floating offshore could not be recovered to identify cause of death. A sixth carcass was reported by a navy pilot but could not be relocated and confirmed. With no more than two dead right whales found on the calving grounds in any previous year, this spate of deaths was an alarming development that captured the attention of Congress and increased pressure on the Fisheries Service and other agencies to be more proactive.

Initially, navy training exercises were suspected as a cause for at least some of those deaths. The US Navy maintains three facilities bordering the calving grounds: the King Bay Submarine Base in southern Georgia, the Mayport Naval Station at the mouth of the St. Johns River, Florida, and the Jacksonville Naval Air Station. Coincident with the whale deaths, the navy had been conducting live-fire gunnery practice from ships 20 nmi east of the critical habitat boundary and aerial bombing practice with 250- to 500-pound bombs farther offshore to the south.

To determine if those exercises were involved, the US Navy and Fisheries Service conducted a joint review in early March to compare their location and timing with carcass sighting locations and necropsy results. The findings revealed no compelling link to any of the right whale deaths. Nevertheless, the commander-in-chief for the Atlantic Fleet issued immediate orders for additional measures to protect right whales. By mid-March, the navy shifted the location of several planned exercises farther from the calving area and suspended all live-fire exercises within 50 miles of the coast. In addition, all navy ships operating in the calving grounds were ordered to post whale lookouts, use "moderate" speed, and travel perpendicular to shore to minimize travel distance and time spent in areas where right whales were most likely to occur.

The developments also led the Marine Mammal Commission, 21 members of Congress, and numerous environmental groups to write the navy and urge that it conduct section 7 consultations with the Fisheries Service similar to those carried out with the coast guard.[16] The navy agreed. Although the consultations were not completed before the start of the 1996 to 1997 calving season, the navy voluntarily instituted additional measures still under discussion in December 1996. Within designated critical habitat off Florida and Georgia, plus an adjoining five-mile buffer area seaward of critical habitat boundaries, the navy limited operations, particularly at night and during periods of poor visibility. It also funded additional whale surveys farther offshore, expanded its role in relaying EWS right whale sightings as noted above, and instructed its captains to use the slowest possible speed consistent with training requirements whenever within 5 nmi of any EWS right whale sighting less than 12 hours old or when a right whale was sighted from their ship.

When the consultations ended, the Fisheries Service provided the navy with its biological opinion in May 1997; the opinion concluded that the exercises and operations would not threaten right whales as long as the measures put in place the previous fall continued.[17] It also recommended that aerial bombing practices be limited to offshore waters warmer than 68°F (20°C), where navy oceanographers had pointed out that right whales were rarely seen. Finally, the Fisheries Service recommended that the navy investigate technological solutions to further examine and mitigate ship-strike risks.

To help identify research options and priorities, the Marine Mammal Commission suggested the navy investigate whether sonar could be used to detect whales from ships or perhaps from an array of sonar buoys lining port access channels.[18] The navy had already started a study to detect vocalizing whales on the calving grounds using hydrophones installed on a coastal pier, but it agreed to consider other research alternatives as well. To do so, the commission recommended a workshop to examine the feasibility of whale detection technologies.

The workshop, held in July 1999, brought together scientists with the Fisheries Service, navy, and commission; ship captains experienced in piloting large vessels; and technicians familiar with available sonar systems. The results were discouraging.[19] Ship captains pointed out that large ships, such as tankers and container ships, traveling at 12 to 15 knots typically needed more than a mile to execute an emergency stop or turn. After adding time to decide on an evasive action, an order for a course change would require detecting whales 2 to 4 miles in front of a ship. Although some commercial sonar systems were powerful enough to detect submarines at distances of a mile or two, detecting whales, particularly those at the surface where wave action disrupts sonar signals, posed a far greater challenge. Moreover, those systems cost several million dollars, required trained experts to operate, and could not reliably distinguish whales from dense schools of fish at ranges beyond a few hundred yards. The participants therefore concluded that available hull-mounted sonar systems were neither technically nor economically feasible for detecting whales.

A similar conclusion was reached for lining port access channels with active sonar or passive acoustic devices. Because of the 20- to 25-mile distance across the calving grounds and the limited sonar detection range, as well as the costs for laying cables from hydrophones to a shore station and for personnel to operate the system around the clock, installing and operating a whale detection system for a single channel were projected to cost tens of millions of dollars. Added to that was concern that the loud sonar sound necessary to detect whales could itself affect right whales and other marine life. Navy experiments to detect vocalizing right whales also suggested that they made calls too infrequently on the calving grounds to be located or tracked with passive listening devices.[20]

The one research option that all participants agreed had merit was testing alarm sounds that might alert whales to approaching vessels. Work on that possibility was undertaken in 2002 by Douglas Nowacek and colleagues at the Woods Hole Oceanographic Institution.[21] They tested responses of feeding right whales to broadcast recordings of ship noise, vocalizing whales, and an acoustic alarm sound using a new tag type called a "D-Tag" (short for Digital Tag). D-tags are attached to whales' backs with suction cups. These tags can be equipped with sensors to record sound levels received by a whale, as well as the animal's changes in pitch, yaw, and depth, which indicate its movements. With those data, scientists can then correlate changes in a whale's motion with broadcast sounds. They found that whereas the whales ignored ship sounds, they responded to the alarm signal by abruptly halting their feeding activity and rapidly rising to the surface where the risk of being hit by the passing ship would be increased. The findings effectively ended most interest in further investigating acoustic alarms.

Taking Stock

By the end of the 1990s, the US Navy and US Coast Guard had adopted constructive mitigation measures that offered a potential model for managing other vessels.[22] However, because commercial and recreational vessels did not involve

federal authorization or funding that required section 7 consultations, those provisions of the Endangered Species Act could not be used to compel operating measures similar to those instituted for navy and coast guard ships. Thus, some groups believed that regulations limiting speed and routing would be necessary to reduce collision risks from nongovernmental vessels.

Speed restrictions, however, were particularly unappealing to mariners, in part because they were considered contrary to age-old traditions that vested sole authority for safe operation of vessels in ship captains. Maritime agencies steeped in those traditions, such as the navy and coast guard, shared that view. Deferring to their judgment, the Fisheries Service continued to rely solely on outreach efforts, still hoping that vessel operators would somehow devise ways to avoid whales once aware of the problem. As right whale ship-strike deaths continued to mount, it was becoming increasingly apparent that this was a misplaced hope (figure 18.3). Therefore, as the 1990s came to a close, the case for moving toward regulations on vessel speed and routing restrictions as the backbone of a new ship-strike reduction strategy began gathering steam—at least in some quarters.

NOTES

1. Philbrick, N. 2000. *In the Heart of the Sea: The Tragedy of the Whaleship* Essex. Viking Press. New York, NY.

2. Chase, O. 1821. Narrative of the most extraordinary and distressing shipwreck of the whaleship *Essex* of Nantucket. *In* TF Heffernan. 1981. *Stove by A Whale: Owen Chase and the* Essex. Wesleyan University Press. Middletown, CT, US, pp. 26–27 (quotation).

3. Parker, H. 1996. *Herman Melville: A Biography, Volume 1, 1819–1851*. Johns Hopkins University Press. Baltimore, MD, US, p. 198.

4. Hakluyt, R. (EJ Payne, ed.). 1907. *Voyages of Hawkins, Frobisher, and Drake: Selected Narratives from the "Principal Navigations" of Hakluyt.* Oxford at the Clarendon Press. London, UK, pp. 141–142 (quotation).

5. Smith, EW. 1978. *Passenger Ships of the World Past and Present.* George H. Dean. Boston, MA, US.

6. Kludas, A. 2000. *Record Breakers of the North Atlantic, Blue Riband Liners, 1838–1952.* Chatham. London, UK.

7. Laist, DW, AK Knowlton, JG Mead, A Collett, and M Podesta. 2001. Collisions between ships and whales. Marine Mammal Science 17(1):35–75.

8. Ibid., pp. 39–40.

9. From JR Twiss Jr., executive director of the Marine Mammal Commission, to WW Fox, director of the Office of Protected Resources, National Marine Fisheries Service, July 15, 1993, on proposed rules to designate critical habitat for northern right whales.

10. Harris, M. 1997. Southeastern U.S. Implementation Team for recovery of the northern right whale to reduce ship strikes. *In* AR Knowlton, SD Kraus, DF Meck, and ML Moonney-Seus. *Shipping / Right Whale Workshop.* Report 97-3. Aquatic Forum Series. New England Aquarium. Boston, MA, US, pp. 86–91.

11. Hain, JHW, SL Ellis, RD Kenney, and CK Slay. 1999. Sightability of right whales in coastal water of the southeastern United States with implications for the aerial monitoring program. *In* GW Garner, JL Laake, BFJ Manly, LL McDonald, and GD Robertson (eds.), *Marine Mammal Survey and Assessment Methods: Proceedings of the Symposium on Surveys, Status, and Trends of Marine Mammals: Seattle, Washington, USA, 25–27 February 1998.* A. A. Balkema. Rotterdam, Netherlands, pp. 191–207.

12. National Marine Fisheries Service. September 15, 1995. (Unpublished record for the US Coast Guard.) National Marine Fisheries Service Endangered Species Act Section 7 Biological Opinion on the Effects of Coast Guard Vessel and Aircraft Activities along the Atlantic Coast on Endangered Marine Species.

13. National Marine Fisheries Service. July 22, 1996. (Unpublished record for the US Coast Guard.) National Marine Fisheries Service Endangered Species Act Section 7 Biological Opinion on the Effects of Coast Guard Vessel and Aircraft Activities along the Atlantic Coast on Endangered Marine Species.

14. Marine Fisheries Service. 1996. North Atlantic right whale protection (proposed rule). 61 Fed. Reg. 41116. (August 7, 1996.)

15. Marine Fisheries Service. 1997. North Atlantic right whale protection (final rule). 62 Fed. Reg. 6729. (February 13, 1994.)

16. From JR Twiss Jr., executive director of the Marine Mammal Commission, to EL Munsell, deputy assistant secretary of the navy (Environment & Safety), March 8, 1996, on navy activities in and around the right whale calving area; and from George Miller, Gerry Studds, and 19 other members of Congress, to WJ Perry, secretary of defense, March 19, 1996, commenting on the deaths of at least five right whales off the coast of Florida in early 1996.

17. National Marine Fisheries Service. May 15, 1997. (Unpublished record for US Navy.) National Marine Fisheries Service Endangered Species Act Section 7 Biological Opinion on the Effects of Navy Vessel Operations and Exercises on Right Whales off the Southeastern United States.

18. From JR Twiss Jr., executive director of the Marine Mammal Commission, to E Munsell, deputy assistant secretary of the navy, November 12, 1997, proposing a meeting on the possible use of active sonar to reduce collisions with right whales.

19. Marine Mammal Commission. 1999. Assessment of the possible use of active acoustics (sonar) to reduce right whale mortalities and injuries from ship strikes. Unpublished summary of an interagency workshop, Crystal City, VA, July 28, 1999.

20. A decade later, however, Chris Clark from Cornell University successfully installed a real-time passive acoustic detection system in the shipping lanes off Boston. A computer algorithm to detect distinctive right whale calls was used to analyze sound recordings that were transmitted via satellite or cell towers back to Cornell, where expert analysts validated the calls. This eliminated the need for costly cables and trained experts to constantly monitor recorded sounds; the system continues to operate today.

21. Nowacek, DP, MP Johnson, and PL Tyack. 2004. North Atlantic right whales (*Eubalaena glacialis*) ignore ships but respond to alerting stimuli. Proceedings of the Royal Society of London, B, Biological Sciences 271:227–231, doi:10.1098/rspb.2003.2570.

22. From JR Twiss Jr., executive director of the Marine Mammal Commission, to EL Munsell, deputy assistant secretary of the navy, July 10, 1997, on navy research and mitigation actions to reduce collisions with right whales.

19

SLOW SPEED AHEAD

IF ONE YEAR stands out as marking a shift in prospects for reducing ship collisions with right whales, it might be 1997. Margaret Mead once offered a bit of wisdom: "Never doubt that a small group of thoughtful, committed citizens can change the world. Indeed, it is the only thing that ever has."[1] In late 1997 such a group began laying the foundation for regulatory measures to reduce ship strikes. Three of those people were Lindy Johnson, Bruce Russell, and Amy Knowlton. They all shared an abiding passion for the ocean, yet each had followed their obsession down diverging career paths.

Lindy Johnson was an eager, outgoing attorney with a special interest in the important, yet obscure, field of commercial shipping. Hired just a few years earlier by NOAA's Office of the General Council for International Law, she was rapidly becoming the agency's resident authority on shipping regulations and had become a regular participant on US delegations to the International Maritime Organization (IMO). Bruce Russell, a commander in the coast guard, had recently retired after 20 years of service, with a stint as chief of the coast guard's Marine Safety and Environmental Protection Program in Washington, DC. Looking for a new challenge, he contracted with the International Fund for Animal Welfare (IFAW) to lead its campaign for more effective measures to reduce right whale ship strikes. He had an in-depth understanding of how ships operated and how the coast guard managed them. Amy Knowlton, an experienced right whale biologist with the New England Aquarium, brought a unique combination of skills. In addition to 15 years of field studies on right whales, she had firsthand knowledge of vessel operations since earning a license to pilot a 100-ton vessel.

Their paths first crossed in April 1997 at a meeting the Right Whale Consortium organized to examine all aspects of the ship-strike problem. Knowlton, one of the meeting organizers, helped cobble together funds from the National Fish and Wildlife Foundation, Exxon Corporation, the Marine Mammal Commission, and the Fisheries Service for a two-day "Shipping / Right Whale Workshop."[2] The meeting brought together 75 experts from industry, academia, the conservation community, and government sectors to brainstorm mitigation options ranging from mariner education to vessel speed and routing.

In the United States these matters typically fall to the coast guard, which in turn follows international "rules of the road" developed by the International Maritime Organization (IMO), a United Nations subsidiary based in London. Established in 1958 to bring a measure of consistency to the standards for ships and shipping across international boundaries,[3] the IMO is an assembly of all nations interested in shipping, which is to say virtually all the world's nations. With no regulatory powers of its own, it instead serves as a forum for negotiating management principles and reviewing measures that member states may adopt under their respective national authorities. Initially, its focus was on navigation safety, with most measures hammered out by its Maritime Safety Committee and Subcommittee on Safety of Navigation. However, as spills of oil and other hazardous cargo became international concerns, the IMO's scope expanded to include environmental protection. To consider those issues, the Marine Environment Protection Committee was formed in 1973.

Over the IMO's first 50 years, some 20 international conventions and many more nonbinding codes and guidelines were approved.[4] Even as late as the 1990s, however, it was rare for IMO measures to focus strictly on environmental protection and no measures had been adopted solely to protect whales. Nevertheless, recognizing the organization's evolving role in environmental protection, participants at the April 1997 workshop flagged IMO measures that might be useful for protecting right whales. Two measures in particular stood out: (1) routing measures (e.g., provisions for various types of vessel traffic lanes and "Areas to Be Avoided") that might steer ships away from important whale habitats; and (2) directives on how ships should operate (e.g., crew training and watch keeping, and mandatory ship reporting systems to alert vessel operators to unusual or hazardous local conditions). It is around those measures that efforts to reduce right whale collision risks would begin to gel.

Shortly after the 1997 workshop, steps were taken to breathe life into its recommendations. Over the next few years, groups other than the Fisheries Service provided most of the money for right whale recovery work. The IFAW alone invested more than $1 million. In addition to contracting with Russell to design a management strategy, it dedicated its own research vessel, *Song of the Whale*, for right whale work; it also provided funding for research itself, including research by Chris Clark at Cornell University to investigate sonar and passive acoustic technology for detecting whales. Anxious to counter complaints by fishermen that all of the regulatory burden for protecting right whales was falling on their shoulders, staff of the Fisheries Service's Northeast Regional Office, particularly Christopher Mantzaris and Patricia Gerrior, directed much of the agency's limited funding to supplement Russell's work.

To open communications with the IMO, the Marine Mammal Commission drafted an information paper for the organization's Marine Environment Protection Committee (MEPC) that described the threat ships posed to right whales and US interests in reducing those threats. The commission planned on asking the coast guard, which represents the United States at IMO meetings, to submit the paper, but after sharing a draft with the Fisheries Service, Lindy Johnson advised that NOAA had similar plans. The two agencies, therefore, jointly revised and sent the paper to the coast guard, which submitted it to the MEPC for its next meeting in July 1997.[5] The paper asked IMO members to advise all vessels calling at US East Coast ports of the new early warning system (EWS) broadcasts reporting recent whale sightings and to be alert for right whales.

Mandatory Ship Reporting Areas

When the IMO paper was submitted, US officials were still unsure what further actions might be taken, but shortly thereafter, Lindy Johnson prevailed on NOAA's leadership to pursue designation of two mandatory ship reporting areas—one for right whale calving grounds off Florida and Georgia, and the other for feeding areas in Cape Cod Bay and the Great South Channel. Authority for establishing ship reporting areas had been approved by the IMO to advise vessel operators of unusual local conditions or operating constraints. It requires nongovernmental ships to call a shore station for cautionary advice to ensure navigation safety. Johnson, however, recognized that it also could be used to alert ships of collision risks with right whales, EWS broadcasts of whale locations, and tips on avoiding whales. Vessel operators would still be left to decide how to use that information, but the call-in requirement provided as much certainty as possible that vessel crews were aware of whale collision risks when it counted most; that is, when they were entering critical right whale habitats. Despite its innocuous nature, the measure put NOAA's resolve for protecting right whales to an unexpected test.

With input from Bruce Russell, the Marine Mammal Commission, implementation teams, and right whale scientists, Johnson began crafting the proposal's details. Boundaries for the southeast reporting area that covered the calving grounds were drawn to include the region's designated critical habitat plus adjacent waters where high numbers of right whales also had been documented. Similarly, the northeast area encompassed critical habitats in Cape Cod Bay and the Great South Channel, plus adjacent waters between the two areas. Because large vessels had been implicated in most right whale deaths caused by collisions, the proposal applied only to ships greater than 300 gross tons. These constituted a vessel class recognized by the IMO that was recently required to carry a new satellite communications system (NAVTEX) that could be used to transmit reporting messages. Upon entering either area, ships would be required to contact a shore station via NAVTEX and provide its name, call number, destination, expected travel route, and speed. In return, a short text message would be sent urging crews to be alert for right whales and to check other sources for recent sightings and tips on avoidance.

As a novel application of the ship reporting measure, there was a need to provide the IMO with an opportunity to review and comment on the proposal. Johnson, with the help of Russell and the Marine Mammal Commission, drafted a paper describing the proposal for a July 1998 meeting of the IMO's Subcommittee on Safety of Navigation. To be placed on its docket, the paper had to be submitted by early May and was therefore first circulated for federal agency review early in 1998 as part of the normal interagency clearance process for any US proposals to international organizations.

Requiring nothing more of vessel operators than sending a short message to a shore station, it hardly seemed an onerous measure. It nevertheless sparked intense controversy. Resistance came not from shipping companies or vessel operators, but from the coast guard and navy. Although their ships were exempt from requirements because they were government vessels, both agencies were concerned about its precedent. The coast guard argued that mandatory reporting was intended solely for navigation safety and that using it to protect a single species of whale was inappropriate. It also raised concerns about enforcement obligations, which would fall to its officer, because of limited resources for boarding ships that failed to report.

The US Navy's Judge Advocate General's Office raised similar concerns, claiming that the action would open a door to the proliferation of mandatory reporting areas around the world. Notwithstanding the navy's exemption from the requirements and prior concurrence with the measure when the IMO initially approved the tool, it now asserted that the location of their own ships could somehow be revealed by the measure and national security would be threatened. As an alternative, the navy proposed asking the IMO to issue a safety of navigation circular (a nonbinding advisory message) that would advise national shipping industry representatives of the need for caution in the two proposed reporting areas and to monitor communication channels for right whale sightings when those areas were being crossed.[6]

At the urging of Johnson, NOAA's leadership maintained that its proposed reporting requirements were necessary to ensure timely advice to mariners as they entered the right whale habitats and were consistent with the IMO's new environmental protection responsibilities. She also countered that concern about revealing the location of US Navy ships was unfounded, given their exemption from reporting. Pointing out that the circumstances requiring right whale protection were highly unusual, she also argued that a proliferation of similar areas was unlikely. The Marine Mammal Commission weighed in, maintaining that short of restricting vessel speeds and routes, mandatory ship reporting was the only meaningful approach and should not be delayed.[7]

The coast guard and navy, however, were unmoved and remained committed to their positions. As a result, during the first three months of 1998, the stalemate precipitated a flurry of communications between officials at the highest levels in NOAA, the Commerce Department, the navy, the coast guard, the State Department, and offices in the White House, including the Council on Environmental Quality and the Marine Mammal Commission. The State Department tried to broker a compromise that merged elements of both sides. It proposed that the IMO issue a safety of navigation circular that asked vessel operators to respect existing requirements for calling ports 24 hours before arrival in order to distribute information on right whale protection and that the coast guard monitor call-in records for compliance. If compliance was low after five years, the State Department proposed that mandatory ship reporting areas be pursued at that time.

As the State Department's proposal was being considered, coast guard opposition began to soften. In late March 1998, 12 members of Congress,[8] as well as the Chamber of Shipping of America,[9] expressed support for NOAA's proposal. In addition, Russell and Johnson convinced coast guard officials that enforcement could be done through routine dockside inspections already conducted by coast guard marine safety officers when vessels arrive in port. By requiring that shore station messages be saved, coast guard officers could simply check those dated messages to verify compliance. After considering those points, the coast guard wrote to the State Department on March 31, advising that it could accept NOAA's proposal as long as it allowed for dockside enforcement.[10]

The navy, on the other hand, stuck fast to its position. NOAA did so as well, arguing that the possibility of a five-year delay in effective compliance under the State Department proposal was unacceptable, given the right whale's plight. The navy therefore elevated the matter to the National Security Council in the White House. In early April, National Security Advisor Sandy Berger summoned involved parties to a meeting to review their positions. With the navy and NOAA at loggerheads and time running out to submit a proposal to the IMO, options were put before President Bill Clinton to make a final decision on April 21— Earth Day 1998.

President Clinton sided with NOAA, and on April 23, he instructed the coast guard to submit NOAA's proposal to the IMO: "We must address this threat so these magnificent creatures can flourish once again."[11] With that, NOAA's proposal for two mandatory ship reporting areas to protect right whales was submitted to the IMO Safety of Navigation subcommittee the following day with changes to reflect the coast guard's preferred enforcement strategy.[12] The IMO subcommittee approved the proposal with no opposition and passed it on to the full Maritime Safety Committee for final review. At that committee's seventieth session, in December 1998, the proposal received final approval, again with no concerns or objections.

Over the next six months the US Coast Guard and the Fisheries Service set up an automated communications system for both the northeast and southeast ship reporting systems (figure 19.1). The combined startup costs for both systems, roughly $300,000, were split between the two agencies and IFAW, with

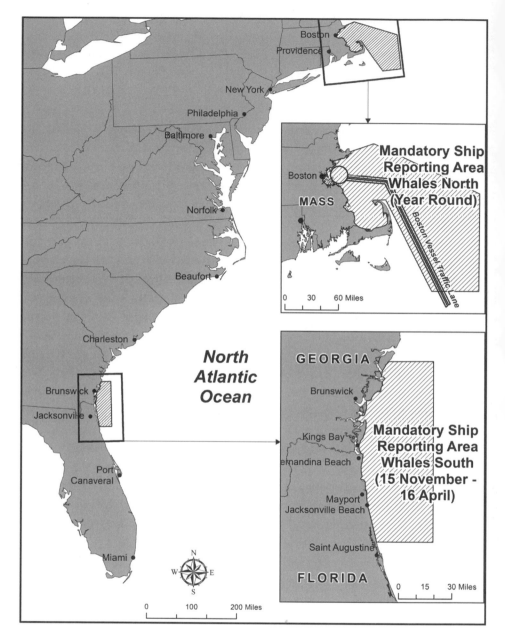

Figure 19.1. Boundaries of Mandatory Ship Reporting zones established in 1999 off the northeast and southeast US coasts. Within these areas ships are required to contact shore stations for information on North Atlantic right whale protection. (Figure by Linda White.)

each contributing a third of the cost. In December 1998, Congress amended section 11 of the Ports and Waterways Act to clarify coast guard enforcement authority, and on June 1, 1999, the coast guard published interim final rules for implementation.[13] A month later the requirement for year-round reporting in the northeast area went into effect. That time frame recognized the occurrence of at least some right whales in that area in all months. Requirements for the southeast area, however, were operational only from mid-November to mid-April during the calving season and went into effect in November 1999. Permanent rules were adopted in November 2001 that incorporated minor changes in the information vessels had to report.[14] With costs for text messages amounting to a few thousand dollars annually and covered by the coast guard and NOAA's Fisheries Service, vessel owners incurred no financial obligations.

In the first two years, compliance with the call-in requirements ranged from 35 to 50 percent for the southeast area, and 50 to 75 percent for the northeast area. Enforcement ramped up slowly. The coast guard began issuing warnings to ships that failed to report in 2001, and in 2002 it began issuing citations. Citations carried potential fines of up to $27,500, but most ranged from $200 to $500. Once fines were imposed, compliance rates rose to an average of 70 to 85 percent.

Looking Beyond Mandatory Ship Reporting

Even as the ship reporting systems were being put in place, some groups doubted that vessel operators would be able to avoid whales, given the limited maneuverability of large ships and the difficulty in sighting the animals. Thus, they began examining other alternatives, particularly speed and routing options. Leading that effort was the Northeast Implementation Team, created by the Fisheries Service in 1994. In 1998 the team formed a ship-strike committee and asked Bruce Russell and Amy Knowlton to serve as co-chairs. The committee was charged with crafting measures to reduce ship collisions with right whales that the implementation team could then recommend to the Fisheries Service.

As the committee began its work, IFAW and the Marine Mammal Commission offered two different approaches, each without knowing the other was recommending alternatives. IFAW, following Russell's counsel, wrote to the coast guard in December 1997 recommending that US regulations for the International Safety Management (ISM) code be modified to include actions that mariners should take to avoid whales.[15] The ISM code defines what constitutes "prudent seamanship." IFAW recommended that it be expanded to include measures, such as posting trained whale lookouts and reducing speed in right whale habitats, or avoiding those areas entirely, and factoring those actions into voyage planning. At the same time, the Marine Mammal Commission wrote to the Fisheries Service recommending development of cooperative agreements with shipping companies using ports adjacent to right whale critical habitats.[16] It suggested seeking corporate directives similar to those issued by the coast guard and navy for using slow speeds and following routes to minimize transit time through right whale habitats. To compensate for lost time, it also advised asking that shipping companies direct their vessels to adjust sailing schedules and use slightly higher speeds on long open-ocean legs between ports.

Though the Fisheries Service agreed the commission's approach seemed useful, and despite an offer by the commission to help fund the initiative, the agency deferred funding for two years, citing budget constraints. In 1999, however, the commission and the Fisheries Service both provided funding to IFAW to develop a partnership program with shipping companies operating along the US East Coast. Led by Bruce Russell, the project was carried out under the aegis of the northeast implementation team's Ship-Strike Committee.

When the Fisheries Service was unable to continue its funding in 2000, IFAW committed over $250,000 to carry on the work. Russell recognized that nonbinding cooperative agreements with shipping companies offered a potential path forward, but he was also keenly aware that such agreements offered limited prospects for success. Experience taught him that shipping companies rarely change operating patterns unless all companies were required to do so, thereby ensuring a level playing field. Otherwise, industry competition could favor companies choosing not to participate, which would place cooperating companies at a disadvantage. Thus, after submitting his report to the Fisheries Service and the commission, Russell recommended changing the approach to require mandatory measures.

As the project's direction shifted to consider mandatory options, research was beginning to offer new insights into factors causing ship strikes. Independent computer simulations by Amy Knowlton and Heather Clyne revealed that hydrodynamic forces created by ships could increase collision risks.[17] They found that objects the size and density of a right whale lying in a ship's path would first be pushed away by the vessel's bow wave, but would then be drawn toward the hull and propellers at the ship's middle and aft portions. The models, however, assumed whales would not react to passing ships. Subsequent studies with physical models of whales and ships in a wave tank revealed similar results, but again assumed whales would not attempt to avoid passing ships.[18] Yet the fact that whales are often caught on the bows of ships clearly indicates whales must have been moving to overcome the forces of bow waves pushing objects away. Another study compared right whale sighting locations to ship routes reported through the new mandatory ship reporting systems. The authors concluded it was impossible for ships using ports adjacent to critical habitats to avoid important right whale habitat, but that some routes seemed to pose at least slightly lower risks.[19]

The most important study, however, was sparked by debate in implementation teams over advice on speed reductions. The Marine Mammal Commission had been urging the use of specific speed limits that were slow enough to allow whales more time to detect and avoid ships. It was concerned that advice calling for use of "reduced" or "slow" speeds was too vague for different vessel types whose normal operating speeds could range from 10 to more than 25 knots. The commission therefore urged that advice recommend speeds below 15 knots. It reasoned that reports from the two coast guard vessels that struck and killed right whales while traveling at speeds of 15 knots demonstrated that 15 knots was not safe when transiting whale habitats.

The recommendation was flatly dismissed by the implementation teams. Scientists on the teams pointed out that information was insufficient to identify what speed was safe, while coast guard and navy representatives objected on grounds that reference to any specific speed was contrary to customary maritime practices that couched advice on speed in vague terms such as "safe speed" or "slow safe speed." They maintained that such terms were necessary to afford vessel captains flexibility to weigh a multitude of constantly changing conditions when deciding what speed was safe. Port officials were particularly vocal in their opposition. They asserted that shipping companies would shift business to other ports where such advice did not apply.

The debate underscored the need for more data on what speeds were safe for whales. Deliberately driving ships at whales to determine collision risks at different speeds was obviously impossible given the risks of injuring or killing whales. However, collision reports by coast guard and navy vessel operators suggested that other accounts might also exist to inform the question of what constitutes a safe speed. The Marine Mammal Commission therefore organized a study led by the author of this book, Amy Knowlton, and Jim Mead to compile records of collisions between motorized vessels and whales. Because of the small number of reported right whale collisions, the scope of the study sought records involving any great whale worldwide.

At that time ship strikes were still considered freak accidents of interest only as odd tales of the sea. The study, however, proved otherwise. Earlier investigators were aware of such collisions, and in 1974, William C. Cummings had requested accounts of encounters between marine animals and vessels through the popular magazines *Yachting* and *Sea Secrets*. Cummings never got around to summarizing the responses he received, but like any good scientist, he saved them. To help with the investigation, he generously provided these to the study team.

Those accounts, along with others collected by the investigators and a search of stranding records, revealed that large whales, particularly fin, right, humpback, and sperm whales, were hit far more frequently than previously recognized.[20] Along the East Coast of the United States—one of the few areas where stranding records had been computerized at the time of the study—a third of all dead fin whales (31 of 92) and right whales (10 of 30), two-thirds of all sei whales (2 of 3), and 5 percent of all minke whales (5 of 105) found stranded between 1975 and 1995 had large propeller slashes or injuries implicating strikes by large ships as their cause of death. Near the Chesapeake Bay, a quarter of all stranded humpback whales (9 of 36) showed such evidence. A fifth of all fin whales found in France between 1972 and 1998 (16 of 72) and in Italy between 1986 and 1997 (8 of 39) also had similar evidence of ship strikes.

Many of the ship-struck rorqual whales, including fin, sei, blue, and minke whales, were discovered draped across the bulbous bows of large ships entering ports (figure 19.2). On the one hand, rorquals carried into port this way were more likely to be discovered, suggesting that the stranding records attributed to ship strikes might be inflated. On the other hand, it seemed that the number of whales struck and pinned to the bow of a ship must be a small fraction of all collisions and that many others must be killed and left adrift, never to be recorded.

The study also uncovered 58 firsthand accounts of motorized vessels hitting whales between 1885 and 2000. They involved all sizes and types of ships, from small outboards to mammoth aircraft carriers. The vast majority of cases causing lethal or serious whale injuries, however, involved ships longer than 260 feet. More than 90 percent of the accounts also stated that whales were not seen before they were hit or were seen only at the last moment when it was too late to take evasive action. Indeed, a quarter of the accounts went unnoticed until whales were found on ship bows when entering port. In one case, the passenger ship *Royal Majesty* left Boston on August 1, 1995, and carried a fin whale 1,100 nmi on its bulbous bow before discovering it on arrival at Bermuda. From engine log notations recording a sudden unexplained vibration and slight speed reduction, it was determined that the whale must have been hit off Cape Cod, just 30 miles into the trip. Given that most whales were either unseen or seen only seconds before contact, the findings raised grave doubts that even attentive pilots could detect and avoid whales.

Finally, the study found 28 accounts with information on both the speed of vessels at the time of collisions and the fate of the whales. Although collision speeds ranged from 6 to 51 knots, all of those causing lethal or serious injuries exceeded 10 knots (11.5 miles per hour), and 90 percent were at speeds above 13 knots (15 mph). (Serious injuries were defined as collisions with observations of blood in the water immediately after a whale was hit, but with no signs of a dead whale.) Some accounts hinted that whales made dives at the last second when ships were within a few hundred yards. Researchers therefore suggested that at least some whales were able to detect and avoid ships, that their ability to do so decreased sharply as ship speeds increased from 10 to 13 knots, and that they had little chance of avoiding ships moving 14 knots or faster. They concluded that vessel speeds of 13 knots or less posed lower risks of hitting and killing whales than faster speeds.

Figure 19.2. A Bryde's whale was found draped across the bulbous bow of the container ship P&O *Nedlloyd Pantanal* as it entered the port of Guayaquil, Ecuador, on December 10, 2004. (Photo courtesy of Fernando Félix.)

Results of those studies were tracked closely by the ship-strike committee. Recognizing that any speed and routing measures would be controversial, Russell held 28 workshops—all funded by IFAW—over a two-year period in US ports from New England to Florida in order to solicit mariner views on possible management strategies. The results bolstered his belief that mandatory measures would be needed. Even seasoned vessel captains counseled that speed and routing measures should be mandatory. They pointed out that mandatory measures were needed to hold all vessels to the same standard and allow operators to estimate exactly how much time they needed to budget for using slower speeds or different routes during voyage planning. Port officials, however, expressed a decidedly different view. They continued opposing regulatory measures, claiming that their ports would be placed at a competitive disadvantage and that ships would move to other ports to avoid such requirements.

After the exhaustive series of regional meetings and extensive debate over measures, the ship-strike committee was unable to resolve differing opinions about measures to recommend. It therefore agreed that Russell and Knowlton should take the differing views into consideration and prepare a final report on behalf of the committee. In August 2001, the report was completed and submitted simultaneously to both the northeast and southeast implementation teams and to the Fisheries Service, which had been pressing for the report's completion.[21] It included an analysis by Knowlton of right whale movements and ship-strike deaths that indicated collision risks existed off all major ports from Boston to northeastern Florida.

The committee's review had led some members to urge a year-round 13-knot speed limit within 30 nmi of shore along the entire East Coast. Knowlton, however, concluded that was excessive. The report therefore instead recommended speed limits for a series of seasonal management zones within 20 to 25 nmi of major ports from Georgia to New England during the right whale migratory season. It also recommended seasonal management areas around right whale feeding and calving grounds designated as critical habitat. The network of management areas afforded protection where ship traffic was heaviest—around port entrances—and assured that no ports would be placed at a competitive disadvantage. The report also recommended that speed limits apply to all vessels 65 feet or longer, a length that the coast guard used as a criterion for some vessel's safety measures and that included the smallest vessel (82 feet long) then known to have killed a right whale. Other recommendations called for speed limits along a segment of the vessel traffic lanes leading into Boston and designating waters east of those lanes in the Great South Channel as an "Area to Be Avoided," an IMO-sanctioned routing measure.

For all other waters off the East Coast, the report recommended a "dynamic" management system to protect temporary right whale aggregations found at unpredictable times and places. This measure was fashioned after a "dynamic area management" system previously adopted by the Fisheries Service to reduce right whale entanglements (chapter 21). Under that scheme, 15-day management zones were established around temporary right whale feeding groups spotted by aerial survey teams. In those areas, which were already being announced over voice radio and other media for fishery management purposes, ships would be required to either slow down or change their route during the effective dates.

Although some ship-strike committee members opposed any speed limits and others supported 12- or 13-knot speeds, as determined by results of the Marine Mammal Commission's study, Russell found analyses of ship hydrodynamics by Heather Cline (also funded by IFAW) more compelling. That study concluded that forces pulling whales toward ships declined significantly at speeds below 10.5 knots,[22] which happened to match the point below which lethal whale collisions seemed to disappear from collision records in the commission's study. After consulting colleagues in the shipping industry and marine engineers to ensure that large vessels could maneuver safely at speeds of 10 knots in most situations, he concluded that speed limits for both seasonal and dynamic management areas should be set at 10 knots.

Before including that as a recommendation in the report, however, Russell, again with IFAW funding, contracted Hauke Kite-Powell, a scientist at the Woods Hole Oceanographic Institution, to evaluate economic costs from increased travel time assuming a 10-knot speed limit for the recommended management areas. For tugs and tows typically traveling at 10 to 12 knots, time delays were negligible. For tankers and bulk cargo carriers that normally operate at maximum speeds of 14 to 16 knots (16–18.5 mph) and make up most of the vessels calling at East Coast ports, the speed limit added 30 minutes to an hour to cross seasonal management areas. Container ships and passenger vessels typically steaming at 20 to 25 knots were the vessels most affected; about two hours could be added to their travel time. However, recognizing that added time could be factored into voyage planning and that speed restrictions applied for only part of the year, it was estimated that the speed limits would have minimal economic impacts on vessel operations.

Finally, the committee report recommended additional studies to flesh out certain proposal details. In addition to further research on the most appropriate effective dates for seasonal management areas, it recommended (1) economic studies, (2) a series of "port access route studies" routinely required by the coast guard to assess navigational safety issues when changing major ship routes, (3) an evaluation of legal authorities for implementing speed restrictions, and (4) a risk assessment.

After reviewing the committee report, the implementation teams wrote to the Fisheries Service. In October 2001, the southeast team commended the subcommittee for its extensive efforts to meet with officials in ports and shipping companies, and confirmed that they unanimously supported consideration of the proposed measures subject to further review of economic impacts.[23] In January 2002, the northeast team wrote that its members had differing views but that most supported the proposal and agreed further economic analyses were needed.[24]

With submission of the report and implementation team comments, the ball was back in the Fisheries Service's court for action.

As the committee report was being completed, prospects for right whale recovery were turning from bad to worse. Eight right whale carcasses, a record high, were discovered in 2001, including at least two hit by ships and two entangled in fishing gear.[25] A new modeling study also concluded that right whale numbers declined during the 1990s and that preventing the death of a single reproductive female per year could tip the balance between the species' recovery and decline.[26] Right whale births also plunged to an all-time low. Whereas an average of 11 births per year had been recorded from the 1980s through the mid-1990s, they fell to 6 in 1998, 4 in 1999, and just 1 in 2000. Equally alarming were observations of thinning blubber layers and unexplained skin lesions on half of all adults; these indicated a declining trend in the health of most whales.[27]

Crafting a Ship-Strike Reduction Strategy

When the Fisheries Service received the committee report, it was faced with a dilemma. Given disturbing information about the right whale population's trend, it could cite the species' desperate condition and expedite steps to adopt or modify the recommended measures; or it could follow a slow, methodical course that ensured all affected parties had every possible opportunity to review and comment on possible restrictions. Given the sharp, unexpected resistance to innocuous ship reporting measures, the agency girded for controversy and opted for a slow, deliberate route. Recognizing approval of at least some measures would require IMO review, the responsibility for crafting ship-strike mitigation measures was retained in the Fisheries Service headquarters, while regional offices were assigned management of fishery impacts.

Lindy Johnson, with help from Greg Silber in the Fisheries Service's Office of Protected Resources, was charged with developing the agency's approach. To ensure that no data, concern, or option was left unconsidered, the agency repeated all the studies and evaluations supported by IFAW and others up to that time and conducted all the additional analyses recommended in the committee report and by the implementation teams. In early 2002, staff from the US Coast Guard and Florida Fish and Wildlife Commission jointly updated analyses of ship routes through critical habitats using the first year of data from the mandatory ship reporting system operations.[28] The results revealed that ships entered reporting areas from all directions at speeds ranging from 5 to 25 knots, with half traveling 14 knots or faster. A later assessment in 2005 that analyzed three years of data found similar results.[29]

In mid-2002, a port-by-port economic analysis estimated financial impacts that might result from speed limits ranging between 6 and 14 knots through the management areas recommended in the committee report, as well as resulting hourly operating costs for different vessel types.[30] Both passenger and container ships were again projected to incur the greatest costs; tankers and bulk carriers were found to sustain little added cost, while tug and barge traffic showed virtually no additional cost. Based on a 10-knot speed limit, cumulative costs incurred by the 33,000 vessel calls per year at affected ports were estimated at $10 million, half of which was borne by the port of New York–New Jersey, due to its high volume of traffic. Average costs for ships normally traveling faster than 10 knots ranged from about $1,700 to $3,500 per ship call.

Port officials criticized the study for omitting costs associated with shipping companies that would drop affected ports from their schedule to avoid restrictions. Others, however, pointed out that the analysis overestimated costs because most ships already slowed to 10 knots several miles offshore to pick up port pilots. The Fisheries Service therefore contracted for another economic study that was completed in 2005. Its results were largely the same. For an industry valued at nearly $300 billion across all the affected ports, it concluded that projected economic impacts of even several million dollars would be minimal.[31] For oceangoing vessels moving cargo worth millions of dollars per port call, it concluded that an added cost of a few thousand dollars at most per call was unlikely to cause ships to switch ports.

The Fisheries Service also repeated the Marine Mammal Commission's study of collision records. The new study, completed in 2003, combed through additional stranding reports, government agency records, and data from other countries for any accounts of collisions with large whales.[32] It found nearly 300 vessel-related stranding deaths, doubling the number in the commission's study. It also added 11 more records with information on ship speeds at the time whales were hit, which again indicated that the vast majority of lethal collisions involved ships traveling 13 knots or faster.

With shipping interests continuing to urge the use of sonar systems, alarms, or other technology to detect and avoid whales, the Fisheries Service also repeated the analysis of options conducted by the US Navy and Marine Mammal Commission in 1999, but with broader participation from shipping interests. The workshop, eventually held in July 2008, came to the same conclusion—there were no readily available technologies for detecting or warding off whales to reduce collision risks.[33]

As its new studies were being undertaken, the Fisheries Service formulated a proposed management approach. In October 2003, two years after receiving the ship-strike committee report, it met with other federal agencies to unveil its preliminary "right whale ship-strike reduction strategy." This plan was virtually identical to that recommended in the 2001 committee report, but with new details on management area boundaries and effective dates. Instead of proposing a specific speed limit, however, other agencies were asked for advice on appropriate speeds. For whatever standard that might be chosen, however, the agency announced its intent to adopt speed limits under the domestic authority that every nation retains to set conditions for allowing vessel entry into their ports. In this case, the Fisheries Service concluded that speed limits safe for right whales constituted an appropriate condition for port access. Other elements of its proposal included measures already in place (mandatory ship reporting areas, distributing education materials, conducting section 7 consultations, and communicating with Canadian officials on right whale conservation plans in both countries).

The response by federal agencies was mixed. The Marine Mammal Commission expressed strong support for the approach and recommended a 12-knot speed limit in seasons of peak right whale occurrence and a more restrictive 10-knot limit when mother-calf pairs might be at risk.[34] The US Navy and US Coast Guard, on the other hand, questioned whether *any* specific speed limit could be imposed without IMO approval and recommended further economic analyses. After considering agency comments, the Fisheries Service decided to follow a rule development process that would afford the public and shipping interests every possible opportunity to comment on considered mitigation strategies.[35] The first step was taken in June 2004, when it published an "advanced notice of proposed rulemaking" and schedule of public meetings in major ports all along the East Coast. The meetings sought comments on the same set of measures presented to federal agencies nine months earlier and effectively repeated those convened five years earlier by Russell. An August deadline was set for suggestions on measures to be considered in the environmental impact statement,[36] but that was extended to mid-November to allow time for public meetings in other ports.[37]

As the Fisheries Service inched ahead, right whale deaths continued to mount. Between August 2001, when the ship-strike committee report was completed, and February 2005, 15 more right whales were found dead. Nearly half were attributed to ship collisions, including 4 within a span of 13 months between January 2004 and February 2005 (figure 19.3). To make matters worse, three of the four ship strikes involved pregnant females with full-term fetuses, and a fourth was a

Figure 19.3. A 53-foot right whale named Stumpy was towed ashore for necropsy on February 7, 2004, after being struck by a ship off Virginia Beach, Virginia. (Image courtesy of Regina Campbell-Malone, Woods Hole Oceanographic Institution, Woods Hole, MA.)

reproductively mature female. Together, they represented at least 5 percent of the population's reproductively mature females.

The spate of deaths elicited another flood of letters to the Fisheries Service urging immediate action. In January 2005, the Marine Mammal Commission recommended the Fisheries Service adopt emergency rules restricting ship speeds within 25 miles of major ports from Jacksonville, Florida, to Portland, Maine.[38] Similar letters were sent by environmental groups, and in March, six US senators wrote to President George W. Bush urging him to instruct the Fisheries Service to accelerate efforts to reduce both ship collisions and entanglements. In May, 11 environmental groups jointly petitioned the secretary of commerce, NOAA, and the Fisheries Service for emergency rules restricting vessel speed. In July, eleven right whale scientists published a statement on the species' dire plight in the nation's most prestigious scientific publication, *Science*, imploring the Fisheries Service to adopt emergency rules to prevent ship collisions.[39]

The Fisheries Service tried to respond without altering its ongoing rule-making schedule. In early 2005 it proposed an interagency "summit" to consider additional actions federal agencies might take. That proposal was soon abandoned when the navy and coast guard objected on the grounds that they were already following Fisheries Service recommendations included in section 7 consultations that the agency had earlier said would be adequate to protect right whales. Therefore, in May, the agency asked the coast guard to modify right whale advisories and mandatory ship reporting messages to request that vessel operators voluntarily use speeds of 12 knots or less in high-use right whale habitats.[40] The coast guard, however, refused, and said that the international convention for advice on vessel speeds used terms such as "safe speed," rather than specific numbers. It also objected on the grounds that such messages would be taken as a coast guard endorsement of speed limits to protect whales, something it had not given.[41]

By June 2005 it was apparent that the Fisheries Service was not going to adopt an emergency rule to restrict vessel speeds. On June 22 it announced plans to prepare an environmental impact statement with five alternatives and asked for public comments.[42] These alternatives included (1) a no action alternative, (2) designating dynamic management areas to protect whale aggregations, (3) establishing seasonal speed restrictions in designated management areas, (4) designating mandatory vessel routes, and (5) some combination of those options. In July, the agency responded to the Marine Mammal Commission's recommendation for emergency rules, and confirmed its intent to reject the approach.[43] It stated that emergency rules required environmental and economic analyses comparable to those for nonemergency rules and that it could not prepare those documents any faster. Thus, notwithstanding the label "emergency rules," the agency maintained it could not adopt them any faster than the rules already being developed under the normal rule-making process.

While the petition for emergency rules presented by environmental groups worked its way through Department of Commerce channels, the Fisheries Service and coast guard continued moving ahead with their respective regulatory and management proposals. In February 2005, the coast guard announced plans for a series of "port access route studies" that would identify new vessel routes to protect right whales off Atlantic coast ports.[44] Two months later, another economic impact assessment was completed by the Fisheries Service; this one found that impacts of proposed actions would amount to just 0.005 percent of the value of trade goods passing through affected ports.[45]

In September, the Fisheries Service formally announced a finding that the petition for emergency rules by environmental groups was "unwarranted" because of the agency's ongoing work on a permanent rule.[46] The environmental groups who had filed the petition were livid. And the coast guard's refusal to modify whale advisories to simply request voluntary use of 12-knot speeds in right whale critical habitats was salt in the wound. Defenders of Wildlife and three other petitioners quickly responded to the Fisheries Service's announcement. They notified the coast guard and Fisheries Service on November 3 of their intent to sue the agencies for failure to use their authority under the Endangered Species Act to protect right whales.[47] Less than a week later, three of those groups filed a lawsuit in the US District Court for the District of Columbia. They charged that the administrator of the Fisheries Service, the secretary of the Department of Commerce, and the commandant of the coast guard were in violation of the Endangered Species Act (*Defenders of Wildlife et al. v. Carlos Gutierrez, Secretary of Commerce, et al.*). In part, the groups asserted that the Fisheries Service's refusal to issue emergency rules was an arbitrary and capricious decision.

Even within the Fisheries Service, staff close to the issue unofficially welcomed the lawsuit as a needed prod to the agency for more aggressive action. The lawsuit achieved some success in speeding action, but for right whales it could not come soon enough. In January 2006, two months after the suit was filed, two more calves were killed on the calving grounds—one had its tail lopped off by the propeller of an unknown ship; the other drowned in a gill net.

Meanwhile, in December 2005, Richard Merrick, chief of the Marine Mammal Division in the Fisheries Service's Northeast Fisheries Science Center, completed an analysis to identify the best times and areas for vessel-related management zones in the Gulf of Maine.[48] That was followed in May 2006 by the coast guard's port access route studies, which proposed new routes for ships using ports adjacent to the calving grounds off Florida and Georgia and feeding areas off Massachusetts.[49] For the calving area, where no designated routes existed previously, the coast guard proposed "precautionary two-way routes" off the ports of Jacksonville, Fernandina, and Brunswick. The new routes directed ships through areas where right whales had been seen the least frequently. This offered at least some prospect for reducing collision risks. Off Massachusetts, it recommended (1) new precautionary routes at the entrance of the Cape Cod Canal; (2) a triangle of precautionary two-way routes (a single lane in which vessels can travel in opposite directions) between the canal, Boston, and Provincetown; and (3) shifting the east-west leg of existing traffic separation lanes (two separated lanes with each dedicated to travel in an opposite direction, similar to a high-speed highway for cars) servicing Boston Harbor to instead pass through an area of relatively low right whale sightings. It also recommended narrowing each lane and the median separation strip on the route's east-west leg from two miles to one to increase the distance from right whale habitat in Cape Cod Bay. Finally, the coast guard proposed designating right whale feeding grounds east of shipping lanes through the Great South Channel as an Area to Be Avoided.

No regulations are necessary to designate shipping lanes or Areas to Be Avoided, and adhering to them is not mandatory. Yet for liability reasons, almost all operators of large vessels follow them religiously. In the event of an accident, following those routes helps demonstrate prudent seamanship for insurance purposes. To implement such routes, the coast guard and Fisheries Service need only instruct publishers of navigational charts, such as the National Ocean Service's Office of Coast Survey, to mark the new routes on their charts. Modifying traffic separation systems and designating an Area to Be Avoided, however is subject to IMO review.

The Fisheries Service Proposal

In June 2006, a month after the coast guard completed its port access route studies, the Fisheries Service published its long-awaited ship-strike reduction proposal.[50] It was now nearly five years since the ship-strike committee submitted its report, and the proposal was essentially identical to its recommendations. At its core were routing measures proposed in the coast guard's port access route studies and a regulation requiring all vessels 65 feet long or longer to use speeds of 10 knots or less in 11 seasonal management areas and dynamic management areas.

The 11 seasonal zones were off all major ports along the species' migratory corridor and around designated critical habitats. Most were 30 nmi arcs around port entrances from Georgia to New York. Although the ship-strike committee report recommended arcs 20 to 25 nmi wide, the service concluded 30 nmi was warranted, based on what was believed to be the distance from shore that right whales migrate. The effective dates for most areas covered three- to five-month periods when right whales were most likely to be present.

As in the committee report, 15-day dynamic management areas were proposed to protect temporary right whale feeding groups outside of seasonal management areas. Criteria for designating those areas were the same as those already being used to establish dynamic area management zones for the reduction of fishery interactions with right whales (chapter 21).[51] The formula for defining management area boundaries produces a roughly 30 × 30 nmi perimeter around sighted groups. Once announced over voice radio and other ship communications media, ships would be required to slow to 10 knots when transiting the areas, or else route around them. If subsequent surveys confirmed whales had left the area before 15 days, the zone could be terminated early; if whales were still there after 15 days, it could be extended an additional 15 days.

The agency set a 60-day comment period on its June proposal, but extended it to mid-October 2006 when the World Shipping Council asked for more evaluation time.[52] When the comment period closed, more than 10,000 comments had been filed. The vast majority were from members of environmental organizations expressing support for the proposal. Opposition was principally from affected ports and a port pilot association.

A Final Rule Limiting Ship Speeds

When comment periods on proposed rules end, agencies normally publish a final decision within 30 to 90 days. For "major rules" with possible economic impacts exceeding $100 million or raising substantial policy issues, final rules must also be cleared by the White House Office of Management and Budget, best known by its abbreviation, OMB. According to the executive order that authorized these reviews, which was signed by President Clinton in 1993 and amended by President Bush in 2002, OMB is to ensure that such rules "are consistent with applicable law [and based] upon a reasoned determination that the benefits of the intended regulation justify its costs."[53] The order also directs that OMB complete the reviews within 90 days and follow a transparent process.

Although the Fisheries Service predicted economic effects well below $100 million, it had to forward the final ship-strike rule to OMB in February 2008 on the grounds that it was a significant policy issue. The final rule was essentially the same as its proposal. The George W. Bush administration, however, had contorted the OMB review process from its intended function as a "counselor" to that of a "gatekeeper."[54] Instead of using it for its original purpose, White House officials used the process to block or change rules they found objectionable for political reasons. In this case, a representative of the World Shipping Council with close ties to Vice President Richard Cheney visited the OMB to register objections to the rule. His concerns were echoed by the Maritime Administration, which represents US shipping interests. In response, the chief of the White House Council of Economic Advisers intervened, and approval of the final rule was blocked.

Over the next six months, OMB repeatedly asked the Fisheries Service for additional information on the rule's justification. As information was provided, Commerce Department officials from Director of the Fisheries Service William Hogarth, to NOAA Administrator Admiral Conrad Lautenbacher, to Secretary of Commerce Carlos Gutierrez met with high-level White House officials and staunchly defended the rule. Their efforts were to no avail. After several rounds of questions and meetings, White House officials were still unwilling to accept the agency's analyses and explanations and began asking for raw data to conduct their own analyses. As the back-and-forth continued, OMB continued to withhold approval even though its 90-day review period had long since elapsed.

As the review dragged on, environmental groups and members of Congress began citing the delay as exemplary of a broader pattern of political interference with science-based decisions on regulatory matters. The charge was part of a growing outcry over what was widely considered an abuse of discretionary authority by White House officials seeking to derail or dilute environmental protection measures in deference to partisan political interests. Similar delays were holding up decisions on other environmental issues, particularly those affecting endangered species, global warming, and air pollution. In August 2007, the situation prompted members of both the House and Senate to write President Bush demanding immediate action on the stalled ship-strike rule, but again to no avail.[55]

The stalemate, however, did not prevent the Fisheries Service and coast guard from proceeding with nonregulatory routing measures not subject to OMB approval. With a simple directive from NOAA administrators to its Office of Coast Survey, new precautionary routes through the calving grounds were added to navigation charts late in 2006. However, to realign the Boston traffic separation scheme and designate the Great South Channel Area To Be Avoided, a paper describing the two actions first had to be submitted to IMO for review at a March 2008 meeting of its Subcommittee on Safety of Navigation.[56] Both actions received approval from the IMO Marine Safety Committee without objection in July 2008 and went into effect in June 2009.

The IMO approval had been eased by a prior precedent-setting approval for similar actions to protect right whales in Canada. In 2002, Transport Canada, the Canadian agency responsible for shipping, and Canada's Department of Fisheries and Oceans asked the IMO to approve shifting vessel traffic separation lanes in the Gulf of Maine off Nova Scotia 3.9 nmi east to protect right whales in the Bay of Fundy.[57] Analyses predicted this could reduce encounters between ships and right whales in their Bay of Fundy feeding grounds by half. When the IMO endorsed the change in December 2002, it was the first time vessel traffic lanes had been moved explicitly to protect whales. After navigation charts were updated, the new route went into effect in the summer of 2003. In July 2007, the two Canadian agencies also asked the IMO to approve a new Area To Be Avoided to protect a right whale feeding ground in the Roseway Basin off the southern tip of Nova Scotia. It too was approved and went into effect in May 2008.[58]

As the Bush administration entered its final year and OMB's review of the ship-strike rule passed its one-year mark, Senators John Kerry and Olympia Snow attempted to force action by introducing a bill entitled the "Ship Strike Reduction Act of 2008" on February 20, 2008. The bill would have required the secretary of commerce to finalize the pending ship-strike regulations; however, it was not taken up and the White House continued to withhold approval as long as it could. In the summer of 2008, with the Bush administration coming to an end, White House officials abandoned their delaying tactics, recognizing that if left to a new administration it would likely be approved as proposed. Therefore, they directed the Fisheries Service to change the final rule in ways not previously considered or subject to public review.

White House officials directed that the width of seasonal management areas along the right whale migratory corridor be reduced from 30 to 20 nmi; that speed limits for temporary dynamic management areas be voluntary instead of mandatory; and that a "sunset" provision be added to termination of the rule after five years. During the five-year period, the Fisheries Service was to study effectiveness of the measures and decide whether they should be continued, modified, or allowed to lapse. As a concession to decreased protection, the Fisheries Service negotiated approval to link three seasonal management areas off ports along the southern migratory corridor into a continuous 20 nmi wide corridor from Brunswick, Georgia to Wilmington, North Carolina.

The new sunset provision squarely ignored advice by Fisheries Service scientists. The reduced width of management areas off Atlantic ports left the outer third of the suspected right whale migratory corridor without protection in an area where drifting carcasses were least likely to be found. Fishery Service scientists also pointed out that five years was not adequate for gathering and evaluating needed data, particularly given the usual amount of time required to build up enforcement and compliance and prepare associated environmental analyses for the new rules. The provision therefore effectively imposed added agency costs amounting to hundreds of thousands of dollars for preparing new rules before a definitive analysis of effectiveness could be completed.

However, resolute support by high-level officials in the Commerce Department preserved the 10-knot speed limit. By then its selected speed limit was further supported with a new study by Canadian scientists Angela Vanderlaan and Chris Taggart.[59] Assuming whales were killed purely as a function of impact force and that the whales made no attempts to avoid ships, they concluded that lethal injuries increased most sharply between vessel speeds of 8.6 and 15 knots and that they would be reduced by 50 percent if speeds were decreased to 11.8 knots. Because there were so few reports of serious or lethal injuries at lower speeds, they were unable to calculate the probability of lethal collisions at speeds below 10 knots with any statistical certainty.

Subject to its imposed changes, OMB granted approval for publishing the final rule and accompanying final environmental impact statement in late August 2008. Although the changes weakened the Fisheries Service's intended action, the 10-knot speed limit in seasonal management zones still offered significant protection. Enormous credit for its retention is due to the staunch support of high-level NOAA officials, particularly Fisheries Service director Dr. William Hogarth and his deputy, Sam Rauch. Special credit, however, also belonged to the agency's staff, particularly Lindy Johnson, who buttressed their position with exhaustive analyses of related data and legal authorities.

The new provisions, however, were now at odds with National Environmental Policy Act requirements for transparent decisions affording opportunity for

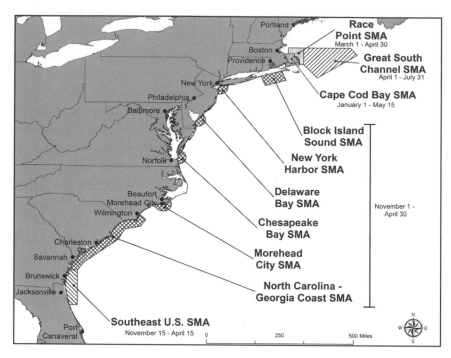

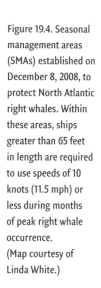

Figure 19.4. Seasonal management areas (SMAs) established on December 8, 2008, to protect North Atlantic right whales. Within these areas, ships greater than 65 feet in length are required to use speeds of 10 knots (11.5 mph) or less during months of peak right whale occurrence. (Map courtesy of Linda White.)

public review. To avoid legal challenges on those grounds, the Fisheries Service added a new preferred alternative to the final environmental impact statement incorporating the measures not previously considered. The final statement was released in late August of 2008,[60] allowing a 30-day public comment period, but with an unwritten understanding by White House officials that there would be no further changes. As soon as the comment period closed on September 29, the Fisheries Service published a final rule on October 10, 2008;[61] it became effective on December 8, a month before President Barack Obama took office (figure 19.4).

Once the final rules were published, attention shifted to implementation. The first task was an intensive outreach campaign to ensure vessel operators were aware of the new requirements. Over the first year, the Fisheries Service and coast guard announced the new requirements through mandatory ship reporting messages, brochures, placards, NOAA weather broadcasts, articles in professional trade magazines, and other media. After saturating affected ports for a year, the coast guard ramped up enforcement. In late 2009, it began issuing written warnings, and in 2011 it started issuing citations subject to fines of up to $5,750 per violation. The first citations were levied on three ships whose repeated violations resulted in cumulative fines totaling between $11,500 and $92,000.[62]

Relief at Last

With the new rule set to expire in December 2013, as soon as the final measures went into effect in December 2008 the Fisheries Service began gathering data to make its decision on whether to then modify, extend, or allow the ship-strike rule to lapse. To allow time for preparing a new rule and associated documents and for soliciting public comments, the Fisheries Service was able to collect data for only two to three years before having to start a new rule-making process. Evaluations of those data were completed in February 2012.[63] As expected, they concluded that time had been insufficient to thoroughly analyze the rule's effectiveness. However, they noted there was "a meager amount of evidence" suggesting vessel-related deaths and injuries had declined. With a previous average discovery rate for ship-struck carcasses of only one or two per year and some years with none, statistically meaningful results were impossible. The analyses had been made even more difficult by reducing the size of management areas off ports from 30 to 20 nmi. Because the outer third of the suspected migratory corridor was unprotected and carcasses could drift in or out of the regulated areas prior to their discovery, additional time would be needed to account for possible ship strikes just beyond 20 nmi from shore.

Nevertheless, the report provided other important insights. First, it examined vessel compliance with the 10-knot speed limit. Without compliance, a fair test of effectiveness was clearly impossible. Fortunately, this was a relatively easy task thanks to a new IMO measure requiring that all ships over 300 gross tons and all passenger vessels regardless of size carry Automatic Identification System (AIS) transponders and receivers for navigational safety. The AIS transponders broadcast signals every 2 to 10 seconds, identifying the ship's name, vessel type, location, speed, and heading. This allows other ships within a perimeter of few tens of kilometers to detect their presence, heading, and speed, even if they cannot be seen. Shoreside receivers can also pick up AIS signals within a range of 20 to 30 nmi, making it possible to record nearshore ship traffic and ship speeds in real time.

Those data revealed that just 32 percent of all vessels traveled at 10 knots or less throughout transits of seasonal management areas during the rule's first year of 2009. After warnings and fines were issued in 2010 and 2011, those percentages increased to 36 and 53 percent, respectively.[64] Like automobiles on a highway, however, most ships slowed to speeds closer to the 10-knot limit. Between 2009 and 2011 the proportion of ships maintaining speeds below 12 knots increased from 58 to 74 percent. The vessels traveling fastest were container ships and passenger vessels, which averaged between 15 and 16 knots, respectively. Tugs and tows, and tankers, which are vessel classes that normally travel well below those speeds, averaged 12 knots or less. The analyses also revealed there was virtually no compliance with the voluntary speed restrictions for any of the temporary dynamic area management zones that were established.

To determine whether collisions with whales declined, Fisheries Service scientists examined the amount of time between documented collision events. They reasoned that there should be an increase in time between collisions if the new rule was effective. However, to boost statistical power for their analyses, they considered collisions with all species of large whales by any vessel anywhere along the East

Coast at all times of the year, even though that could include collisions that the rules would not have prevented. With just three years of data from post-rule experience compared to three years of pre-rule data, they could detect no statistical difference.

Finally, the Fisheries Service reexamined the economic impacts using actual vessel speeds through management zones recorded by AIS receivers. Compared to the initial projections of $137.3 million per year made in 2008,[65] the new estimates of total economic costs ranged between $52.4 and $79 million per year, varying largely due to fluctuating fuel costs.[66] There was no evidence of shipping companies dropping ports due to speed restrictions. Even those reduced estimates, however, were likely inflated. A more recent study not considered in their research revealed that some container companies had adopted a "slow steaming" policy for entire transits between ports that produced a 20 percent reduction in total costs as a result of fuel savings.[67] A container ship carrying 8,000 containers, for example, was found to reduce fuel consumption by more than half by slowing transit speed from 21 to 15 knots. Thus, true economic costs of compliance with management zones are almost certainly lower than those estimated.

Based on its analyses, the Fisheries Service concluded that vessel speed restrictions should be continued as the core element of its ship-strike reduction strategy and that further studies should be done to evaluate their effectiveness.[68] It also noted that further consideration should be given to expanding the size of existing seasonal management areas and designating additional seasonal management areas in places, such as Jeffrey's Ledge and Jordan Basin in the Gulf of Maine, where dynamic management areas had been designated repeatedly.

In early 2012 the agency began work on a new rule to replace the one set to expire in December 2013. Environmental groups hoping to add provisions dropped from the original 2006 proposal urged the agency to publish its new proposal in 2012. However, officials in the Obama administration had placed a hold on any potentially controversial regulations until after the November 2012 elections. As a result, release of the proposed rule was delayed until June of 2013.[69] With little time to consider controversial changes, the agency simply proposed eliminating the sunset clause, thereby extending the existing rule indefinitely with no other changes. Once adequate information on its effectiveness was gathered, the agency pledged to reconsider whether and how the regulation might be changed.

The proposal was again opposed by the same segments of the shipping industry that opposed the original action in 2006, principally those representing ports and port pilots. The American Pilots' Association claimed that large vessels could not operate safely in dredged channels due to reduced maneuverability at speeds of 10 knots.[70] Although the rules allow vessel operators to use higher speeds if conditions warrant them and an explanation is noted in the ship's log, the association recommended exempting all dredged shipping channels from the speed restrictions; they argued that doing so would have a negligible effect on collision risk because dredged channels represented less than 1 percent of the regulated areas.

The Marine Mammal Commission and others, however, supported elimination of the sunset provision. The commission also provided preliminary results from a new study that suggested the regulations had indeed been effective.[71] It found that between December 2008, when the speed restrictions went into effect, and October 2013, no right whale deaths attributed to ships had been found inside or within 45 nmi of any effective seasonal management zone. This was twice as long as the longest period with no such carcasses since 1990 when comparable data were first collected. It also represented a statistically significant reduction in documented right whale collisions.[72]

After considering those and other comments, along with the encouraging new findings, the Fisheries Service adopted a final rule on December 9, 2013, eliminating the sunset provision while retaining all other requirements in the 2008 regulations.[73] This effectively extended its provisions indefinitely. The rule, however, still had another hurdle to cross. To approve its adoption, the OMB directed that the agency consider the recommendations by the American Pilots' Association to be a petition for amending the rule by exempting all dredged channels; this was to be taken up as a separate action in 2014. Thus, on January 30, 2014, the Fisheries Service requested comments on the association's "petition" to exempt all dredged channels off major ports from the speed restrictions.[74]

Comments by environmental groups resolutely opposed the change proposed in the petition, pointing out that the navigation concerns raised by the association seemed questionable, given that the coast guard had established an identical 10-knot speed limit for large vessels transiting other areas, such as the Cape Cod Canal, which were subject to strong winds and currents. The Marine Mammal Commission opposed the petition as well, noting that the existing provisions already allowed vessels the option of higher speeds when captains deemed it necessary for navigation safety.[75] The commission also remarked that it was likely many of the ship-struck right whales found near those channels before the 2008 rules went into effect were hit by the ships then presumably using them; this was probably the case even though the channels themselves represented less than 1 percent of the seasonal management areas. After 18 months of discussions and high-level negotiations with the coast guard, OMB, and the Maritime Administration, Fisheries Service leadership led by Deputy Assistant Administrator Sam Rauch again maintained staunch support for the rule. On October 15, 2015, the agency rejected the petition and thus left the 2013 rule intact.[76]

Although some ship strikes are still bound to occur around port entrances, the effectiveness of the speed restrictions to date have exceeded the expectations of even their most ardent supporters. As of January 1, 2016, there had still been no ship-struck right whale carcasses found in or near any active seasonal management areas since the speed rules went into effect seven years earlier. However, there was a report of one right whale hit and seriously injured in one of those areas by a recreational boat less than 65 feet long, exempt from the speed restrictions.

Despite a tortuous and lengthy rule-making process, the measures chosen by the Fisheries Service to reduce ship strikes were based on the best available scientific information and a well-reasoned scientific rationale. Once measures were selected, agency leaders and staff displayed a tenacious resolve to give them a chance despite intense political pressure. Their apparent effectiveness has been all the more important because the species' other major threat—entanglement in commercial fishing gear—has shown no signs of abating and has even increased. As will be seen in the following chapters, reducing entanglement-related deaths has been even more difficult, and the agency's resolve more questionable.

NOTES

1. Nancy C. Lutkehaus. 2008. *Margaret Mead: The Making of an American Icon*. Princeton University Press. Princeton, NJ, US, p. 261 (quotation).

2. Knowlton, AR, SD Kraus, DF Meck, and ML Mooney-Seus. 1997. *Shipping / Right Whale Workshop*. Report 97-3. New England Aquarium Research Department. Boston, MA, US.

3. Johnson, LS. 2004. *Coastal State Regulation of International Shipping*. Oceana Publications. Oxford University Press, Oxford, UK.

4. Ibid.

5. US Delegation to the IMO. Ship strikes of endangered North Atlantic right whales. MEPC 40/INF.9. June 30, 1997. An information paper (unpublished) submitted by the United States for Agenda Item 20 of the 40th Session of the Marine Environment Protection Committee in London, UK.

6. Memorandum from Rear Admiral JD Hutson, Department of Defense representative for Ocean Policy Affairs, to the Office of the General Council, NOAA, February 18, 1988, on the NOAA proposal for establishing mandatory ship reporting systems to reduce ship strikes with northern right whales.

7. From JR Twiss Jr., executive director of the Marine Mammal Commission, to Admiral RE Kramek, commandant, US Coast Guard, March 6, 1998, on need for mandatory ship reporting systems; and from JR Twiss Jr., executive director of the Marine Mammal Commission, to Admiral JD Hutson, representative for Ocean Policy Affairs, Department of Defense, March 6, 1998, in support of establishing proposed mandatory ship reporting systems.

8. From Senator Edward M. Kennedy and 11 other members of Congress, to Secretary of Defense William S. Cohen, February 27, 1998, in support of the proposal for establishing mandatory ship reporting areas.

9. From K Metcalf, director of Maritime Affairs, Chamber of Shipping of America, to KA McGinty, chair, Council on Environmental Quality, April 9, 1998, in support of establishing mandatory ship reporting systems.

10. From Rear Admiral ER Riutta, assistant commandant for Operations, US Coast Guard, to MB West, deputy assistant Secretary for Oceans and Space, Department of State, March 31, 1998, on alternative approaches to establishing mandatory ship reporting systems.

11. President William J. Clinton. 1998. Statement on protecting the northern right whale. Thursday, April 23, 1998. 34 WCPD 658. Daily Compilation of Presidential Documents. US Government Printing Office. Washington, DC.

12. US Delegation to the IMO. Ship reporting systems off the eastern coast of the United States. NAV 44/3/1. April 24, 1998. An action paper submitted by the United States for agenda item 3 of the 44th Session of the Sub-Committee on Safety of Navigation, London, UK.

13. US Coast Guard. 1999. Mandatory ship reporting systems. 64 Fed. Reg. 29229. (June 1, 1999.)

14. US Coast Guard. 2001. Mandatory ship reporting. 66 Fed. Reg. 58066. (Nov. 20, 2001.)

15. From TP Moliterno, director of Field Operations, International Fund for Animal Welfare, to the US Coast Guard Headquarters, Marine Safety Council, December 29, 1997, on proposed rules for the International Safety Management Code.

16. From JR Twiss Jr., executive director of the Marine Mammal Commission, to RA Schmitten, assistant administrator for Fisheries, National Marine Fisheries Service, December 23, 1997, recommending measures to improve the conservation of right whales along the US Atlantic coast.

17. Knowlton, AR, FT Korsmeyer, JE Kerwin, HY Wu, and B Haynes. 1995. *The Hydrodynamic Effects of Large Vessels on Right Whales.* Final report to the National Marine Fisheries Service, Northeast Fisheries Science Center. Contract No. 40ANFF400534. Woods Hole, MA, US. Knowlton, AR, FT Korsmeyer, and B Haynes. 1998. *The Hydrodynamic Effects of Large Vessels on Right Whales, Phase II.* Final report to the National Marine Fisheries Service, Northeast Fisheries Science Center. Contract No. 46EANFF600004. Woods Hole, MA, US. Clyne, H. 1999. Computer Simulations of Interactions between the North Atlantic Right Whale (*Eubalaena glacialis*) and Shipping. Master's thesis, Napier University.

18. Silber, GK, J Slutsky, and S Bettridge. 2010. Hydrodynamics of a ship/whale collision. Journal of Experimental Marine Biology and Ecology 391:10–19.

19. Silber, GK, LI Ward, R Clarke, KI Schumacher, and AJ Smith. 2002. *Ship Traffic Patterns in Right Whale Critical Habitat: Year One of the Mandatory Ship Reporting System.* NOAA Technical Memorandum NMFS-OPR-20. National Marine Fisheries Service. Silver Spring, MD, US.

20. Laist, DW, AR Knowlton, JG Mead, AS Collett, and M Podesta. 2001. Collisions between ships and whales. Marine Mammal Science 17(1):35–75.

21. Russell, BA. 2001. *Recommended Measures to Reduce Ship Strikes of North Atlantic Right Whales.* Report submitted to the National Marine Fisheries Service by the Northeast and Southeast Implementation Teams for the Recovery of the North Atlantic Right Whale. US Office of Protected Resources. Silver Spring, MD.

22. Clyne, Computer Simulations (n. 17).

23. From BJ Zoodsma, chair of the Southeast US Right Whale Recovery Plan Implementation Team, to WT Hogarth, assistant administrator for Fisheries, National Marine Fisheries Service, October 18, 2001, commenting on the report entitled, *Recommended Measures to Reduce Ship Strikes of North Atlantic Right Whales* (n. 21).

24. From T French, chair of the Northeast Implementation Team, to WT Hogarth, assistant administrator for Fisheries, National Marine Fisheries Service, January 29, 2002, commenting on the report entitled, *Recommended Measures to Reduce Ship Strikes of North Atlantic Right Whales* (n. 21).

25. Marine Mammal Commission. 2002. *Marine Mammal Commission Annual Report: Calendar Year 2001.* Bethesda MD, US, pp. 9–29.

26. Fugiwara, M, and H Caswell. 2001. Demography of the North Atlantic right whale. Nature 144(29):537–541.

27. Hamilton, PK, and MK Marx. 2005. Skin lesions on North Atlantic right whales: categories, prevalence, and change in occurrence in the 1990s. Diseases of Aquatic Organisms 68(1):71–82.

28. Silber, GK, LI Ward, R Clarke, KL Schumacher, and AJ Smith. 2002. *Ship Traffic Patterns in Right Whale Critical Habitat: Year One of the Mandatory Ship Reporting System.* NOAA Technical Memorandum NMFS-OPR-20. Office of Protected Resources. National Marine Fisheries Service. Silver Spring, MD, US.

29. Ward-Geiger, LI, GK Silber, RD Baumstark, and TL Pulfer. 2005. Characterization of ship traffic in right whale critical habitat. Coastal Management Journal 33:236–278.

30. Kite-Powell, HL, and P Hogland. 2002. *Economic Aspects of Right Whale Ship Strike*

Management Measures. Final Report to the National Marine Fisheries Service. Order no. 40EMNF100235. Office of Protected Species, National Marine Fisheries Service. Silver Spring, MD, US. The estimated economic costs to vessel operators cited here and in subsequent economic analyses referenced in this book may be greatly inflated given that many vessel operators have adopted at their own initiative a practice of "slow steaming" to reduce fuel costs and operational expenses. See Maloni, M, JA Paul and DM Gilgor. 2013. Slow steaming impacts on ocean carriers and shippers. Marine Economics & Logistics 15:151–171, doi:10.1057/mel.2013.2.

31. Nathan Associates. 2005. *Economic Analyses for the Environmental Assessment of the North Atlantic Right Whale Ship Strike Reduction Strategy.* Final report prepared for the National Marine Fisheries Service, National Oceanic and Atmospheric Administration. Silver Spring, MD, US.

32. Jensen, A, and GK Silber. 2003. *Large Whale Ship Strike Data Base.* NOAA Technical Memorandum NMFS-OPR-25. US Department of Commerce. Bethesda, MD, US.

33. Silber, GK, S Bettridge, and D Cottingham. 2008. *Report of a Workshop to Identify and Assess Technologies to Reduce Ship Strikes of Large Whales: Providence, Rhode Island, 8–10 July 2008.* NOAA Technical Memorandum NMFS-OPR-42. Office of Protected Species, National Marine Fisheries Service. Silver Spring, MD, US.

34. From D Cottingham, executive director, Marine Mammal Commission, to WT Hogarth, assistant administrator for Fisheries, National Marine Fisheries Service, October 23, 2003, on plans for developing a ship-strike reduction strategy.

35. From WT Hogarth, assistant administrator for Fisheries, National Marine Fisheries Service, to D Cottingham, executive director, Marine Mammal Commission, February 4, 2004, on plans for developing a ship-strike reduction strategy.

36. National Marine Fisheries Service. 2004. Endangered fish and wildlife; advanced notice of proposed rulemaking (ANPR) for right whale ship strike reduction. 69 Fed. Reg. 30857. (June 1, 2004.)

37. National Marine Fisheries Service. 2004. Endangered fish and wildlife; advanced notice of proposed rulemaking (ANPR) for right whale ship strike reduction; extension of public comment period. 69 Fed. Reg. 55135. (September 13, 2004.)

38. From D Cottingham, executive director, Marine Mammal Commission, to WT Hogarth, assistant administrator for Fisheries, National Marine Fisheries Service, January 12, 2005, on the need for emergency rules to reduce ship collision risks.

39. Kraus, SD, MW Brown, H Caswell, CW Clark, M Fujiwara, PK Hamilton, RD Kenney, et al. 2005. North Atlantic right whales in crisis. Science 309:591–592.

40. From WT Hogarth, assistant administrator for Fisheries, National Marine Fisheries Service, to Admiral TH Collins, commandant, US Coast Guard, May 9, 2005, asking that right whale advisories issued by the coast guard recommend use of vessel speeds of 12 knots or lower.

41. From Admiral TH Collins, commandant, US Coast Guard, to WT Hogarth, assistant administrator for Fisheries, National Marine Fisheries Service, June 9, 2005, concerning the recommended use of vessel speeds of 12 knots or lower to protect right whales.

42. National Marine Fisheries Service. 2005. Notice of intent to prepare an environmental impact statement and conduct public scoping on a right whale ship strike reduction strategy. 70 Fed. Reg. FR 36121. (June 22, 2005.)

43. From WT Hogarth, assistant administrator for Fisheries, National Marine Fisheries Service, to D Cottingham, executive director, Marine Mammal Commission, July 1, 2005, on a decision not to implement emergency rules to reduce right whale ship strikes.

44. US Coast Guard. 2005. Port access routes study of potential vessel routing measures to reduce vessel strikes of North Atlantic right whales. 70 Fed. Reg. 8312. (February 18, 2005.)

45. Nathan Associates, *Economic Analyses* (n. 31), p. 87.

46. National Marine Fisheries Service. 2005. Petition to initiate emergency rulemaking to

prevent the extinction of North Atlantic right whales: final determination. 70 Fed. Reg. 56884. (September 29, 2005.)

47. From JR Lovvorn, Humane Society of the of the United States; M Senatore, Defenders of Wildlife; S Weaver, The Ocean Conservancy; and C Butler-Stroud, Whale and Dolphin Conservation Fund, to the secretary of the Department of Homeland Security, secretary of the Department of Commerce, commandant of the US Coast Guard, and administrator for Fisheries, National Marine Fisheries Service, November 3, 2005, regarding notice of violations of the Endangered Species Act regarding the endangered North Atlantic right whale.

48. Merrick, RL. 2005. *Seasonal Management Areas to Reduce Ship Strikes of Northern Right Whales in the Gulf of Maine.* Northeast Fisheries Science Center Reference Document 05-19. National Marine Fisheries Service. Woods Hole, MA, US.

49. US Coast Guard. 2006. Port access route studies of potential vessel routing measures to reduce vessel strikes of North Atlantic right whales. 71 Fed. Reg. 29876. (May 24, 2006.). US Coast Guard. 2006. *Port Access Route Study to Analyze Potential Vessel Routing Measures for Reducing Vessel (Ship) strikes of North Atlantic Right Whales.* USCG-2005-20380-86. Office of Navigation Systems, Waterways Management Directorate. Washington, DC.

50. National Marine Fisheries Service. 2006a. Proposed rule to implement speed restrictions to reduce the threat of ship collisions with North Atlantic right whales. 71 Fed. Reg. 36399. (June 26, 2006.). National Marine Fisheries Service. 2006b. *Draft Environmental Impact Statement to Implement the Operational Measures of the North Atlantic Right Whale Ship Strike Reduction Strategy.* Office of Protected Resources. Silver Spring, MD, US.

51. Clapham, P, and R Pace. 2001. *Defining Triggers for Temporary Area Closures to Protect Right Whales from Entanglements: Issues and Options.* Northeast Fisheries Science Center Reference Document 01-06. National Marine Fisheries Service. Woods Hole, MA, US.

52. National Marine Fisheries Service. 2006c. Notice of extension of the public comment period for proposed rules to implement speed restrictions to reduce the threat of ship collisions. 71 Fed. Reg. 46440. (August 14, 2006.)

53. Executive Order no. 13258, February 26, 2002, amending Executive Order 12866, on regulatory planning and review. 67 Fed. Reg. 9385. (February 28, 2002.)

54. US General Accounting Office. 2003. *OMB's Role in Reviews of Agencies' Draft Rules and Transparency of Those Reviews.* GAO-03-929. US General Accounting Office. Washington, DC.

55. From Representatives Nick J. Rahal II, Madeleine Z. Bordalla, Thomas H. Allen, Wayne T. Gilchrest, Frank Pallone Jr., and Jim Saxton, to President George W. Bush, August 6, 2007, calling for immediate action to protect right whales from the threat of vessel strikes; and from Senators John F. Kerry, Olympia J. Snowe, and Edward M. Kennedy, to President George W. Bush, August 6, 2007, calling for immediate action to finalize the ship-strike reduction rule.

56. US Coast Guard. 2008. *Routing of Ships, Ship Reporting, and Related Matters: Area to be Avoided "In the Great South Channel."* Document submitted by the United States to the 54th Session of the Sub-Committee on Safety of Navigation, International Maritime Organization. March 25, 2008. London, UK.

57. Brown MW, RD Kenney, CT Taggart, ASM Vanderlaan, J Owen, JB Ring, et al. 2001. *Measures to Reduce the Potential for Interactions between the Endangered North Atlantic Right Whale and Large Vessels in the Waters of Atlantic Canada.* Regional Canadian Marine Advisory Council. Halifax, Nova Scotia.

58. Silber, GK, ASM Vanderlaan, AT Arceredillo, L Johnson, CT Taggart, MW Brown, S Bettridge, and R Sagarminaga. 2012. The role of the International Maritime Organization in reducing vessel threat to whales: process, options, action and effectiveness. Marine Policy 36(6):1221–1233.

59. Vanderlaan, ASM, and CT Taggart. 2007. Vessel collisions with whales: the probability of lethal injury based on vessel speed. Marine Mammal Science 23(1):144–156.

60. National Marine Fisheries Service. 2008a. *Final Environmental Impact Statement to Implement Vessel Operational Measures to Reduce Ship Strikes to North Atlantic Right Whales.* Office of Protected Resources, National Oceanic and Atmospheric Administration. Silver Spring, MD, US.

61. National Marine Fisheries Service. 2008b. Final rule to implement speed restrictions to reduce the threat of ship collisions with North Atlantic right whales. 73 Fed. Reg. 60173–60191. (October 10, 2008.)

62. National Oceanic and Atmospheric Administration. 2012. Three Vessels Charged with Violating Right Whale Ship Strike Rule Pay Penalties. NOAA press release, January 10, 2012.

63. Silber, GK, and S Bettridge. 2012. *An Assessment of the Final Rule to Implement Vessel Speed Restrictions to Reduce the Threat of Vessel Collisions with North Atlantic Right Whales.* NOAA Technical Memorandum NMFS-OPR-18. Office of Protected Resources. National Marine Fisheries Service. Silver Spring, MD, US.

64. Ibid., p. 14.

65. Nathan Associates. 2008. *Economic Analysis for the Environmental Impact Statement of the North Atlantic Right Whale Ship Strike Reduction Strategy.* Report prepared for the Office of Protected Resources. National Marine Fisheries Service. Silver Spring, MD, US.

66. Silber and Bettridge, *An Assessment* (n. 63), pp. 29–31.

67. Maloni et al., Slow steaming (n. 30).

68. Silber and Bettridge, *An Assessment* (n. 63), p. 39.

69. National Marine Fisheries Service. 2013a. Proposed rule to eliminate the expiration date contained in the final rule to remove the sunset provision of the final rule implementing speed restrictions to reduce the threat of ship collisions with North Atlantic right whales. 78 Fed. Reg. 34024-34030. (June 6, 2013.)

70. American Pilots' Association to N. LeBoeuf, chief of Marine Mammal and Sea Turtle Conservation Division, National Marine Fisheries Service, July 31, 2013, on the proposed rule to eliminate the expiration date contained in the final rule to reduce the threat of ship collision with North Atlantic right whales.

71. From RJ Lent, executive director, Marine Mammal Commission, to N LeBoeuf, chief of Marine Mammal and Sea Turtle Conservation Division, National Marine Fisheries Service, August 5, 2013, on proposed rules to reduce ship collision risks for right whales.

72. Laist, DW, AR Knowlton, and D Pendleton. 2014. Effectiveness of speed limitations to protect North Atlantic right whales. Endangered Species Research 23:133–147. As of January 1, 2016, there had still been no deaths attributed to large vessels found in or near any of the seasonal management areas.

73. National Marine Fisheries Service. 2013b. Proposed rule to eliminate the expiration date contained in the final rule to reduce the threat of ship collisions with North Atlantic right whales. 78 Fed. Reg. 73726–73736. (December 9, 2013.)

74. National Marine Fisheries Service. 2014. Petition for rulemaking to exclude federally-maintained dredged entrance channels and boarding areas for ports from New York to Jacksonville from vessel speed restrictions. 79 Fed. Reg. 4883–4884. (January 30, 2014.)

75. From RJ Lent, executive director, Marine Mammal Commission, to DS Wieting, chief of Marine Mammal and Sea Turtle Conservation Division, National Marine Fisheries Service, March 4, 2014, on petition to exempt dredged channels from speed restrictions to reduce ship collision risks for right whales.

76. National Marine Fisheries Service. 2015. Finding for a petition to exclude federally-maintained dredged port channels from New York to Jacksonville from vessel speed restrictions designed to reduce vessel collision with North Atlantic right whales. 80 Fed. Reg. 62008. (October 15, 2015.)

20

ENTANGLEMENT

FEW, IF ANY, issues so critical to the survival of an endangered species have received so much attention yet proven to be so intractable for resource managers, so infuriating for industry groups, and so frustrating for conservation advocates as the entanglement of North Atlantic right whales. The problem is deceptively simple. In the course of their migrations and while feeding, right whales, like other whales, blunder into fishing gear set by commercial fishermen. When they do, lines can jam between baleen plates or wrap around flippers and tails as startled whales instinctively spin or thrash in response to the unexpected contact (figures 20.1 to 20.3). If not shed quickly, lines can impede feeding, slice deep cuts into skin, and sap energy due to increased drag.[1] Over the course of months or even years, the debilitating effects of entangled fishing gear can wear whales down and cause a slow, agonizing death.

The odds of such encounters may seem vanishingly small for a population of just a few hundred whales in the vast ocean, yet they occur with astonishing frequency. As soon as scientists began photographing right whales for identification purposes in the mid-1970s, evidence began to mount. When the first international meeting on right whales was held in June 1983, eight North Atlantic right whales had already been photographed carrying burdens of line or webbing.[2] By

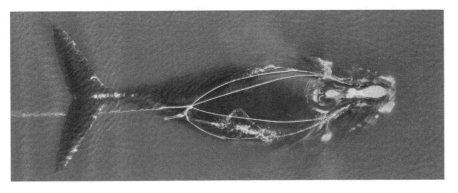

Figure 20.1. A juvenile right whale (whale #3701) found entangled on December 26, 2008, southeast of St. Augustine, Florida. The whale was successfully disentangled the following day. (Image courtesy of the Florida Fish and Wildlife Conservation Commission, taken under NOAA Permit Number 594-1759.)

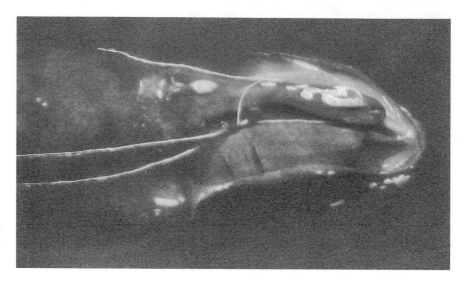

Figure 20.2. An adult female right whale named Wart (whale #1140) found entangled in Cape Cod Bay on March 6, 2008. The whale was successfully disentangled two years later in the Great South Channel. (Image by Peter Duley, courtesy of the Northeast Fisheries Science Center, Woods Hole, MA, taken under NOAA permit number 775-1875.)

the end of the 1990s, 25 more were seen entangled, and between 2000 and 2011 another 65 new entanglements were documented.[3] When a 30-year review was completed in 2012 examining entanglement scars on photographed right whales, 83 percent of all well-photographed individuals were found to have been entangled at least once. An average of 25 percent of all whales acquired new entanglement scars every year.[4]

Although deaths due to entanglement were less than half of those caused by collisions with vessels in the 1990s, scientists believe that the carcasses resulting from entanglement are less likely to be found than ship-strike victims. Entangled whales lose fat reserves over the course of their long, slow deaths and are therefore more likely to sink and go undetected. Despite this bias in discovery rate, entanglement deaths have been increasing at an alarming rate, which raises ominous prospects for such a small population. Whereas just 2 right whale entanglement deaths were confirmed in the 1990s, there were 7 in the first decade of the 2000s and 10 between 2010 and 2014. Since 2005, entanglement has been the leading cause of human-related deaths among all observed right whale carcasses.

Where and in precisely what fisheries the entanglements occur is rarely identified with certainty. The fishing gear in which whales are caught is normally left untended for days or weeks at a time, and because entangled animals typically carry it off with them, fishermen usually never know whether or when their gear was involved. Moreover, lines removed from whales are mostly tattered, nondescript fragments that could have come from almost any gear type involving rope. In the rare cases when distinctive gear parts or license numbers on buoys have been recovered from entangled right whales, they have almost always been traced to either gill-net or trap fisheries, particularly the lobster fishery. Because most mitigation options have focused on ways to make gear parts less likely to entangle whales, a brief review of the involved equipment and fisheries would be helpful before considering management struggles.

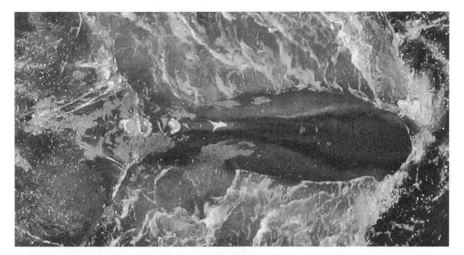

Figure 20.3. An entangled right whale (whale #2470) found in the Great South Channel on May 13, 2010, with severe injuries on its flukes and tail stock; it was trailing 130 feet of line. Views are from the air (*top*) and vessel (*bottom*). The whale was successfully disentangled the same day. (*Top*, image by Tim Cole, courtesy of the Northeast Fisheries Science Center. Woods Hole, MA, taken under NOAA permit 775-1875; *bottom*, image courtesy of the Center for Coastal Studies, Provincetown, MA, taken under NOAA permit 932-1489.)

Gill Nets, Traps, and Fishery Management

The largest gill-net fisheries on the US East Coast target species of "groundfish" living on the sea floor, principally cod, haddock, flounder, dogfish, and monkfish, as well as a few species living at midwater or shallow depths, such as mackerel and sharks. In the 1980s, some 500 gill-netters, most based in New England, set 10,000 miles of netting every day during peak fishing seasons. As overfishing forced drastic cuts in catch quotas, attrition in the industry reduced their numbers by half by 2010.

Although gill-net characteristics, such as mesh size and net height, can differ widely among fisheries, there are basic similarities. Most nets consist of 10 to 30 individual net panels, each some 50 to 100 feet long, and tied together into a "string." The webbing in each panel—the part that actually snares fish by their gills—is suspended between a "float line," with a series of small buoyant floats running across the top, and a "lead line," with heavy weights running along the bottom. The individual net panels are tied together with "bridles," or pairs of lines a few feet long that link float line to float line, and lead line to lead line. When the net panels are strung together, a string of nets normally stretches a half-mile to two miles in length (figure 20.4).

Gill nets can be set at any depth and fished in different ways depending on the species sought. Some nets are left to drift with the current marked at either end by buoys; others are fixed in place with anchors tied to both the float line and lead line of the first and last panels, with short "groundlines" at either end of the string. Buoy lines (also called endlines or vertical lines) run from the anchors to either a single buoy of varying size or to a "surface system" composed of a large spherical buoy, or "polyball," up to 30 inches in diameter and tied to a separate "highflier." Consisting of a disc-shaped or rectangular float and a long vertical rod that extends

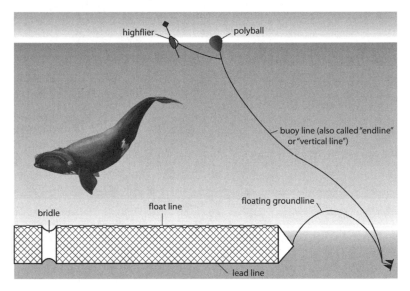

Figure 20.4. Typical configurations of commercial gill nets (top) and lobster traps (bottom) used along the US East Coast. (Image modified from a figure provided courtesy of Provincetown Center for Coastal Studies, Provincetown, MA.)

above the surface and sports a flag or radar reflector, highfliers make the gear easier for passing vessels and fishermen to spot.

The major trap fisheries on the East Coast target lobsters, crabs, sea bass, conchs, and hagfishes using box-like traps, or "pots," set on the sea floor and marked with a buoy or surface system connected to a trap by a buoy line (figure 20.4). Depending on the fishery and water depth, one or more traps may be strung together into "trawls" of up to 40 or more traps. When trawls begin to exceed three to five traps, they are usually marked with buoys at either end. The traps are linked together either with groundlines running directly from one trap to another in short trawls or by means of a separate groundline, with each trap tied to it by a short "branch line" or "gangion."

Buoy lines are thought to pose the greatest entanglement risk for whales, but groundlines can also be hazardous. Most trap fishermen prefer to use "floating" line for groundlines because it floats above the bottom. This reduces snags on rocky outcrops and chafing that could otherwise lead to more frequent line replacement. When trawls are fished, traps are typically set 60 to 120 feet apart. As a result, floating groundlines can arch tens of feet up into the water column between traps, where it can snag passing whales.

The American lobster fishery is by far the largest, most valuable trap fishery on the US East Coast. While gill-net fisheries were declining in the final decades of the 1900s, the lobster fishery was booming with no limits on either numbers of lobsters that could be caught or traps that could be fished. In 1983, when the first management plan for lobster fishing was adopted, annual landings worth $86 million were caught using some 2.2 million traps deployed by 10,000 vessels.[5] By 2006, its value quadrupled to nearly $400 million per year. Although lobster boats declined by nearly half, the number of licensed traps more than doubled. In 2006, fear of overfishing led fishery managers to impose an 800-trap limit per fisherman, which was far more than all but a few could fish. Yet even then, more than 3 million traps were deployed annually. This represented 90 percent of traps set in all trap fisheries along the East Coast. (Because the cost of trap licenses is low, fishermen often purchase them for traps they do not intend to use as a hedge against possible future limits on the fishery. Their extra licenses might be sold at a premium or ensure an ability to purchase more in the future should managers impose stricter limits on the number of fishermen or traps to prevent overfishing.) When combined with lobster fishing in Canada and other trap fisheries in the range of North Atlantic right whales, the number of traps licensed annually that could potentially entangle right whales approaches 8 million.

Responsibility for managing fisheries in US waters is split between federal and state agencies. Whereas states manage fishing within 3 miles of shore, the National Marine Fisheries Service has authority from that point seaward to the edge of the US Exclusive Economic Zone (EEZ) that ends 200 nautical miles (230 mi) from shore. Overall, the EEZ for the United States covers an expanse equal to one-fifth of the entire country's land area, thus making it the largest EEZ of any coastal nation.

It also includes some of the world's most productive fishing grounds. In 2014 alone, 9.3 billion pounds of fish and shellfish landed from the US EEZ valued at $5.4 billion.[6] With so much at stake, it is hardly surprising that decisions on allocating fishing rights and catch quotas are matters of intense interest and political pressure.

Fishing in federal waters is regulated according to management plans required by the Fishery Conservation and Management Act of 1976. Today the act is called the Magnuson-Stevens Act after two senators, Warren Magnuson of Washington and Ted Stevens of Alaska, who worked tirelessly for decades to fine-tune its provisions. Responsibility for approving and implementing fishery plans falls to the Fisheries Service's regional offices. To prepare them, the agency relies heavily on two groups: regional fishery management councils and the agency's own regional fisheries science centers. The councils are advisory bodies established by the Magnuson-Stevens Act and composed of fishermen, fishery managers, and fishery scientists appointed by the director of the Fisheries Service and governors of coastal states. Councils are charged with drafting recommended fishery management plans and periodic amendments submitted to Fishery Service regional offices for approval. Fishery science centers, on the other hand, are subunits of the regional offices. They gather and analyze data on the abundance and trends of fish stocks, amounts of fish caught, and rates at which nontarget species are caught; councils and regional offices use this information to develop and update fishery plans.

When the Fisheries Act passed in 1976, fishing in US waters was dominated by foreign factory ships that were rapidly stripping fish stocks bare. The act's immediate goal was therefore to rein in foreign fishing and replace it with more sustainable harvests by US vessels. Within 15 years, foreign fishing was all but banished, with US vessels taking its place. In its zeal to promote US fishing, however, the Fisheries Service overshot its goal of establishing "maximum sustainable yields" from fish stocks, and overfishing continued.

Estimating fish abundance is an imperfect science at best. Uncertainty around estimates leaves fishery managers subject to intense pressure from fishing lobbies and politicians seeking the "maximum" catch possible. With key officials in the Fisheries Service typically appointed because of their ties to regional fishing communities, they routinely bent to pressure for larger catch quotas and gave fishermen any benefit of the doubt about the size of fish stocks, the amount of fish that could be safely caught, the effectiveness of management measures, and other uncertainties. As a result, by the early 1990s many of the nation's most lucrative fish stocks were overfished and declining.

The First Attempt to Mitigate Entanglement

When the right whale recovery plan was adopted in 1991, one of the fisheries in crisis was that of New England groundfish, which was managed under the "Northeast Multispecies Fishery Management Plan." Groundfish are caught by both trawlers and gill nets. The latter included the majority of East Coast gill-netters with some 350 participating vessels, most operating in New England. In 1991, gill nets were thought to pose the greatest entanglement risk for right whales. Thus, in 1993, when an assessment of groundfish abundance revealed that cod and haddock stocks were in danger of complete collapse, the Fisheries Service's Northeast Regional Office gave the New England Fisheries Management Council the unenviable task of slashing catch levels by 50 percent. To do so, the council began working on an amendment to the groundfish management plan.

The plan's overhaul seemed a good opportunity to craft management regulations that would simultaneously protect groundfish and right whales, as well as the humpback whales and harbor porpoises that were also caught in high numbers in gill nets. (At the time, more than 1,000 harbor porpoises were being taken annually in New England gill nets, and their population in the Gulf of Maine had been proposed for listing under the Endangered Species Act.) The council was therefore charged with incorporating measures to reduce bycatch of right whales, as well as humpbacks and harbor porpoises as part of its fishery plan amendment.

The New England Fisheries Management Council decided that the best way to reduce overfishing with the least economic impact would be an across-the-board, 50 percent reduction in fishing days at sea, to be phased in over five years. For the first year it proposed reducing allowed days at sea by four days per month. Apparently assuming that any reduction in fishing effort would satisfy necessary reductions of right whale entanglements, no numerical goals were set for decreasing that bycatch. The only measure specifically linked to right whale protection was a recommendation to continue an existing seasonal fishery closure established years earlier to protect spawning haddock between February and May in an area south of the Great South Channel (Haddock Closure Area 1).[7] That area happened to overlap a small part of a right whale critical habitat in the Great South Channel whose designation was pending at the time. The council claimed that it had planned to eliminate the closure but had decided to retain it to protect right whales that aggregated in the area in spring. Thus, in effect, the council proposed changing the purpose of the closure from protecting spawning haddock to protecting right whales, but with no change in its boundaries, effective dates, or other provisions.

Whereas fishermen objected strenuously to any reduction in days at sea, others believed that the measures fell far short of protecting either declining fish stocks or right whales. The Marine Mammal Commission pointed out that most of the seasonal haddock spawning closure lay outside boundaries of the pending right whale critical habitat where most whales occurred and that its effective dates did not match right whales peak occurrence in the area in May and June.[8] It therefore recommended changing closure boundaries to match those of the pending Great South Channel critical habitat and extending its effective dates through June. It also recommended designating a similar closure for the proposed Cape Cod Bay critical habitat in spring. To the extent those changes reduced fishing opportunities otherwise allowed, the commission suggested that the council adjust other fishing restrictions to compensate for lost fishing prospects, such as opening parts of the haddock spawning area outside of the critical habitat that the council said it had planned to open anyway.

Because the amendment substantially revised management of a fishery that threatened endangered species, including right whales, the Fisheries Service was obligated to conduct a consultation on its new measures pursuant to section 7 of the Endangered Species Act. When such consultations involve fishery management plans, it raises something of a conflict of interest. That is, Fisheries Service officials with approval authority for fisheries plans (e.g., directors of the agency's regional offices) also oversee the staff charged with carrying out section 7 consultations. This dual responsibility opens a door for subverting consultation evaluations to favor desired outcomes subject to strong political pressure.

In November 1993, the Fisheries Service completed its biological opinion on the New England Council's proposed amendment. It concluded that fishing under the new management measures would not jeopardize right whales, but offered a list of measures it considered prudent. In part, it recommended that the council "seek actions, such as time-area closures . . . to reduce entanglement of endangered large whales."[9] The council, however, was already facing a constituency of fishermen angry over reductions in days at sea and was not anxious to further alienate them with additional closures. Therefore, with a biological opinion concluding that the amendment would not jeopardize right whales, the council decided that the existing haddock spawning area closure satisfied the opinion's recommendation for time-area closures and made no further changes.

When the council submitted its final recommended amendment for adoption, the director of the Fisheries Service's Northeast Regional Office approved it with no further changes, thereby tacitly agreeing that agency recommendations for time-area closures regarding right whale protection had been satisfied. Thus, the first effort to reduce right whale entanglements came to naught, with no new measures specifically for the protection of endangered whales.

By December 1994, it became apparent that the new measures were insufficient to reduce overfishing of groundfish stocks. The Fisheries Service therefore adopted emergency rules closing three other fish spawning areas, pending another amendment to its groundfish fishery management plan.[10] Although none of these areas included right whale critical habitat, a closure off New Hampshire and southern Maine over a submerged offshore ridge called Jeffreys Ledge would later be identified as important feeding habitat for the species.

Two key factors obstructing meaningful right whale protection during this process would continue to hobble recovery efforts: (1) the composition of the fishing industry; and (2) conflicting Fisheries Service mandates for maximizing sustainable fish catches under the Fisheries Act, while minimizing marine mammal bycatch under the Marine Mammal Protection and Endangered Species Acts. Trap and gill-net fisheries are composed of thousands of small, independently owned and operated ventures run with narrow profit margins. As a result, conservation measures, including seasonal time-area closures that have a negligible effect on overall landings, could still put a few individual fishermen out of business if they have limited options to fish elsewhere. This possibility pitted seasonal closures to improve survival prospects for right whales against the livelihood of individual fishermen with life savings invested in boats and equipment, rather than against the industry as a whole.

Confronted with this dilemma, the Fisheries Service stopped short of measures that might disproportionately affect even a few individual fishermen, even if their overall contribution to industry landings was insignificant. This effectively eliminated one of the few promising methods (seasonal area closures) for reducing right whale entanglements. To justify this decision, the agency used its discretion to decide what constitutes effective whale protection and followed the deceptive tactic that agencies sometimes employ by citing faulty, incomplete, or uncritical scientific analyses and unproven assumptions. In that way, they can justify politically preferred actions.[11] Alternatively, to deflect actions they wish to avoid, they can refer it to another body for further consideration.

As noted in chapter 17, when critical habitat for right whales was designated in the spring of 1994, the Fisheries Service postponed restrictions on hazardous fishing gear pending advice from its two new regional implementation teams. By the end of 1995, the two teams were still debating the issue. In the interim, however, at least some additional protection was provided when Florida voters approved a referendum banning gill nets within the state's three-mile jurisdictional limit. Combined with a similar ban adopted earlier by Georgia, nearshore portions of the right whale's calving grounds were thus protected from gill nets.

A New Management System

In late 1994, Congress amended the Marine Mammal Protection Act, creating a new system for managing marine mammal–fishery interactions nationwide. The new system directed the Fisheries Service to convene "take reduction teams" for the purpose of drafting "take reduction plans" for fisheries that frequently catch marine mammals from "stocks" (a term synonymous with populations) classified as "strategic." These include any species listed as endangered or threatened under the Endangered Species Act or whose bycatch in fisheries exceeds their "potential biological removal" (PBR) level. The PBR level is a conservative estimate of the number of marine mammals that can be removed from a population annually (not including natural mortality) with little risk that the population will decrease below its optimum sustainable level. To convert this concept into operational practice, the Fisheries Service developed a formula to calculate PBR levels for individual marine mammal populations based on estimates of their minimum abundance and maximum growth rate, and a recovery factor to add an extra measure of safety depending on the extent to which a species is endangered or depleted.[12]

When Congress authorized the creation of take reduction teams, it envisioned something along the lines of locking concerned parties (fishermen, conservationists, scientists, and officials of involved government agencies) in a room until they reached consensus on a set of measures to reduce the bycatch levels to required goals. As an immediate goal, Congress stipulated that take reduction plans should reduce fishery-related mortality and serious injury of marine mammals below their PBR level within six months of implementation. For the longer term, the act required that within five years, deaths and serious injuries be reduced to a "zero mortality rate goal," which the Fisheries Service has interpreted to be 10 percent or less of PBR. To develop recommended plans, teams were allowed six months to thrash out a consensus on measures that would meet the statutory goals. The Fisheries Service then had nine months to publish a draft plan, solicit public comments, and issue final regulations.[13] If a team failed to reach a consensus on needed measures, the Fisheries Service was obligated to develop its own plan to meet those goals. If an adopted plan failed to meet its immediate or long-term goals, the team was to be reconvened to develop further measures.

The time frames Congress laid out were entirely unrealistic. It normally takes at least two years to prepare such regulations and all the necessary documents and then afford time for public comment required by other laws. Similarly, achieving compliance with new regulations and collecting data to assess whether goals are being met within six months is virtually impossible. Nevertheless, in most cases the act's directives have had their intended effect of speeding agreement on potentially effective take reduction measures. The most notable exception, however, has been the team charged with reducing entanglements of North Atlantic right whales.

While the Fisheries Service was planning steps to implement the act's new provisions, its northeast right whale implementation team completed analysis of fishery management needs as requested by the agency and wrote to the director in July 1996.[14] The team had concluded that the best way to mitigate whale entanglements with the least impact on fisheries was to reduce hazardous fishing gear at times and in areas where right whales were most abundant and fishing activity was relatively low. Recognizing that these conditions described the new right whale critical habitats, the team recommended that the Fisheries Service immediately adopt a regulation prohibiting hazardous fishing in the Great South Channel critical habitat during months of peak right whale occurrence. For critical habitats including state waters (Cape Cod Bay and southeast calving grounds), it recommended that the agency consult with state agencies to develop similar regulations.

By the time the Fisheries Service replied to its implementation team in November 1996, the agency had already begun applying the new bycatch reduction provisions under the 1994 amendments to the Marine Mammal Protection Act. It had developed a formula for calculating PBR and convened a new "Atlantic Large Whale Take Reduction Team." The agency also had announced plans to reinitiate section 7 consultations on all fisheries thought to entangle right whales. The director of the Fisheries Service therefore advised both implementation teams that it would not be executing their recommendations because the agency was now relying on its new take reduction team and the next round of section 7 consultations for advice on fishery measures.[15]

Responsibility for supporting the Atlantic Large Whale Take Reduction Team and implementing the related take reduction plan was assigned to the Fisheries Service's Northeast Regional Office (recently renamed the Greater Atlantic Regional Office). However, because right whales also occur along the southeast coast, the agency's Southeast Regional Office was to be consulted on matters affecting fisheries and right whales from Virginia southward. After invitations were sent to prospective team members in early August, the team's 37 members met for the first time on September 16 and 17, 1996. Anticipating contentious discussions, a professional conflict-resolution firm was hired to moderate discussions. The team was charged with reducing entanglement risks for all endangered whales along the East Coast (right, humpback, fin, sei, and sperm whales), but virtually all attention focused on right whales due to their highly endangered status.

By the team's first meeting, the list of suspected fisheries had grown to include the lobster fishery, as well as gill-net fisheries for groundfish, monkfish, dogfish, and sharks. The team therefore included representatives of each of those fisheries from different areas, as well as conservation organizations, whale research groups, and government agencies. The miniscule size of the right whale population and its low reproductive rate produced an equally miniscule PBR of 0.4 whales per year. Thus, the team's statutory goals were to reduce right whale entanglement-related deaths and serious injuries to less than two whales every 5 years within 6 months of plan implementation, and to less than one whale every 20 years within 5 years.

In December 1996, with the team still locked in its initial six-month deliberation, the state of Massachusetts and the Fisheries Service completed separate analyses of measures needed to reduce right whale entanglements. The state's analysis was triggered by a lawsuit filed by environmental activist Max Strahan in the District Court for Massachusetts on April 21, 1996 (*Strahan v. Cox*). He contended that the state's Division of Fisheries had violated Endangered Species Act protection requirements by issuing permits to fish in state waters with lobster traps and gill nets that were known to entangle and kill right whales and other endangered whales. The judge in the case, Douglas P. Woodlock, agreed. He concluded that entanglement risks in Massachusetts waters were greatest in the Cape Cod Bay critical habitat and ordered the state to develop a plan restricting or eliminating fixed fishing gear from the critical habitat.[16]

The state immediately formed a working group to draft a plan. After it was finished, the state reported back to the court in December.[17] The resulting plan sought to reduce entanglement risks while minimizing economic impacts to fishermen by imposing new conditions on fishing in Cape Cod Bay between January and mid-May when right whale numbers were greatest. Although no gill-net fishing occurred in the bay during those months, the state pledged to nevertheless prohibit it to prevent any future shifts in fishing patterns. For the handful of winter lobstermen fishing in the bay, the state argued that reducing the amount of line they deployed would reduce entanglement risks to a negligible level.

The state of Massachusetts therefore proposed requiring that winter lobstermen cluster their pots into four-trap trawls instead of the one or two traps per buoy normally fished, in order to cut their number of buoy lines in half. It also proposed requiring that groundlines between lobster traps be made of sinking line instead of floating line; this would eliminate loops rising between traps that could entangle whales. Finally, the state pledged that it would develop and require a 150-pound weak link for use between buoys and buoy lines. They reasoned that a weak link with a lower breaking strength than that of the attached line might increase the chances that a whale could shed the buoy and then the lines in which it had become entangled. The court found the plan to be in good faith and accepted it. The state therefore put the measures—except for weak links whose design was still being developed—into effect in January 1997, in time for the start of Cape Cod Bay's winter whale season.

Massachusetts's response to the court was tracked closely by the Fisheries Service as it worked on its new section 7 consultations for gill-net and lobster fisheries. In separate biological opinions on each fishery completed in December 1996, the agency concluded that the fisheries in their current state were likely to jeopardize right whales.[18] To avoid such risk, the agency's biological opinions recommended measures similar to those in the state's plan, but on a broader geographic scale. As with the state's action to prevent gill nets in its portion of the Cape Cod Bay critical habitat, for critical habitat in federal waters, the opinions recommended prohibiting gear types that posed entanglement risks for whales but were not then used in those areas, to prevent their possible future use. Conversely, the biological opinions recommended allowing potentially hazardous gear types that were already being used in the critical habitats, provided that they were modified to reduce whale entanglement risks.

Accordingly, for the Cape Cod Bay critical habitat, the biological opinions recommended extending the same measures required by the state in its part of the habitat into federal parts of the habitat. That is, they proposed prohibiting gill nets throughout the designated critical habitat from January through mid-May, but allowing lobster traps equipped with the weak links, sinking line, and trawls no less than four traps long. Conversely, for the Great South Channel critical habitat they recommended prohibiting lobster traps during its whale season (April to June) "unless gear or alternative fishing practices are developed that eliminate the likelihood of entanglement,[19] while allowing gill-net fishing to continue in the one area where it already occurred. This area was a narrow strip along the southeast edge of the Great South Channel critical habitat that came to be called the "sliver."

Although the opinions offered no explanation as to why remaining entanglement risks from lobster fishing in Cape Cod Bay were justified, two reasons were cited for leaving the sliver area open to gill nets in the Great South Channel critical habitat.[20] First, the opinion on the gill-net fishery stated that because only 2 percent of all right whale sightings in that critical habitat were from the sliver area, entanglement risks would be low. The sliver area, however, was also only a little more than 2 percent or so of the critical habitat's area and no evaluations were presented to compare risks in the sliver *versus* other parts of the critical habitat. Second, it asserted that because gill-net fishing in the sliver area occurred at depths of 35 to 40 fathoms (210 to 240 feet) and most right whales in the Great South Channel were seen in waters deeper than 50 fathoms (300 feet), gill-netting in the sliver posed little or no entanglement risk. It failed to note, however, that right whales feed at all depths, including those from 35 to 40 fathoms.

Release of the Massachusetts plan and the Fisheries Service's biological opinions brought a swift end to contentious debate within the take reduction team over time-area closures in critical habitats. With the Fisheries Service announcing what it considered safe in critical habitat, which included virtually all existing fishing provided it used untested gear modifications, all incentive for team fishermen to negotiate over closing those areas evaporated. Fishermen then maintained that any time-area closure, no matter how few fishermen were affected, would simply displace fishing gear to closure perimeters and create a "wall" of gear that would increase entanglement risks for any whales entering and leaving closure areas. Given the fishermen's unyielding position, conservationists and scientists on the team made no further attempt to discuss closures, believing it would stymie negotiations over other measures, even though closures provided one of the most logical prospects for reducing entanglement risks with the least economic impact.

This left only three approaches for team discussion: (1) disentangling whales, (2) removing derelict fishing gear from whale habitats, and (3) gear modifications to prevent or reduce the severity of entanglements. All seemed potentially useful, but none offered much promise for achieving statutory goals of reducing deaths and injuries below the PBR. There was broad support for increasing disentanglement efforts; however, scientists experienced in disentangling whales were quick to point out that this did nothing to reduce the cause of the problem, placed the lives of responders at risk, was expensive, and at best treated only the very small number of entangled whales they could reach.

All sides also supported removing derelict fishing gear. This had a clear benefit for many marine species, but it was not believed to offer a significant prospect for reducing large whale entanglements.[21] When fishing gear is lost or abandoned, it generally balls up in a snarled mass by the time it settles to the sea floor, and this makes it very unlikely to entangle passing whales. Thus, despite its many other values, derelict gear cleanups, which involve removing lost gear resting on the bottom or beaches, were not expected to do much to reduce whale entanglement risks.

This left only gear modifications. This strategy was appealing to all sides. Modifications could be applied everywhere whales occurred, they would allow continued fishing, and they distributed mitigation costs evenly among all fishermen. Options, however, first had to pass muster as being practical and affordable for fishermen, and, second, generally relied on questionable theories. With no information on how whales became entangled and no way to test modifications on live whales, gear research was limited to designing prototypes and testing their performance and acceptability for fishing, rather than their effectiveness in reducing entanglement risks. Short of using them everywhere and waiting to see if entanglements decreased, little could be done to test their mitigation value. As a result, team discussions revolved around speculative debate over whether and how modifications might work, with no data or experience to assess their merits. Numerous modifications were suggested, but tabled. Most were little more than concepts requiring significant research and development.

The modifications that were initially considered included fishing without buoys and their connecting lines, and using weak buoy lines or weak links at various spots on fishing gear. Scientists and conservationists supported options that eliminated lines in the water column and particularly devices or methods that eliminated buoy lines, which would clearly reduce entanglement risk. Fishermen, however, quickly dismissed that approach as impractical. Grappling for trawls of traps with no buoy lines was possible in shallow water, but was too slow and labor intensive for fishermen tending hundreds of traps. Systems that kept buoys and buoy lines coiled on the sea floor until fishermen signaled for their release with a sonic device were also rejected on grounds of cost and reliability. Fishermen also pointed out that surface buoys were needed to let other fishermen know where gear was set so new gear would not be set on top of it and trawlers would not drag their nets or dredges across trap lines.

Discussions therefore turned to the weak links that the Massachusetts Division of Fisheries had already begun testing for use in Cape Cod Bay. Fishermen expressed conditional support for their use subject to costs and the assurance they would not fail during normal use, thereby increasing gear loss. They ruled out weak links at the junction of buoy lines and either traps or gill nets, given the risk of breaking when gear was hauled. This left use of the devices to connections between surface buoys and their lines. If weak links released buoys under the pull of a whale, it was suggested the animals might somehow shed the remaining gear or at least bear less drag that could cause lines to cut into their skin. Cost concerns were eased when designs costing a few cents each were proposed, but potential gear loss remained an issue. Fishermen also noted that acceptable breaking strengths would need to differ depending on sea conditions, water depths, and trawl lengths. Conservationists and scientists expressed tepid support for weak links; they noted that most entangled whales were found entangled in lines lacking buoys, which suggested that the release of buoys may not aid in shedding gear.

When the team's six-month deliberation period ended in early February 1997, the meeting facilitator drafted a summary of discussion points. After incorporating comments from team members, a final report was submitted to the Fisheries Service on behalf of the team.[22] Although consensus was reached on a few recommendations, team conservationists and scientists did not believe they would meet the required goals. Areas of agreement included expanding whale disentanglement work, removing derelict gear, increasing research on possible gear modifications, placing distinctive marks at intervals along lines in order to identify the fisheries and fishing areas responsible for gear found on entangled whales, and preparing outreach materials to inform fishermen of the problem. Also supported were the requirements for lobster fishing that the state of Massachusetts instituted for Cape Cod Bay (sinking groundlines, weak links, and four-trap trawls).

With regard to fishing closures, consensus was limited to those measures recommended in the Fisheries Service's biological opinions and adopted by the state of Massachusetts. The report noted, however, that because the agreed closures in critical habitats allowed virtually all ongoing lobster and gill-net fishing to continue, they were unlikely to reduce whale entanglements. Neither conservationists nor scientists, however, offered any other proposals for closures.

A Rocky Road to the First Atlantic Large Whale Take Reduction Plan

With the team unable to agree on a full set of measures likely to meet statutory goals, the Fisheries Service was left to decide what additional protection would be necessary after taking team deliberations into consideration. Its decision was announced in early April 1997, when it adopted two final rules and proposed a third. The two final rules—one for gill nets and one for lobster traps—focused on fishing inside right whale critical habitats. The proposed third rule included gear modifications for both lobster traps and gill nets that were to be required in all East Coast areas outside of those two critical habitats. Together, the three rules formed the core of the agency's new "Atlantic Large Whale Take Reduction Plan." All rules were issued under the authority of the Marine Mammal Protection Act, rather than under fishery management plans authorized by the Magnuson-Stevens Act, so as to keep provisions subject to the take reduction team process rather than the fishery management council process.

The first was a final rule published on April 1; it established restrictions on gill-netting in the two right whale critical habitats off Massachusetts. Following recommendations in the biological opinion, it banned gill nets in the Cape Cod Bay critical habitat from January through mid-May, and in the Great South Channel critical habitat—except for the sliver area—from April through June.[23] Some comments on the rule urged the expansion of this restriction to include a seasonal ban on gill nets throughout the entire Great South Channel critical habitat, including the sliver area, but the Fisheries Service rejected those suggestions. It erroneously claimed that because 97 percent of all right whale sightings in the critical habitat had occurred outside of the sliver area, the ban would reduce entanglement risks in the critical habitat by 97 percent.[24] Because the measure changed neither the distribution nor amount of fishing gear in the critical habitat, the agency's assertion was spurious.

On April 4, the Fisheries Service published an emergency rule for lobster fishing in the same two critical habitats, again following recommendations in its biological opinions. Whereas lobster fishing in the Cape Cod Bay critical habitat was allowed to continue subject to gear modifications identified in the state plan, fishing in the entire Great South Channel (including the sliver area) was banned from April 1 to June 30 subject to a caveat. This caveat left open the possibility of allowing lobster fishing to resume if the agency determined gear modifications would be adequate to avoid whale entanglements.[25]

The third part of new take reduction plan regulations was a proposed rule published on April 7. This presented a complex network of management areas and gear modifications similar to those adopted for the Cape Cod Bay critical habitat.[26] For each management area, the Fisheries Service proposed varying sets of requirements for sinking groundlines on traps and gill nets, buoy lines composed entirely of sinking line, and weak links with breaking strengths from 150 to 500 pounds between buoys and their lines. It also suggested a requirement for marking buoy lines midway along their length with a stripe of one of five different colors to help identify their source when found on whales; these marks would distinguish near-shore and offshore lobster traps from gill nets set in the northeast, mid-Atlantic, and southeast regions. Finally, it proposed a contingency measure to protect right whales in critical habitats. If any right whale was found dead or seriously injured as a result of entanglement in a gill net or lobster trap set in compliance with requirements for those areas, that gear type was to be banned from the critical habitat unless new measures were identified to ensure it would not happen again. Nonregulatory measures completing the take reduction plan also called for increasing funding to disentangle whales, remove derelict fishing gear, and investigate other gear modifications.

The agency's April 7 proposal set off a firestorm of opposition from both fishermen and conservationists. More than 13,000 comments were submitted. Fishermen vehemently objected to sinking groundlines, arguing they were impractical in many fishing conditions and their cost would put them out of business and cause more traps to be lost. They therefore flooded offices of their congressional representatives with complaints. Conservationists were no less angry over the complete reliance on unproven gear modifications and the failure to reduce fishing effort in right whale critical habitats.

After considering the comments, the agency agreed with fishermen that certain measures were impractical and published an "interim" final rule (a rule the agency expected to modify once additional information was gathered) in July 1997 that substantially differed from its proposal.[27] The interim rule effectively eliminated the need for most fishermen to make any changes to their gear, while offering no new measures to compensate for eliminated modifications. The agency abandoned requirements for specific gear modifications in its various management areas in lieu of a "menu" of six modification options. Fishermen had to select either one or two options from that list, depending on where they fished. Whereas two options were required in order to fish in the Great South Channel critical habitat and a Stellwagen Bank / Jeffreys Ledge management area off New Hampshire and northeastern Massachusetts, only one option was required for all other areas except the Cape Cod Bay critical habitat, where modifications required by the state of Massachusetts remained in place.

For lobster gear, the menu of options included use of buoy lines with a diameter of 7/16 inch or less in inshore areas, and 3/4 inch or less in offshore waters. The Service asserted this was useful because thicker line would be more difficult for whales to break and therefore increase entanglement risks. However, because virtually all lobstermen and gill-netters were already using buoy lines at those diameters, the rule effectively required no gear changes except in the two management areas where a second option was required. The remaining options included weak links that would break under 1,100 pounds of force in nearshore waters and 3,780 pounds of force in offshore waters, or the use of sinking line for either the buoy lines or groundlines. A separate but similar menu of options was adopted for gill nets. It too included use of buoy lines of 7/16-inch diameter or less, sinking buoy lines, and 1,100-pound weak links at the buoy, but also listed the use of weak links between net panels and two alternative anchoring systems to provide resistance against which whales could pull to break free.

Although the revision substantially weakened proposed protection, the Fisheries Service issued a new biological opinion on July 15, 1997, concluding that the new measures would provide effective protection for right whales and other endangered whales.[28] The agency further concluded that the new regulations and nonregulatory measures would achieve the statutory goal of reducing right whale entanglements to near zero within six months.[29]

Notwithstanding the agency's pledge to increase funding for disentanglement work and gear modification research, as well as its new obligations for enforcement and outreach, it was in no position to pay for those expenses. In November 1996, the Marine Mammal Commission held one of a series of reviews of the right whale recovery program and concluded that agency funding, then about $600,000 per year for addressing both ship strikes and entanglement issues, was woefully inadequate. It determined that at least $3 million per year would be required to carry out just the most essential tasks. The commission wrote to the Fisheries Service in December 1996, proposing that part of the funding needs be met through the creation of a nongovernmental right whale fund administered by the National Fish and Wildlife Foundation to solicit private funding and that the Fisheries Service request at least $1.25 million in fiscal year 1997 to meet remaining priorities for right whale recovery work.[30]

At that time, however, Congress was aggressively cutting federal agency budgets to trim a ballooning federal deficit. The Fisheries Service therefore requested no new funds for right whale recovery. When the agency eventually replied to the commission in October 1997, it advised that it looked forward to working with them to develop the right whale fund and that it would increase its budget request for right whales to $1 million in each of fiscal years 1998 and 1999.[31] In 1998, its request was granted and the agency increased funding for gear studies and disentanglement responders to $130,000 and $145,000, respectively. However, with the nation's economy dipping into a recession and charitable donations declining, the Large Whale Fund struggled to get started. It was not until 2001 that it was able to make its first contribution of $250,000, which went to support whale disentanglement work along the East Coast.

To complete its Large Whale Take Reduction Plan, the Fisheries Service still needed to replace its interim rule with a final rule. After testing some new modification options and reconvening the take reduction team early in 1999 to ask for its advice, a final rule was published in February 1999.[32] The final rule retained all of the management areas in the interim rule with only minor changes to its menu of gear modification options. Responding to complaints from fishermen that marking lines with a colored stripe was too difficult, the final rule also eliminated gear-marking requirements in all areas except critical habitats.

Adoption of the final rule on February 22, 1999, marked the completion of work to develop what would be the first version of the Atlantic Large Whale Take Reduction Plan. By then it was slightly more than two years after the first meeting of the Atlantic Large Whale Take Reduction Team. Completing the various sets of rules and supporting documents in that time frame had been a monumental bureaucratic task, particularly given the controversy they created. In practice, however, its only effect was to require a few gear modifications of questionable value in two right whale critical habitats and the Stellwagen Bank / Jeffreys Ledge management area. In Cape Cod Bay, lobster gear had to incorporate sinking groundlines and 1,200-pound weak links on buoys, and cluster at least four pots per buoy during periods of peak right whale occurrence. In the Great South Channel critical habitat and the Stellwagen Bank / Jeffreys Ledge management area, it effectively required seasonal use of newly developed weak links, the easiest and least expensive option on the menu of gear options.

Conservationists and right whale scientists held out no hope for the new measures. Even some staff of the Fisheries Service candidly admitted that the new rules would be ineffective. Those doubts were soon validated. Between early 1999 and early 2000, one and possibly two dead right whales were found entangled in fishing gear. Four new entanglements involving live right whales were also discovered in 1999, only one of which could be successfully disentangled. At the same time, concerns over the plan's weaknesses were magnified by ominous signs in the right whale population's status and trend.

NOTES

1. Van der Hoop, JM, P Corkeron, J Kenny, S Landry, D Morin, and MJ Moore. 2015. Drag from fishing gear entangling North Atlantic right whales. Marine Mammal Science 32(2):619–642, doi:10.1111/mms.12292.

2. Reeves, RR. 1983. *Report of Workshop on the Status of Right Whales. Boston, Mass, USA. 17–13 June 1983.* Report provided to the International Whaling Commission. Cambridge, UK, p. 29.

3. Marine Mammal Commission. 2001–2011. *Marine Mammal Commission Annual Reports to Congress.* Bethesda, MD, US. Knowlton, AR, and SD Kraus. 2001. Mortality and serious injury of northern right whales (*Eubalaena glacialis*) in the western North Atlantic Ocean. *In* PB Best, JL Bannister, RL Brownell Jr., and GP Donovan. Right Whales: Worldwide Status. Special issue 2. Journal of Cetacean Research and Management, pp. 193–208.

4. Knowlton, AR, PK Hamilton, MK Marx, HM Pettis, and SD Kraus. 2012. Monitoring North Atlantic right whale *Eubalaena glacialis* entanglement rates: a 30 yr retrospective. Marine Ecology Progress Series 466:293–302.

5. New England Fisheries Management Council. 1983. *American Lobster Fishery Management Plan.* Saugus, MA, US, p. 8.

6. National Marine Fisheries Service. 2015. *Fisheries of the United States, 2014.* NOAA Current Fishery Statistics No. 2014. US Department of Commerce. Washington, DC, p. vi, https://www.st.nmfs.noaa.gov/commercial-fisheries/fus/fus14/index.

7. New England Fisheries Management Council. 1993. *Amendment 5 to the Multifisheries Fishery Management Plan.* Vol. 1. Newburyport, MA.

8. From JR Twiss Jr., executive director, Marine Mammal Commission, to WW Fox, assistant administrator for Fisheries, National Marine Fisheries Service, November 15, 1993, on proposed amendment 5 to Multispecies Fishery Management Plan.

9. National Marine Fisheries Service. 1993. Consultation and Conference in Accordance with Section 7(a) of the Endangered Species Act Regarding Proposed Management Activities Conducted under Amendment 5 to the Northeast Multispecies Fishery Management Plan. November 30, 1993. Unpublished report. Silver Spring, MD, US.

10. National Marine Fisheries Service. 1994. Northeast multispecies fishery; emergency interim rule. 59 Fed. Reg. 63926. (December 12, 1994.)

11. Artelle, KA, JD Reynolds, PC Paquet, and CT Darimont. 2014. Letter: When science-based management isn't. Science 343:1311.

12. The formula for calculating PBR is (NMIN) (½ RMAX) (FR), where NMIN is the minimum estimated size of a population, ½ RMAX is one-half the maximum rate at which a population could be expected to increase, and FR is a recovery factor ranging between 1 and 0.1. A recovery factor of 1 is used for populations believed to be in good shape, and its designation results in no further reduction in the estimated safe removal number; a recovery factor of 0.5 is for populations listed as threatened, and 0.1 is for endangered species, reducing PBR by 50 and 90 percent, respectively.

13. Marine Mammal Protection Act of 1972, Pub. L. No. 92-522, 86 Stat. 1027 (1972). Section 118 (f)(7).

14. From TW French, chair of the New England Whale Recovery Plan Implementation Team, to RA Schmitten, assistant administrator for Fisheries, National Marine Fisheries Service, July 26, 1996, on actions to reduce the entanglement of right whales and humpback whales in fishing gear.

15. From RA Schmitten, assistant administrator for Fisheries, National Marine Fisheries Service, to TW French, chair of the New England Whale Recovery Plan Implementation Team, November 5, 1996, on team recommendations to reduce entanglement of right whales and humpback whales.

16. Strahan v. Coxe, 939 F. Supp. 963 (D. Mass, 1996), US District Court.

17. McKiernan, D. 1997. State presents plan to reduce entanglement risks for right whales. Division of Marine Fisheries News 17, January–March, p. 1.

18. National Marine Fisheries Service. 1996a. *Endangered Species Act Section 7 Biological Opinion on Re-initiation of Consultations Regarding Current and Proposed Management Activities Conducted under the Northeast Multispecies Fishery Management Plan.* December 13, 1996. Northeast Regional Office. Gloucester, MA, US. National Marine Fisheries Service. 1996b. *Endangered Species Act Section 7 Biological Opinion on Re-initiation of Consultations Regarding Current and Proposed Management Activities Conducted under the American Lobster Fishery Management Plan.* December 13, 1996. Northeast Regional Office. Gloucester, MA, US.

19. National Marine Fisheries Service, 1996a (n. 18), pp. 19–20.

20. Ibid., pp. 17–18.

21. Laist, DW. 1996. Impacts of marine debris: entanglement of marine life in marine debris including a comprehensive list of species with entanglement and ingestion records. *In* JM Coe and DB Rogers (eds.), *Marine Debris: Sources, Impacts, and Solutions.* Springer Series on Environmental Management. Springer. New York, NY, pp. 99–139, 115.

22. Keystone Policy Center. 1997. Draft, Large Whale Take Reduction Plan. Unpublished report submitted to the National Marine Fisheries Service, Northeast Regional Office. February 1, 1997. Gloucester, MA, US.

23. National Marine Fisheries Service. 1997a. Final rule on fisheries of the northeastern United States, northeast multispecies fishery; framework amendment 23. 62 Fed. Reg. 15425. (April 1, 1997.)

24. Ibid., p. 15427.

25. National Marine Fisheries Service. 1997b. Emergency interim rule on North Atlantic right whale protection. 62 Fed. Reg. 16108. (April 4, 1997.)

26. National Marine Fisheries Service. 1997c. Proposed rule and request for comments on taking marine mammals incidental to commercial fishing operations; Atlantic Large Whale Take Reduction Plan regulations. 62 Fed. Reg. 16519. (April 7, 1997.)

27. National Marine Fisheries Service. 1997d. Interim final rule for taking marine mammals incidental to commercial fishing operations; Atlantic Large Whale Take Reduction Plan regulations. 62 Fed. Reg. 39157. (July 22, 1997.)

28. National Marine Fisheries Service. 1997e. *Endangered Species Act Section 7 Biological Opinion on the Atlantic Large Whale Take Reduction Plan.* July 15, 1997. Northeast Regional Office. Gloucester, MA, US.

29. National Marine Fisheries Service. 1997f. *Environmental Assessment and Regulatory Impact Review of the Atlantic Large Whale Take Reduction Plan and Implementing Regulations.* Northeast Regional Office. Gloucester, MA, US.

30. From JR Twiss Jr., executive director, Marine Mammal Commission, to RA Schmitten, assistant administrator for Fisheries, National Marine Fisheries Service, December 12, 1996, on priority needs for North Atlantic right whales.

31. From RA Schmitten, assistant administrator for Fisheries, National Marine Fisheries Service, to JR Twiss Jr., executive director, Marine Mammal Commission, October 16, 1997, on recovery priorities for North Atlantic right whales.

32. National Marine Fisheries Service. 1999. Taking marine mammals incidental to commercial fishing operations; Atlantic Large Whale Take Reduction Plan regulations. 64 Fed. Reg. 7529. (February 16, 1999.)

21

OH, WHAT A
TANGLED WEB

AS THE 1990s came to a close, survival prospects for North Atlantic right whales were turning from bad to worse. Birth rates were plummeting, mysterious white lesions on the skin of adults signified that much of the population was in poor health, and carcass discovery rates, including those attributed to both entanglement and ship strikes, were on the rise. All signs pointed to a species slipping toward extinction. The alarming findings provoked a torrent of appeals for immediate action by the National Marine Fisheries Service.

Congress responded by quadrupling the Fisheries Service's right whale budget request to $4.1 million for fiscal year 2000. In doing so, it directed that $750,000 be used to develop and test whale-safe fishing gear. As it had done at times in the past, however, the Fisheries Service diverted part of the appropriation to unrelated purposes—in this case, salaries and administrative costs incurred by budget cuts to other programs. Angered at the misallocation, yet aware of species' plight, the Senate Appropriations Committee issued a stern warning to agency officials even as it increased right whale recovery funds to $5 million for fiscal year 2001. Its appropriations report for that year stated that it "was dismayed at the agency's lack of responsiveness to the Committee's direction on the allocation of right whale funds in fiscal year 2000"; it ordered the agency to report to them "no later than November 30, 2000, with a spending plan for fiscal year 2001."[1]

Despite the funding increase, hope for prompt action on entanglement risks faded quickly. According to one involved participant, the agency's response, a continuous rule-making process, amounted to a "rope-a-dope" strategy that resulted in a bewildering web of seasonal management areas with different sets of measures. It seemed more intent on ensuring that no fishing opportunity would be lost than on protecting whales. For the next decade, agency attention remained riveted on gear modifications, most of which were of questionable value and largely

unenforceable. As requirements were adopted and changed between 2000 and 2010, nearly 60 new right whale entanglements were documented, and carcasses attributed to the problem increased to nearly a quarter of all known deaths (10 of 43) (figure 21.1). Another half dozen whales disappeared from sighting records and were assumed dead after last being seen entangled or with severe entanglement wounds. When the agency's predicted effectiveness failed to materialize, the cumbersome rule-making process would be repeated three times with no better results; this left all sides frustrated.

First Revision of the Take Reduction Plan: DAMs, SAMs, and Gear Modifications

The first revisions to the take reduction plan were made in early 2000, barely a year after the initial plan was completed. One of the Fisheries Service's first steps was to expand the Atlantic Large Whale Take Reduction Team to 50 seats by adding fishermen and fishing industry representatives targeting various species from additional ports not previously represented on the team. When the new team met in February 2000, it was asked to reexamine gear modifications and time-area fishing closures.[2] Fishermen on the team again made it clear they would not consider seasonal closures. As a fallback alternative, team scientists and conservationists offered a new concept—short-term closures established on an ad hoc basis to protect temporary aggregations of feeding right whales discovered by researchers at unpredictable times and locations. Referred to as a "dynamic area management," or DAM, strategy, closures would be announced over voice radio and remain in effect for two weeks or until whales dispersed.

368

When fishermen opposed that idea as well, team deliberations turned to two gear modifications proposed by the Fisheries Service: (1) reducing the breaking strength of weak links in hopes of making it easier for whales to shed buoys; and (2) reducing slack line in buoy systems, assuming such lines might somehow contribute to entanglement risks. Both measures had some intuitive logic but received only tepid support from team members. Fishermen were concerned weak links would fail under normal use and gear loss would escalate, while scientists questioned the underlying premise. Though weak links may help whales shed buoys, they first had to become entangled to exert a breaking force, and most whales were seen entwined in lines with no buoys. Moreover, disentanglement teams had found that buoys or telemetry tags attached to trailing lines sometimes helped pull lines free, suggesting that releasing buoys was not always advantageous. Nevertheless, with no data to evaluate either modification, scientists and conservationists were willing to give them a try.

Notwithstanding doubts about those two modifications, there was broad optimism that some technological fix with better justification could be devised. Thus, in a rare moment of strong consensus, the team recommended that the Fisheries Service support an aggressive gear research program in cooperation with fishermen. The most promising modification at that time was a proposal to replace floating groundlines with "neutrally buoyant lines" on trap trawls, thereby eliminating the arching rise between traps. It was suggested that line with a density equal to that of sea water might hover just above the bottom where it would neither entangle passing whales nor chafe and snag on bottom obstructions. Lobstermen on the team, principally those from Maine, believed that such lines would fray quickly due to wear against rocks, particularly in areas with strong tides typical off Maine. This, they argued, would cause the loss of more traps to snags and require more frequent line replacement. For a lobsterman fishing up to 800 traps, swapping out all groundlines could cost more than $5,000, a significant expense for a small owner-operated business. Nevertheless, most team members supported research to test neutrally buoyant groundlines.

Given the increased size of the team, discussions became even more unwieldy. In addition, new fishery representatives brought differing perspectives about fishing practices and conditions in their areas. The Fisheries Service therefore split the team into two subgroups—one for New England and the other for waters off mid-Atlantic and southeastern states. In early May 2000 the two subgroups continued discussions independently.

As evidence of the right whale population's decline mounted, Max Strahan of GreenWorld filed a lawsuit charging the Fisheries Service and Maine Division of Marine Resources with violating the Endangered Species and Marine Mammal Protection Acts by authorizing activities—in this case, fishing with traps and gill nets—known to kill and injure right whales and other endangered whales. His suit was followed by two others, one by the Humane Society of the United States and the other by the Conservation Law Foundation; these two were subsequently merged into a single complaint (*Humane Society of the United States et al. v. Norman Mineta, Secretary of Commerce et al.*). The combined suit charged the secretary of commerce with violating the two acts by issuing fishing permits without adequate protection measures that resulted in the death and injury of right whales. All suits cited evidence of continuing fishery-related deaths of right whales and humpback whales, despite the initial take reduction plan.

Under pressure from the courts and others quarters to strengthen take reduction measures, the Fisheries Service reinitiated section 7 consultations on the lobster, northeast groundfish, monkfish, and spiny dogfish fisheries. To settle the lawsuit, the agency agreed to complete a new rule-making action by the fall of 2001 that, in part, would authorize the designation of DAM zones around temporary right whale aggregations, and seasonal area management (SAM) zones in predictable high-use right whale habitats. The promised rule-making was to follow another two-step process involving an interim rule to strengthen gear modifications, followed by a final rule to establish DAMs and SAMs and refine gear modifications.

The interim rule was adopted in December 2000; it skipped the normal public review process on the grounds that "it would be contrary to the public interest to delay action for prior public notice and opportunity to comment."[3] The rule established eight management areas for lobster and trap fisheries and four for gillnet fisheries that covered almost all New England waters with varying requirements for gear modifications. Citing support from the take reduction team, each area required use of weak links with breaking strengths ranging from 500 to 3,780 pounds and vertical buoy lines composed entirely of dense, heavy line that would minimize slack. (Fishermen sometimes splice in lengths of floating line at the tops or bottoms of their buoy lines to prevent snags or make it easier to grab hold of them for hauling.) For most management areas, fishermen were also required to choose one or more modifications from a menu of six options similar to those under the previous rule. In some management areas, different combinations of measures were required in different seasons. For all areas, a practice used by some lobstermen called "wet storage" (leaving traps on the sea floor during nonfishing seasons instead of storing them on land) was banned.

Once the interim rule was adopted, the Fisheries Service turned attention to dynamic and seasonal area management measures. In early 2001, scientists at the agency's Northeast Fisheries Science Center completed a study to develop criteria for establishing temporary DAM zones. From past aerial surveys, they found that a sighting of four or more right whales in an area of 100 nmi^2 (115 square miles) was a reliable indicator that a larger group of feeding right whales would be present within 15 nmi of the sighting location for two weeks.[4] Using those criteria, in March 2001, the agency began broadcasting right whale advisories as it continued working on its new rule. The advisories asked fishermen to voluntarily remove any gear already set in those areas and avoid setting new gear for two weeks, or until it was confirmed whales had left those areas.

In mid-June 2001, the agency completed its new round of section 7 consultations on the fisheries of concern.[5] The resulting biological opinions concluded that the management measures in place for all four fisheries were insufficient to avoid jeopardizing right whales and therefore listed what are known as "reasonable and prudent alternatives." The recommended alternatives called for designating both short-term DAM zones and new SAM zones that either prohibited fishing gear or limited gear to that which had been "proven safe" for whales. They also recommended expanding the take reduction plan to cover all East Coast trap fisheries and increasing support for gear modification research.

Immediately after completing its biological opinions, the Fisheries Service reconvened the take reduction team to ask for its advice on how to implement the recommended measures. Agency officials explained that they were proposing to modify the interim rules in order to authorize the designation of DAMs and to establish a seasonal SAM zone in the Great South Channel critical habitat and surrounding waters. The agency noted that it planned to designate 15-day DAM zones similar to the whale advisories it had begun announcing in March, but now with mandatory rather than voluntary measures. The SAM zones would comprise a broad swath of continental shelf east of Massachusetts where seasonal right whale occurrence was predictable. The question put to the team was whether regulations for DAM and SAM zones should prohibit fishing entirely or allow it with modified gear deemed safe for whales. If the latter, it was asked what modifications could be considered safe for whales.

Most members agreed that some modifications—particularly use of neutrally buoyant groundlines—would help reduce entanglement risks, but there was no agreement on any modifications that could be considered fully "whale safe."[6] Scientists and conservationists argued that short of eliminating all lines from the water column with ropeless fishing gear, no other options could be considered as "proven safe" for whales. Fishermen considered this unrealistic. Similarly, there was no agreement on closing DAMs or SAMs to all fishing. Whereas closing these zones had the support of scientists and conservationists, fishermen considered that option a nonstarter. As a result, decisions on those matters were again left to the Fisheries Service.

CHURCHILL

While the Fisheries Service was preparing its final rules, a three-month drama was unfolding involving an attempt to rescue a severely entangled right whale named "Churchill." The course of events was tracked closely by national and international media. The entangled whale was first spotted by an aerial survey team 75 miles east of Cape Cod on June 8, 2001. It had a tight wrap of line around its upper jaw that had sliced a deep gash into its rostrum. The entanglement was judged life threatening and a response team was dispatched the next day. The whale's elusive behavior made it impossible to remove any gear, but a satellite tag was attached to a trailing line to track its movements for later rescue attempts. Over the next two months, Churchill remained out of rescue range or managed to elude responders. As successive attempts failed, daily updates by the national media reported the whale's latest location and status, as well as rescue plans.

As the weeks wore on, Churchill's condition deteriorated noticeably. Convinced his demise was imminent, a team of marine mammal veterinarians decided the only hope was to sedate the whale so rescuers could get close enough to cut the lines free. Immobilizing an animal the size of a right whale in the open ocean had never been attempted, but any chance of saving one of the world's most endangered animals was worth a try. The veterinarians therefore designed and manufactured a 2-foot-long syringe with a 1-foot stainless steel needle and a gas-activated plunger for injection, and drew up a plan to administer the drug. An opportunity to put their plans into action finally arose in late June off Massachusetts.

After loading the barrel of their syringe with 4,000 milligrams of the powerful sedative midazolam, the giant needle was affixed to the end of a 30-foot cantilevered pole mounted to the bow of an inflatable work boat. Rescuers maneuvered cautiously up to the whale, and when the device was suspended over the whale's back, they let the pole drop (figure 21.2).[7] The injection was a success, but the dosage was not. Despite its weakened state, the whale remained as elusive as ever. Over the course of two more attempts, the dosage was gradually increased but without success. On August 30, the date of the last attempt, the whale appeared

Figure 21.2. A Center for Coastal Studies response team attempts to sedate a right whale named Churchill (#1102) with a specially designed syringe dropped from a cantilevered pole that is attached to the bow of a rescue boat. (Image courtesy of Center for Coastal Studies, Provincetown, MA, taken under NOAA permit 932-1489.)

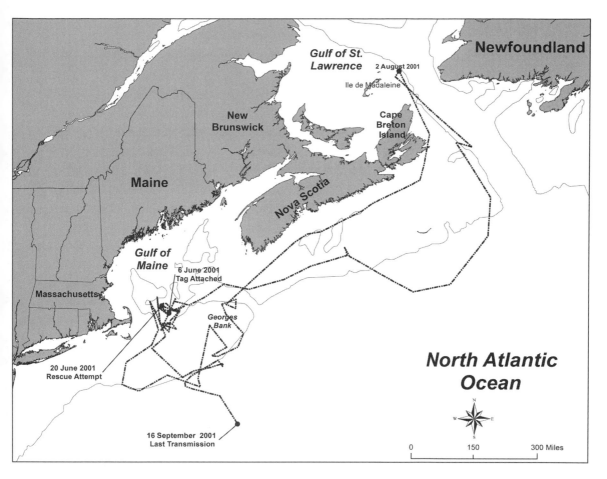

Figure 21.3. The track of an entangled right whale named Churchill (whale #1102) as recorded by a telemetry tag that was tied to a trailing line during unsuccessful rescue efforts between June 9 and September 16, 2001. (Data courtesy of the Center for Coastal Studies, Provincetown, MA; figure courtesy of Linda White.)

to slow slightly, but not enough for rescuers to cut the line free. By then, Churchill was even thinner. His color had faded to ghostly gray and whale lice covered much of his body.

Over the course of three months, Churchill had traveled more than 5,500 miles from Cape Cod to the Magdalen Islands in the Gulf of St Lawrence and back (figure 21.3). Transmissions from his satellite tag finally stopped on September 16, 100 days after it was attached and two weeks after the last sedation attempt. The last signal came from 460 miles east of Cape May, New Jersey, where the whale presumably died and sank. Although the sedation attempt had failed, methods were subsequently revised and tried again in mid-January 2011 on another badly entangled right whale off Florida. Using a specially designed pneumatic dart gun to deliver the syringe and a new combination of drugs, the whale slowed just enough to allow responders to remove all of the entangling gear.[8] Unfortunately, that animal was found dead several weeks later from wounds inflicted by the lines before they could be removed. Nevertheless, the successful sedation effort added a new tool to the arsenal of rescue options.

ONE STEP FORWARD, TWO STEPS BACK

Public attention on Churchill added even more pressure to improve right whale protection. Over the first three days of October 2001, two weeks after Churchill presumably died, the Fisheries Service published three separate rules to modify its Atlantic Large Whale Take Reduction Plan. The first, published on October 1, proposed making management areas listed in its December 2000 interim rule permanent, but with some changes to the menu of gear modification options.[9] Most changes were minor (e.g., reducing weak-link breaking strengths). One, however, proposed eliminating the option for using buoy lines less than 7/16-inch diameter beginning in 2003. At that time, it would be replaced by an option for using sinking or neutrally buoyant groundlines in the northern inshore lobster management zone off Maine. With new manufacturing methods to strengthen line developed during the 1990s, the agency acknowledged that line thickness was not a valid indicator of breaking strength and therefore was not a useful mitigation measure. It nevertheless deferred the change, citing a need for further research on neutrally buoyant line and the lack of other replacement options.

On October 2, a second rule proposed procedures for designating the 15-day DAM zones.[10] Before each zone could become effective, the procedures called for publishing a notice of the coordinates and requirements in the Federal Register, which is the official record of federal rules. Although such notices did nothing to inform fishermen of the new zone, it lengthened the time between sighting a whale aggregation and the zone's effective date from the initially envisioned 3 days to 5 days or more. The proposal also listed three possible management options for designated zones: (1) removing all traps and gill nets and banning the setting of new gear during effective dates, (2) removing all gear not equipped with specified gear modifications, or (3) asking fishermen to voluntarily remove gear and refrain from setting new gear. Decisions on which requirement to apply were to be made on a case-by-case basis, depending on weather conditions, how much gear was in the area, and other fishery management provisions in effect at the time of designation. If whales left a designated zone before the expiration date, the zone could be terminated early; if whales continued to be present after 15 days, it could be extended another 15 days with the same or different boundaries.

Finally, on October 3, the agency also published an advanced notice for its proposed third rule to establish SAM zones in the Great South Channel.[11] This was followed by issuing a proposed rule in late November with further details.[12] It suggested two adjacent SAM zones spanning a roughly 100-mile-wide band east of Cape Cod to the 200-nmi limit of US jurisdiction.[13] The western third of this swath closest to Cape Cod (called SAM West) was to be effective in March and April; the eastern two-thirds (SAM East) would be effective from May through July to reflect a typical shift in whale distribution. To fish in those areas, lobster traps would have to be equipped with the same modifications required for the Cape Cod Bay critical habitat (sinking or neutrally buoyant groundlines and weak links). Gill nets would need to have (1) weak links at set locations along the float line and between each net panel, (2) a specified anchoring system that whales could pull against to break weak links, and (3) weak links on surface buoys and associated high fliers.

Several groups commenting on the proposals found the measures insufficient. The Marine Mammal Commission opposed the two-year delay in deleting the 7/16-inch diameter line from the menu of mitigation options. Noting that the Fisheries Service had already found the option ineffective, it argued that the option should be removed immediately.[14] Several conservation groups also challenged the agency's conclusion that gear modifications in SAM zones were safe for whales. They supported the use of sinking or neutrally buoyant groundlines, but pointed out that both broken and unbroken weak links had been found on entangled whales, which demonstrated that they were not effective at preventing buoy line entanglements. Accordingly, reliance on weak links to prevent endline entanglements was contrary to the agency's own biological opinion recommending that fishing in SAM zones be either prohibited or limited to gear proven effective at preventing entanglements. They therefore recommended that SAM zones be seasonally closed to all traps and gill nets until modifications could be found that were proven effective. Another comment recommended that sinking or neutrally buoyant groundlines should be required on all traps along the entire East Coast whether inside or outside of SAM zones by 2003.[15]

Fishing interests argued that more time was needed to evaluate the effectiveness of measures already put in place. They also objected to gear marking requirements, calling them far too onerous and costly, and claimed that weak links increased trap loss, thereby increasing chances of whales being entangled in derelict gear.

After reviewing thousands of public comments, the Fisheries Service adopted final rules in early January 2002. Only a few changes were made to its original proposal. The menu of optional gear modifications for management areas was expanded to include neutrally buoyant or sinking groundlines. However, because 7/16-inch buoy lines, which fishermen normally used anyway, was retained as an option until 2003, few would need to make any changes to their gear before that year.[16] In the interim, the Fisheries Service advised that it would explore ways to expand the use of neutrally buoyant groundlines through tests with fishermen, discussions with rope manufacturers, and advice from the take reduction team.

The final rule for SAM zones allowed fishing to continue subject to gear modifications, including weak links and either sinking or neutrally buoyant groundlines.[17] To rationalize continued reliance on weak links, the agency explained that the recommendation in its biological opinion for preventing the use of gear not proven effective at avoiding entanglements meant only that modifications had to have a chance of reducing entanglements, but not necessarily preventing them. Notwithstanding the lack of evidence that a weak link had benefited any whale, the agency advised that unbroken weak links on entangled whales simply indicated they had not been entangled in the "right" way, and thus did not prove that none had been saved. In support of its decision to reject recommendations for closing SAM zones to all gear, the agency echoed arguments by fishermen, stating that moving fishing gear to closure perimeters would increase entanglement risks for whales entering and leaving the area. No analyses of gear density were offered to justify its conclusion or explain why the measure was somehow believed effective for protecting fish stocks for fishery management purposes, but not for safeguarding whales.

The action proving most contentious and least effective, however, was the rule authorizing DAM zones.[18] The Fisheries Service designated the first two DAM zones in the spring of 2002 in the southern Gulf of Maine, one covering 1,100 square miles and the other 1,900 square miles.[19] The requirement for first publishing notices in the Federal Register caused a two-week lag between the time whales were seen and zone implementation. When the first two zones went into effect, fishermen were required to remove their gear from the zones within 2 days and refrain from setting any new gear for 15 days. The announcements elicited an immediate, vehement outcry from fishermen, who protested that it was impossible to safely remove all traps and nets on such short notice given unpredictable weather conditions and limited space aboard their small boats to stow gear.

In response, the Fisheries Service rescinded the second DAM zone before it was due to expire and thereafter asked only for voluntary efforts to remove gear when new zones were announced. No monitoring was done to determine whether or to what extent fishermen complied, but given their strenuous opposition to the measure, it was broadly assumed that few, if any, did comply. Further undercutting effectiveness of DAM zones were new policies adopted by the agency. One was a decision not to establish DAM zones in critical habitats where other restrictions applied, even if those measures were less protective or not in effect at the time. The other required that initial sightings of whale aggregations be validated by a second survey; this would delay effective dates by up to three weeks.

Throughout the final months of 2002, fishermen seethed over DAM zones even though they were voluntary. Conservationists were no less irate due to the long delays in zone designations and the dubious value of voluntary measures. In an attempt to ease the controversy, the agency proposed amending the rule in March 2003 to allow fishing in DAM zones if traps and gill nets used a single end-line and either neutrally buoyant or sinking groundlines to minimize the amounts of line whales could encounter in the water column.[20] In doing so, the agency hoped that fishermen would begin using the modifications on all their gear so they could continue fishing whenever and wherever DAM zones might be designated.

In April 2003, the Fisheries Service reconvened its Large Whale Take Reduction Team to review the proposed changes and begin identifying new mitigation measures for a second major overhaul of the plan. After adding several more representatives of fisheries, the team had grown to nearly 60 people, more than two-thirds of whom were fishermen or staff of fishery management organizations. When they met, widely differing views were again expressed and no consensus was reached on the agency's proposed changes.

After considering the team's disparate suggestions, as well as public comments, the agency adopted a final rule in late August 2003.[21] The Fisheries Service decided to allow fishing gear in all future DAM zones, provided it was equipped with all of the modifications identified in its January proposal, except for one. In response to objections by fishermen, the requirement for using one endline to mark trap trawls and gill nets was retracted. Like past requirements, the rules were almost impossible to enforce. Coast guard policies precluded hauling fishing gear without the owner present or a Fisheries Service enforcement officer onboard. And because the agency has no staff dedicated to at-sea enforcement, coast guard officers were unable to check for infractions not readily visible at the surface, which included all gear modifications. Cooperative agreements with state marine patrols became the principal line of enforcement; but even then, with hundreds of thousands of traps and gill nets scattered across hundreds of thousands of square miles of open ocean, inspecting meaningful numbers was impossible, particularly when it came to those more than a few miles from shore.

One final change to its first major overhaul of the plan was adopted in 2002 for gill nets off Florida and Georgia. Although both states had banned these from state waters, they continued to be allowed in federal waters beyond the three-mile limit of state jurisdiction. To protect right whale mothers and calves on the calving grounds, the 1999 take reduction plan established a "Restricted Area" covering federal waters from the Georgia–South Carolina border to Cape Canaveral, Florida. There were two gill-net fisheries then active in the area—one for sharks and the other for mackerel. The Fishery Service deemed both safe for whales because they occurred south of the core calving area, and fishermen remained within sight of their nets. Fishing boats also had to carry a fishery observer at all times, avoid sets when right whales were seen within three miles of their vessels, and call disentanglement teams immediately if a whale became entangled. Realizing that whales could not be seen or disentangled at night, a new rule was adopted in September 2002 that banned nighttime sets in the restricted area during the right whale calving season (November 15 to March 31).[22]

As the Fisheries Service worked on its DAM rules in early 2001, the Ocean Conservancy, a nonprofit environmental group, petitioned the agency to expand right whale critical habitat boundaries. The petition cited whale survey data collected since 1994 when the critical habitats were first designated that revealed concentrations of whales adjacent to the designated areas. In November the agency agreed the petition had merit and requested public comments.[23] Eighteen months later, however, it rejected the petition.[24] Although hundreds of letters of support had been submitted, the Fisheries Service decided the petition did not include adequate information explaining why adjacent areas were important to right whales, something it had recently started to require for designation. The agency therefore committed itself to conduct research on the question and postponed boundary revisions until some unspecified future date when those studies were completed. However, it would take more than a decade, another petition, and another lawsuit before the Fisheries Service would meet its pledge.

By late 2002, doubts about the effectiveness of new take reduction measures were confirmed. In mid-October 2002, the carcass of a badly injured right whale was found one month after it had been disentangled. Fishing gear removed from the whale included a 600-pound weak link required by the new gear modification requirements. In addition, between 2002 and 2003, at least eight more right whales were found with severe entanglements considered potentially life threatening.[25]

At the Marine Mammal Commission's annual meeting in late October 2002, Fisheries Service officials acknowledged that the take reduction plan was inadequate and advised that it planned to reconvene the team to begin work on a second major plan revision. The commission repeated its earlier recommendations that the agency consider prohibiting fishing gear that posed entanglement risks in right whale critical habitats during peak whale seasons, including gear that relied on weak links to prevent entanglements in endlines. It also recommended that all trap fisheries along the East Coast be required to use sinking or neutrally buoyant groundlines by January 2004.[26]

Second Revision of the Take Reduction Plan: Broad-Based Gear Modifications with Sinking Groundlines

Work on a second major plan revision began at a take reduction team meeting in April 2003, even as the Fisheries Service was still putting final touches on its first revision. Agency officials informed the team that it planned to prepare a draft environmental impact statement and proposed rules by mid-2004, with final rules by early 2005.[27] The team agreed that the new plan needed to reduce entanglement risks from both buoy lines and groundlines, but there was no consensus on how to do so. As at previous meetings, discussions focused on gear modifications, but now with a goal of applying them more uniformly along the entire East Coast.

Grist for those discussions was provided by results from an expanded gear research program made possible by increased congressional funding. Between 2002 and 2005, the Fisheries Service spent $2 million on workshops and studies to brainstorm, design, and test new gear modification ideas. To help carry this out, the agency hired two gear specialists to work with fishermen, gear manufacturers, whale scientists, and conservationists. In addition to investigating new designs and applications for weak links, consideration was given to slippery rope, stiff rope, weak rope, knotless rope, glowing rope, rope of different colors, fat-soluble rope that would dissolve when contacting whale skin, electrified rope, weak tag lines to pull up a heavier hauling line, time-tension line cutters to cut rope under the constant strain of swimming whales, mechanical line splicers to avoid bulging knots that might prevent lines from slipping through whale baleen, and streamlined buoys less likely to cause entanglement.

Other studies tested ways to reduce line in the water column with the assumption that less line would mean fewer entanglements. The agency tested pop-up buoys released by electronic signals and galvanic time-release links to keep buoys and coiled buoy lines on the sea floor in baskets attached to traps until gear was ready for hauling. The use of one instead of two buoy lines to mark trawls and gill nets was tested, and experiments were also conducted with groundlines made of neutrally buoyant line instead of floating line.[28] Still other studies used load sensors to measure the minimum breaking strength of line needed to haul different kinds of gear in different situations, the pulling strength of whales, and ways to mark gear to identify its source when whales were disentangled.

As the team sifted through the results, the one modification that continued to stand out as most promising was neutrally buoyant groundline. By 2003, field tests demonstrated that "neutrally buoyant" line had to be just slightly more dense than the density of seawater in order to eliminate hazardous loops between traps, but also that it was more likely to snag on rocks and abrade. This led Maine lobstermen to ask if there were areas where floating groundlines could be used without posing an entanglement risk. Despite all the research done on right whales, their behavior when submerged was still one of the least known aspects of their biology.

Scientists on the team noted that right whales feeding in the Bay of Fundy sometimes surfaced with mud on their heads, indicating they had been diving to the bottom; but with little else to go on, they could only guess at when and how often groundlines might entangle whales. Assuming right whales skimmed upright along the sea floor when feeding, they speculated that lines two feet or less from the bottom might simply slip beneath them. They did not consider the possibility discussed earlier (chapter 5) that right whales may feed upside down with their narrower upper rostrum scraping the bottom. If so, lines even a foot or two off the bottom could catch in the mouth.

Nevertheless, drawing on the scientists' speculation, Maine lobstermen seized on a new concept—"low profile" groundlines that would rise no more than two feet above the bottom to avoid both whale entanglements and line abrasion. To create such a profile, they suggested placing weights at intervals along floating groundlines. Opinions on the merits of "low profile" and "neutrally buoyant" line differed, but there was little disagreement that reducing or eliminating looping groundlines between traps would reduce entanglement risks.

Shortly after the team's meeting, the Fisheries Service asked for public comments on alternative management strategies to analyze in its forthcoming environmental impact statement.[29] By the time of the next round of team meetings, however, the only new modifications that continued to receive support from at least some fishermen were neutrally buoyant and low-profile groundlines. Whereas conservationists and scientists urged use of such groundlines on all traps in all areas, fishing representatives argued for exemptions wherever bottom snags were possible, particularly along the rocky coast of Maine.

With rocky bottoms less common in lobster fishing areas off Massachusetts, however, representatives of the Massachusetts Lobstermen's Association (MLA) expressed a willingness to use neutrally buoyant groundlines. To convert this interest into practice, the International Fund for Animal Welfare (IFAW) offered to fund a gear buy-back program to defray costs for the state's lobstermen to switch from floating to neutrally buoyant groundlines. The IFAW and Massachusetts Lobstermen's Association then lobbied Congress to supplement IFAW funds. In response, Representative William D. Delahunt and Senator Edward M. Kennedy earmarked a one-time congressional appropriation of $685,000 to the Fisheries Service for that purpose. The IFAW contributed $300,000 and, after offering to administer the buy-back program at no charge, most of the Fisheries Service's appropriated funds were transferred to them.[30]

Beginning in September 2004, nearly $1 million was distributed in the form of vouchers to Massachusetts lobstermen to purchase new groundline with a density slightly greater than seawater in exchange for an equal amount of old floating groundline. The vouchers covered three-quarters of the cost of new line, while fishermen paid the rest. Over the next year, some 300 Massachusetts lobstermen—approximately a quarter of the state's licensed commercial lobstermen—took advantage of the program, paying an average of $1,000 to $1,500 as their share. By the end of 2005, 3,250 miles of floating groundline off Massachusetts had been replaced with neutrally buoyant groundlines. The Massachusetts Division of Fisheries subsequently changed state lobster fishing regulations to require sinking groundlines on all traps set in state waters as of January 1, 2007.[31]

As the groundline buy-back program got underway in February 2005, the Fisheries Service completed and circulated a draft environmental impact statement analyzing alternatives for its second revision of the take reduction plan.[32] With take reduction team members continuing to express different opinions on virtually all issues, decisions on mitigation measures were again left to the Fisheries Service. The environmental impact statement proposed six options for "broad-based" gear modifications, including two "preferred alternatives." All but the no action alternative (meaning only the existing measures apply) proposed replacing the DAMs, SAMs, and the menu of gear options for management zones under the first plan revision. Instead, the alternatives identified specific sets of gear modifications for each of ten different fisheries and fishing regions that encompassed virtually all East Coast trap and gill-net fisheries.

Although each alternative identified different combinations of modifications for the ten management units, the underlying theme for all areas was a requirement to use sinking or neutrally buoyant groundline on all traps by 2008. Despite agreement by the take reduction team to address risks of both groundlines and buoy lines, no new measures were included to prevent entanglements in the endlines. To reduce those risks for trap fisheries, the Fisheries Service proposed continued reliance on weak links even though there was no evidence they had prevented even a single right whale death and entanglement rates had actually increased since they were first required under the initial plan. For gill nets, the

alternatives increased the number of weak links on net panels. Sinking or neutrally buoyant groundlines were also required for gill nets even though their short
lengths at either end of a string were a negligible entanglement risk compared to
trap fisheries and the nets themselves.

In April 2005, the Fisheries Service reconvened the take reduction team to
ask for comments on the various alternatives. The team continued to have widely
disparate views. Indeed, positions differed so much that, rather than trying to sum-
marize them in the meeting report, team members were asked to submit their
views individually as part of the public comment process. Conservation groups
complained that all of the alternatives still relied on weak links, thereby failing
to address entanglement risks posed by buoy lines. Others urged consideration
of additional alternatives to compensate for ineffective weak links. The Marine
Mammal Commission was particularly critical.[33] Having repeatedly recommended consideration of seasonal closures in critical habitats, it found that the
agency's failure to include any such option was inconsistent with guidelines that
require environmental impact statements to rigorously evaluate all reasonable
alternatives.[34] Environmental groups also argued that deferral of requirements for
sinking or neutrally buoyant groundlines until 2008 was far too long and inconsistent with the agency's statutory obligation to reduce entanglements below
required levels within six months of plan implementation.

Fishermen were no less critical. To them, the 2008 deadline for eliminating
floating groundlines was far too soon. They argued that more time was needed
for research on ways to avoid groundline snags and abrasion and that more areas
where snags were possible, particularly areas with rocky bottoms and wrecks,
should be exempt from neutrally buoyant groundline requirements. They also
complained that fishermen were being targeted unfairly with onerous regulations,
while vessel operators had yet to be subject to any restrictions.

A month after the comment period closed in mid-May, the Fisheries Service
requested comments on a voluminous proposed rule to implement a preferred
alternative.[35] For all its complexity, its central element was a requirement that
almost all traps and gill nets along the East Coast use neutrally buoyant or sinking
groundlines by January 1, 2008. Except for a few exempted areas in coastal bays
behind headlands and in waters deeper than 1,600 feet (beyond the dive depth of
right whales), neutrally buoyant or sinking groundline was to be used everywhere
from the shoreline to the outer edge of the 200-nmi US Exclusive Economic Zone.
For gill nets, it required an increase in the number of weak links on net panels.
The requirements were to be effective year-round off New England, between September and May off Mid-Atlantic states, and from mid-November to mid-April
in calving grounds off Georgia and Florida. Weak links continued to be the only
measure to prevent buoy line entanglements.

Although legally obligated to reduce entanglement deaths from all causes, the
Fisheries Service decided to defer efforts to mitigate entanglements in "vertical"
lines (buoy lines and endlines leading from the bottom to the surface), pledging

to take up that issue in a future rule-making action—a delay that would take nearly 10 years. To justify its decision, the agency claimed that most take reduction team members had recommended the deferral.[36] The agency's statement was misleading at best. Deferring measures to mitigate vertical line risks was never put before the team as an option and no team scientists or conservationists ever endorsed such an approach. The agency went on to justify deferral of vertical line measures by stating that the team had offered no useful advice and had recommended further research on possible endline gear modifications. While the team had recommended further research on both groundlines and buoy line modifications, in neither case had it agreed this was grounds for deferring management action.

Nevertheless, the Fisheries Service thereby restricted the scope of its second revision to groundline measures and effectively disregarded the cause of perhaps three-quarters of all right whale entanglements, which occur in vertical buoy lines.[37] Ironically, its narrowed scope also disregarded what the agency itself had cited as the reason for revising the take reduction plan—a right whale found dead in October 2002 due to wounds caused by a lobster trap buoy line with a 600-pound weak link that complied with previous plan provisions.[38]

As upset as conservationists were by the proposal's limited scope, Maine lobstermen were even more outraged. In a firestorm of protest, thousands of state lobstermen hotly opposed the mandate to switch to neutrally buoyant or sinking groundlines. They repeated assertions that snags, associated gear loss, and the need to replace abraded line more often would spell economic disaster for small, independent fishermen along Maine's rockbound coast. In some cases, their anger boiled over into threats of physical harm against government officials. To address the controversy, the Maine Department of Marine Resources demanded that all state waters be exempt from the rule on grounds that right whales rarely occurred in Maine waters and further research on groundline alternatives was needed.

Although no major right whale habitats occur in nearshore Maine, the extent to which right whales pass through state waters is poorly understood. Almost no whale surveys have been conducted in Maine waters. What little is known is from opportunistic sightings, entanglement records, and a telemetry study that tracked the movements of half a dozen right whales in the northern Gulf of Maine. The telemetry study revealed that several of the whales visited Maine waters for at least brief periods.[39] Because of the extremely high concentration of lobster traps in the state's waters—in excess of 2 million traps during the peak summer lobster season—even infrequent right whale visits could pose great risk. Indeed, at the time of the proposed rule, a third of the entangled right whales that could be traced back to fishermen from license numbers on removed gear (3 of 9 cases) involved lobstermen working along the Maine coast.

The proposed exemption, however, still left thousands of Maine lobstermen subject to new groundline requirements and did nothing to assuage their ire. To press their case, they flooded the offices of Maine's influential congressional delegation, then led by Senators Olympia Snow and Susan Collins, with complaints.

In response, the legislators met with the secretary of commerce and director of the Fisheries Service, urging the agency to reconsider its decision. The agency agreed.

EMERGENCY RULES FOR RIGHT WHALE CALVING GROUNDS

As the Fisheries Service reconsidered its groundline rule in the fall of 2005, a right whale calf was found dead off Florida after being entangled in a gill net. The calf was first seen with its mother by a right whale survey team in late December off Georgia, with no signs of being entangled. It was seen with its mother three more times in early January off Florida, but by then photos revealed linear scars and a possible buoy trailing behind its flukes. The mother and calf were last seen together off Florida on January 11. Less than two weeks later, the mother was spotted alone in an agitated state. The following day a private vessel owner reported a dead whale floating 16 miles off Jacksonville Beach, Florida, not far from where the agitated mother had been seen. It proved to be a calf and was towed ashore that day for necropsy.

The dead calf was found just outside the region's critical habitat boundary but within the Southeast US Restricted Area that had been established to manage gill-net fishing when the take reduction plan was first adopted in 1999. Regulations for the area imposed certain requirements on what were then the only two gill-net fisheries operating in the region—one for sharks and the other for mackerel. When the calf's carcass was towed ashore, no lines or netting were attached, but the calf's tail stock and back had numerous cuts and diamond-shaped impressions in the skin, consistent with marks caused by gill nets. All sightings of the calf had been within the Southeast US Restricted Area and, after considering all the evidence, the Fisheries Service concluded that the calf died due to entanglement in a gill net in the restricted area during its November 15 to April 1 effective dates.

Contingency measures require that the director of the Fisheries Service immediately close the restricted area to gill-net fishing if a right whale is killed or seriously injured "as a result of an entanglement by gillnet gear allowed to be used in that area."[40] Once closed, the area is to remain closed unless new measures are developed to prevent another entanglement death or injury. On January 30, shortly after the necropsy results concluded the calf had died due to a gill-net entanglement, environmental groups wrote the director of the Fisheries Service insisting the area be closed.[41] After reviewing information, the agency also concluded that the calf died of entanglement in a gill net set inside the restricted area during its effective period. Therefore, on February 16, it issued an emergency rule closing the restricted area to all gill-net fishing for the remainder of the calving season. At the same time, it scheduled a meeting of the take reduction team's southeast regional subgroup to consider whether measures could be developed to reopen the area.

The subgroup met in April in St. Augustine, Florida,[42] to review information on the calf's death and three gill-net fisheries known to have been operating in the area at that time—the shark and mackerel fisheries, and a new unregulated gill-net fishery for small fish called whiting. The shark and mackerel fisheries were conducted principally to the south of St. Augustine in times and areas where right whales were less common; they were subject to various restricted area requirements mentioned earlier that minimized entanglement risks. However, because the whiting fishery was new, no regulations had been developed for its management.

The whiting fishery, which was based in the heart of the calving grounds 10 to 20 miles south of the Florida-Georgia border, was initially suspected of being responsible for the entanglement. However, the small mesh size of its nets did not match the larger net marks on the calf. In a remarkable bit of candor, a shark fisherman on the subgroup confided that the calf had been caught by a shark gill-netter operating off northern Florida who had not been carrying an observer as required and who had cut all his gear from the whale in an attempt to conceal the entanglement. The large mesh marks on the calf were consistent with shark gill nets. Because the offender was fishing illegally, fishermen on the team argued that criteria for closing the area had not been met (the responsible gear was not gear permitted in the area) and therefore asked that the area be reopened. The fisherman reporting the illegal operation, however, refused to identify the offender and the information therefore could not be verified. Commercial fishing invariably involves bad apples who disregard the law using honest fishermen for cover. Increasing enforcement patrols at public expense to allow fishing to continue seemed inappropriate to conservationists on the team and they therefore opposed reopening the restricted area without further restrictions.

After more deliberations, the subgroup recommended additional measures to restrict shark and mackerel fishing to areas and times even farther south (about 10 miles north of Cape Canaveral) where right whales occur less frequently. The whiting fishermen offered to accept almost any restrictions that would allow them to continue fishing, including carrying observers on all boats. However, because their target fish occurred only in the heart of the calving grounds and because the agency's observer program was already underfunded, scientists and conservationists did not believe that any measures allowing the fishery would be safe for mother-calf pairs.

Final approval of regulations for the protection of right whales rests with the director of the Fisheries Service, but it is rare for advice from responsible regional offices to be overturned. The staff of each regional office weighs competing responsibilities for the management of fishing and protection of endangered species differently. The Northeast Regional Office had lead responsibility for the Atlantic Large Whale Take Reduction Plan and consistently favored fishing interests; however, staff of the Southeast Regional Office considered closure of the northern part of the restricted area appropriate, despite differing views raised by the Northeast Office.

The agency's director was slow to make a final decision. With the emergency closure having expired and time running out to put new measures in place for the 2006 winter calving season, the Marine Mammal Commission wrote the Fisheries

Service urging that it act quickly to close the area to all gill-net fishing not recommended by the southeast subgroup of the take reduction team.[43] Shortly thereafter, the Humane Society filed a notice of intent to sue the agency if a new rule was not adopted by the start of the calving season.[44]

On November 15, the first day of the Southeast US Restricted Area's regulatory season, the Fisheries Service published another emergency rule for the coming winter's calving season[45] as well as a proposed permanent rule for future seasons.[46] Both actions followed the advice of the Southeast Regional Office and the take reduction team's southeast subgroup. This included restricting shark and mackerel fishing to waters south of St. Augustine and prohibiting all gill-net fishing north of that point, including operation of the new whiting fishery. A final permanent rule with the same measures was adopted in June 2007.[47]

MARINE MAMMAL COMMISSION REVIEW

Since the early 1990s, the Marine Mammal Commission had held reviews of the right whale protection program every few years to provide advice to the Fisheries Service and its recovery partners. In March of 2006, the commission reexamined the right whale program as part of a congressional request to review the cost-effectiveness of protection programs for the most endangered marine mammals. To conduct its review, the commission formed a panel of marine mammal scientists and invited the Fisheries Service and other involved agencies and organizations to describe their activities and plans for right whale recovery work. The panel's report was provided to all involved agencies and groups in early 2007.[48]

Whereas the panel was impressed by progress on research to assess and monitor right whale biology and ecology, it pointed out that the Fisheries Service's bureaucratic approach had effectively stymied progress on management actions to reduce entanglements. The panel concluded that the adaptive management approach used by the agency had been ineffective and was inappropriate for a species as highly endangered as right whales. It advised turning the approach on its head. Instead of gradually increasing measures with questionable prospects for success, the panel concluded it would be far more cost-effective and fitting to apply science-based management measures with the highest probability of success and then scale back measures as new information or better options became available.

Despite more than a decade of take reduction team meetings, the panel found that the team had consistently failed to reach consensus on needed measures and that the convoluted regulations adopted by the Fisheries Service based on its advice had failed to reduce observed right whale entanglements and deaths. Although one measure, sinking groundlines, had a convincing rationale, plans for requiring its use were proceeding far too slowly, while other options were receiving too little investment (e.g., technology to eliminate buoy lines) or were underutilized (e.g., seasonal fishing closures in high-use right whale habitats). Because all options proposed by the agency relied on weak links to prevent buoy line entanglements, the panel found it difficult to imagine any scenario in which proposed measures would meet required goals. The panel therefore concluded that the expense of convening a 60-member take reduction team that could not agree on needed actions was not a cost-effective investment and that the statutory provisions authorizing take reduction teams had never envisioned such a prolonged, open-ended process for reducing right whale entanglements.

The panel therefore recommended that the Large Whale Take Reduction Team be replaced by a less costly scientific body, such as a small recovery team composed of individuals with direct knowledge of right whale biology, involved fisheries, and whale entanglement issues. It also recommended that fixed gear in major right whale aggregation areas, including designated critical habitats, be limited to that which had been demonstrated as being whale safe before it was approved. To move in that direction as quickly as possible, the panel recommended prohibiting gear with vertical lines in known aggregation areas during peak right whale seasons and requiring sinking or neutrally buoyant groundlines in all other areas. The panel noted this would encourage the creativity of fishermen to develop ways of catching lobsters and finfish without using hazardous buoy lines in times and areas where whales were most abundant. The panel's recommendations, however, fell on deaf ears. The Fisheries Service proceeded on its previous course of action with no changes.

FINAL RULES FOR BROAD-BASED GEAR MODIFICATIONS

As 2006 came to a close and work on a final broad-based groundline rule continued, the Fisheries Service reconvened the take reduction team to provide its members with an update on the results of gear research and to begin discussing ways to mitigate vertical line entanglement risks as part of a third major revision of the take reduction plan. However, with controversy over the broad-based groundline requirements still swirling, the team spent most of its time reexamining the pending rule in light of new research results on neutrally buoyant and low-profile line. The new studies revealed that to avoid arching groundlines between traps, line density had to be greater than that of sea water, thereby rendering the term "neutrally buoyant line" a misnomer. Scientists and conservationists on the team urged that the term be abandoned in lieu of "sinking line" and that it be required for groundlines everywhere. Fishermen on the team, however, urged that local and state fishing interests be afforded an opportunity to submit proposals for "low profile" groundline and identify where it should be used, the rationale for its use, and the line density or height off the bottom that should be allowed.

Following the team's meeting, Maine lobstermen and the Maine Department of Marine Resources continued their adamant opposition to sinking groundline. The Fisheries Service, in consultation with the Maine Department of Marine Resources, therefore entered into direct negotiations with fishing interests, outside of the take reduction team process, to consider additional areas off Maine that could be exempted from groundline requirements. Concurrent with those negotiations, the agency provided the Maine Lobstermen's Association a $2 million grant from funds earmarked for the agency by Senator Olympia Snow to organize a floating groundline buy-back program for Maine fishermen, similar to the earlier program for Massachusetts' lobstermen.

With no sign of when the agency planned to move ahead with its "broad-based" groundline rule that had been proposed more than a year and a half earlier, the Humane Society of the United States and the Ocean Conservancy filed a lawsuit in February 2007 (*Humane Society of the US et al. v. Carlos M. Gutierrez, Secretary of Commerce et al.*). It alleged that the Fisheries Service's indefinite rule-making delay violated obligations under the Marine Mammal Protection and Endangered Species Acts to protect endangered whales. The court agreed and the Fisheries Service pledged to publish a rule by October 1, 2007.

On August 17, 2007, the agency released a final environmental impact statement on its broad-based gear modifications. It identified a new preferred option that closely mirrored its previous proposal, but with a significantly expanded exemption area for the Maine coast. The new exemption area included 70 percent of the waters within the state's three-mile jurisdiction. Although the analysis included no information on the number of traps likely to be exempted from requirements, it probably included at least 1 to 2 million, amounting to nearly half of all those set along the East Coast.

Because the expanded exemption zone had not been subject to previous public review, the agency provided a 30-day comment period on its final statement. To justify the expanded exemption zone, the Fisheries Service stated that, upon reexamining right whale sighting data, it now found that right whales were so scarce in the enlarged exemption zone that whales were unlikely to encounter groundlines. It also noted that because there was no evidence that right whales foraged and dove to the bottom in the area, they were unlikely to encounter floating groundlines. Finally, given perceptions of fishermen that right whales rarely occurred along the Maine coast, the agency noted that groundline restrictions in the area would undermine the willingness of fishermen to comply with take reduction measures.[49]

After reviewing the agency's new exemption area and rationale, the Marine Mammal Commission and conservation groups opposed its expansion. They pointed out that the agency failed to account for evidence of past right whale entanglements in the expanded zone. They also noted that the absence of sighting records when few or no sighting surveys had been conducted along the Maine coast was not evidence that whales did not occur in the area. The commission again noted the agency had failed to meet its obligation to consider a full range of possible mitigation alternatives, including seasonal fishing closures in high-use right whale habitats that could compensate for buoy line entanglements. Maine lobstermen, on the other hand, opposed the new exemption boundary claiming it was still too narrow. They recommended extending it even farther offshore, stating that the absence of right whale sightings applied to all state waters as well as adjacent federal waters.

While it took nearly two years to consider comments on its 2005 draft environmental impact statement, under the pressure of litigation the agency completed its analyses of comments on the final environmental impact statement in four days. On September 21, 2007, it signed a record of decision rejecting all recommendations for further change. In response to comments that past entanglements had been documented in the exemption zone, the agency stated they considered the number of entangled whales traced to Maine lobstermen was too small to justify a reliable trend.[50] It also dismissed the need to consider closures, stating that it considered options in its environmental impact statement to be a comprehensive set of alternatives. Calls to expand the exemption area were dismissed in a similarly terse manner. The agency stated that expanding the exemption area would increase entanglement risks and prevent achievement of its legal mandates. In neither case did the agency support its conclusions with a thorough review of available data.

On October 5, 2007, two weeks after signing the record of decision, a 90-page final rule was published to implement broad-based groundline requirements in what was the second major revision of the Atlantic Large Whale Take Reduction Plan.[51] The new rule required sinking groundlines in all nonexempt areas. Most of its ancillary requirements (e.g., revised weak link breaking strengths) became effective on November 5, 30 days after the rule's publication. Requirements for sinking groundlines, however, were deferred from January 1, 2008, the date initially proposed in 2005, to October 5, 2008, to provide more time for fishermen to purchase sinking line and replace their floating groundline.

At the end of April 2008, the Fisheries Service reconvened the take reduction team to review the status of the new rules and restart deliberations on the next plan revision to mitigate endline entanglements. Most of the meeting, however, again focused on new groundline rules, with fishermen requesting additional exemptions along the coast Maine, in offshore waters, in the mid-Atlantic region, and off North Carolina and Florida. None of the requests received consensus support and the Fisheries Service subsequently rejected them.

That fall, however, the Maine Lobstermen's Association wrote the Fisheries Service asking for a further delay in groundline requirements off Maine.[52] The letter claimed that the October effective date fell in the middle of an important fishing season and that it would disrupt fishing if lines had to be changed at that time. They also argued that it was unclear how the agency planned to enforce the measure and that fishermen needed more time to switch out all their groundlines. To the consternation of conservation groups who believed there had been more than ample notice of the impending change given that it had been pending as a proposed rule since 2005, the agency amended the rule to further delay the starting date by six months to April 2009.[53] In doing so, it let DAM and SAM requirements lapse, even though the new groundline requirements intended to replace them had not yet gone into effect. The Humane Society and Defenders of Wildlife filed a lawsuit to compel the agency to reinstate the lapsed rules and extend them until the new groundlines came into effect (*Humane Society of the US et al. v. Carlos M. Gutierrez, Secretary of Commerce et al.*). The court agreed and at its direction, the Fisheries Service adopted a rule in early October 2008 that restored the lapsed provisions and continued them through April 2009.[54]

To further the replacement of floating groundlines, Congress appropriated an additional $3 million to help Maine lobstermen purchase new sinking lines.[55] This time the program was administered jointly by the Maine Department of Marine Resources and the Gulf of Maine Lobsterman's Association. During the course of 24 exchange events in 2009 and 2010, $2.3 million was provided to 2,000 lobstermen and 2 million pounds of floating groundline was collected and sent to recycling centers.[56]

When the sinking groundline rule finally went into effect on April 5, 2009, it had been six years since the groundline rule-making process began. Although nearly half of all lobster traps had been exempted from the requirements, it was the first modification with a convincing rationale that had a chance of reducing right whale entanglements. When it went into effect, the provisions for DAMs and SAMs and the menu of optional gear modifications under the take reduction plan's first major revision expired. The second revision also incorporated the new restrictions on gill-net fishing in the calving grounds, which significantly improved protection for whales in that area. Assuming that perhaps a quarter of all right whale entanglements had been caused by groundlines[57] and that the traps in exempted areas along the Maine coast posed at least some risk of entanglement, the new groundline and gill-net provisions might have reduced right whale entanglement risks by perhaps 10 to 15 percent.

Third Revision of the Take Reduction Plan: A Vertical Line Rule

By the time new groundline requirements went into effect in the spring of 2009, the take reduction team had yet to devote any meaningful attention to endline measures. Up to then, the Fisheries Service had relied on unsupported claims that weak links would mitigate entanglement risks, even though they were being found on more and more entangled whales. To restart discussions of ways to prevent entanglements in vertical buoy lines, the Fisheries Service reconvened meetings of the full team or its separate regional subgroups half a dozen times between the spring of 2009 and early 2012. At meetings in April 2009, agency officials laid out its schedule for another major revision of the take reduction plan—a "vertical line" rule to mitigate endline entanglements. Its schedule called for another five-year rule-making process to be completed by 2014.

Recognizing that research had identified no new technological options and there was no evidence that weak links provided protection, the team agreed to consider ways of reducing amounts of line in the water column. Three methods were suggested during meetings in 2009: (1) reducing trap numbers, (2) imposing caps on the number of allowed endlines, and (3) grouping more traps per endline (a strategy called "trawling up"). At the urging of team scientists and conservationists, the northeast subgroup also considered setting aside areas in which no vertical lines would be permitted in order to encourage experimentation with ropeless fishing methods.

To establish a quantitative goal for line reductions, scientists and conservationists also proposed decreasing the number of lobster traps by 50 to 60 percent. This suggestion was based in part on studies that revealed lobster traps to be little more than feeding stations for the crustaceans. Video recordings of deployed traps showed lobsters of all sizes entering to feed on bait and leaving at will. This suggested that catch levels depended more on the density of lobsters in the area than the number of traps.[58] This was further supported by experimental fishing at Monhegan Island, Maine, where fishermen found they could catch almost as many lobsters with a fraction of the traps.[59] Between 2005 and 2007, lobstermen on the island agreed to fish 500 traps in some areas and 167 traps in other areas of comparable size and habitat for two two-month periods. To their amazement, the total catch in areas with 167 traps was just 15 percent less than areas with 500 traps.[60] In light of those findings, island fishermen agreed to reduce the number of traps they set by half. Although their total catch was slightly lower, savings in fuel, bait, equipment, and fishing time actually increased their net profits.

Fishermen on the team, however, opposed the approach, pointing out that the situation at Monhegan Island was unique because there was an unwritten understanding among lobstermen that those waters were off-limits to nonislanders. Team fishermen argued that it would be impossible to control trap densities in other fishing areas. With no other alternatives offered, there was no agreement on an overall goal for reducing trap or endline numbers. Instead, fishermen in different states agreed to submit proposals to reduce endline numbers to an extent they considered practical. To evaluate proposals, the team recommended that a "co-occurrence model" be developed to estimate the likelihood of whales encountering endlines in different times and areas. It recognized that line encounter rates were not necessarily the same as entanglement risks. Scientists on the team pointed out that the likelihood of death or serious injury in a short buoy line in shallow water with a single trap was far smaller than entanglement in a long line in deep water with a heavy string of 20 traps. However, no method could be identified to account for such differences. Nevertheless, developing a model to estimate encounter rates seemed valuable, and the Fisheries Service's Northeast Regional Office hired a consultant, Industrial Economics, Inc., for that purpose.

Conceptually, the model was straightforward. The entire East Coast Exclusive Economic Zone from Maine to Florida was divided into a grid of 10 minute squares of latitude and longitude (roughly 10 square miles). For each square or "cell," the consultant was to estimate monthly numbers of endlines and whale densities. A measure of encounter rates could then be calculated for each cell in each month by multiplying the density of each whale species by the number of endlines to produce a co-occurrence score reflecting a measure of line encounter rates. Regional and overall encounter risks could then be calculated by adding co-occurrence scores for each cell. By the same token, the model could identify times and areas where encounter rates were most likely and management needs were greatest for each whale species. Reductions in co-occurrence scores achieved by reducing endline numbers could then be calculated as a measure of whale protection.

Generating reliable data for the model, however, proved to be challenging. Fishermen had never been required to record the number of endlines they fished and this varied by fisherman, fishing location, and season. The team urged the Fisheries Service to contact state, regional, and federal fishery managers to obtain relevant data, but the results were modest at best. Only the state of Massachusetts required lobstermen in state waters to record endline numbers and trawl lengths. The state of Maine also had surveyed some lobstermen asking how many traps they thought they set per string by area and time of year, but for other states and federal waters only qualitative estimates by agency managers were available. This left the consultant with the enormous task of estimating endline numbers from data on the number of fishing permits in different areas, the number of licensed traps or nets per fisherman, and a disparate array of qualitative advice about how many traps per trawl were set in different months and areas. Initially, data were compiled for the 2008 fishing year, but were later updated through 2010.

Estimates of whale numbers were more readily available, but these data had their own problems. At first, the Fisheries Service provided the consultant with data from standardized whale surveys conducted by its own scientists for just a few areas. Those surveys, however, covered only a relatively small area, and team scientists found the resulting whale density maps inconsistent with what was known from more extensive data about right whale occurrence in high-use habitats. They therefore recommended using "sightings per unit of effort" (SPUE) data archived by the Right Whale Consortium from aerial and shipboard surveys dating back to the 1970s.

That data did a far better job describing the distribution of right whales as well as other large whales, but still had a troubling weakness. Even with 30 years of survey effort, many areas for quite a few months of the year had either not been surveyed or had been surveyed for only a few hours with no sightings. As a result, many cells had no right whale sightings and produced whale densities of zero. Because co-occurrence scores were calculated by multiplying whale densities by endline numbers, cells with no sightings resulted in co-occurrence scores of zero, regardless of how many endlines were present. Opportunistic sightings made independently of surveys, however, demonstrated that right whales occurred at least occasionally in almost all areas. Thus, the model underestimated entanglement risks by an unknown extent. This was particularly troubling off Maine where gear densities were exceptionally high, but sparse survey effort yielded no whale sightings. Team scientists asked that the contractor estimate more realistic minimal whale densities for cells showing a zero whale density. Despite their request, no steps were taken to estimate minimal whale densities to ensure that the model was not underestimating risks.

Members of the team also raised concern over the lack of information with which to assess the plan's effectiveness. Ropes removed from entangled whales offered the greatest potential for shedding light on sources of the gear and whether associated risks were being reduced; however, the rope-marking requirements initially proposed in 1997 had been reduced to apply only to gear set in critical habitats. The marking requirements also called for only a single colored band on each buoy line. Some team members therefore urged requiring a more elaborate and comprehensive system in which lines would be marked in all areas at more frequent intervals using paint or tape to create a colored band. They noted that a simple color coding system could help identify involved fisheries (trap vs. gill net), gear parts (groundline vs. vertical line), and geographic regions. Some fishermen, however, opposed any expansion in marking requirements, repeating claims that it was too difficult a task given the amount of lines they used. As a result, the team was unable to agree on any marking proposals, which would mean that the source of gear removed from entangled whales would likely continue to be unidentifiable (figure 21.4).

On June 14, 2011, the Fisheries Service began the "scoping" process for the new rule by requesting comments on options to reduce endline entanglements.[61] Respondents from the fishing community generally supported the status quo option (no new measures) or expanding existing exemption zones. They argued that (1) they never saw right whales in the areas in which they fished,

Figure 21.4. A disentanglement team with the Florida Fish and Wildlife Research Institute grapples for a line trailing the mouth of an entangled juvenile right whale (#4057) off Jacksonville, Florida, in February 2014. (Image courtesy of Florida Fish and Wildlife Commission, taken under NOAA permit number 932-1489.)

(2) information was insufficient to justify any new regulations, (3) increasing the number of traps per trawl would create safety problems, and (4) regulations should focus on ship-strike mortalities rather than penalizing fishermen. Conservationists, on the other hand, urged that the regulations focus on right whales more than other whales, that measures be adopted before 2014, and that gear marking requirements be expanded.

The Marine Mammal Commission provided some of the most extensive comments.[62] It recommended modifying the co-occurrence model to account for right whales in unsurveyed or minimally surveyed areas with no sightings. The commission again urged that all fishing with vertical lines be prohibited in right whale critical habitats. It also recommended that lines be marked at intervals comparable to lengths of line removed from whales, and that fishermen be required to record the number and location of endlines fished in order to provide data for monitoring their numbers and locations over time.

At the team's last face-to-face meeting before publication of proposed rules in January 2012, five proposals for reducing endline numbers were reviewed;[63] one each came from the states of Maine and New Hampshire, one came from lobstermen in central Maine, and two came from the Fisheries Service. The proposals from the Fisheries Service included region-wide approaches, with one for New England and the other for mid-Atlantic and southeast states. All proposals applied to multiple geographic management zones (up to 35 zones in the Fisheries Service's northeast proposal). Each proposal included (1) a different set of management zones with minimum trawl lengths ranging from 2 to 20 traps for per zone, (2) requirements for one endline on trawls with fewer than a specified number of traps (usually 3 to 5 traps), and (3) areas exempt from all vertical line requirements. For most proposed management areas, the co-occurrence model was used to estimate reductions in endlines and encounter rates. However, the team was provided with no overall estimate of risk reduction for all areas, including both exempt and non-exempt areas. Moreover, with no agreed-upon goals for quantifying needed endline reductions, each team member was left to make his or her own qualitative judgment about whether proposed measures would achieve the statutory aim of reducing entanglement deaths and serious injuries to less than one whale every two years.

Team scientists and conservationists were particularly troubled by the validity of model results. No steps had been taken to address their recommendations for adjusting risk estimates in cells with zero whale densities. Moreover, they were upset about assumptions made to exempt areas along the Maine coast where more than half of all endlines occurred. Although the team had not agreed to it, the Fisheries Service had already decided to exempt the same area that had been excused under its earlier groundline rule. It apparently did so to avoid another outburst of opposition from Maine lobstermen. Thus, co-occurrence analyses were presented only for "regulated waters" that excluded risk estimates for the areas exempted from earlier groundline requirements.

New England fishermen were most disturbed by proposals to increase trawl lengths in nearshore nonexempt areas. Because of limited deck space on their small boats, they pointed out that they could not safely handle longer trawls. They also objected to the use of only one endline on even short trawls, citing fears of increased trap losses and the possibility of gear being set on top of that already deployed, since location of the earlier traps would not be as easily determined. In addition, Maine fishermen continued to believe the exempted area off their coast was too narrow. Fishermen in other states raised concerns that it was unfair that their own nearshore waters were not to be exempted as well. Moreover, when results of the co-occurrence model revealed that 90 percent of all endlines along the East Coast were deployed off New England, fishermen in mid-Atlantic and southeastern coastal states argued that the entanglement issue was effectively a New England problem and that regulations should be limited to that region.

By the end of the January 2012 meeting, all sides found the proposals flawed and needing change. Conservationists and scientists concluded they did not provide enough protection for whales and that time-area closures in high-use right whale habitats offered the only readily available alternative. Fishermen urged exemptions for more areas and shorter trawl lengths. With no agreement on needed measures and no funding to reconvene the team for another in-person meeting, the Fisheries Service advised that it would consider and analyze any new proposals submitted by team members over the coming month and then hold a conference call to review them.

In the weeks that followed, six new proposals were submitted; these were discussed during a conference call in February 2012.[64] Separate groups of team scientists and conservationists each submitted suggestions for a total of six seasonal closures, some of which overlapped. They included proposed closures in Cape Cod Bay, the Great South Channel, Jeffreys Ledge, Stellwagen Bank, and Jordan Basin in the central Gulf of Maine. Fishery groups submitted three proposals for exempting additional state waters and establishing alternative management zones with different minimum trawl lengths. Maine lobstermen proposed three zones parallel to the shore that were 3, 6, and 12 miles out, with different minimum trap limits based on the typical vessel sizes fishing in each strip. To address failure of the co-occurrence model to account for whales in cells with few or no surveys, the Marine Mammal Commission funded a team scientist to develop a method that the Fisheries Service's consultant could use to modify their model and account for that deficiency.[65] The resulting report and correction methods were endorsed and submitted by all team scientists and conservationists as a separate proposal.

Following the meeting, each proposal was analyzed using the co-occurrence model for presentation to the team's northeast and mid-Atlantic/southeast subgroups during separate conference calls in mid-April.[66] The team again failed to endorse any individual or combination of proposals, so the Fisheries Service decided to consider most of them in its forthcoming environmental impact statement. With regard to modifying the co-occurrence model, however, the agency advised it was too late to incorporate changes in its model, but that it would undertake a sensitivity analysis to determine the extent to which such a change might alter model results.

On July 13, 2013, the Fisheries Service released a draft environmental impact statement for its vertical line rule. Three days later it published the proposed regulations.[67] Different approaches were proposed for the northeast, mid-Atlantic, and southeast regions. For the northeast region, the core requirements included four parts: (1) 25 new management areas, each specifying varying trawl lengths of 1 to 20 traps and weak link breaking strengths of 600 to 2,200 pounds; (2) use of one endline for trawls of five traps or less; (3) three new seasonal closures applicable to lobster fishing but not gill nets (Jordon Basin, Jeffreys Ledge, and an expanded Cape Cod Bay–outer Cape zone); and (4) marking all endlines at the top, middle, and bottom according to an expanded color coding system to distinguish between lines from traps in three different regions, and those from gill nets in two regions.

For the mid-Atlantic and southeast regions, proposed measures effectively maintained the status quo. In the mid-Atlantic (New York to North Carolina), the agency proposed requiring 600-pound weak links and marking buoy lines on traps and gill nets with distinctive colors. No restrictions on trawl length were proposed. In the southeast region, the agency proposed a new restricted area for trap fisheries (e.g., sea bass, whelks, and hagfish) with boundaries identical to the existing gill-net restricted area. In the new trap fishery restricted area, buoys would have to be equipped with 200- to 600-pound weak links, buoy lines would have to be marked with a distinctive color, and multiple-trap trawls would be prohibited. Whereas the agency sought to reduce endlines in the northeast region by clustering traps on buoys, in the southeast it decided that the added weight of multiple trap trawls could drown or injure right whale calves and therefore required that every trap have its own buoy with no multi-trap trawls.

The proposal elicited over 40,000 comments.[68] The vast majority were form letters submitted by members of environmental groups. They expressed support for most measures, but objected to the exemption of state waters in Maine and recommended refinements in the gear-marking system to distinguish more fishing areas from one another. Form letters from fishing organizations supported the proposed exemption areas and urged others, but strongly opposed any gear-marking requirements and all time-area closures. Over 500 letters provided separate comments from individuals, agencies, and organizations.

Most fishermen submitting individual comments supported proposed trawl lengths of 15 to 20 traps in waters 12 miles or more offshore, but because most trawls in those areas were already at least that long, the requirements meant little or no change. However, Maine lobstermen fishing closer to shore strongly objected to requirements for using 3 to 10 trap trawls in proposed zones within the 12-mile limit but beyond the state's regulatory exemption line. They stated that hauling trawls of that length safely aboard small boats (often less than 25 feet long) was impossible due to limited deck space and the risk of capsizing in rough weather. The Maine Department of Marine Resources echoed those concerns and added that the cost and labor required to convert single traps into trawls would be too great. It recommended a new zone between 3 and 6 miles from shore, with either a shorter trawl minimum or an exemption for small boats. It also proposed additional exemption areas around offshore islands. Citing similar concerns, the state of New Hampshire sought exemptions for fishermen in its state waters.

Almost all fishermen strongly opposed additional gear marking; they considered it unnecessary, unlikely to produce useful results, and too difficult to accomplish. In contrast, conservationists and scientists argued that gear marking was essential for evaluating the effectiveness of mitigation measures and that it should be even more detailed. They urged that the marking system for New England use additional colors or bicolor codes to distinguish more fishing areas, fisheries, gear parts, and fishing trawl lengths, all to identify the source of entanglements more clearly. They also recommended that the marks be enlarged and placed at more frequent intervals to improve detection on line fragments removed from whales or seen by aerial observers.

Almost all fishermen strenuously opposed the proposed seasonal closures, particularly those in the Jordon Basin and at Jeffreys Ledge. The Maine Department of Marine Resources noted that the co-occurrence model predicted only a small reduction in entanglement risks from those closures compared to the minimum trawl length requirements, while estimated costs of foregone catch from closures could be 10 times higher than costs for trawling up. However, they offered no suggestions to compensate for the protection benefits of the closures they opposed.

Several groups, particularly the Marine Mammal Commission, the Massachusetts Division of Fisheries, the New England Aquarium, and the Humane Society challenged the validity of the co-occurrence model. Like one of the three independent peer reviewers that the Fisheries Service asked to review their model, they concluded that it was not ready for use as a management tool. They pointed out that no steps had been taken to validate either the model or its estimates of endline numbers, and that none of its estimates included statistical confidence limits for assessing reliability. Doubts were heightened when preliminary results of another model for the coast of Maine suggested that the agency's model overestimated conservation benefits by as much as 50 percent. Many reviewers therefore recommended that the agency collect more reliable data on the number and location of endlines and that an analysis be performed to validate the model before relying on its results to make management decisions. Citing examples of both right whale and humpback whale deaths in gill nets, several groups also urged that the proposed seasonal closures apply to gill nets as well as traps.

Some state agencies even recommended abandoning the trawling-up strategy altogether or at least supplementing it with other measures. State agencies in Georgia, Massachusetts, and Rhode Island noted the inconsistent Fisheries Service rationale for measures in the northeast and southeast. They pointed out that whereas proposed measures for the southeast sought to prevent multitrap trawls on the grounds that longer, heavier trawls would cause more significant injuries for whales that became entangled, particularly for small calves, the agency argued that in the northeast, longer trawls would reduce entanglement risks by decreasing endline numbers. The states of Georgia and Rhode Island therefore recommended that instead of increasing trawl lengths, overall trap numbers should be decreased in order to reduce line numbers.

Similar concerns were echoed by the New England Aquarium. It noted that analyses of past entanglement trends revealed an increase in severe entanglement injuries and deaths coincident with increases in line strength since the 1990s.[69] Accordingly, it recommended limiting the breaking strengths of buoy lines to 1,500 pounds north of Cape Hatteras, North Carolina, and 1,000 pounds south of that point, in order to protect calves.

A FINAL PLAN FOR VERTICAL LINES

In May 2014, the Northeast Regional Office of the Fisheries Service released a final environmental impact statement.[70] After a 30-day comment period, it published a final rule on June 27.[71] Similar to past rule-making actions, the Northeast Office adopted many recommendations and requests by fishing interests that increased risks to whales, while rejecting those urged by scientists and conservations that strengthened protection.

The final rule eliminated the seasonal closures the Northeast Office had initially proposed for the Jordan Basin and Jeffreys Ledge on grounds that their model predicted conservation benefits were small, while economic impacts could be substantial. At the same time, it adopted a Massachusetts Bay Restricted Area that was smaller than initially proposed (a closure effective January 1 to April 30), even though its predicted conservation benefits were comparable to those of the other closures it was now rejecting. The agency also expanded exempted areas to include all state waters in New Hampshire and areas within a quarter mile of three additional offshore islands in Maine. Minimum trawl lengths beyond the exemption line in Maine state waters and for all state waters in Rhode Island and Massachusetts were reduced to two traps in light of safety concerns for small boat operators. Also eliminated were gear-marking requirements for Maine lobstermen in the exemption zone; here, the Fisheries Service decided it was too difficult for fishermen to use separate sets of marks when fishing inside and outside of that zone. Otherwise, the proposed marking system was unchanged. Finally, effective dates for the new rule were deferred to allow more time for fishermen to bring their gear into compliance. Fishermen in the southeast had until November

2014 to comply, whereas those in the northeast had until June 2015.

Even though whale protection measures in the final rule were significantly weakened, the agency predicted a 37.9 percent decrease in entanglement risks in "regulated" areas other than exempted waters, as based on its dubious co-occurrence model. The Fisheries Service exercised its discretion to conclude that this was adequate to achieve the protection required by the Marine Mammal Protection and Endangered Species Acts.

Similar to events following the second revision, fishermen and state agencies registered strong objections and sought further exemptions and changes. This time, four states submitted formal exemption requests to the Northeast Regional Office pursuant to guidelines established as part of the Atlantic large whale take reduction process. Under those guidelines, requests are to be reviewed by the team and must include alternative measures to compensate for any resulting reduction in protection. Upon receiving the requests, the Fisheries Service reconvened the take reduction team.

Massachusetts asked for a change in the effective dates for seasonal fishing closure covering the Cape Cod Critical habitat and adjacent areas (the Massachusetts Restricted Area). They sought to shorten the January 1 to April 30 closure dates by eliminating the month of January in order to allow an economically important fishing season for local lobstermen to continue into January. The state requested that the closure start on the first of February instead of the first of January; this request was addressed during a team conference call in early October 2014.[72] To compensate for diminished whale protection, the state proposed expanding closure boundaries to include additional areas southeast of Cape Cod, where there was a significant co-occurrence risk for right whales and fishing gear in April. The co-occurrence model suggested the change would provide protection comparable to the longer closure period. With more reliable data on endlines in Massachusetts waters than elsewhere, the team concluded the change would not reduce right whale protection and agreed to support the request. In response, the Fisheries Service amended the plan in mid-December 2014 and changed the area's boundaries and the start date to February 1.[73]

The states of Maine, New Hampshire, Massachusetts, and Rhode Island also requested additional exemptions from trawling-up requirements for all or parts of their state waters. For those requests, no compensatory protection measures were identified. To consider the proposals, the take reduction team was reconvened for an in-person meeting in January 2015.[74] While the states maintained the new exemptions would have a negligible effect on whale protection, conservation groups proposed reinstating at least one of the seasonal closures (either Jeffreys Ledge or Jordan Basin) that the agency had withdrawn in its final action as a compensatory measure. When fishermen adamantly opposed any closures, conservationists withdrew their proposal in lieu of a requirement for fishermen to mark their gear in both areas with a unique mark; they hoped that this would at least enable the identification of any whales that became entangled in gear in those areas. Most, but not all, team members agreed to the exemption requests with that addition, even though it offered no compensatory protection.

Following the team's meeting, the Fisheries Service published a proposed amendment on March 19, 2015, to adopt all the state exemption requests to the new take reduction plan rules, and asked for comments.[75] The notice stated that the proposed exemptions had followed the Atlantic take reduction team exemption request process and would have only minimal effects on whale protection. This claim was based on co-occurrence model results that projected a decrease in projected conservation benefits of 8.2 percent. That is, instead of reducing co-occurrence scores by 37.9 percent in "regulated" waters under the final rule adopted in June 2014, it would reduce them to 29.7 percent with the new exemptions.

The Marine Mammal Commission strongly opposed the proposed exemptions.[76] It noted that the 8.2 percent reduction in right whale protection amounted to a nearly 20 percent reduction compared to the final rules adopted in June 2014, which hardly seemed minimal. Considering the agency's record of consistently overestimating the effectiveness of its protection measures and the questionable reliability of the co-occurrence model, the commission considered this reduction in projected effectiveness highly inappropriate. It also noted that the states had not followed the established exemption guidelines and procedures because neither state exemption requests nor the agency's proposed rule identified measures to compensate for the reduced protection. The commission therefore recommended that the agency defer the rule until compensatory protection measures were developed. They also asked the Fisheries Service to explain why it considered such a large decrease in expected protection to have a minimal effect, and why it considered the requests consistent with the established exemption request guidelines.

On May 28, 2015, the agency adopted the final rules implementing the amendment as proposed, but with no response to the commission's comments and questions.[77] The new exemptions completed the third major revision of the Atlantic Large Whale Take Reduction Plan. But again, it left little hope for reducing right whale entanglement deaths and injuries.

After 20 years of contentious struggle and the expenditure of tens of millions of dollars to reduce right whale entanglements, there had been no measureable reduction in entanglements or the resulting deaths. Although some useful measures had been implemented—requirements for sinking groundlines, a seasonal ban on hazardous fishing gear in Cape Cod Bay, and further restrictions on gill-netting in the species calving ground—prospects for success under the latest plan revision were still dim. The new plan relied almost entirely on reducing gear encounter rates by clustering more pots on fewer endlines off New England. However, with such poor data on the numbers of endlines, exemptions for over half of all traps, and an almost impossible enforcement task, a meaningful reduction in the numbers of buoy lines sufficient to significantly reduce entanglement risks seemed unlikely. Moreover, by increasing the number of traps per trawl, and thus their overall weight, the new trawling-up strategy raised the specter of actually increasing the risk of serious injury and death for any whales that did become entangled.

Perhaps the greatest accomplishment over this long process has been the Fisheries Service's movement toward protection measures based more on scientific reasoning with less reliance on unproven speculation about requirements such as weak links, line thickness, and slack in buoy lines. Nevertheless, there is little doubt that yet another major modification of the take reduction plan will be needed if entanglement deaths are to be reduced. When it is revised—hopefully sooner than later—meaningful progress will require a reexamination of the management strategy put in place in 2015 that required the clustering of more traps on fewer endlines in light of alternative gear modifications. Recent research has shown promising results for new modifications, particularly (1) weak or "whale-safe" rope that breaks before entangled whales can be seriously injured or killed, and (2) improved designs for "ropeless fishing" devices (elimination of vertical buoy lines) that store buoy lines at the bottom between catch and retrieval operations. A combination of these two approaches applied to different segments of the relevant fisheries may well provide greater whale protection than the current management strategy that creates longer, heavier trawls that are more likely to injure entangled animals.

Even more important will be a stronger commitment by the Fisheries Service to implement effective measures in a timelier manner. Seasonal fishing closures in high-use right whale habitats will remain a critical component of the take reduction plan and will need to be strengthened, at least until gear modifications are proven effective. This and any other management options (e.g., reducing trap numbers) that affect fishing effort should also be developed as part of a more unified ecosystem management perspective. This perspective should simultaneously recognize the benefits of measures for conserving not only right whales but also stocks of target fish and shellfish by integrating them into both take reduction plans *and* fishery management plans. In this way, restrictions on fishing opportunities that also contribute to fishery conservation could be counted as conservation credits in management plans and allow easing of other plan restrictions.

In addition, the Fisheries Service must pay far more attention than it has in the past to monitoring and data collection needs at the same time any new take reduction plan measures are implemented. A plan that cannot monitor or assess the adequacy of its measures is not an adequate plan. The importance of data needs (gear marking to identify the fisheries and areas responsible for entanglements and data on the number of endlines, number of traps per trawl, and breaking strength of buoy lines fished in different areas at different times of the year to assess entanglement risks) and the provisions to collect them cannot be persistently disregarded or put off as afterthoughts, as they have been over the first 20 years of planning.

NOTES

1. Senate Appropriations Committee. September 8, 2000. Subcommittee on Commerce, Justice, State, and the Judiciary and Related Agencies Appropriation Bill, 2001. US Government Printing Office. Washington, DC, pp. 117–118 (quotation).

2. RESOLVE. 2000. *Atlantic Large Whale Take Reduction Team: Final Summary of Deliberations, Recommendations, and Next Steps from the 22–24 February 2002 Meeting.* Report provided to the Northeast Regional Office, National Marine Fisheries Service, Gloucester, MA.

3. National Marine Fisheries Service. 2000. Taking marine mammals incidental to commercial fishing operations; Atlantic Large Whale Take Reduction Plan Regulations; Interim final rule. 65 Fed. Reg. 80368. (December 21, 2000.)

4. Clapham, P, and R Pace. 2001. *Defining Triggers for Temporary Area Closures to Protect Right Whales from Entanglements: Issues and Options.* Northeast Fisheries Science Center Reference Document 01-06. National Marine Fisheries Service. Woods Hole, MA, US.

5. National Marine Fisheries Service. June 14, 2001. *Endangered Species Act Section 7 Consultation Biological Opinions on: (1) Re-initiation of Consultation on the Federal Lobster Management Plan; (2) Authorization of Fisheries under the Northeast Multispecies Fishery Management Plan; (3) Authorization of Fisheries under the Monkfish Fishery Management Plan; and (4) Authorization of Fisheries under the Spiny Dogfish Fishery Management Plan.* Northeast Regional Office. Gloucester, MA, US.

6. RESOLVE. 2001. *Meeting Summary: Atlantic Large Whale Take Reduction Team Meeting, June 27–28, 2001.* Report provided to the Northeast Regional Office, National Marine Fisheries Service. Gloucester, MA, US.

7. Moore, M, M Walsh, J Bailey, D Brunson, F Gulland, S Landry, D Mattila, et al. 2010. Sedation at sea of entangled North Atlantic right whales (*Eubalaena glacialis*) to enhance disentanglement. PLoS ONE 5(3):e9597, doi:10.1371/journal.pone.0009597.

8. Ibid.

9. National Marine Fisheries Service. 2001a. Taking of marine mammals incidental to commercial fishing operations; Atlantic Large Whale Take Reduction Plan regulations: gear modifications. Proposed rules; request for comments. 66 Fed. Reg. 49896–49909. (October 1, 2001.)

10. National Marine Fisheries Service. 2001b. Taking of marine mammals incidental to commercial fishing operations; Atlantic Large Whale Take Reduction Plan regulations; Proposed rules. Dynamic Area Management. 66 Fed. Reg. 50160–50163. (October 2, 2001.)

11. National Marine Fisheries Service. 2001c. Marine mammals; Atlantic Large Whale Take Reduction Plan regulations: seasonal area management program; advanced notice of proposed rulemaking; notice of intent to prepare an environmental impact statement; request for comments. 65 Fed. Reg. 50390. (October 2, 2001.)

12. National Marine Fisheries Service. 2001d. Taking of marine mammals incidental to commercial fishing operations; Atlantic Large Whale Take Reduction Plan regulations: Seasonal Area Management; Proposed rule; request for comments. 65 Fed. Reg. 59394–59404. (November 28, 2001.)

13. Merrick, RL, PJ Clapham, TVN Cole, and RM Pace III. 2001. *Seasonal Management Zones for North Atlantic Right Whale Conservation.* Northeast Fisheries Science Center Document 01-14. National Marine Fisheries Service. Woods Hole, MA, US.

14. From RH Mattlin, executive director of the Marine Mammal Commission, to MA Colligan, chief of Protected Resources Division, Northeast Regional Office, National Marine Fisheries Service, October 31, 2001, on three proposed rulemaking actions to reduce entanglement risks for North Atlantic right whales.

15. From RH Mattlin, executive director of the Marine Mammal Commission to MA Colligan, chief of Protected Resources Division, Northeast Regional Office, National Marine Fisheries

Service, December 13, 2001, on the proposed rule for establishing seasonal management areas to protect North Atlantic right whales.

16. National Marine Fisheries Service. 2002a. Taking of marine mammals incidental to commercial fishing operations; Atlantic Large Whale Take Reduction Plan regulations: gear modifications; Final rule. 66 Fed. Reg. 1300–1314. (January 10, 2002.)

17. National Marine Fisheries Service. 2002b. Taking of marine mammals incidental to commercial fishing operations; Atlantic Large Whale Take Reduction Plan regulations: dynamic area management; Final rule. 67 Fed. Reg. 1133–1142. (January 9, 2002.)

18. National Marine Fisheries Service. 2002c. Taking of marine mammals incidental to commercial fishing operations; Atlantic Large Whale Take Reduction Plan regulations: seasonal area management; Final rule. 67 Fed. Reg. 1142–1160. (January 9, 2002.)

19. National Marine Fisheries Service. 2002d. Taking of marine mammals incidental to commercial fishing operations; Atlantic Large Whale Take Reduction Plan; Temporary rule; temporary area and gear restrictions. 67 Fed. Reg. 20699–20701. (April 26, 2002.). National Marine Fisheries Service. 2002e. Taking of marine mammals incidental to commercial fishing operations; Atlantic Large Whale Take Reduction Plan; Temporary rule. 67 Fed. Reg. 44092–44093. (July 1, 2002.)

20. National Marine Fisheries Service. 2003a. Proposed rule for the taking of marine mammals incidental to commercial fishing operations; Atlantic Large Whale Take Reduction Plan regulations: dynamic area management. 68 Fed. Reg. 10195–10199. (March 4, 2003.)

21. National Marine Fisheries Service. 2003b. Taking of marine mammals incidental to commercial fishing operations; Atlantic Large Whale Take Reduction Plan regulations: dynamic area management; Final rule. 68 Fed. Reg. 51159–51201. (August 26, 2003.)

22. National Marine Fisheries Service. 2002f. Taking of marine mammals incidental to commercial fishing operations; Atlantic Large Whale Take Reduction Plan regulations; Final rule; gillnet fishing in the right whale calving grounds. 67 Fed. Reg. 59471–50477. (September 23, 2002.)

23. National Marine Fisheries Service. 2002g. Endangered and threatened species; Finding for a petition to revise critical habitat for northern right whales; Notice of 90-day finding. 67 Fed. Reg. 69708–60710. (November 19, 2002.)

24. National Marine Fisheries Service. 2003c. Endangered and threatened species: Finding for a petition to revise critical habitat for northern right whales; Response to petition; final determination. 68 Fed. Reg. 51758–51763. (August 28, 2003.)

25. Marine Mammal Commission. 2003d. *Marine Mammal Commission Annual Report to Congress 2002*. Bethesda, MD, US, pp. 23–24. Marine Mammal Commission. 2004a. *Marine Mammal Commission Annual Report to Congress 2003*. Bethesda, MD, US, pp. 15–16.

26. From JE Reynolds III, chairman, Marine Mammal Commission, to WT Hogarth, assistant administrator for Fisheries, National Marine Fisheries Service, November 27, 2002, on the conservation issues for North Atlantic and North Pacific right whales.

27. National Marine Fisheries Service. 2003e. *Atlantic Large Whale Take Reduction Team Meeting, Providence, RI, April 28–30, Meeting Summary*. Northeast Regional Office. Gloucester, MA, US.

28. Donovan, G. 2005. *Report of the Workshop on Modification of Fishing Gear to Reduce Large Whale Entanglement: 13–15 October 2004, North Falmouth, Massachusetts*. Report provided to the National Marine Fisheries Service, Gloucester, MA, US. Knowlton, AR. 2005. *A Review of Gear Modifications and Fishing Practices Aimed at Reducing the Level and Frequency of Entanglements with Right and Humpback Whales in the Western North Atlantic*. Report by the New England Aquarium to the National Marine Fisheries Service for Contract No. 40EMNF300159. Northeast Regional Office, National Marine Fisheries Service. Gloucester, MA, US.

29. National Marine Fisheries Service. 2003f. Marine mammals: Notice of intent to prepare an environmental impact statement for the Atlantic Large Whale Take Reduction Plan. 68 Fed. Reg. 38676–38678. (June 30, 2003.)

30. Burke, E, and D McKiernan. 2006. *Massachusetts Division of Fisheries Right Whale Conservation Program: 2005 Projects and Accomplishments.* Report submitted to the National Marine Fisheries Service. Massachusetts Division of Fisheries. Boston, MA, US.

31. Burke, E, and D McKiernan. 2007. *Massachusetts Division of Fisheries Right Whale Conservation Program: 2006 Projects and Accomplishments.* Report submitted to the National Marine Fisheries Service. Massachusetts Division of Fisheries. Boston, MA, US.

32. National Marine Fisheries Service. 2005a. *Draft Environmental Impact Statement for Amending the Atlantic Large Whale Take Reduction Plan: Broad-based Gear Modifications.* Northeast Regional Office. Gloucester, MA, US.

33. From D Cottingham, executive director of the Marine Mammal Commission, to M Colligan, assistant regional administrator for Protected Resources, National Marine Fisheries Service, Gloucester, MA, May 12, 2005, commenting on the Draft Environmental Impact Statement for amending the Atlantic Large Whale Take Reduction Plan.

34. US Code of Federal Regulations, Part 40 CFR §15-2.14.

35. National Marine Fisheries Service. 2005b. Taking of marine mammals incidental to commercial fishing operations Atlantic Large Whale Take Reduction Plan regulations; Proposed rule; request for comments. 70 Fed. Reg. 35894–35944. (June 21, 2005.)

36. Ibid., p. 35897.

37. Johnson, A, G Salvador, J Kenney, J Robbins, S Kraus, S Landry, and P Clapham. 2005. Fishing gear involved in entanglements of right whale and humpback whale. Marine Mammal Science 21(4):635–645.

38. National Marine Fisheries Service. 2005b (n. 35), pp. 35895–35896.

39. Mate, BM, SL Nieukirk, and SD Kraus. 1997. Satellite-monitored movements of the North Atlantic right whale. Journal of Wildlife Management 61:1393–1405.

40. US Code of Federal Regulations Part 50 CFR §229.32(g)(1).

41. From SB Young, Marine Issues field director, Humane Society of the United States, to WT Hogarth, assistant administrator for Fisheries, National Oceanic and Atmospheric Administration, January 30, 2006, requesting immediate closure of gill-net fishing in the Southeast US Restricted Area.

42. Ellenberg Associates. 2006. *Key Outcomes for the Southeast Subgroup of the Atlantic Large Whale Take Reduction Team, April 11–12, 2006.* Prepared for the National Marine Fisheries Service. St. Augustine, FL, US.

43. From TJ Ragen, executive director of the Marine Mammal Commission, to WT Hogarth, assistant administrator for Fisheries, National Oceanic and Atmospheric Administration, September 26, 2006, recommending adoption of an emergency rule to prohibit gill-net fishing in the Southeast US Restricted Area.

44. From PR Lane, senior attorney, Humane Society of the United States, to WT Hogarth, assistant administrator for Fisheries, National Oceanic and Atmospheric Administration, October 2, 2006, regarding litigation concerning emergency closure of the Southeast US Restricted Area to gill-net fishing gear to protect North Atlantic right whales.

45. National Marine Fisheries Service. 2006. Right whale protection; Southeast U.S. gillnet closure; Emergency rule. 71 Fed. Reg. 66469–66482. (November 15, 2006.)

46. National Marine Fisheries Service. 2007a. Taking of marine mammals incidental to commercial fishing operations; Atlantic Large Whale Take Reduction Plan; Proposed rule; request for comments. 71 Fed. Reg. 66482–66495. (November 15, 2006.)

47. National Marine Fisheries Service. 2007b. Taking of marine mammals incidental to commercial fishing operations; Atlantic Large Whale Take Reduction Plan regulations; Final rule. 72 Fed. Reg. 57104–57194. (October 5, 2007.)

48. Reeves, RR, AJ Reed, L Lowry, SK Katona, and DJ Boness. 2007. *Report of the North*

Atlantic Right Whale Program Review: 13-17 March 2006, Woods Hole, Massachusetts. Marine Mammal Commission. Bethesda, MD, US. This was one of five reports prepared in response to a directive from Congress to the Marine Mammal Commission to assess the effectiveness of protection programs for the most endangered marine mammals in US waters.

49. National Marine Fisheries Service. 2007c. *Final Environmental Impact Statement for Amending the Atlantic Large Whale Take Reduction Plan: Broad-based Gear Modifications.* Northeast Regional Office. Gloucester MA, US, pp. 5-48 to 5-49.

50. National Marine Fisheries Service. 2007d. *Record of Decision: Final Environmental Impact Statement; Amendments to the Atlantic Large Whale Take Reduction Plan.* September 21, 2007. Northeast Region. Gloucester, MA, US.

51. National Marine Fisheries Service. 2007e. Taking of marine mammals incidental to commercial fishing operations; Atlantic Large Whale Take Reduction Plan. 72 Fed. Reg. 34632–34643. (June 25, 2007.)

52. From P McCarron, executive director, Maine Lobstermen's Association, to JW Balsiger, assistant administrator for Fisheries, National Oceanic and Atmospheric Administration, April 24, 2008, regarding request for temporary deferral of enforcement of final rule implementing the Atlantic Large Whale Take Reduction Plan in Maine coastal waters.

53. National Marine Fisheries Service. 2008a. Taking of marine mammals incidental to commercial fishing operations; Atlantic Large Whale Take Reduction Plan regulations; Final rule. 73 Fed. Reg. 51228–51242. (September 2, 2008.)

54. National Marine Fisheries Service. 2008b. Notice; Taking of marine mammals incidental to commercial fishing operations; Atlantic Large Whale Take Reduction Plan regulations. 73 Fed. Reg. 58942–58943. (October 8, 2008.)

55. National Marine Fisheries Service. 2009. Notice of funding availability: northeast region fishing gear exchange project. 74 Fed. Reg. 34552–34555. (July 16, 2009.)

56. Maine Department of Marine Resources. [n.d.] *Maine Fishing Gear Exchange and Research Program: Final Report, October 1, 2009–December 31, 2011.* Report submitted to the National Marine Fisheries Service, Northeast Regional Office. Gloucester, MA, US.

57. Johnson et al., Fishing gear (n. 37), p. 641, table 2.

58. Jury, SH, H Howell, DF O'Grady, and WH Watson III. 2001. Lobster trap video: in situ surveillance of the behavior of *Homarus americanus* in and around traps. Marine and Freshwater Research 52:1125–1132.

59. Myers, RA, SA Boudreau, RD Kenney, MJ Moore, AA Rosenberg, SA Sherrill-Mix, and B Worm. 2006. Saving endangered whales at no cost. Current Biology 17(1):R10–R11, http://dx.doi.org/10.1016/j.cub.2006.11.045; and Woodward, C. 2005. *The Lobster Coast: Rebels, Rusticators, and the Struggle for a Forgotten Frontier.* Viking Penguin. New York, NY.

60. Schreiber, L. 2008. Can fewer traps and lines help whales and overall catch? Fishermen's Voice 13(4), http://www.fishermensvoice.com/archives/0408canfewertrapshe%23144518.html. Accessed June 1, 2014.

61. National Marine Fisheries Service. 2011. Marine mammals: notice of intent to prepare and environmental impact statement for the Atlantic Large Whale Take Reduction Plan. 76 Fed. Reg. 43654–34656. (June 14, 2011.)

62. From TJ Ragen, executive director of the Marine Mammal Commission, to M Colligan, associate regional administrator for Protected Resources, National Marine Fisheries Service, September 12, 2011, regarding Atlantic Large Whale Take Reduction Plan scoping.

63. CONCUR. 2012a. *Atlantic Large Whale Take Reduction Team: January 9–13, Provincetown, RI.* Prepared for the Northeast Regional Office, National Marine Fisheries Service. Gloucester, MA, US.

64. Northeast Regional Office. 2012. ALWTRT conference call, February 15, 2012. Summary Report. National Marine Fisheries Service. Gloucester, MA.

65. Kenny, R. 2012. *Estimating Minimum SPUE Values for Right Whales and Humpback Whales in Northeast Areas with Low Survey Effort: An Analysis Prepared for the Atlantic Large Whale Take Reduction Team.* Marine Mammal Commission. Bethesda, MD.

66. CONCUR. 2012b. Atlantic Large Whale Take Reduction Team: Northeast Subgroup Webinar April 12, 2012, 10 a.m. to noon; Mid-Atlantic/Southeast Subgroup Webinar, 1 p.m. to 3 p.m. Summary report prepared for the Northeast Regional Office, National Marine Fisheries Service. Gloucester, MA, US.

67. National Marine Fisheries Service. 2013. Taking of marine mammals incidental to commercial fishing operations; Atlantic Large Whale Take Reduction Plan regulations; Proposed rule; request for comments. 73 Fed. Reg. 42654–42675. (July 16, 2013.)

68. Industrial Economics, and National Marine Fisheries Service. 2014. *Final Environmental Impact Statement for Amending the Atlantic Large Whale Take Reduction Plan: Vertical Line Rule.* Vol. 2. Northeast Regional Office, National Marine Fisheries Service. Gloucester, MA, US.

69. Knowlton, AR, J Robins, S Landry, H McKenna, SD Kraus, and T Warner. 2012. *Implications of Fishing Gear Strength on Large Whale Entanglements.* Final report submitted by the Consortium for Wildlife Bycatch Reduction to the National Marine Fisheries Service in partial fulfillment of Contract # NA09NMF4520413. National Marine Fisheries Service. Gloucester, MA, US.

70. Industrial Economics, and National Marine Fisheries Service, *Final Environmental* (n. 68).

71. National Marine Fisheries Service. 2014a. Taking of marine mammals incidental to commercial fishing operations; Atlantic Large Whale Take Reduction Plan regulations; Final rule. 74 Fed. Reg. 36586–36621. (June 27, 2014.)

72. Greater Atlantic Regional Office. 2014. Atlantic large whale take reduction team meeting teleconference: October 1, 2014, 1 p.m.–5 p.m. Meeting summary. National Marine Fisheries Service. Gloucester, MA, US.

73. National Marine Fisheries Service. 2014b. Taking of marine mammals incidental to commercial fishing operations; Atlantic Coastal Fisheries Cooperative Management Act provisions; American lobster fishery Final rule. 79 Fed. Reg. 73848–73852. (December 12, 2014.)

74. CONCUR. 2015. *Atlantic Large Whale Take Reduction Team Meeting January 12–14, 2015, Providence, RI.* Prepared for the Greater Atlantic Regional Office, National Marine Fisheries Service. Gloucester, MA, US.

75. National Marine Fisheries Service. 2014c. Final rule: taking of marine mammals incidental to commercial fishing operations; Atlantic Coastal Fisheries Cooperative Management Act provisions; American lobster fishery. 80 Fed. Reg. 14345–14356. (March 19, 2015.)

76. From RJ Lent, executive director of the Marine Mammal Commission, to K Damon-Randal, assistant regional administrator for Protected Resources, National Marine Fisheries Service, April 20, 2015, on amendments to exempt areas from requirements of the Atlantic Large Whale Take Reduction Plan and add new gear marking requirements.

77. National Marine Fisheries Service. 2014d. Taking of marine mammals incidental to commercial fishing operations; Atlantic Large Whale Take Reduction Plan regulations; Final rule. 80 Fed. Reg. 30367–30379. (May 28, 2015.)

22

TEN THOUSAND
RIGHT WHALES

FEW MARINE SPECIES have influenced the course of human history and our relationship with the sea as much as North Atlantic right whales. For more than a thousand years, their pursuit provided a livelihood for many and riches for a few. Their past abuse and current plight is as much a reminder of the risk of species extinctions, even in vast oceans, as it is a yardstick of our commitment to manage activities in ways that preserve our natural maritime heritage. After commercial whalers wreaked their havoc and hunting right whales was mercifully banned in the 1930s, the species slowly began inching its way back from the brink of extinction. By the 1980s, as dedicated recovery efforts were taking root, new obstacles to their survival were developing in the form of larger and faster ships, and the proliferation of stronger, longer-lasting fishing gear. Although the deaths they caused were unintended, collisions and entanglement have been no less lethal than the whaler's harpoon, and the frequency of their occurrence has been sufficient to overwhelm the species' low calving rate and minuscule population size.

A baby boom in the first decade of the 2000s markedly improved the North Atlantic right whale's recovery prospects. The number of births leapt from an average of 11 per year in the 1980s and 1990s to over 20 during the first decade of the new millennium (figure 22.1). This enabled the species' population to grow from the upper 300s in 2000 to nearly 500 whales by 2010. After 20 years of stagnant population growth in the late twentieth century, this was a welcome development. Yet the right whale's future is still far from secure. Even between 2000 and 2010, their annual growth rate was anemic compared to most other large whale populations, including those of the southern right whale. Moreover, between 2010 and 2015, calf production fell to an average of 15 per year, closer to the pre-2000 average than the relatively good years in the early 2000s. If calving rates are linked to prey availability, which in turn is governed by fickle oceanographic conditions and

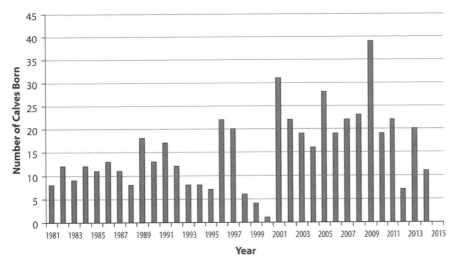

Figure 22.1. Number of observed North Atlantic right whale calves, 1981–2014.

climate change, as many scientists believe, reproduction rates may again slide back to the dismal levels of the late 1900s. What may be a harbinger of such a decline is the recent abandonment of many high-use feeding areas that right whales have used for decades, coincident with the slowing birth rate between 2011 and 2015. The species' shift to new foraging grounds in areas yet to be discovered suggests that prey abundance in many of their long-time feeding haunts has dipped at least temporarily to levels unable to support right whales in some areas.

Sustained recovery of North Atlantic right whales will therefore continue to depend on minimizing human-related deaths and injuries. There are encouraging signs that ship collisions, the leading cause of death in the late 1900s and early 2000s, have been brought under control by new speed restrictions and routing measures. Although the 10-year process to develop and implement measures was excruciatingly slow for such an urgent conservation issue, the new requirements are based on a well-reasoned, science-based rationale and required only a single rule-making process to develop an effective solution. Between December 2008 and the end of 2015, only two deaths attributable to ship strikes were found in US waters, neither of which was found in an established speed zone or close enough to one to think they might have drifted from within a speed zone. The Fisheries Service deserves enormous credit for implementing these restrictions despite opposition from some politically influential segments of the shipping industry.

The same cannot be said for reducing deaths from fishing gear. This has proven to be an extraordinarily difficult management issue with no single solution, and there has been no sign of progress in reducing entanglement deaths since management efforts began in the early 1990s. Indeed, between 2000 and 2015, the incidence of entanglement deaths and serious injuries has increased to the point where they are now the largest known cause of North Atlantic right whale deaths.[1] They have effectively outstripped the conservation gains realized by reduced vessel–whale collisions. Whereas eight carcasses attributed to entanglement were found in the first decade of the 2000s, nine were discovered between just 2010 and 2014. If this trend continues, and considering that perhaps an equal or even greater number of entanglement deaths are unrecorded, the spurt of population growth in the early 2000s may turn out to have been a short-lived blessing.

The development of protection measures throughout the take reduction plan's 20-year rule-making history has rarely been based on a well-reasoned, science-based rationale. The process also has not reflected a good-faith effort to address all critical aspects of the issue. Moreover, it has already required three major rule-making overhauls and nearly a dozen more minor adjustments. The result has been a patchwork quilt of largely unenforceable measures that still has little hope of significantly reducing entanglement risks. This dismal history has been marked by Fisheries Service leadership with little apparent political resolve for addressing difficult issues that are no less critical for right whale conservation than those for minimizing ship-strike deaths.

Preventing these fishery-related injuries and deaths continues to be one of the most difficult and frustrating marine mammal conservation issues in the United States. Although several potentially useful measures have been identified (time-area closures, sinking groundlines, and reducing the number of vertical lines), they have been stridently opposed by fishermen, thus putting fishery managers in an awkward spot. As independently minded small business operators inherently suspicious of government regulators, trap fishermen are quick to unite in opposition to any measures that might threaten any of their livelihoods.

Fisheries Service administrators are well aware of their dual responsibilities for protecting right whales and ensuring fishermen an opportunity to ply their trade. However, the agency's Northeast Regional Office, which is in charge of both reducing whale entanglements and managing most of the fisheries that cause them, has been loath to approve any measures that might adversely affect an individual fisherman disproportionately (closures) and has instead relied almost entirely on largely ineffective measures broadly applicable to all fishermen (weak links, minimum trawl lengths designed to make slight reductions in endline numbers). The Fisheries Service's commitment to ensuring no loss of fishing opportunities has hobbled use of the few measures most likely to reduce whale entanglement risks. And recognizing the agency's record of bending toward the position of the fishing industry, there has been little incentive for fishermen to negotiate on difficult issues, thereby rendering the take reduction team process ineffective.

Researchers are investigating some potentially effective new gear modifications. Fishing with pop-up buoys that keep buoys and lines on the sea floor between retrieval operations has been shown to be technically possible, and with further research to improve designs, the buoys could become a more acceptable piece of equipment for at least some trap fishing. Their high cost, roughly between $1,500 and $2,500 per buoy, would almost certainly them rule out for nearshore fishermen using large numbers of short trawls, but they could be economically feasible for larger offshore operations employing fewer and longer trawls. If requirements for their use are phased in over a period of years and large-scale production can reduce the cost per unit, they could become economically possible. This still leaves an important and unsettled issue—how to ensure that fishermen know where gear is already set so they can avoid deploying their equipment on top of it or dragging bottom dredges and net trawls across trap lines. Further research on technological solutions to detect set gear without buoys or new management arrangements to avoid gear conflicts may be feasible.

Another leading prospect is the use of weak rope. Recent studies led by Amy Knowlton at the New England Aquarium and her colleagues suggests this could be a very effective conservation measure, particularly for small fishing operations working in relatively shallow areas with short trawls.[2] Before the mid-1990s, when ropes were weaker, it appears that right whales were more likely to break entangling lines and shed gear before sustaining serious and lethal wounds. For fishing in shallow areas, the appeal of stronger rope has not been its improved ability for hauling gear but rather its extended life and the less frequent need for replacement.

By testing the breaking strength of ropes removed from whales between 1990 and 2010, researchers have concluded that serious and potentially lethal right whale injuries could be reduced by nearly 75 percent if fishermen used lines with breaking strengths of less than 1,700 pounds.[3] The breaking strength of lines now manufactured for fisheries averages more than 3,500 pounds and can range up to 12,000 pounds. For traps set in relatively shallow waters, a breaking strength of 1,700 pounds is more than ample for hauling gear, particularly when trawls are short. Thus, instead of attempting to reduce the number of buoy lines by grouping more traps together, a safer and more effective strategy for fishing from small boats in inshore water might be to use shorter trawls with weaker, whale-safe endlines.

If further research on whale-safe rope and pop-up buoys can resolve at least some of the thornier technical issues, the current Large Whale Take Reduction Plan might be modified into an approach with six basic elements. These would include (1) a combination of shorter trawls and whale-safe buoy lines (those strong enough to haul gear but weak enough for whales to break if entangled) in shallow waters less than about 165 feet (50 meters) deep, where small boats with relatively short trawls operate; (2) pop-up buoys in waters greater than about 330 feet (100 meters), where large boats fish with fewer and longer trawls; (3) sinking groundlines; (4) expanded time-area closures; (5) expanded line marking (including marks on groundlines) to better identify the source of fishing gear found on whales; and (6) record keeping on the number and location of endlines to better evaluate entanglement risks. Only three of these elements (sinking groundlines, time-area closures, and gear marking) are addressed in the current take reduction plan. Two of those (time-area closures and gear marking) have been used far too sparingly to be effective.

Alternatively—or perhaps in concert with the strategy outlined above—reducing entanglement deaths may require an entirely new management paradigm, one that would include measures, such as time-area closures and limits on fishing effort, to seamlessly and simultaneously protect both right whales *and* fisheries. Such an approach would have the best chance of success if responsibility for drafting those measures was vested in a small group with expertise in the science and management of both fisheries and whales, rather than in the take reduction team. Recognizing the team's poor track record of reaching consensus on difficult issues, a smaller group without direct economic interests in measures would likely be more successful in agreeing on those measures that could be effective components of both fishery management plans and the take reduction plan. Measures identified by this smaller group could then be presented to the take reduction team and modified, but only through team consensus. Identical measures included in both fishery management and take reduction plans would free administrators from being placed in the position of approving some of the more controversial whale protection measures independently from, and in addition to, those necessary to ensure sustainable fishing. That difficulty for fishery administrators has worked against the adoption of important take reduction measures in the past.

This approach, however, could require amending the Marine Mammal Protection and Magnuson-Stevens Acts to authorize a group other than the take reduction teams and fishery councils to draft measures that overlap conservation goals for both whales and target fishery stocks. Such authorization could be limited to take reduction teams that consistently fail to reach agreement on a full set of bycatch reduction measures or that fail to achieve statutory goals within a set period of time (e.g., 10 years). Such a contingency would also provide added incentive needed for team members to reach consensus on difficult issues.

While further work is urgently needed on finding ways to reduce right whale entanglement risks, resource managers will also need to contend with other emerging threats that either separately or in combination could throw new roadblocks into the path of whale recovery. In March of 2010, plans were announced to lift a moratorium on offshore oil and gas development off mid- and south Atlantic states that had been in place since the 1980s.[4] Although President Obama reversed this decision in March 2016 when he withdrew areas off the Atlantic coast from the offshore oil and gas leasing schedule, seismic exploration for oil and gas reserves in the region is continuing, at least for now. And with such valuable resources still available, political decisions governing their exploitation are bound to shift yet again as economic conditions and administrations change. When they do, it will once more open the door for offshore drilling with its inevitable accompaniment of oil spills. With the potential for spilled oil to drift hundreds, or even thousands of miles, it could have devastating effects on the plankton swarms that right whales depend on for food.

In the interim, as seismic exploration continues, loud sounds from air guns towed behind vessels crisscrossing the continental shelf in search of oil-bearing geologic structures could displace whales from preferred habitats, drown out the vocalizations through which mothers and calves and other whales communicate with one another, and possibly cause temporary or even permanent hearing loss.[5]

Increases in shipping, seismic exploration, construction, and other sources of anthropogenic sound have already elevated levels of ocean noise in some areas 100-fold since the 1960s, and a similar increase is expected in most oceans in coming decades as those activities continue to increase.[6]

Incentives to reduce carbon emissions from electric utilities have also spurred a rush to erect offshore wind farms all along the eastern seaboard.[7] The first of these facilities, a project called "Cape Wind," with 130 turbines off the southern coast of Massachusetts, entered the construction phase in 2015. Close on its heels are similar proposals for areas off states from Maine to Florida. Because most proposed sites are between 10 and 50 miles from shore—far enough from shore to be largely invisible from land but close enough for cost-effective transmission of electricity—many wind energy fields could lie directly in the path of migrating right whales. Although the effects of turbine sound and electromagnetic fields emanating from transmission cables on large whales are still uncertain, there is to date no evidence that their operation is likely to have adverse effects.[8] Indeed, navigation hazards created by turbine towers could benefit migrating right whales by creating a corridor relatively free of large ships and might reduce collision risks. On the other hand, if facilities end up rerouting vessel traffic closer to shore into the preferred right whale migratory corridor, ship-strike risks could increase significantly. A more immediate although temporary concern is pile driving during the construction phase. Years of pounding with jackhammer-like noise to install footings for turbine towers up to 400 feet tall could alter right whale migratory routes and habitat-use patterns. Transiting construction and service vessels would also increase ship-strike risks.

Plans by the US Navy's Atlantic Fleet to expand warfare training and munitions testing may also threaten right whales. Of particular concern is the development of a new antisubmarine warfare testing range just 15 to 20 nautical miles seaward of the critical habitat designated for calving and nursing right whales off northeast Florida. Although the navy is well aware of the risks to right whales and has a record of being very responsive to protection needs, the impacts of noise from powerful sonars, ships, and low-flying helicopters used in training exercises to search out hiding submarines are still poorly understood; they could interfere with right whale hearing or displace them from calving habitat. Navy ships exempt from speed restrictions transiting to and from the range could also increase the risk of strikes.

Looming over effects of all of these activities is climate change. With more than 90 percent of the heat from global warming absorbed by the oceans,[9] rising ocean temperatures could alter the timing and location of copepod swarms, which in turn could affect the distribution and movements of foraging whales. Changing temperatures could also shift the location of calving grounds and migratory corridors. As noted earlier, right whales have recently abandoned many of their preferred summer feeding grounds used for decades. Where they have gone to feed instead and whether this is just a short-term shift in habitat is unknown, but many scientists believe it is yet another sign of the effects of global warming and could have long-term implications. Because the speed with which such changes might occur is uncertain, scientists and managers will need to monitor right whales closely to determine if adjustments are required in the effective dates and location of management measures, such as seasonal speed restrictions or fishery closures.

The above list of potential threats is not intended to suggest that right whales face a gloomy future unless all ocean activities are stopped cold. Rather, they illustrate the need for continued vigilance and thoughtful, adaptive management interventions. To meet this challenge, scientists have never been better positioned than they are now to gather and analyze data for making informed management decisions. The growing array of technological wizardry and available data they can bring to bear is nothing short of extraordinary.

The foundation of scientists' analytic capability has been and almost certainly will remain the right whale identification catalogue. After 35 years of data collection, virtually every North Atlantic right whale is now individually identifiable by callosity patterns and scars. For many individuals, there also is associated data on age, sex, seasonal habitat-use patterns, calving interval, history of entanglements and ship strikes, and general health. In addition, there are biopsy samples for most animals that provide a means of genetically determining parents and filial relationships. These biopsies also enable identification of individuals even after carcass decomposition has obliterated the callosity patterns normally used for that purpose. All this information offers a powerful tool for tracking trends in habitat-use patterns, behavior, and interactions with human activity, relating them to human activities and population risks, and crafting mitigation options.

A second pillar of right whale research that will likely remain important for at least the near future are aerial and shipboard surveys. These are now the best way to count whales for abundance estimates and to take the photographs necessary for analyses of survival rates, health assessments, and general evaluations of population status. They also are the primary means for detecting entangled whales and carcasses drifting offshore. Aerial surveys, however, are expensive, fail to identify most of the whales present at the time of a survey, and are not without risk to researchers. Despite great attention to safety, three researchers and their pilot were killed in a plane crash while surveying right whales off the coast of Georgia in January 2003. In the not-too-distant future, however, other technologies, such as passive acoustics and drones, might be used to collect much of the data now gathered by aerial and vessel observers.

A third cornerstone of whale research in general has been tagging studies. Although the list of technological options being built into tags grows more impressive with each passing year, the means of attaching tags to whales has been a major obstacle for this line of study. Currently there are two basic approaches. One involves implanting tags into the skin and underlying tissue using a crossbow or air rifle. Initial designs were cylindrical metal tubes roughly a foot long and an inch or two in diameter (but now somewhat smaller) that housed electronics and batteries. Anchored with pronged barbs at one end and an antenna extending from the other end above the skin, these tags were designed to track long-term movements. Most tags fall out within a few weeks or months, but while implanted, they beam location data back to the scientists' computers several times a day via passing satellites when whales are at the surface.

No large whales are known to have died from tag implants, and stone and metal objects of comparable size from whaling days at least 100 years ago are still being found in bowhead whales killed by Native subsistence whalers. However, because of divots and swelling left on some animals after tags fall off, many North Atlantic right whale researchers have opposed their use since the early 1990s. They are nevertheless still used with great success on other large whale species, including southern and North Pacific right whales. Because of opposition to these tags on North Atlantic right whales, one of the best tools for defining travel routes, discovering unknown habitats, and evaluating where human interaction risks are greatest has seldom been used on this species. In recent years, however, researchers have been testing a new and less invasive tag design—the "limpet" tag. Barely two inches in diameter, it rests on top of the skin and is anchored in place with a pair of barbs extending just a few inches into the underlying blubber. This greatly reduces the risk of injury. So far, these tags have remained attached no longer than a few weeks, but they are already beginning to fill in missing information on migratory routes. With smaller tags reducing animal welfare concerns, their expanded use on North Atlantic right whales may soon resolve some of the most vexing and important questions about their migrations and changing habitat-use patterns.

A second type of tag attachment involves suction cups to affix various types of sensors to a whale's back. Because there is no penetration of the skin with the suction cup, this approach poses virtually no risk of injury. However, tags applied with this technique remain on whales for only a few hours before dislodging and floating to the surface. The resulting data are stored in digital form on a memory chip in the tag itself, which explains their common name of D-Tag. These tags have provided the first detailed view of what whales do and sense when they dive beneath the surface. The amazing array of sensors attached to whales with suction cups include video cameras; hydrophones to record whale vocalizations and what they hear; pressure sensors to trace depth of dive profiles; motion activators to detect fluke beats and changes in orientation as animals roll side to side; a compass to record direction of travel; and a clock and GPS to place all collected data in time and space. By calculating swim speeds, rates of decent and ascent, time spent at different depths, and changes in orientation, a virtual image can be created describing exactly how whales move, the depths at which they feed, their vocalizations, and how they react to sounds and other perturbations in their environment.

Further brightening future research prospects are several exciting technologies that seem to be on the cusp of offering valuable new tools for answering long-standing questions about right whale travel routes, occurrence rates in poorly surveyed areas, and unknown high-use habitats. Among these are drones equipped with cameras and other sensors. Drones hold great promise for monitoring habitat-use patterns and collecting whale identification photos more efficiently, with less risk to researchers and at less cost. Drones have already been put to use for specialized purposes, such as revealing configurations of ropes on entangled whales to help response teams free them. They could soon be used to monitor whale behavior near approaching ships or for warning operators of seismic arrays, drilling rigs, or other equipment when whales are within hazardous zones around work areas.

Perhaps even more promising is passive acoustic monitoring (PAM) technology that uses hydrophones to detect whale vocalizations distinctive to right whales. How often right whales vocalize can vary by individual and behavior (traveling, feeding, mating, nursing, etc.), but hydrophones have the advantage of recording data 24 hours a day in all weather conditions, whether whales are at the surface or at depth. And with a detection range of two to five miles, hydrophones have routinely documented right whales where and when visual observers on aerial or shipboard whale surveys have failed to do so. In the process, they have revealed the presence of right whales at times and places that have contradicted conventional wisdom, such as their occasional presence in the calving grounds off Georgia even in summer.

Several types of acoustic recorders have been developed. One incorporates hydrophones into buoys that can be anchored at a single location and at any depth for months at a time. When placed in an array with overlapping detection ranges, recorded calls can be triangulated to locate, count, and track individual whales moving through the array. Because most buoys archive sound data using onboard memory chips, they need to be retrieved periodically to download data before it can be analyzed. Hydrophones have also been incorporated into underwater gliders able to roam vast ocean areas for months on end, recording marine mammal vocalizations in times or areas too difficult or expensive to reach with ships or planes. Still other hydrophones are deployed as "towed arrays" behind ships on whale surveys to detect vocalizing animals that can be heard but not seen by observers.

Whereas most buoys require periodic retrieval to collect their data, some gliders have been programed to surface periodically and send their archived data back to scientists via satellite links. Some surface buoys have also been designed to transfer their data in near real time via satellite links, or if close enough to shore, over cell phone towers. As mentioned earlier (chapter 19), Chris Clark at the Cornell University Bioacoustics Research Program pioneered this "smart" system when he installed a network of 10 buoys lining a 70-mile stretch of the Boston Traffic Separation System to detect right whale calls. Within minutes, results are posted on a web site that vessel operators, as well as the general public, can access over the internet to identify when whale calls were last detected at each buoy. Armed with this information, vessel operators can determine when and where to reduce their speed to prevent potential ship strikes.

Of course, passive acoustic devices record all sounds, including those from transiting ships, seismic surveys, pile driving, and sonar pings, as well as natural sounds from moving ice, thunder, storm waves, earthquakes, and other marine life. Using sophisticated analytic techniques capable of synthesizing massive amounts of data from continuous recordings by multiple hydrophones, scientists are learning how to characterize soundscapes across regional seas and bays, and even entire ocean basins. As those methods are honed, it will be possible to monitor noise levels and trends of sound sources at different frequencies over time. As scientists improve methods to distinguish calls produced by right whales and other marine species, and as costs fall for analyzing massive acoustic data sets, prospects will open for detecting right whales and their habitat-use patterns in response to different types of maritime activities. Perhaps eventually, this will allow the identification of individual whales and even estimate local whale abundance.[10]

Another important new source of data not directly related to whales but nonetheless important for managing ship strikes is the expanding automatic identification system (AIS) used to track vessel movements. AIS transponders on equipped vessels automatically broadcast signals via satellite and VHF systems every 10 seconds to identify their location, speed, heading, length, type, and name. Initially proposed as a navigational safety measure to prevent vessels from colliding with one another, the system was first required in 2002 by the International Maritime Organization for commercial ships greater than 300 gross tons and all commercial passenger vessels. Since then, other vessels have begun carrying AIS transmitters and additional satellites have been put into orbit to relay their signals in all major shipping areas. As a result, tracking ship traffic throughout the world's oceans has improved enormously with information now accessible to anyone over the internet in real time. This wealth of data provides a previously unimagined ability to compare vessel traffic patterns and whale distribution in order to evaluate collision risks and mitigation options.

As these and other technologies continue to advance, the potential for uncovering new secrets about right whale biology and ecology is enormous. By the same token, they offer hope for vast improvements in understanding the vulnerability of right whales to human activities and ways to mitigate impacts. Both are vital ingredients for a well-managed recovery program, but are only as useful as the political will of responsible officials to use them. Here, readers of this book and the public at large can make a real difference. Ocean resources within US jurisdiction, be they fish, whales, coral reefs, oil and gas deposits, recreational opportunities, transportation corridors, or some other feature, are assets held in the public trust. They belong as much to someone living in Kansas who has never seen the ocean but might hope to one day as they do to someone who lives on the coast and makes their living from the sea. Although it may seem futile and trite, expressing thoughtful views on ocean management issues through letters on proposed actions, participation in public hearings, and support for like-minded interest groups can and has changed the outcome of decisions affecting North Atlantic right whales.

What may be most encouraging of all, however, is the growing number of talented, committed young scientists and resource managers eager to invest careers in developing and applying science-based information to contentious high-stakes political struggles between ocean exploitation and conservation. Every year, the number of students interested in marine mammal conservation increases, and colleges and universities offer more and more courses in this and related subjects to meet the demand. As these students join the work force and rise through the ranks to leadership positions, there will be a new opportunity to secure the species' future. At the same time, new technologies and information and broader views of the species' place in the marine ecosystem will open new possibilities for applying ecosystem management principles; these will do a far better job than has been done so far to integrate measures for both species conservation *and* resource use. With such a multitasking approach, the children of today may well live to see the day when 10,000 right whales once again grace the North Atlantic Ocean. And North Atlantic right whales could rise from the depths of former whaling grounds like a phoenix from the ashes of ruin to once again enliven coastal waters of not only North America but also Europe and northwestern Africa and resume the role they once played in creating a rich and productive ocean ecosystem.

NOTES

1. Knowlton, AR, PK Hamilton, MK Marx, HM Pettis, and SD Kraus. 2012. Monitoring North Atlantic right whale *Eubalaena glacialis* entanglement rates: a 30-year retrospective. Marine Ecology Progress Series 466:293–302.

2. Knowlton, AR, J Robbins, S Landry, HA McKenna, SD Kraus, and TB Werner. 2016. Effects of fishing rope strength on the severity of large whale entanglements. Conservation Biology 2:318–326.

3. Ibid.

4. Obama, B. March 31, 2010. Remarks by the president on energy security at Andrews Air Force Base, 3/31/2010. Office of the Press Secretary. The White House. Washington, DC, https://www.whitehouse.gov/the-press-office/remarks-president-energy-security-andrews-air-force-base-3312010.

5. Gordon, J, D Gillespie, J Potter, A Frentzis, MP Simpkins, R Swift, and D Thompson. 2003/2004. A review of effects of seismic surveys on marine mammals. Marine Technology Society Journal 37(4):16–34. Nowacek, DP, CW Clark, D Mann, PJO Miller, HC Rosenbaum, JS Golden, M Jasny, et al. 2015. Marine seismic surveys and ocean noise: time for coordinated and prudent planning. Frontiers in Ecology 13(7):378–386.

6. McDonald, MA, JA Hildebrand, and SM Wiggins. 2006. Increases in deep ocean ambient noise in the North Pacific west of San Nicolas, California. Journal of the Acoustic Society of America 3:65–70. Hatch, L, C Clark, R Merrick, S Van Parijs, D Ponirakis, et al. 2008. Characterizing the relative contributions of large vessels to total ocean noise fields: a case study using the Gerry E. Studds Stellwagen Bank National Marine Sanctuary. Environmental Management 42:735–752.

7. Russo, G. 2014. Wind power tests the waters. Nature 513:478–480.

8. Baily, H, KL Brooks, and PM Thompson. 2014. Assessing environmental impacts of offshore wind farms: lessons learned and recommendations for the future. Aquatic Biosystems 10:8, doi:10.1186/2046-9063-10-8.

9. Church, JA, NJ White, LF Konikow, CM Domingues, JG Cogley, et al. 2011. Revisiting the Earth's sea-level and energy budgets from 1961 to 2008. Geophysical Research Letters 38, doi:10.1029/2011GL048794.

10. McCordic, JA, and SE Parks. 2015. Individually distinctive parameters in the upcall of the North Atlantic right whale (*Eubalaena glacialis*). Journal of the Acoustical Society of America 137, 2196 (2015), doi.org/10.1121/1.4919983; and Širović, A, A Rice, E Chou, JA Hildebrand, SM Wiggins, and MA Roch. 2015. Seven years of blue whale and fin whale call abundance in the Southern California Bight. Endangered Species Research 28:61–76.

APPENDIX

Table A.1. Estimates of the number of right whales and bowhead whales landed (not including those struck and lost) by Basque whalers based along the Strait of Belle Isle from 1530 to 1610. (Numbers not in parentheses are from cited references; numbers in parentheses are new estimates developed for this book based on cited assumptions; * = vessels returning half empty; bbs = barrels.)

Assumptions	1530–1579 Right Whales Catch/Year	Total Catch	1530–1579 Bowhead Whales Catch/Year	Total Catch	1580–1610 Right Whales Catch/Year	Total Catch	1580–1610 Bowhead Whales Catch/Year	Total Catch	1530–1610 Right Whales Total Catch	1530–1610 Bowheads Total Catch
Aguilar (1986) • 100% right whales • 20–30 ships/yr • 1,000 bbs of oil/ship • 85 bbs of oil / right whale	300–500	(15,000–25,000)	—	—	300–500	(9,300–15,500)	—	—	24,000–40,000 (24,300–40,500)	—
Barkham (1984) • 100% right whales • 20–30 ships/yr • 15,000 bbs of oil/yr • 50 bbs of oil / right whale	300	(15,000)	—	—	(150*)	(4,650)	—	—	20,000 (19,650)	—
Revised Estimate 1: If all oil was from bowheads • 20–30 ships/yr • 1,000 bbs of oil/ship • Full load 1530–1579 • ½ load 1580–1610 • 70–90 bbs/bowhead	—	—	(222–429)	(11,111–21,429)	—	—	(111–214)	(3,444–6,643)	—	(14,555–28,072)
Revised Estimate 2: If ½ the oil was from right whales and ½ from bowheads • 20–30 ships/yr • 1,000 bbs of oil/ship • Full load 1530–1579 • ½ load 1580–1610 • 70–90 bbs/bowhead • 42–45 bbs / right whale	(222–357)	(11,111–17,857)	(111–214)	(5,556–10,714)	(111–179)	(3,444–5,536)	(83–107)	(1,722–3,321)	(14,555–23,393)	(7,278–14,035)

Table A.2. Qualitative estimates of North Atlantic right whales killed (including those struck and lost) by colonial and US whalers, from 1660 to 1930. (S = Shore-based whalers; O = Offshore sloops and small vessels patrolling near shore waters beyond the range of shore whalers; P = Pelagic vessels stripping blubber at sea beyond local waters.)

	1660s		1670s		1680s		1690s		1700s		1710s		1720s		1730–1800		1801–1850		1851–1930	
	S	O/P	S	O/P	S	O/P	S	O/P	S	O/P	S	O/P	S	O/P	S	O/P	S	O/P	S	O/P
Long Island	40	5	80	100	400	300	400	400	400	400	400	400	100	200	90	80	80	50	80	0
Cape Cod	5	0	40	0	150	0	350	0	400	50	40	200	30	100	20	20	10	10	5	5
Nantucket	0	0	0	0	0	0	10	20	300	140	300	200	400	120	80	120	20	30	5	10
Martha's Vineyard / S. New England	0	0	0	0	0	0	5	5	10	10	15	15	20	20	10	20	5	20	0	10
Cape Anne	0	0	0	0	0	0	0	10	0	20	0	10	0	10	0	10	0	0	0	0
New Jersey	0	0	0	0	0	0[1]	100	0[1]	100	0[1]	50	0[1]	50	0[1]	20	0[1]	10	0[1]	5	0
North Carolina	0	0	0	0	0	0[1]	0	0[1]	0	0[1]	30	0[1]	80	0[1]	80	0[1]	50	0[1]	40	0
Southeast US	0	0	0	0	0	0	0	0	0	0	0	0	0	0	0	0	0	0	0	30
Total	50		220		850		1,390		1,830		2,020		1,130		550		285		190	

[1]Additional offshore/pelagic whaling was carried out in this region at this time by New England whalers, but those catches are included in totals for the regions (Long Island, Nantucket, Cape Cod and/or Martha's Vineyard) where whale products were landed.

INDEX

Page numbers in bold refer to figures and tables.

Acosta, Jose de, 165–66, **166**

acoustics: acoustic alarms, 305, 313, 330; impacts, 292; passive acoustic monitoring (PAM), 50, 313, 318, 414–15

Adams, John, 103

Adams, John Quincy, 245

Ælfric Bata, 94–95

aerial surveys, 254, 273, 283–85, 287, 292, 307–8, 310, 370, 396, 412, 414

Afonso I of Portugal, King, 115

Afonso II of Portugal, King, 115

Aguilar, Alex, 115–16, 125–26, 136, 141, **420**

Albert the Great (Albertus Magnus), 95–97

Aleutian Islands, 72–75, **73**, 85

Alfonso XI of Castile, King, 112

Alfred the Great, 88–89, 94, 159

allee effects, 255

Allen, Glover M., 48

Allen, Joel Asaph, 21, 28

AlÞing, annals of, 142–43

Ambulocetidae, **57**, 59–61, **Plate 1**

American Pilots' Association, 240–41

Antarctica: circumpolar current, 58–63; paleoclimate changes, 56–58, 61–63; whale fossils, 62–63; whaling, 267

archaeocetes, 56–61

archeological sites: Amaknak Bridge, **73**, 74–78; Cape Krusenstern, **73**, 74, 77; Île aux Basques, **121**, 125, 140; L'Anse Amour, 76–77, 121; L'Anse aux Meadows, 121, 124; Margaret Bay, **73**, 75–77; Un'en'en **73**, 78, 79

archeology: Arctic Small Tool Tradition, 75, 80; maritime tradition, 74–75, 80, 160; middens, 74–75, 77–78, 80; Old Whaling Culture, 74–77; semi-subterranean houses, 74–75, 77–78; whaling implements, 64, 65, 74, 75–77, **77**, 79–80, 81, 88, 90, 169

Archer, Gabriel, 168

Arctic Natives, 18, 81, 90, 93, 105, 140–41, 149. *See also* Eskimos; Inuit; Thule

Arif, Muhammad, 59

Army Corps of Engineers, US, **280**, 287, 292, 307, 337

Artiodactyla, 20, 56

Atlantic Large Whale Take Reduction Plan: first revision, 368–71, 374–78; original, 361–64; second revision, 378–89; third revision, 389–99

Atlantic Large Whale Take Reduction Team, 356–400

Automatic Identification System (AIS), 329, 415

Azores, 86, 253–54

Balaenidae (right whale family): *Balaena biscayensis*, 26; *Balaena cisarctica*, 27, **27**; *Balaena glacialis*, 21, 24, 28; *Balaena montalionis*, 63; *Balaena ricei*, 63; *Balaena tarentina*, 27; *Balaenella brachyrhynu*, 63; diversification, 61–63; *Eubalaena shinshuensis*, 63–64; feeding methods, 34–35, 51; *Fucaia buelli*, 62; *Morenocetus parvus*, 63; taxonomy, 20, 23–29, **57**; toothed, 62

Balaenopteridae (rorquals), 34, **57**, 63

baleen (whalebone): economic value and uses, 43–44, 109, 133; entanglement, 347, 378; filtering dynamics, 31–36, **35**, 43–44, **44**, 49, **253**; length, 18, 29, 44, 45, 139; protobaleen, 62; structure and composition, 43–44

Barentsz, Willem, 152, 153, 157

Barkham, Selma Huxley, 122, 124, 132, 136, **420**

barrels of oil: capacity, 107, 136; construction, 126–27, 132–33, 199, 204, 217–18; number per bowhead, 131, 141, 206, 209, **420**; number per right whale, 46, 131, 141, 185, 187, 201, 206, 225, 242–43, **420**

barricas, 107, 126, 132

Basilosauridae, **57**, 59, 60, 61, **Plate 1**

Basques: cod fishing, 103, 111–13, 123–26, 129, 140; homeland, 85, 102, 103, 111–12; language, 101–2; relations with Natives, 129, 140–41

Basque whaling: basis for success, 102–3; cooperative agreements, 131, 139; Davis Strait, 135, 140–41, 144; end of dominance, 133–35, 155–57; financial arrangements, 127, 135; Gulf of Maine, 139–40, 168–69; harpoons (Arpoi), 104–5, **105**, **143**; off Iceland, 111–13, 141–44; marketing networks, 103, 109, 113, 133, 135; off Norway, 144, 159; offshore, 110–14; pelagic, 114–15; shore, 103–10; shore stations, 107, **108**, **123**, 128, 130, 139–40, 142;

Basque whaling (*continued*): off Spitsbergen,
157; Strait of Belle Isle, 18, 120, **121**, 122–33;
taxation, 109, 112; town seals, 109, **110**, 116;
watchtowers, 103, 116; whale boats (chalupas),
103–5, **106**, 132, **plate 5**

Basque whaling ports, **85**: Bayonne, 103, 109,
124, 133; Bermeo, 109, 113; Biarritz, 109;
Bilbao, 133; Fuenteffabia, 109; La Rochelle,
129; Lequeitio, 109, 112–13, 115, 116; San
Sebastian, 26–27, 103, **104**, 109, 131–32,
133, 135, 144, 157; St. Jean de Luz, 133, 155;
Zumaya, 105. *See also* Red Bay

Basque whaling ships: cargo capacity, 126, 127, 142;
crew, 127–28; economic importance, 126–27;
provisioning, 122, 126–27, 142; ship-building
skills, 111–12; types, 111, **114**

Bayeux Tapestry, 84

Bay of Fundy, **121**, 251, 252, 273, 284, 285, 336

Bear Island, 150, 151, 152, 153, 156

Bering Land Bridge, 72, 73

Bering Sea, 50, 72, **73**, 75–85, 255

Bering Strait, 72, **73**, 75

biological opinions (section 7 consultations):
fisheries, 354, 356–57, 358–63, 371, 375;
legislative directive, 278, 295; US Navy exercises, 311–12; vessel operations, 309–10, 312,
313–14, 331–32

Biscay Bay, 16, 26, **85**, 103, 115, 127, 129, 142, 144

BLM (Bureau of Land Management), **280**, 281,
284–85. *See also* Minerals Management Service

blowholes, 38, 50, 60, 155, 166

blubber: processing, 107–9, **123**, 130, 138–40, 155,
157–58, 160, 181–82, 187, 198, 201, 228, **241**,
253, 269–71, **plate 5**; storage, 111, 158; thickness, 106, 185

blue whales (*Balaenoptera musculus*), 17, 19, 34, 94,
135–37, 267, 270, 325

bomb lance, **209**, 209–10, 242, 253

Bonnaterre, Pierre Joseph, 25, 28

Boston, 183, 188, 197–201, **198**, 210, 221, 229, **294**,
309, 322, 326–27, 333–34, 336, **338**, 414

bowhead whales (*Balaena mysticetus*): catch levels,
142, 151, **420**; common names, 17–19, 23, 29;
Davis Strait, 137–38, 144, 206, 209, 223; feeding grounds, 92–93, 137, 139, 143, 149, 151–
52; filter feeding, 33–35, **35**, 44–45, 48–51; off
Greenland, 50, 81; Iceland, 91–93, 143; off Jan
Mayan Island, 152; life span, 64, **65**; migration,
72, 92, 125, 137–38; Native hunting, 78–81, **79**;
population size, 258; prey, 50; in Spitsbergen,
149–52, 156, 159–62; Strait of Belle Isle, 120,
125, 131–42; taxonomy, 23, **25**, 25–30, 63–64

Bradford, William, 169, 176

Braham, Howard, 258

Brickell, John, 236

Brisson, Matruin-Jacques, 23

buoy lines (end lines and vertical lines): breaking
strength, 363, 373–74, 397; configuration,
349–50; entanglement risks, 350–51, 360, 382,
387–88, 390. *See also under* fishing gear

Bureau of Land Management (BLM), **280**, 281,
284–85

Bush, George W., 322, 335–36

Butler, Samuel, 222, 232–33

Cabot, John, 124

Calanus finmarchicus, **45**, 49–50

Caldwell, David and Melba, 273, 283

callosities, 37; appearance, 17, 37–38, **43**; function,
48–49, 51; habitat for whale lice, 37, 66; identifying individual whales, 37–38, 273, 283

calves, right whale: calving grounds, 66, 84, 86–87,
93, 221, 244, 252–53, 255, **280**, 287–94, **294**;
mother-calf bond, 37, 86–87, 116, 247; number,
329, 406–7, **407**

Cantabrian Sea. *See* Biscay Bay

Cape Cod, **198**; Barnstable, 197–99, 202–5,
222–23; catch records, 203, **421**; decline of
whaling, 205–10; Dennis, 212n45; Eastham,
169, 177, 197–200, 202, 205–7; first hunting
of live whales, 200–201; offshore/pelagic whaling, 204, 206–7; peak whaling era, 201–5; and
Pilgrims, 169–72, 175–77; Provincetown, 1,
3–4, 5, 176, 196, 204, 206–10; salvage of drift
whales, 196–201, 203, 204; Sandwich, 197–99,
201–4, 208; taxes, 198–200, 202; Truro, 204;
Wellfleet, 169, 204–5; whale carcass disposal,
205; whale inspectors, 202; Yarmouth, 198–
204, 208, 218

Cape Cod Bay, 49, 137, 168, **198**, 204, 206, 209,
251, 273, 282–83, 285, 293, **294**, 319, **338**,
353–63, 394–95, 399

Cape Farwell, **85**, 253

Cape Fear, 232, 234, **235**, 238–**39**

Cape Hatteras, 228, **235**, 236, 239, 240

Capellini, Giovanni, 27

Cape Lookout, **235**, **239**. *See also under* North
Carolina

Cape May: catch records, 223–24, 251, **421**; early
Dutch whaling, 172–75; settlement by Long
Island whalers, 189, 223–25

Caribbean monk seals (*Monachus tropicalis*), 65,
67, 267

Carlson, Carol, 4

Carmen, Caleb, 223

Carteret, Peter, 229–30

Cartier, Jacques, 129

Center for Coastal Studies, 1–15, 286–87, **372**

Central American land bridge, 56–58, 64–67

Central American Seaway, 65–67

Ceratodactyla, 20

Cetacean Turtle Assessment Program (CeTAP), **28**, 284–85, 287

Charles I of England, King, 175, 185, 216, 228–29

Charleston, South Carolina, 229, **239**, 240, 245, 246, **338**

Chesapeake Bay, 174, 232, 325, **338**

Chukotka Peninsula, 72, **73**, 78, **79**

Churchill (whale), 371–74

Cintra Bay, 252–53

Clapp, William, 204

climate change, 64, 83, 149, 252, 407, 411–12

Clinton, William J., 321, 335

Coast Guard, US: and disentanglement, 1, 9–12; fishery management, 377; vessel management, 287, 292, 305–7, 309–10, 313–14, 317, 318–23, 327–29, 331–34, 336, 338

Collett, Robert, 272

Columbus, Christopher, 112, 123–24

Congress, US, **280**; fishery management actions, 355–56, 382–83, 385, 389; funding, 282, 286–87, 289, 291, 295, **296**, 363, 367, 378, 380; legislation, 276–77, 322, 355–56; vessel management actions, 312, 320–21, 336

Convention for the Regulation of Whaling, 269–72

convergent evolution, 36–37, 47–48

co-occurrence model, 390–91, 393–94, 398–99

Cope, Edward D., 27, **27**

copepods, 45, 49–50, 283, 411

Coues, Elliot, 339–340

critical habitat, 294–95; designation for right whales, **280**, 284, 291, 292–95, **295**, 306; and fishery interactions, 354–59, 361–64, 371, 374, 376–78, 383, 386, 392–93, 398; legal authority, 278; and ship strikes, 306, 310–12, 319, 323–24, 327, 329, 333–34

Crockford, Susan J., 80

Cumbaa, Steven L., 136

Cummings, William C., 325

Cuvier, Georges, 25–26

cyamids: genetic analyses, 66–**67**; as indicator of health, 38, 50–51; relationship to whale host and callosities, 37–38, 48

Darwin, Charles, 37, 40–41, 122

Davis Strait, 80–81, 135, 137, 140–41, 144. *See also under* bowhead whales

Dehm, Richard, 59

Delaware Bay, 27, 171–75, 223–24, 261, 309, **338**

derelict fishing gear, 359, 361, 362, 375

Dietz, Johann, 151, 160

disentanglement: of first whale, Ibis, 4–8; as mitigation measure, 14–15, 288, 359–64, 377–79, **392**; of Nantucket, 11–14; and sedation, 372–73

DNA, 55, 259–60

Dorudontinae, 60

Doty, Josiah, 236

drones, 412, 413–14

D-tags, 49, 313, 413

Dudley, Paul, 185

Dürer, Albrecht, 21, **22**

Dutch Harbor, **73**, 74–75

Dutch West Indies Company, 172

Dutch whaling: basis for success, 160; Davis Strait, 144; Delaware Bay, 171–75; Jan Mayan Island, 159; Spitsbergen, 135, 138, 140, 151–52, 157–58, 160–61

dynamic area management (DAM): fisheries, 368, 370–77, 380, 388–89; vessel traffic, 232, 234, 237, 239–40, 327–28

Early Warning System (EWS): in biological opinions, 310, 312; formation, **280**, 306–8; and Mandatory Ship Reporting, 319–20

East Hampton, New York, 181, 183, 186–87, 190

Edgartown, **198**, 222

Edge, Thomas, 130, 154–56

Edwards, Everett, 169

Edward III of England, King, 229

Einarsson, Björn, 93–94

Endangered Species Act, 276, 278, **280**. *See also* biological opinions; critical habitat

endlines. *See* buoy lines

enforcement: fishery restrictions, 363, 368–69, 377, 384, 388, 399; vessel restrictions, 311, 320–23, 328

entanglement: cod traps, 2–3, 210; first records, 153, 293; gillnets, 4–3, 348, 353, 383–84; trap/pot gear, 348, 351, 382; trends, 295–**96**, 347–48, 397, 407–8

Eomysticetidae, 63

Eschricht, Daniel Frederik, 26–27, 138–39

Eschrichtiidae, 33, **57**

Eskimos, 93, 103

Eskimo whaling, 64, **65**, 74, 77, 78

Essex, sinking of, 300–302

Evans, William E., 290

evolution: analytic methods, 29, 55; balaenid whales, including right whales, 62–68; first whales, 56–61; modern whale families, 61–62; speciation factors, 64; whale propulsion, 60–61

Exclusive Economic Zone (EEZ), 351–52, 381, 390

Faroe Islands, 84, **85**, 88, 89, 253–54

Feduccia, Alan, 47

Ferdinand III of Spain, King, 109

fin whales (*Balaenoptera physalus*): feeding methods, 34; ship strikes, 325–26; as target of early whalers, 17, 94, 267 fisheries: groundfish, 349, 352–54, 370; lobster, 348, 351, 357; shark, 349, 357, 377, 383–85; whiting, 384–85

fisheries management: conflicts of interest, 281, 354; gear marking, 362, 364, 375, 392–97, 400, 409; gill net bans, 355, 361; groundline buy-back programs, 380, 387, 389; time-area closures, 354–55, 359, 361, 368, 381, 385, 387–88, 394–98, 408–9, 412; trawling-up, 390–400. *See also* Atlantic Large Whale Take Reduction Plan

Fisheries Management Councils, 252, 352–55, 400, 410

fishery management plans: legislative directive, 352, 361, 400, 410; Northeast Multispecies Fishery Management Plan, 352–55

fishing gear: breaking strength, 363, 373–74, 397; buoy line entanglement risks, 350–51, 360, 382, 387–88, 390; buoy lines, end lines, and vertical lines, 349–**50**, **351**, 358–63, 370, 374–99; floating line, **347**, 350, 351, 358, 370, 379; gill nets, 4, 333, 348–**50**, 351–95; groundline, **350**–52, 358, 361–64, 369, 371, 378–83, 386–89, 399; traps and pots, **350**, 351–52, 354–55, 360, 362, 369–71, 374–82, 387, 389–400

fishing gear modifications: funding, 363–69, 378; low profile groundline, 379, 386; neutrally buoyant groundline, 369–82, 386; pop-up buoys, 360, 379, 408–9; ropeless (on call pop-up buoy), 360, 371, 390, 400; sinking groundline, 358, 361–64, 374–76, 378, 380–82, 385–86, 388–89, 399, 408–9; as take reduction strategy, 359–60, 362, 364, 367–83, 386–89; types considered, 378–79; weak links, 358, 360–64, 369–70, 374–75, 378, 380–82, 385, 388–90, 395, 400, 408; weak rope, 360, 378, 409

flamingos (Phoenicopteridae): and convergent evolution, 36–37, 40–42, 47–51; diet and feeding behavior, 41–42; inverted feeding, 50–51; similarities to right whales, **36**–37, 40–42, 46–48

Flemish, 94–97, 102, 104, 151, 160

Florida Fish and Wildlife Commission, 289, 329

Flower, William H., 26–27, 56

Foreland Sound, 153–**54**, 156, 158

Fotherby, Robert, 130, 157, **plate 4**

Freeman, Fredrick, 198, 205

Frobisher, Martin, 302

Gardiner, Lyon, 179, 180, 181

Gasco, Francisco, 27

Gaskin, David, 258

genetic analyses: Basque whaling effects, 136–37, 258–60; cyamids, 66–67; gray whale population structure, 86; right whale taxonomy, 66; Saqqaq origin, 80

genetic diversity, 259–60

Georges Bank, **151**, 206, 247, 374

George II of England, King, 238

Georgiacetus vogtlensis, 61

Gerrior, Patricia, 318

Gesner, Konrad, 20–21, **22**, **23**, **24**, **105**, **108**, **109**

Giddings, John Lewis, 74, 77

Gloucester, 4, 7

Gmelin, Johann Friedrich, 23

Godyn, Samuel, 172–73

Gosnold, Bartholomew, 139–40, 168, 171

Gould, Stephen J., 141, 148

Grand Banks, **121**, 124

Grand Bay, 18, 120–29, 131, 137, 158. *See also* Strait of Belle Isle

Gray, John Edward, 26–27

gray whales, Atlantic population (*Escherictius robustus*), 18, 217

gray whales, North Pacific populations (*Escherictius robustus*): captive feeding, 35; eastern, 84–85, 255–56; feeding method, 33; migration, 85; western, 84–85, 255

Greater Atlantic Regional Fisheries Office. *See* Northeast Regional Office for NMFS

Great South Channel: critical habitat, 293–**94**, 310; early records, **121**, 206, 251–52; feeding area, 282, 284, 287; fishery management, 353, 356, 358–59, 362–64, 371, 374, 394; ship strike reduction, 319, 327, 334, 336, **338**

Greenland, **85**, **150**: confusion with Spitsbergen, 153; first human occupants, 74, 79–81; Norse colonies, **85**, 86, 88, 92–94, 124; paleoclimate, 62; whales and whaling, 120, 138, 140, 142, 144, 158, 161, 206, 253, 302

Greenland whale (Greenland right whale), 18, 23, 25–26, 28, 91–93, 138, 176. *See also* bowhead whales

Greenland Sea, 144, **150**, 151, 161

Guldburg, Gustav A., 28

Gulf of Maine: fisheries management, 376, 382, 394; possible Basque whaling, **121**, 137, 139, 165, 168–69; right whale habitat, 251, 273–74, **280**, 282, 287–88; vessel management, 333, 336, 340

Gulf of Mexico, 66, 86

Gulf of St. Lawrence, 18, 120, **121**, 128–29, 137–40, 165, 223, **373**

Gulf Stream, 65–66, 127, 133, 151

Guðmundsson, Jón, 142

habitat: historical vs. present, 251–52, 262, 292; preferences, 84, 177, 287; protection measures, 291, 310, 318–20, 323–24, 330–34; range, 84, 86–87, 152; research, 283–85, 287–89. *See also* critical habitat; fisheries management: time-area closures

Hain, James, 284

Häkon Häkonsson of Norway, King, 91, **plate 2**

Harlan, Richard, 60

Hartford, Connecticut, 180, 182

Hawaiian monk seals (*Monachus schauinslandi*), 67, 290

health assessments, 38, 51, 296, 329, 367, 412

Higgenson, Francis, 177

History of Orosius, 88

Hofman, Robert J., 281

Hogarth, William, 335, 337

Holocene, 68, 73–74, 79

Horn Sound, 153, **154**, 158, 160

Hudson, Henry, 153–54, 157, 171–72

Hudson River, 179, 190

Hughes, Humphrey, 229

Humane Society of the United States, 369–70, 385, 387, 389, 396

humpback whales (*Megaptera novaeangliae*): catch of, 19, 77, 171, 208, 240, 253, 267; disentanglement, 2–8; feeding methods, 34, 49; fishery interactions, 353, 357, 370, 396; ship strikes, 310, 325; singing, 5

Hussey, Christopher, 219

hyoid bone, 77, 88

Iceland, **85**, **150**; medieval and Renaissance, 18, 86–94, 97, 111–13, 117; modern, 86, 253–**54**; 17th and 18th century, 23, 25, 44, 142–44, 151–52, 157–58, 161

ICES (International Council for the Exploration of the Sea), 269–70

Implementation Teams for the Right Whale Recovery Plan, **280**, 292, 294, 306–8, 319, 323–24, 327–28, 355–56

Indohyus, 58–59, **plate 1**

International Fund for Animal Welfare (IFAW), 317–18, 321, 323–24, 327–29, 380

International Maritime Organization (IMO), 317–22, 327, 329, 331, 334, 336, 339

Inuit, 78

Iron Age, 84, 88

Isasti, Lope de, 141

IWC (International Whaling Commission), 270–72, **280**, 285–86

Jan Mayen Island: Basques, 142, 144, 159, 161; discovery, 152–53; Dutch, **150**, 151–52, 159–60, 161, 172, 174

James I of England, King, 154, 175, 185

Jamestown, Virginia, 171–72, 174, 176, 179

Jeffreys Ledge, 354, 363–64, 394–98

John of England, King, 109

Johnson, Lindy S., 317, 319–21, 329, 337

Jones, M. Rice, 139

Josephson, Elizabeth, 261

Jough, Hans, 175

Jumieges, 103

Katona, Steven K., 282

Kellogg, Remington, 59

Kenney, Robert D., 284, 287, 293

King's Mirror, 191–94, **plate 2**

Kite-Powell, Hauke, 328

Klein, Jacob Theodor, 21, 23

Knowlton, Amy R., 247, 317, 323–25, 327, 409

Kraus, Scott D., 284–85

krill, 34, 45, 49–50, 62–63

Kritzler, Harry, 272

Labrador, 18, 77, 79, 86–87, 117, 120–**21**, 129–30, 206, 254, **plate 3**

Lacépède, Comte de, 25

Lamarck, Jean Baptiste, 41

Larson, Laurence Marcellus, 91–93

Lautenbacher, Conrad, 335

Lawson, John, 232

lawsuits: Basque, 102, 122, 127, 129–32; colonial, 197, 201, 222–23, 231, 234; fishery impacts, 357, 369–70, 387, 389; vessel impacts, 309–10, 333

League of Nations, 269

Leaming, Christopher and Thomas, 224–25

Lien, Jon, 2–3

Linnaeus, Carl, 19–21, 23–24, 29–30

Little, Elizabeth, 330

Little Ice Age, 83, 93, 120, 137–38, 152, 179, 251

lobster fisheries, 348, 351, 357

Long Island, New York: catch records, 188, **421**; coastal voyages, 189, 204, 206, 220, 221, 222, 223; colonial orders and ordinances, 180–82; deeds and contracts with Native residents, 180–81; duties and taxes, 182–83, 185–88; first settlements, 180; marketing of oil, 186, 188; salvage of drift whales, 180–82; shore whaling companies, 182–83, **184**, 187–89; whale-boat wars, 189–90; whaling after American Revolution, 190–93; whaling methods, 184–85

Loper, James, 183, 217–18

Macy, Obed, **215**, 217–19

Macy, Thomas, 215–16

Magellan, Ferdinand, 112

Magnus, Olaus, 21, **23**, 24

Magnuson-Stevens Fishery Conservation and Management Act, 276, 352, 361, 410

Maine Department of Marine Resources, 382, 386, 389, 396

Maine Lobstermen's Association, 387–88

manatees (*Manatus sp.*), 55, 156, 266, 269, 277–78, 306

Mandatory Ship Reporting system (MSR), 318–23, **322**

Manigault, Gabriel. E., 245

Mantzaris, Christopher, 318

marine debris, 290. *See also* derelict fishing gear

Marineland of Florida, 273, 283, 286

Marine Mammal Commission: actions related to fishery interactions, 375, 378, 381, 384, 385–87, 393–94, 396, 399; actions related to ship strikes, 305–6, 312, 317–34, 353, 363; legislative formation, 277–78; role in initiating right whale recovery program, **280**, 281–93

Marine Mammal Protection Act: administration, 277, 356–57; adoption and provisions, 276–78, **280**; 1994 amendments governing fishery interactions, 355–56, 361

Marmaduke, Thomas, 156

Martha's Vineyard, **198**, 215, 216, 221–23, 232, **421**

Massachusetts Bay Colony, 175, 177, 180, 196–97, 201, 203–4, 215, 221

Massachusetts Division of Fisheries, 357–58, 360, 380, 396

Massachusetts Environmental Trust, 289

Massachusetts Lobstermen's Association, 380

Mate, Bruce R., 85–86, 288

Mather, Cotton, 203

Mattila, David, 2–14, 7

Maury, Matthew Fontaine, 28, 257

Mayflower (ship), 169, **170**, 176

Mayflower (whale), 242–**43**, **244**

Mayhew, Thomas, 215, 222

Mayo, Charles "Stormy," III, 2–14, 49, 283–84

Mead, James G., 283, 325

medieval optimum (medieval warm period), 80, 83, 92

Melville, Herman, 38, 42, 268–69, 302, **303**

Merrick, Richard, 333

Merritt, Frank, 245, **247**, 272

Mey, Cornelis Jacobszoon, 172

Minerals Management Service (MMS), 284, 288, 289, 292. *See also* Bureau of Land Management

minke whales (*Balaenoptera acutorostrata*), 35, 235

Miocene, 59, 62–64

Mitchell, Edward, 240, 242, 258

Monomoy Island, 10, 12, **198**

Montaukett Indians, 180–81

mortality, North Atlantic right whale: annual levels and trends, 295–**97**, 304–5, 311, 329, 330, 341, 385, 399; and fishing gear, 347–48, **368**, 370, 385, 397, 399; and ship strikes, 304, 319, 325–**26**, 327, **331**, 339, 341

Mowat, Farley, 258

Mulford, Samuel, 186–87, 192

Müller, Orthone Friderico, 23

Muscovy Company, 153–57, 160–61, 196

Mysticeti (baleen whales): feeding methods, 33–35, 44, 62; oldest known, *Llanocetus denticrenatus*, 62; taxonomic group, 20, 33, 60; toothed, 62; toothless, **43**, 63

Nansen, Fridtjof, 88–89

Nantucket (whale), 9–14, **10**, **13**

Nantucket Island, **198**, **215**: catch records, 214, 219–20, **421**; drift whales, salvage of, 216–17; first hunting of live whale, 217; involvement of Native whalers, 169; marketing of oil, 221; offshore/pelagic whaling, 189, 192, 219; settlement, 214–16; shore whaling, 183, 203, 216–21

Nantucket sleigh ride, 106, 300. *See also* whaling methods: tow

National Fish and Wildlife Foundation, 289, 317–18, 363–64

National Large Whale Conservation Fund, 289, 363–64

National Marine Fisheries Service (NMFS and Fisheries Service), **280**; critical habitat petitions, 293; disentanglement authorization, 2, 14–15; funding for right whales, 282, 286–87, 289, 292, 295–**96**, 318, 323–24, 363–64, 367, 378, 380, 389; mitigating fishery interactions, 352–400; mitigating vessel interactions, 318–42; recovery planning, 190–92, 258; responsibility for marine mammals, 277, 278, 281–82, 295, 341–42, 351–52

Native whaling accounts: archaeological evidence, lack of, 169–70; English, 166–70, 184–85; Spanish, 165–66, **167**. *See also* archeological sites; archeology

Navy, US, **280**, 287, 292, 305–7, 323–25, 331–32

Neave, David J., 273–74

necropsies, 288, 295, 305, 312, **314**, **331**, 383

Neuhauser, Hans, 283–84

New Amsterdam, 172, 174–75, 185, 216

New England Aquarium: Bay of Fundy feeding ground, 283; calving ground surveys, 287, 307; comments on fisheries management, 396, 397; photo-identification catalogue, 14, 287 (*see also under* right whales, North Atlantic); right whale research initiative, 286–87; Shipping / Right Whale Workshop, 306; workshop on status of right whales, **280**, 285–86. *See also* Knowlton, Amy R.

New England Fisheries Management Council, 352–55

Newfoundland, **121**: Basque cod fishing, 124–25; Norse settlement, 124; whale habitat, 18, 86–87, 137–38; whaling, 120, 138, 172, 207, 209, 220, 254

New Jersey: Barnegat Bay, 225; catch records, 223, **421**; court at Burlington, 223; Egg Harbor, 225; Seven Mile Beach, 225. *See also* Cape May

New Netherland, 172, 174–75, 179, 186, 216

New Plymouth. *See* Plymouth

NOAA (National Oceanographic and Atmospheric Administration): assignment of marine mammal authority, 277; initial IMO Information paper, 317; mandatory ship reporting proposal, 319–23; vessel speed and routing measures, 332, 335–38

Noordsche Company, 157–61, 172

Nordkapers, 17–26, 151, **253**

Normandy, 95, 115

Normans, 94–97, 104, 115, 201

Norse, the: burial artifacts, 88; contact with Arctic Natives, 90, 93–94; in Greenland, 88, 93; *Grettir* saga of, 87; legal codes, 89–91; longboats and knarrers, 83–84; spear whaling, 90–91, 93–94

North Atlantic right whale. *See* right whales, North Atlantic

North Atlantic Right Whale Consortium, 285–93, 317, 391

North Atlantic Right Whale Recovery Plan, 258, **280**, 290–93, 295, 305, 307, 311

North Atlantic Right Whale Recovery Team, **280**, 290–92, 293

North Carolina, **235**, **239**: Beaufort, 232, 234, 236, 238, **239**, 243; Bogue Banks, **239**–40; Cape Fear, 232, **234**, 238–**39**; Cape Hatteras, 228, **235**, 236, **239**, 240; Cape Lookout, 232–42, **235**, **239**; catch records, 234, 236–37, 240–43, **421**; colonial charters, 228–29, 230; Diamond City, **239**, **240**, 240–43, **243**; encouragement of whaling, 230–34, 237–38; licenses, 229, 232–**33**, 236–37, 244–45; naming whales, 242–43; New England whalers, 228, 232–37, 245; Ocracoke Island, 236–37; Outer Banks, 222, 228–30, 232, 234–39, **235**, **239**; salvage, drift whale, 229–31, 236–37; Shackleford Banks, 234–36, 238–42, **239**; shore whaling, start of, 232, 235; taxes, 230–31, 233–34, 236–38; whaling decline, 240–43

Northeast Implementation Team (NEIT), **280**, 323–24, 328

Northeast Multispecies (Groundfish) Fishery Management Plan, 352–54

Northeast Regional Office for NMFS, 318, 353–55, 357, 384, 390

northern fur seal (*Callorhinus ursinus*), 269

North Sea, 66, **85**, 94, **150**

Norway, **85**, **150**: Basque whaling, 142, 144, 157, 159; North Cape, 17, **85**, 86, 88–89, **150**, 152–54, 161; petroglyphs, 88; whale habitat, 21, 23–24, 25, 86–87; whaling by northern

Europeans, 16–17, 28, 151, 159–60, 161, 253

Nova Scotia, **121**: Basque presence, 125, 129; vessel traffic lanes, 336; whale habitat, 86, 117, 273, 284, 287–88; whaling, 88–90, 117, 139, 169, 220, 251, 254

Nowacek, Douglas, 113

Obergfell, Friedlinde, 58

Odelica, Clemente de, 129

Odontoceti, 60–62

offshore oil and gas development, 281, 284, 289, 410

Ohthere, of Hålogaland, 88–89

oil: commercial uses, 109; processing methods, 107–8, 111, 140, 156, 158, **191**, 219, **241**, **plate 4**, **plate 8**

Oligocene, 61–63

Olson, Storrs L., 47

OMB (Office of Management and Budget), 335–337, 341

Orkney Islands, **85**, 88

Orosius, Paulus, 88

Osborn, John, 207

Our, John, 9, 12, **13**

Överhogdal tapestries, 84

Paddock, Ichabod, 203, 218, 222

Pakicetidae, **57**, 59, **plate 1**

Paleo-Eskimos, 75–76, 78–79

Parks Canada, 122

Payne, Roger, 283

Peach, Robbin, 289

Penn, William, 189, 223

petitions: critical habitat, **280**, 293, 377–78; exemptions from fishery restrictions, 398–99; exemptions from speed restrictions, 341; for speed restrictions, 332–33; for vessel approach limits, 310–11

petroglyphs, 87–88

Philip II of Spain, King, 134–35

Phoenicopteridae. *See* flamingos

photo-identification, 14, 86, 247, 273, 287, 290, 291, 307, 412

Pleistocene, 68, 72–73

Plymouth, **198**; Colony, 3, 169, 175–76, 196–203, 215, 222, 236; court at, 197, 199, 202; town, 176–77

Poole, Jonas, 253–56

population models, 256–57

population size, North Atlantic right whale: Basque whaling effects, 259–60, 262, **420**; colonial whaling effects, 261–62, **421**; after commercial whaling, 254; current, 254, 279, 406; pre-exploitation, 257–58, 262–63

Port Access Route Studies, 328, 333–34

Portuguese whaling, 115, 117

Potential Biological Removal level (PBR), 15n2, 355, 357, 359

Potvin, Jean, 34, 44

Prescott, John H., 285–86

Price-Fairfield, Carol, 284

Protocetidae, **57**, 59, 60–61

Provincetown, 1, 2–6, 176, 196, **198**, 204, 206–10, 333–34

Pynde, Edward, 223

Qazwīnī, al-, 95

Quebec, 86, **121**, 128

Ranga Rao, Anne, 58

Raoellidae, **57**, 58

Rattray, Jeannette, 169

Rauch, Samuel D., III, 337, 341

Ray, John, 20

Records, John, 334

Red Bay, Labrador, **121**, 122, **123**, 124, 129–32, 136, 138–39, 254, 258–60, **plate 5**

Reeves, Randal R., 240, 242, 261

regulations: critical habitat, 293–94; first revised take reduction plan, 370, 374–77; mandatory ship reporting, 329–33; original take reduction plan, 361–64; third revised take reduction plan (vertical lines), 382, 386–88, 395–99; vessel approach limit, 310–11

regulations, second revised take reduction plan (broad-based use of groundline): broad-based groundlines, 381–82, 387–89; effective date deferral, 388; gill net fishing on calving grounds, 383–85; temporary reinstatement of dynamic and seasonal zones, 388–89

regulations, vessel speed: compliance, 338–39; economic impacts, 328, 329–30, 331, 333, 335, 340; effectiveness, 339–41; petitions, 32–33; seasonal management areas, 334–38; sunset provision, 337, 341–42

Remingtonocetidae, **57**, 59

research: acoustics, 50, 312–13, 412, 414–15; aerial surveys, 254, 273, 283–84, 287, 292, 307–8, 328, 370, 391, 396, 412; drones, 412, 413–14; genetics, 17, 29, 55, 66–67, 77, 86, 136–37, 255, 258–60; necropsies, 288, 295, 305, 312, **314**, **331**, 383; shipboard surveys, 283, 284–85, 287, 291, 412; telemetry, 280, 288, 292, 369, 371–72, **373**, 382, 412–13; vessel routing assessments, 324, 329, 336. *See also* fishing gear modifications; population models

Rhode Island, University of, 283, 284, 286, 287, 293

right whale feeding methods: continuous ram, 34, 44–46; filtration, 34–**35**, 44; foraging, 48–51; prey, **45**, 49, 283, 411; skim, **35**–36, **43**, 48–**49**

right whales, North Atlantic (*Balaena glacialis*): calving grounds, eastern North Atlantic, 68, 84, 86–87, 252; calving grounds, western North Atlantic, 66, 68, 84, 86–87, 221, 244, **280**, 283, 285, 292–**94**, 306, 319; catch levels, 115–16, 151, 240–42, **246**, **253**, 261–62, **420**, **421**; common names, 16–17, 18–19, 23, 29; economic importance, 18, 116, 126–27, 134, 205, 234, 237, 267; extinction risks, 193, 254–55, 266, 279, 367, 406; feeding grounds, eastern North Atlantic, 86, 93, 139–41, 161, 151–52, 253; feeding grounds, western North Atlantic, 68, 86, 110, 113, 206, 254, 283, 284, 287–88, 292–93, 306, 319, 334, 336, 411; genetic analyses, 29, 66–67, 77, 136–37, 258–60, 412; movements and migration, 1, 38, 66, 84, 86, 161, 285, 288, 292, 327, 373, 382, 412–13; photo-identification catalogue, 14, 247, 273, 287, 290, 291, 307, 412; population (stock) structure, 84–87; prey, 45, 49–50; reproduction, 38, 64, 86–87, 285, 288, 329, 406–7; speciation, 67; taxonomy of, 17, 20–30. *See also* biological opinions; mortality; population size

right whales, North Pacific (*Eubalaena japonica*): abundance, 255; distribution, 255, 272; illegal whaling, 255; separation from North Atlantic right whale, 28–29, 64–67

right whales, southern (*Eubalaena australis*), 29, 66, 67, 260, 272, 277, 283, 406

ringed seal (*Phoca hispida*), 179

rorqual whales, 34, 35, 36, 89, 94, 136

Roseway Basin, 336

Rosier, Thomas, 167–69

Russell, Bruce, 317–21, 323–24, 327–28, 331

Sakhalin Island, 85–86

Sallowes, Allan, 157

Sanderson, S. Laurie, 47

San Juan (ship), 122, 131–32

Saqqaq, 79–80

Savage, John, 117–18

Scanlan, James J., 97

Schellenger, Abraham, 192

Schellenger, Jacop, 183, 192

Schevill, William E., 247, 272

Scoresby, William, Jr., 25–26

Scotland, **150**, 153, 160, 254, 272

scrag whale. 240. *See also* gray whales

sealskin floats, 78

section 7 consultations. *See* biological opinions

Segura, Doming de, 129

seismic exploration, 410–11, 414–15

Shackleford, John, 234–36

Shackleford Banks, 234–36, 238–42, **239**

Shetland Islands, **85**, 88, **250**, 253–54

shipping lanes, 305, 306, 315–16n20, 318, **322**, 327, 333–34, 336, 414

Shipping / Right Whale Workshop, 306, 318

ship speed. *See* vessel speed

Ship Strike Committee, 232–24, 327–29, 330, 331, 334

Ship Strike Reduction Strategy, 314, 329–34, 340. *See also* regulations, vessel speed

ship strikes: annual levels and trends, 295–**96**, 304–5, 311, 329, 330, 341; bow caught, 325–**26**, **plate 9**; historical records, 304; hydrodynamic effects, 324, 328; injuries, nature of, 304, **305**, 337; mortality, 304, 319, 325–**26**, 327, **331**, 339, 341; public outreach, 304–8, 338; speed at time of collision, 326, 337; technology to detect and avoid, 305, 312–13, 330

Silber, Gregory K., 329

Simpson, Marcus B., Jr., and Sally W., 237

Sirenians, 55. *See also* West Indian manatees

Slijper, Everhard J., 88

Smeerenberg whaling station, **154**, 158

Smith, John, 171, 172

Smith, Martha Tunsstall, 188

Smith, Timothy D., 161

Smith, William "Tangier," 188

Snow, Olympia, 336, 382–83, 387

sonar, 292, 305, 312–13, 318, 330, 411

Southampton, New York, 180–84, 186–87, 189–90, 229

Southeast Implementation Team (SEIT), **280**, 306

southern right whales. *See* right whales, southern

Southworth, Constant, 199

Spitsbergen, **150**, **154**; Basque whalers, 155–58; bay whaling, 149–50, 152–61; catch records, 151, 156, 158, 161; discovery, 152–54, 157; Dutch whalers, 157–58, 160–61, 172; English whalers, 154–58, 196, **plate 4**; first years of whaling, 154–56; species caught, 150–51, 155–56, 161

Starbuck, Alexander, 185

Starbuck, Edward, 215–16

Starbuck, Mary, 220

Staten Island, 172, 187

Steingrimsfjörður, 142

Strahan, Richard Max, 308–10, 357, 369

Strait of Belle Isle, 18, 87, 120, **121**, 125, 128, 136–42, 258–60

Stuyvesant, Peter, 175

Suárez, Jose Leon, 269

Subcommittee on Safety of Navigation. *See* International Maritime Organization

surveys, shipboard, 283, 284–85, 287, 291, 412

Szabo, Vicki E., 96

Taggart, Christopher T., 337

taxonomy, 19–21; Balaenidae, 20, 23–29, **57**; bowhead, 23, **25**, 25–30, 63–64; right whale, North Atlantic, 17, 20–30

telemetry, 280, 288, 292, 369, 371–72, **373**, 382, 412–13

Tethys Sea, 58, 60

Thewissen, Hans, 58–59

Thule, 74, 78, 80–81, 93, 140

toggling harpoon, 76–77, 104

tongue, right whale, 34, **35**, **43**, 44, 45–46, 77, 103, 107–9, 111, **253**

Torre, Nicolas de la, 129

True, Fredrick W., 28

tryworks: onboard, 103, **114**; portable, 111, 140, 156–57; shore-based, 107, **123**, 128, 142, 158–59, **191**, 192, 204, **241**, **plate 4**

Twiss, John R., Jr., 281

'Udhrī, al-, 95–97, 113

Umiak, 76, 78

Unalaska, **73**, 74–75

Vanderlaan, Angela S. M., 337

vertical lines. *See* buoy lines

vessel management: ATBAs (Areas to Be Avoided), 318, 327, 334, 336; shipping lanes, 305, 318, 327, 334, 336, 407; voluntary actions, 306–8, 311, 319, 324–25, 332–33, 337, 339. *See also* regulations, vessel speed

vessel noise, 292, 313, 411

vessel routing assessments, 324, 329, 336

vessel speed: history of, 303–4; relationship to ship strikes, 9, 324–26, 330, 337

Vikings. 83–84, 101, 124. *See also* Norse, the

Vries, David Pieterszoon de, 171–75, **173**

walmanni, 95

walruses (*Odobenus rosmarus*), 72, 76–79, 89, 139, 153–58, 277–78

Wampanoags, 219

Wassersug, Richard, 47

Watkins, William A., 247, 273

Werth, Alexander, 34, 44

Westfjords Peninsula, 142–44

West Indian manatees (*Trichechus manatus*), 156, 266, 269, 306

West Indies, 126, 165–66

Weymouth, George, 140, 166–68, 171

whaleboats, 78, 103–4, **106**, **132**, 182, 189–90, **plate 5**

whalebone. *See* baleen

whale lice. *See* cyamids

whaler's creed, 201–2

whales, early illustrations, 21, **22**, **23**, **24**, **25**, **27**, **plate 6**

whaling charts, 257, **plate** 7

whaling methods: carcass salvage, 89–90, 103, 166, 181, 182, 196–99, 203–4, 222, 228–29, 236–37; offshore/pelagic, 110–17, 130, 138–40, 144, 157–61, 189, 221, 260, **421**; onboard tryworks, **114**, 157; shore, 103–10, 129–30, 158–59, 187–88, 190, 202–5, 208, 218–19; shore processing stations, 157–59, **191**–93, 204, 252–54, **plate 3, plate 4, plate 8** (*see also under* Basque whaling); spear, 90–91, 93–94, 97; tow, 90, 96–97, 105, 106, 300

Willis, Josephus, Sr., 243

wind energy facilities, 411

Winn, Howard E., 283–86

Winthrop, John, 196, 203

Winthrop, John, Jr., 179, 186

Wolley, Charles, 184–85

Woodcocke, Nicolas, 153, 155–57

Woods Hole Oceanographic Institute (WHOI), 247, 273, 286, 287, 313, 328

World Shipping Council, 335

World Wildlife Fund, 286

Wright, B. S., 273–74

Wyandanch (Montaukett chief), 180–81

Zwaanendael, 72–74